Coccidiosis of Man and Domestic Animals

Editor

Peter L. Long
D. W. Brooks Distinguished Professor
Department of Poultry Science
University of Georgia
Athens, Georgia

CRC Press
Taylor & Francis Group
Boca Raton London New York

CRC Press is an imprint of the
Taylor & Francis Group, an **informa** business

First published 1990 by CRC Press
Taylor & Francis Group
6000 Broken Sound Parkway NW, Suite 300
Boca Raton, FL 33487-2742

Reissued 2018 by CRC Press

Library of Congress Cataloging-in-Publication Data

Coccidiosis of man and domestic animals/editor, Peter L. Long.
p. cm.
Includes bibliographical references.
ISBN 0-8493-6269-5
1. Coccidiosis. 2. Coccidiosis in animals. 3. Coccidia.
I. Long, Peter L.
QR201.C59C63 1990
593.1'9--dc20 90-1364

Publisher's Note
The publisher has gone to great lengths to ensure the quality of this reprint but points out that some imperfections in the original copies may be apparent.

Disclaimer
The publisher has made every effort to trace copyright holders and welcomes correspondence from those they have been unable to contact.

ISBN 13: 978-1-315-89164-4 (hbk)
ISBN 13: 978-1-351-07074-4 (ebk)

Visit the Taylor & Francis Web site at http://www.taylorandfrancis.com and the
CRC Press Web site at http://www.crcpress.com

PREFACE

The first book on coccidiosis which dealt in detail with the full range of biological characteristics of coccidia was *Coccidia and Coccidiosis of Domesticated Game and Laboratory Animals and of Man* by E. R. Becker (1934, Iowa State College Press, Ames, Iowa). There have been numerous books since then in English including those by Davies, Joyner and Kendall (1963, Oliver and Boyd, Edinburgh and London) and Pellerdy (1965; 1974, Verlag Paul Parey, Berlin and Hamburg). In 1970 the late Dr. D. M. Hammond and I considered that the subject matter was too great for a single author and accordingly we invited ten coccidiologists to contribute to *The Coccidia*, Hammond and Long (1973, University Park Press, Baltimore), a book devoted to the biology of coccidian parasites. I tried to bring information in this book up to date by producing another book *The Biology of the Coccidia* (P. L. Long, Ed., 1982, University Park Press, Baltimore). This book had 14 authors and co-authors and covered a wider range of topics than the 1973 work. The production of that book was a monumental task sufficient to discourage editing a new book of this type. Unfortunately, the book went out of print quite quickly due to the failure of the publishing house, at a time when demand for the work continued.

Since 1982 many interesting discoveries have been made including the recognition of *Cryptosporidium* as a common coccidium of animals and man. There has also been a vast number of publications on *Toxoplasma* and *Sarcocystis* sufficient for the appearance of several books devoted entirely to these coccidia including Dubey and Beattie (1989, CRC Press, Boca Raton, Florida); Dubey, Speer and Fayer (1989, CRC Press, Boca Raton, Florida).

The present book deals with the biology of a wide range of coccidia of numerous genera including *Eimeria, Isospora, Sarcocystis, Toxoplasma, Caryospora*, and *Cryptosporidium*. It is interesting that the genus *Cryptosporidium* was first described by Tyzzer in 1907 but was not widely studied until about 10 years ago. It is now recognized as an important parasite of man and domestic animals.

I believe that the subject of coccidia and coccidiosis has been of varied and continuing interest and there is still a need for a book on the biology of the coccidia in which all coccidia are discussed under the same cover. It is also necessary for all the latest information to be available to those with general interest in this group of parasites.

In the present work I have departed from the format of the two earlier works in which I have been involved. Here we have special chapters on each of the important groups of coccidia containing up-to-date information but aimed at avoiding excessive duplication of information given in other books. We have 17 relatively short chapters with a total of 26 contributors. Communications between me and the authors (and between authors) has been considerable, but helped by modern communication technology (e.g., facsimile machines).

I think that the book will be valuable for advanced undergraduates, graduate students, research workers, and teachers in biology, especially useful for parasitologists teaching the subject and essential for coccidiologists doing research on coccidia. The book has also an appeal for physicians, veterinarians, and zoologists needing an update of information in the general field of coccidiosis.

Finally, I wish to thank Myra Jane Blount for doing a vast amount of secretarial and communication work on this project.

THE EDITOR

Peter L. Long, D.Sc., Ph.D., is D. W. Brooks Distinguished Professor in the Department of Poultry Science at the University of Georgia at Athens, Georgia, U.S.

Dr. Long received his Ph.D. from Brunel University, London, England, in 1969 and D.Sc. degree from the same university in 1977. Dr. Long was engaged in research, first as research technician and research associate with Dr. Clifford Horton-Smith, one of the pioneers of coccidiosis research, at the Houghton Poultry Research Station, Cambridge, England. Apart from a year's sabbatical leave at Cornell University in 1968, working with Dr. P. P. Levine, and short periods with F.A.O. of United Nations in the Middle East, he spent most of his working life at the Houghton Laboratory. He was Head of the Department of Parasitology at Houghton from 1972 to 1979. He joined the staff at the University of Georgia in 1979 and has been there ever since. He is a member of the American and British Societies of Parasitology, Society of Protozoologists, Poultry Science Association and an Honorary Member of the British Veterinary Poultry Association. He is currently on the review boards of the *Journal of Protozoology* and *Poultry Science*.

Dr. Long has published more than 200 scientific papers on avian parasites (mainly on coccidia), edited (or co-edited) 5 books and written chapters for 14 books. He has received numerous awards for his work including the 'Tom Newman' and 'The Gordon memorial' medals for poultry research, The Merck Award from the Poultry Science Association, and the Creative Research Medal from the University of Georgia. He has received the D.A.A.D. traveling award to work with Dr. Erich Scholtyseck at the University of Bonn, Germany, on two occasions. He has given scientific papers at meetings in about 18 different countries. His current research interests are host and site specificity of coccidia, immunity and immunization against coccidiosis.

CONTRIBUTORS

Patricia C. Allen
Livestock and Poultry Sciences Institute
U.S. Department of Agriculture
Protozoan Diseases Laboratory
Beltsville, Maryland

Patricia C. Augustine
Livestock and Poultry Sciences Institute
U.S. Department of Agriculture
Protozoan Diseases Laboratory
Beltsville, Maryland

S. J. Ball
Honorary Research Fellow
Department of Biosciences
Polytechnic of East London
London, England

Byron L. Blagburn
Associate Professor
Department of Pathobiology
College of Veterinary Medicine
Auburn University
Auburn, Alabama

Richard J. Cawthorn
Associate Professor
Department of Pathology and Microbiology
Atlantic Veterinary College
University of Prince Edward Island
Charlottetown, Prince Edward Island, Canada

W. L. Current
Senior Research Scientist
Infectious Disease Research
Lilly Research Laboratories
Adjunct Professor of Pathology
Indiana University School of Medicine
Indianapolis, Indiana

Harry D. Danforth
Livestock and Poultry Sciences Institute
U.S. Department of Agriculture
Protozoan Diseases Laboratory
Beltsville, Maryland

Mary Agnes Fernando
Professor
Department of Pathology
University of Guelph
Guelph, Ontario, Canada

Martin W. Gregory
Department of Parasitology
Central Veterinary Laboratory
New Haw, Weybridge, Surrey, England

Thomas K. Jeffers
Director, Animal Science Discovery Research
Lilly Research Laboratories
Greenfield, Indiana

Alan M. Johnson
Chief Hospital Scientist and Senior Lecturer
Department of Clinical Microbiology
Flinders Medical Centre
Flinders University of South Australia
Bedford Park, South Australia

Michael H. Kogut
Assistant Professor
Department of Botany and Microbiology
University of Arkansas
Fayetteville, Arkansas

David S. Lindsay
Research Fellow
Department of Pathobiology
College of Veterinary Medicine
Auburn University
Auburn, Alabama

Peter L. Long
D.W. Brooks Distinguished Professor
Department of Poultry Science
University of Georgia
Athens, Georgia

Larry R. McDougald
Professor
Department of Poultry Science
University of Georgia
Athens, Georgia

Robert Michael Pittilo
Senior Lecturer in Life Sciences
School of Life Sciences
Kingston Polytechnic
Surrey, England

M. Elaine Rose
Head, Department of Parasitology
Houghton Laboratory
Institute for Animal Health
Huntingdon, Cambridgeshire, England

Michael D. Ruff
Livestock and Poultry Sciences Institute
U.S. Department of Agriculture
Protozoan Diseases Laboratory
Beltsville, Maryland

Dennis M. Schmatz
Associate Director
Department of Biochemical Parasitology
Merck Sharp & Dohme Research Laboratories
Rahway, New Jersey

Martin W. Shirley
Division of Molecular Biology
Institute for Animal Health
Houghton Laboratory
Huntingdon, Cambridgeshire, England

C. A. Speer
Professor and Head
Veterinary Molecular Biology
Marsh Laboratory
Montana State University
Bozeman, Montana

Richard G. Strout
Professor
Department of Animal and Nutritional Sciences
University of New Hampshire
Durham, New Hampshire

Christine A. Sundermann
Assistant Professor
Department of Zoology and Wildlife Sciences
Auburn University
Auburn, Alabama

Steve J. Upton
Assistant Professor
Division of Biology
Kansas State University
Manhattan, Kansas

D. Wakelin
Professor and Head
Department of Zoology
University of Nottingham
Nottingham, England

TABLE OF CONTENTS

Chapter 1

TAXONOMY AND LIFE CYCLES

William L. Current, Steve J. Upton, and Peter L. Long

TABLE OF CONTENTS

I. INTRODUCTION

The coccidia comprise a diverse group of obligate intracellular protozoa of vertebrates, with a few species infecting invertebrate hosts. These parasites have been placed in the phylum Apicomplexa, characterized by the presence of an assemblage of organelles within the anterior end of the invasive stages. This apical complex facilitates the entry of the parasite into host cells. Most named species infecting vertebrates are homoxenous (one host in life cycle) or facultatively heteroxenous (may also have dormant merozoite or sporozoite stages capable of initiating an endogenous cycle) and most develop within epithelial cells of the intestine. Other species causing disease in man and domesticated animals are obligatory heteroxenous parasites, having an intestinal phase of development (asexual and sexual, or sexual only) in one host and extraintestinal development (usually asexual) in an intermediate host. The coccidia of medical and veterinary importance, the focus of this book, are found within three of the more than nine families comprising the true coccidia (suborder Eimeriorina).

The majority of named species of coccidia within the family Eimeriidae are intestinal parasites belonging to the two genera *Eimeria* and *Isospora*. Because infections with homoxenous *Eimeria* and *Isospora* spp. are transmitted by environmentally resistant oocysts released in the feces of infected animals and because poultry and livestock are often kept in large numbers in confinement pens, intestinal coccidiosis has become an important disease of poultry and livestock throughout the world.[1] Intestinal coccidiosis, caused by *Eimeria* spp., is of such importance to poultry production that almost all commercially raised chickens are given anticoccidial drugs in the feed. Without such feed additives, the poultry industry as we know it today would probably not exist. Species of *Eimeria* also cause major economic losses in livestock, especially cattle and to a lesser extent sheep and goats. Species of homoxenous or facultatively heteroxenous *Isospora* spp. are found in a variety of avian and mammalian hosts. Several species produce diarrheal disease in dogs and cats, and one species, *I. suis*, is the cause of neonatal swine coccidiosis, a diarrheal disease of major concern to the swine production industry.[2] Another species, *I. belli*, has long been recognized as an infrequent cause of gastroenteritis and diarrhea acquired by humans residing or traveling in endemic areas. Recently, this parasite has been recognized as a cause of severe unremitting diarrhea in persons with the acquired immune deficiency syndrome (AIDS) and is especially prevalent in patients living in developing countries of the tropics.[3] Also within the family Eimeriidae are species of another coccidian genus, *Caryospora*, some of which have intestinal stages in reptiles or raptors and tissue-dwelling stages in mammals.[4] At least one species of *Caryospora* can produce clinical disease in dogs with compromised immunity.

The family Cryptosporidiidae is represented by small homoxenous coccidia assigned to a single genus, *Cryptosporidium*. One species, *C. parvum*, is now recognized as an important cause of diarrheal illness in several mammalian hosts, including man. In immunocompromised patients, especially those with AIDS, *C. parvum* may produce a prolonged, life-threatening, cholera-like illness.[5] This parasite may also be an unrecognized cause of respiratory illness in the immunocompromised host. Another species, *C. baileyi*, can produce severe respiratory disease in chickens and turkeys. Other species, including *C. meleagridis*, may cause enteritis and diarrhea in commercially reared poultry.[6]

The family Sarcocystidae is represented in this book by obligatory heteroxenous coccidia assigned to six genera: *Sarcocystis, Frenkelia, Toxoplasma, Besnoitia, Hammondia,* and *Neospora*. Both *Sarcocystis* and *Frenkelia* belong to the subfamily Sarcocystinae, coccidia that have sexual stages and formation of oocysts only in the intestine of the definitive host and asexual stages only in various tissues of the intermediate host. One asexual stage (the bradyzoite) of these two genera occurs within tissue cysts and is the life cycle form that transmits infections from intermediate to definitive hosts. Bradyzoites of these two genera are formed from metrocytes, pre-merozoite mother cells. Extraintestinal endogenous stages of *Sarcocystis* spp. are frequent causes of disease in livestock and more than 45 documented cases representing at least seven distinct species have been associated with nonintestinal infections in humans.[7] Intestinal disease has also been reported in humans when they serve as the definitive host for at least three species of *Sarcocystis*.

The other four genera of the family Sarcocystidae have been placed in a different subfamily (Toxoplasmatinae) because life cycles of these parasites include asexual stages that can transmit the infection from one intermediate host to another and because metrocytes are not found in association with bradyzoites. *Toxoplasma gondii* has long been recognized as a tissue-dwelling parasite capable of causing disease in man. In immunocompetent humans, *T. gondii* infections are usually mild or asymptomatic and result in the development of immunity and in the formation of tissue cysts (most in the central nervous system) that can persist for life. However, when the parasite is present in an immunocompromised host, such as a fetus (by transplacental transmission) or a patient with AIDS, it can produce a life-threatening infection. Approximately 30% of the AIDS patients that are seropositive for *T. gondii* will eventually develop toxoplasmic encephalitis because they lack immune surveillance mechanisms that prevent establishment of acute infections following cyst rupture.[8] *Toxoplasma gondii* is also an important cause of abortion and morbidity in livestock, especially sheep and goats.[9] Closely related to *T. gondii* are coccidian parasites of the genus *Besnoitia*, coccidian parasites with asexual, tissue-dwelling stages that have a predilection for invading fibroblasts. In some regions of the world, such as South Africa, China, and the Mediterranean countries, one species *(Besnoitia besnoiti)* may cause significant morbidity in cattle.[10] At the time of this writing, the definitive host(s) of this parasite has not been clearly defined. Species of *Hammondia* have asexual, tissue-dwelling stages in mammals that can transmit the infection to canids or felids, the definitive hosts.[10] In 1988, a coccidian closely related to *T. gondii* was shown to produce a fatal toxoplasmosis-like infection in dogs.[11] The role of this parasite, *Neospora caninum*, as a cause of disease in dogs and other domesticated mammals is under intensive investigation. At the time of this writing, the definitive host of this parasite is also unknown.

From this introduction it should be clear to the reader that the coccidia are represented by a large number of organisms, some of which have an impact on human health and on man's ability to produce poultry and livestock. This chapter also addresses the history, taxonomy, and life cycles of coccidia of medical and veterinary importance.

II. HISTORY

The early history of our knowledge of the coccidia has been reviewed extensively[1,12] and, as pointed out by Levine,[12] it is so tangled in the history of other protozoa (and helminths) that is is impossible to give a straightforward chronological account. It is beyond the scope of this chapter to revisit this early history; however, a brief discussion of some important events, beginning with the major advances in our understanding of these parasites made possible by the electron microscopic studies of the 1960s and 1970s,[13,14] will provide some valuable insight into our present understanding of the taxonomy and life cycles of the coccidia of medical and veterinary importance.

Research published in 1965 and 1967 initiated a change in our overall concept of the coccidia and the relationship of these parasites to the other organisms classified under the so-called "Sporozoa", an unnatural assemblage of diverse groups of protozoan parasites. Electron microscopy revealed that the extraintestinal merozoites (zoites) of *Toxoplasma gondii* and *Sarcocystis* spp. were ultrastructurally similar to those of the intestinal-dwelling eimerian species, suggesting that these two genera were coccidia, even prior to elucidation of the coccidian nature of their life cycles.[15,16] The first of these observations was soon followed by Hutchison's discovery of *Toxoplasma* infectivity in feline feces.[17] He first suspected transmission of *Toxoplasma* through the eggs of the nematode, *Toxocara cati*, in a manner similar to transmission of the protozoan *Histomonas* in the eggs of *Heterakis*, a nematode infecting turkeys. The nematode egg theory of transmission was discounted after *Toxoplasma* infectivity was dissociated from *T. cati* eggs.[18] In 1970, just 5 years after Hutchison's discovery, our knowledge of the life cycle of *T. gondii* was completed by the finding of sexual stages of the parasite in the gut of the cat and by the demonstration that oocysts are shed in the feces of felids.[19,20] Two years later, in 1972, the coccidian nature of the life cycle of *Sarcocystis* was revealed when Fayer[21] reported that bradyzoites (asexual stages in tissue cysts) obtained from an infected grackel developed into coccidian gametes and oocyst-like stages in cultured cells. During the same year Rommel et al.[22] independently found oocysts in the feces of a cat fed *Sarcocystis* cysts obtained from muscle tissue of sheep. Thus, during a 10-year period that began in the early 1960s, our concept of the coccidia changed dramatically from that of homoxenous, strictly enteric pathogens that are transmitted only by oocysts to include heteroxenous parasites with life cycle stages that can invade many organs of the host and can transmit the infection to another host by carnivorism.

Approximately 10 years following the discoveries that both *T. gondii* and *Sarcocystis* spp. were coccidian parasites, another coccidian gained recognition as an important pathogen of man and domesticated animals. In 1982, our concept of *Cryptosporidium* spp. began to change from that of a rare organism infecting mammals into that of important, widespread causes of diarrheal illness in several animals species, including humans.[23] *Cryptosporidium parvum* is now recognized as a common cause of self-limited diarrheal illness in immunocompetent persons and as a cause of prolonged, life-threatening, cholera-like illness in immunocompromised persons, especially patients with AIDS. Reports of infections of the respiratory tract, pancreatic ducts, and biliary tree demonstrate that the developmental stages of this protozoan are not confined to the gastrointestinal tract and suggest that *C. parvum* may also be an under-reported cause of respiratory and biliary tract disease, especially in the immune deficient host.[5]

III. TAXONOMY

As a direct result of the rapid advances in our understanding of the fine structure of parasitic protozoa that occurred in the early 1960s,[13,14] the phylum Apicomplexa Levine 1970 was established in an attempt to make some sense of the "Sporozoa" that contained diverse groups of protozoan parasites.[24] The phylum Apicomplexa brings together all protozoa that possess an **apical complex**, an assemblage of organelles at the anterior end of certain life cycle stages that facilitate attachment to or entry into host cells. These organelles (Figure 1) include: one or more electron dense **polar rings**; a **conoid** formed by several spirally coiled microtubules inside the polar ring; a number of **rhoptries** (toxonemes, paired organelles, lankesterellonemes, eimerianemes, dense bodies) — electron dense, tubular, or saccular organelles, often enlarged posteriorly, extending back from the anterior region inside the conoid; a number of **micronemes** (sarconemes, convoluted tubules, rod-shaped granules) — elongate, electron-dense organelles extending longitudinally in the anterior part of the cell, perhaps attached to the rhoptries; and a number of **subpellicular microtubules** (subpellicular microfibrils) — slender, electron-dense hollow structures extending back just beneath the pellicle from a polar ring. Thus, the Apicomplexa as we know it today includes not only the true coccidia (suborder Eimeriorina) but also the malarial parasites (suborder Haemosporina) of man and other animals, the piroplasms of domesticated and wild animals, the heteroxenous hemogregarines, and the gregarines of invertebrates. A more thorough treatment of the taxonomy and biology of the more than 4000 named species and over 300 named genera in the Apicomplexa can be found in books dealing with the protozoa[24] or

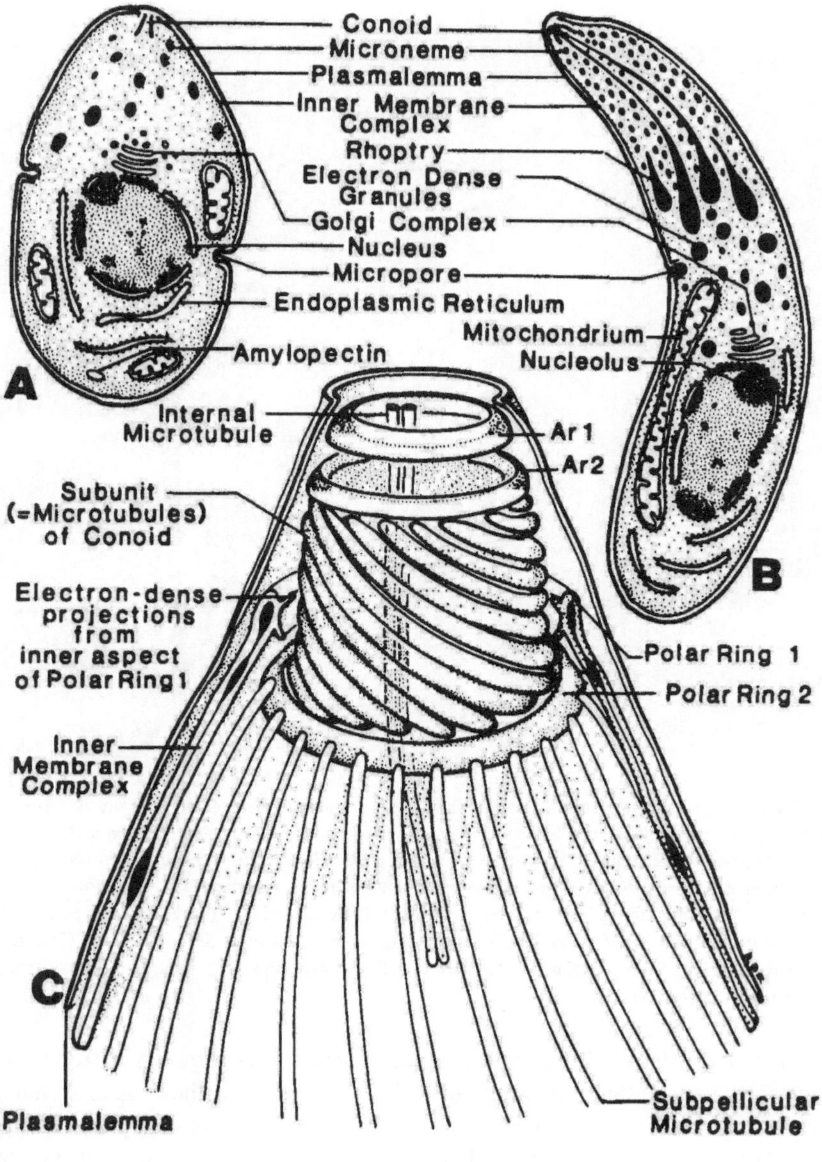

FIGURE 1. Diagrammatic representation of a metrocyte (A) and a merozoite (B) of *Sarcocystis* sp. with an enlargement of the apical complex (C) showing the organelles common to invasive stages of most coccidia. (From Dubey, J. P., Speer, C. A., and Fayer, R., *Sarcocystis of Animals and Man*, CRC Press, Boca Raton, FL, 1988. With permission.)

more specifically with the phylum itself.[25] A more recent phylogenetic analysis of the apicomplexan class Sporozoea (Sporozoasida)[26] suggests that previous treatments[24,25] of relationships of the coccidia to other groups such as the piroplasms and hemogregarines will have to be reconsidered; however, this does not affect the relationships among the true coccidia, especially those of medical and veterinary importance.

The coccidia of medical and veterinary importance are contained in three different families of the suborder Eimeriorina (the true coccidia). The taxonomy of the true coccidia has been the subject of considerable controversy and change during the past two decades, and as new information concerning the morphology, life cycles, and molecular genetics become available, the changes and controversy will continue. One ongoing controversy involves species of *Isospora (I. felis, I. rivolta, I. canis,* and *I. ohioensis*) which are

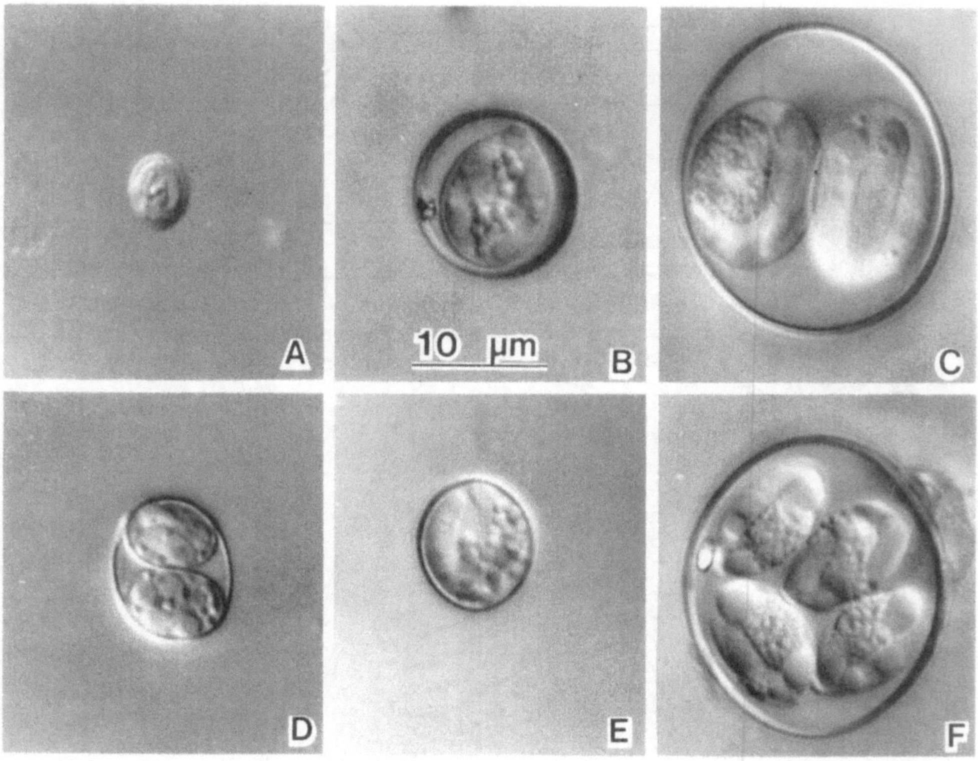

FIGURE 2. Nomarski interference contrast photomicrographs showing structure and comparative sizes of sporulated oocysts of some coccidia of medical and veterinary importance. (A) *Cryptosporidium parvum* (oocyst without sporocysts and with four naked sporozoites) from a goat. (B) *Caryospora bigenetica* (oocyst with one sporocyst containing eight sporozoites) from an Eastern diamondback rattle snake. (C) *Isospora rivolta* (oocyst with two sporocysts each containing four sporozoites) from a domesticated cat. (D) *Toxoplasma gondii* (oocyst with two sporocysts each with four sporozoites) from a domesticated cat. (E) *Sarcocystis* sp. (free sporocyst containing four sporozoites) from a snake, *Crotalus atrox*. (F) *Eimeria nieschulzi* (oocyst with four sporocysts each containing two sporozoites) from a rat.

facultatively heteroxenous and form cyst-like stages containing single sporozoites (hypnozoites) in the intermediate host. Dubey,[10] in 1977, proposed that these heteroxenous isosporans should be placed in a separate genus, *Levineia*, and independently that same year, Frenkel[27] proposed they should be placed in a new genus, *Cystoisospora*. Although both types of *Isospora* have obvious differences in their life cycles, we have adopted the conservative approach of placing both types in the single genus *Isospora*. The most recent classifications of the coccidia are ones in which both oocyst structure and aspects of the parasite life cycle are reflected in the placement of different genera. A comparison of the different types of oocysts found in coccidia of medical and veterinary importance is presented in Figure 2. Table 1 presents the classification scheme used in this book for the coccidia of medical and veterinary importance; it is basically the one proposed by Levine.[24,25]

IV. LIFE CYCLES

Coccidia have life cycle stages that occur inside the host(s), termed **endogenous stages. Exogenous stages** occur outside the host, and involve maturation of oocysts (sporogony). As mentioned above, certain species of coccidia are **homoxenous**, with all endogenous stages occurring within one host. Other species are **heteroxenous**, with certain endogenous stages occurring within the definitive host and other stages occurring within the intermediate

TABLE 1
Taxonomic Classification of the Coccidia of Medical and Veterinary Importance

Phylum Apicomplexa Levine, 1970

Apical complex present at some stage, usually consisting of polar ring(s), rhoptries, micronemes, conoid, and subpellicular microtubules; micropore(s) generally present at some stage; sexuality, when present, by syngamy; all species parasitic; two or three classes (depending on taxonomic scheme); about 4000 named species.

Class Sporozoasida Leukart, 1879

Conoid usually present, forming a complete cone; reproduction usually both asexual and sexual; oocysts with sporozoites, the infective stage resulting from sporogony; locomotion by gliding, undulation, flexion or flagella; flagella, when present, only on microgametes; homoxenous or heteroxenous; two subclasses; about 4000 named species.

Subclass Coccidiasina Leukart, 1879

Gamonts usually present, normally small and intracellular; conoid not modified into mucron or epimerite; syzygy usually absent, if present involves gametes; gametes usually anisogametes; life cycles normally consist of merogony, gamogony, and sporogony; most species parasites of vertebrates; four orders.

Order Eucoccidiorida Léger and Doboscq, 1910

Merogony, gamogony, and sporogony present; in vertebrates or invertebrates; two suborders, the Eimeriorina (the true coccidia) and the Haemosporina (containing the malarial parasites).

Suborder Eimeriorina Léger, 1911

Macrogamete and microgamete develop independently; syzygy generally absent; microgamont usually produces numerous microgametes; zygote usually not motile; homoxenous or heteroxenous; 13 families.

Family Eimeriidae Minchin, 1903

Oocysts with 0, 1, 2, 4, or more sporocysts, each with one or more sporozoites; merogony and gamogony usually within same host; homoxenous or facultatively heteroxenous; microgametes with two or three flagella; in vertebrates or invertebrates; 18 genera.

Genus *Caryospora* Léger, 1911

Oocysts with a single sporocyst, each with eight sporozoites; homoxenous or facultatively heteroxenous; in reptiles and birds; about 41 named species; Type Species *Caryospora simplex* Léger, 1911.

Genus *Eimeria* Schneider, 1875

Oocysts with four sporocysts, each with two sporozoites; homoxenous, although some species may have sporozoites and/or merozoites that may become dormant intestinally or extraintestinally and possess the capability of later producing patent infections; a few species have extraintestinal developmental stages; in vertebrates, rarely invertebrates; about 1200 named species; Type Species *Eimeria falciformis* (Eimer, 1870) Schneider, 1875.

Genus *Isospora* Schneider, 1881

Oocysts with two sporocysts, each with four sporozoites; homoxenous, although some species may have sporozoites and/or merozoites that may become dormant intestinally or extraintestinally and possess the capability of later producing patent infections; some species. (*Cystoisospora* Frenkel 1977) may have an additional intermediate host in which dormant sporozoites can transmit the infection to the definitive carnivore host; in vertebrates, rarely in invertebrates; sporogony inside or outside of host; about 250 named species; Type Species *Isospora rara* Schneider, 1881.

Family Cryptosporidiidae Tyzzer, 1907

Oocysts small, with no sporocysts and four naked sporozoites; development just under host cell membrane so that the parasite projects into the lumen; developmental stages form a feeder organelle or attachment organelle that anchors the parasite to the base of the parasitophorous vacuole; developmental stages of some species lacking mitochondria; microgametes without flagella; homoxenous; in vertebrates; sporogony endogenous; one genus.

Genus *Cryptosporidium* Tyzzer, 1907

With characteristics of the family; more than 20 named species, only about six of which are valid; Type Species *Cryptosporidium muris* Tyzzer, 1907.

Family Sarcocystidae Poche, 1913

Heteroxenous, with most species producing oocysts containing two sporocysts, each with four sporozoites; syzygy absent; intestinal in definitive host; asexual stages in various tissues of intermediate host; two subfamilies.

Subfamily Sarcocystinae Poche, 1913

Obligatory heteroxenous; asexual multiplication in intermediate host (a prey animal); sexual reproduction and sporogony in intestine of vertebrate host (a predator); first generation meronts in vascular endothelium; last generation meronts in striated muscle or in central nervous system contain metrocytes that form bradyzoites; three genera.

TABLE 1 (continued)

Genus *Frenkelia* Biocca, 1968
Last generation meronts typically in central nervous system; merozoites elongate; definitive hosts raptors; intermediate hosts rodents; about two named species; Type Species *Frenkelia microti* (Findlay and Middleton, 1934) Biocca, 1968.
Genus *Sarcocystis* Lankester, 1882
Last generation meronts typically in striated muscle, some in neural tissue; merozoites elongate, about 125 named species; Type Species *Sarcocystis miescheriana* (Kühn, 1865) Labbé, 1899.
Subfamily Toxoplasatinae Biocca, 1956
Complete life cycle obligatory heteroxenous, but asexual stages transmissible from one intermediate host to another; metrocytes not formed; sporogony exogenous; four genera.
Genus *Besnoitia* Henry, 1913
Meronts in fibroblasts and other cells; host nuclei within meront wall; tissue cyst wall thick; definitive hosts mammals; intermediate hosts mammals and reptiles; about seven named species; Type Species *Besnoitia besnoiti* (Marotel, 1913) Henry, 1913.
Genus *Hammondia* Frenkel, 1974
Meronts in numerous cell types; oocysts not infectious to definitive host; definitive host felids and canids; intermediate hosts mammals; about three named species; Type Species *Hammondia hammondi* Frenkel, 1974.
Genus *Neospora* Dubey, Carpenter, Speer, Topper, and Uggla, 1988
Oocysts and intestinal stages unknown; zoites divide by endodyogeny in numerous tissues, especially brain and spinal cord, usually without formation of parasitophorous vacuole; tissue cysts thick-walled, found generally in the central nervous system; in canids; about one named species; Type Species *Neospora caninum* Dubey, Carpenter, Speer, and Uggla, 1988
Genus *Toxoplasma* Nicolle and Manceaux, 1908
Meronts in numerous cell types; host cell nucleus outside meront wall; cysts typically spherical or subspherical; definitive hosts felids; intermediate hosts consist of numerous species of vertebrates; one known valid species but about five additional species are named based on cysts found in intermediate hosts (several may prove to represent non-*Toxoplasma* spp.); Type Species *Toxoplasma gondii* (Nicolle and Manceaux, 1908)

host. For heteroxenous coccidia, the **definitive host** is the one in which sexual stages of the endogenous life cycle occur; the **intermediate host** supports only endogenous asexual stages. Regardless of whether a particular species is homoxenous or heteroxenous, all known coccidia progress sequentially through a series of life cycle stages that ultimately result in the formation of oocysts, environmentally resistant forms that can transmit the infection to the next susceptible host. Prior to specific discussions of the coccidia addressed in this book, life cycle events common to the coccidia of medical and veterinary importance will be reviewed in an attempt to define the terminology associated with these complex protozoan parasites. Additional details of the structure and biology of coccidia as they progress through these different life cycle events have been reviewed.[5,7,9,10,27-29]

A. SPOROGONY

Oocysts are the exogenous stages that are usually shed in the feces of a definitive host. Depending on the species, environmentally resistant oocysts may undergo sporogony within or outside the host. **Sporogony** is the process by which a one celled **sporont** (zygote) within the oocyst wall undergoes a series of divisions to form **sporozoites**, which may lie free within the oocyst wall or which may be contained within **sporocysts** (Figure 3). For those species whose oocysts sporulate outside the host, oxygen, moisture, and optimum temperatures are important. Only sporulated oocysts, those containing fully formed sporozoites, are infective to the definitive and/or intermediate hosts.

B. EXCYSTATION AND HOST CELL ENTRY

After being ingested by a definitive or intermediate host, the sporulated oocysts undergo the process of **excystation**, the release of infective sporozoites. In most of the coccidian species examined to date, excystation can occur *in vitro* by exposing oocysts or sporocysts

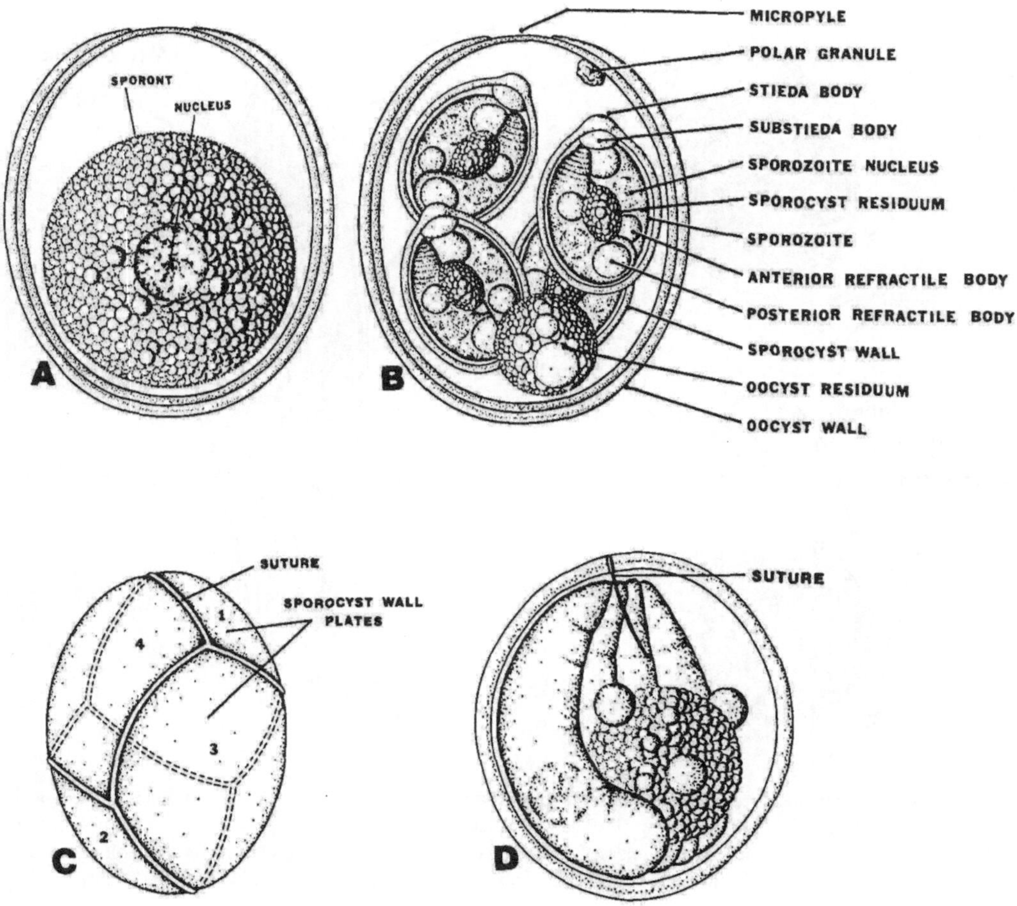

FIGURE 3. Diagrammatic representation of oocysts and sporocysts of several coccidian genera. (A) Unsporulated oocyst of *Eimeria* sp. (B) Sporulated oocysts of *Eimeria* sp. containing four sporocysts, each with two sporozoites. (C) Sporocyst of *Sarcocystis* sp. showing arrangement of wall plates and sutures. The sporocyst wall encloses four sporozoites (redrawn from Reference 30). (D) Oocyst of *Cryptosporidium parvum* showing the suture that, when dissolved, opens up to release sporozoites.

to conditions that simulate the gastrointestinal environment of the host, i.e., reducing conditions, CO_2, host body temperature, bile salts, and trypsin. Such investigations have suggested that there are two major steps in excystation. The first step, an alteration of oocyst wall permeability, can be triggered by exposure to host body temperature and CO_2. This step may also be accomplished by sodium hypochlorite treatment or by physical means such as the grinding process that occurs in the gizzard of an avian host or in a tissue grinder. The second step is the release of sporozoites from sporocysts by the action of pancreatic enzymes and/or bile salts. Sporozoites of many eimeriid coccidia escape through an opening in one pole of the sporocyst that is formed by degradation of a plug, the **Stieda body** (Figure 3). It is believed that trypsin degrades the Stieda body and that bile salts stimulate sporozoite motility. Other coccidia have sporocysts whose walls are composed of four **plates joined by sutures** (Figure 3). The wall of *Cryptosporidium* spp. oocysts contains a single suture that traverses approximately one third the circumference of the oocyst (Figure 3). Trypsin and/or bile salts cause dissolution of the sutures, allowing sporozoites to escape between the collapsed plates. Once free in the intestinal lumen, the motile sporozoites actively penetrate into a host cell, usually enterocytes or cells of the lamina propria. It is believed

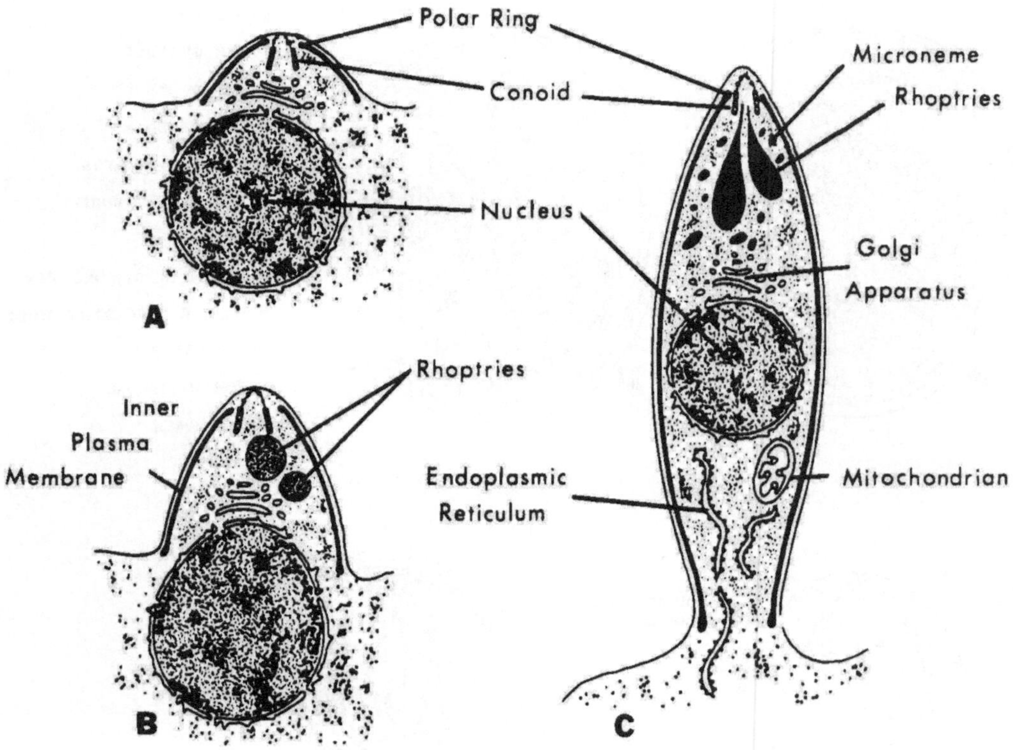

FIGURE 4. Diagrammatic representation of merogony (ectomerogony). (A) Nucleus and organelles of the apical complex begin to "bud" from the surface. (B) The inner plasma membrane continues to form. (C) Soon after this late stage of merogony, the merozoite is released from the surface of the meront. (Adapted from Scholtyseck, E., in *The Coccidia*, Hammond, D. M. and Long, P. L., Eds., University Park Press, Baltimore, 1973, chap. 4.)

that organelles of the apical complex (Figure 1) of the invasive stages (sporozoites and merozoites) are involved in the active penetration of host cells.

C. MEROGONY (SCHIZOGONY)

Once inside the host cell, sporozoites round up and transform into a **uninucleate meront** (trophozoite, uninucleate schizont). **Merogony**, the asexual proliferative phase of most coccidia, is initiated when several mitotic nuclear divisions occur and is completed when elongate merozoites are released from the surface of the meront by multiple fission (Figure 4). These **merozoites** then leave the host cell to enter other host cells and form one or more additional meront generations. Most coccidian species usually form a characteristic number of merozoites in each meront generation and most have a characteristic number (usually two to four) of asexual generations. Certain coccidia, especially some species of *Isospora* and those within the family Sarcocystidae, have some asexual stages that multiply by endodyogeny (Figure 5). **Endodyogeny** is the division process by which two daughter organisms are formed within the mother cell, resulting in the ultimate destruction of the mother cell. **Zoites** (tissue-dwelling merozoites) formed by endodyogeny that proliferate rapidly, often forming colonies or pseudocysts, are called **tachyzoites** and are the tissue stages associated with acute disease. Zoites which proliferate more slowly by endodyogeny within resistant tissue cysts are called **bradyzoites**, and it is this life cycle stage that may persist indefinitely within the host.

D. GAMOGONY

Merozoites of the final generation of merogony enter host cells and initiate the sexual

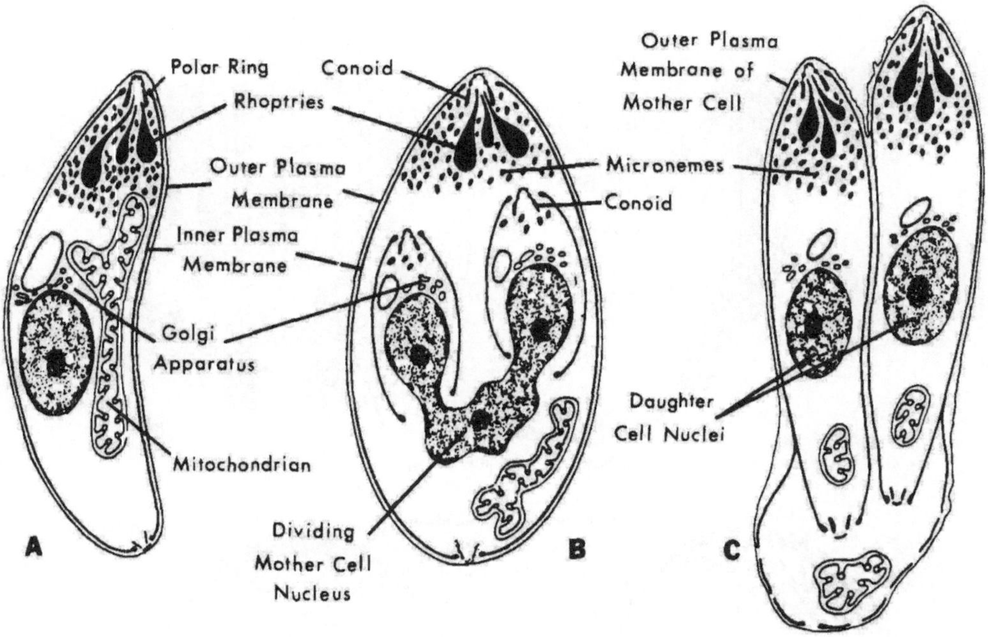

FIGURE 5. Diagrammatic representation of endodyogeny. (A) Merozoite-shaped meront prior to division. (B) Initiation of division to form two daughter merozoites within the mother cell (meront). (C) Late stage of division showing that the outer plasma membrane of the mother cell forms the outer plasma membrane of the separating daughter cells. (Adapted from Scholtyseck, E., in *The Coccidia*, Hammond, D. M. and Long, P. L., Eds., University Park Press, Baltimore, 1973, chap. 4.)

portion of the endogenous cycle (**gamogony**) by developing into male or female gamonts, the microgamonts and macrogamonts, respectively. Some merozoites become **microgamonts** that undergo repeated nuclear divisions followed by cytoplasmic divisions. Following nuclear division, several to many **microgametes** (Figure 6) bud from the surface of the microgamont. All known coccidian microgametes, except those of the genus *Cryptosporidium*, are fusiform, contain a nucleus (some may have two), and have two or three flagella for locomotion; those of *Cryptosporidium* do not have flagella and move by gliding. Other merozoites become **macrogamonts** which do not undergo nuclear divisions but increase several times in size (Figure 6). This increase in size of the macrogamont includes the proliferation of various organelles, including the **wall-forming bodies** that are involved in the subsequent formation of an oocyst wall. A macrogamete is fertilized by a microgamete to form a **zygote**, which then forms at its margin an environmentally resistant **oocyst wall** (Figure 6). The oocyst is released from its host cell and shed into the environment. Oocysts of some species may undergo sporogony within the host whereas oocysts of other species will not sporulate until they are exposed to the proper environmental conditions outside the host.

E. SPECIFIC LIFE CYCLES

A comparison of the known life cycles of coccidia of medical and veterinary importance, including endogenous and exogenous stages, routes of transmission, and number of hosts is presented diagrammatically in Figure 7 and the life cycle of each genus is discussed below.

1. *Eimeria* spp. — Homoxenous Life Cycle (Figure 7a)

As presented, species of *Eimeria* are usually homoxenous and have an endogenous intestinal cycle with asexual stages that proliferate by multiple fission (merogony) followed

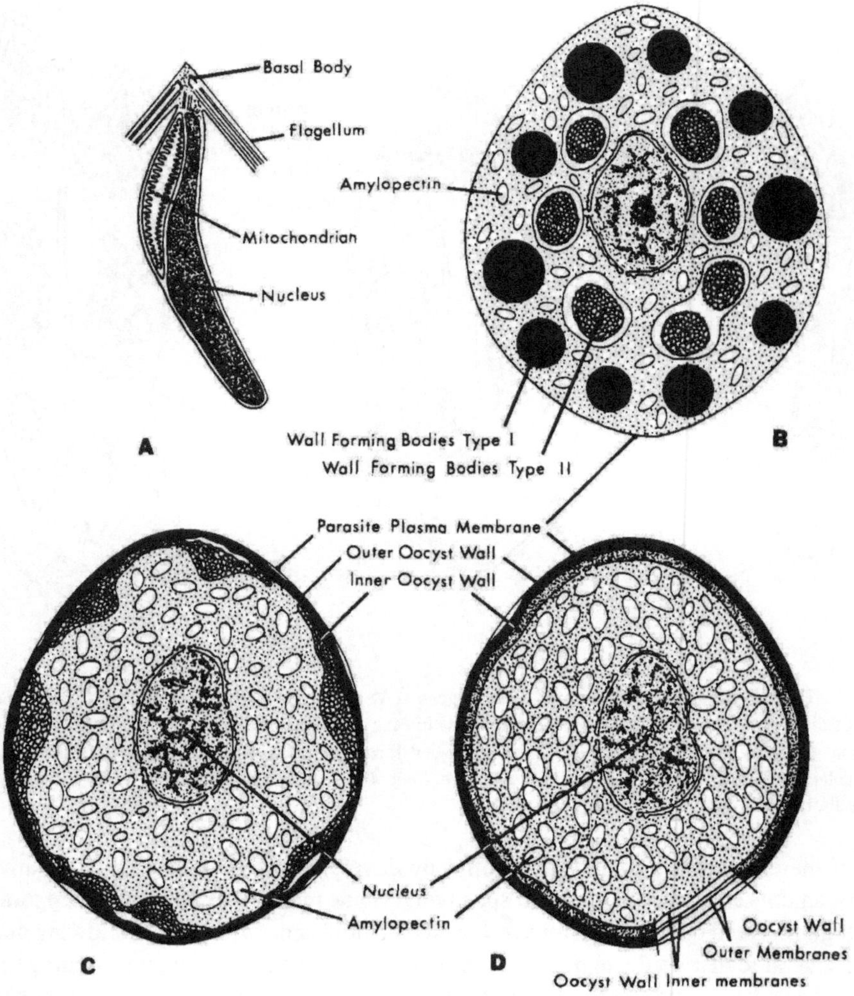

FIGURE 6. Diagrammatic representations of sexual stages of coccidia. (A) Flagellated, motile microgamete showing just the anterior portion of the two flagella which normally extend beyond the posterior of the cell. (B) Macrogamont showing wall forming bodies that coalesce between membranes at the surface of the parasite to form the outer and inner oocyst walls. (C) Developing oocyst with outer oocyst wall complete and the inner wall being formed. (D) Unsporulated oocyst showing completely formed outer and inner oocyst walls surrounding a one-celled sporont. (Adapted from Scholtyseck, E., The *Coccydia,* Hammond, D. M. and Long, P. L., Eds., University Park Press, Baltimore, 1982.)

by sexual stages (gamogony) and oocyst formation. Unsporulated oocysts are passed from the host and undergo sporogony in the environment to form sporulated oocysts containing four sporocysts, each with two sporozoites. The life cycle is completed when a nonimmune host ingests the infective, sporulated oocysts. Some species of *Eimeria* have adapted their life cycles to include development in extraintestinal sites such as the liver, gall bladder, kidneys, and even in the placenta, peritoneal cavity and uterus. Sporozoites and merozoites of some *Eimeria* spp. of rodents also appear to be capable of entering extraintestinal tissues, becoming dormant and reinitiating infection at later intervals. Additional details of the eimerian life cycle are presented in Chapter 4 and in the review of Hammond.[29]

2. *Isospora* spp. — Homoxenous Life Cycle (Figure 7b)

Most species of *Isospora* have life cycles and host specificity similar to those of *Eimeria* spp. The major differences between the two genera include oocyst structure (oocysts of

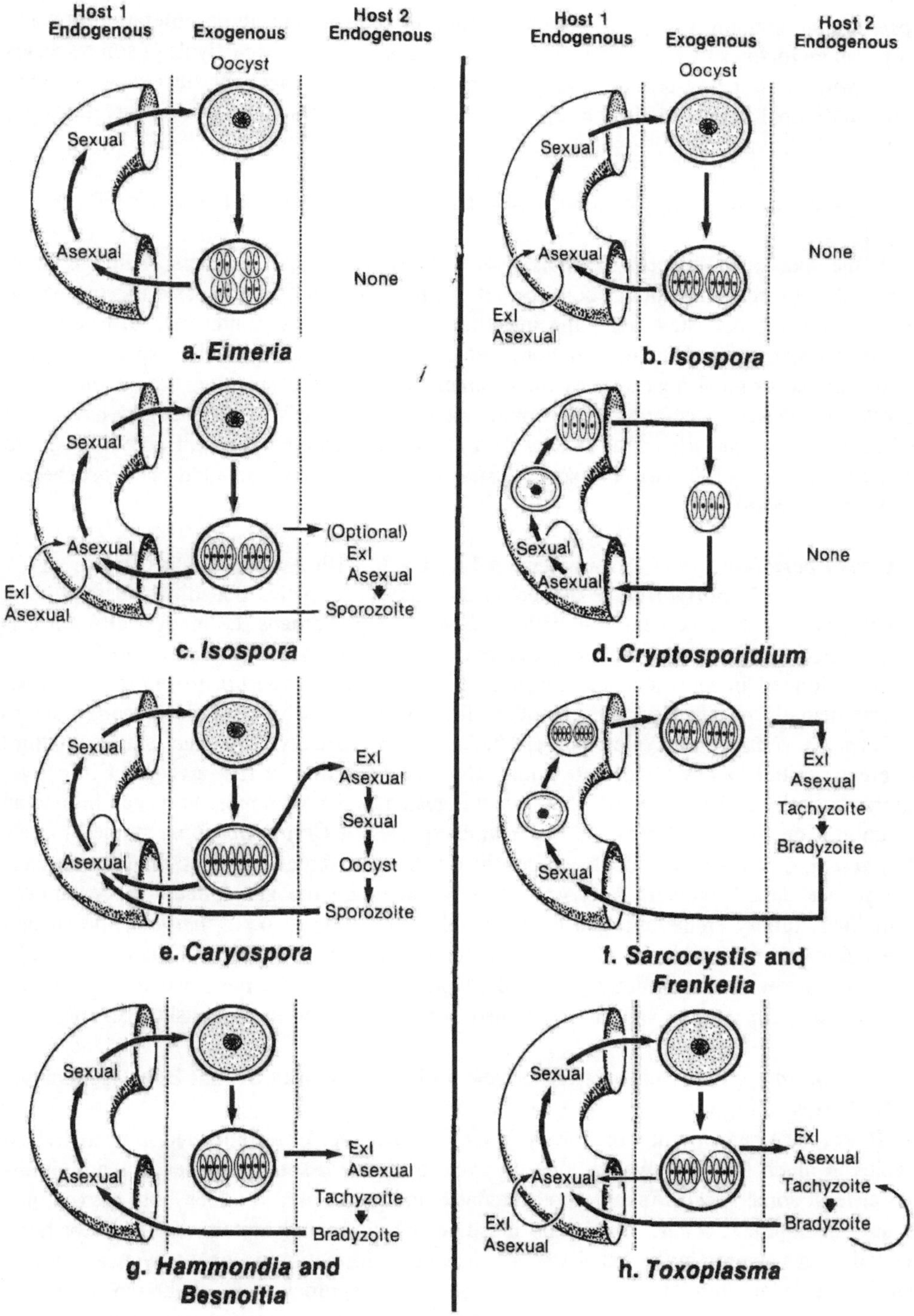

FIGURE 7. Simplified life cycles of coccidian genera of medical and veterinary importance. For each genus, a diagrammatic representation of the intestinal tract appears on the left under "Host 1" (definitive host), oocyst structure is shown in the center under "Exogenous", and the extraintestinal stages (ExI) that develop in the intermediate host(s) are listed in order of appearance at the right under "Host 2". In some genera, extraintestinal stages (ExI) also occur in the definitive host. (Portions adapted from Fayer, R. and Dubey, J. P., *Int. J. Parasitol.*, 17, 615, 1987.)

Isospora spp. contain two sporocysts each with four sporozoites), the proliferation of some isosporan merozoites by endodyogeny, and that portions of the asexual cycle of some species of *Isospora* (now being termed *Atoxoplasma*) occur in extraintestinal sites such as blood cells. Additional details of the life cycle of *Isospora* spp. producing enteric infections are contained in Chapter 5.

3. *Isospora* spp. — Homoxenous Cycle with Optional Heteroxenous Cycle (Figure 7c)

Some species of *Isospora* (***Cystoisospora***) differ from most members of the genus in their ability to infect additional host species. In the additional host species the sporozoite excysts from the oocyst, crosses the intestinal wall, enters a nonintestinal host cell, and remains dormant until, through predation, it enters the gut of the original host species. Within the gut of the original host species the dormant sporozoite (cystozoite, dormozoite, hypnozoite) is released, penetrates an intestinal epithelial cell, and initiates endogenous asexual and sexual development just as if it were a sporozoite released from an oocyst. A more detailed discussion of the *Isospora* spp. (*Cystoisospora*) that utilize an additional host species has been published.[28]

4. *Cryptosporidium* spp. — Homoxenous Life Cycle with Autoinfection (Figure 7d)

Species of *Cryptosporidium* studied to date have life cycles that differ in four basic aspects from the homoxenous life cycle of *Eimeria* spp. Perhaps the biggest difference is the presence of thin-walled, autoinfective oocysts in some species of *Cryptosporidium* that allow prolonged infections in the absence of repeated oral exposure to the thick-walled, environmentally resistant oocysts. Another difference is the ability of some meront stages of *Cryptosporidium* to recycle, thus producing larger numbers of asexual stages. A third difference is that oocysts sporulate within the host, resulting in the passage of infective, environmentally resistant oocysts that, when ingested by a nonimmune host, can initiate an infection. The fourth difference is that several species of *Cryptosporidium* studied to date (*C. parvum, C. baileyi*) are not always confined to the gastrointestinal tract, but can complete endogenous development in a variety of sites, including the gut epithelium, respiratory epithelium, biliary epithelium and conjunctival epithelium. Also, *C. parvum*, the species believed to be responsible for all well documented cases of cryptosporidiosis in mammals, lacks the narrow host specificity exhibited by species of *Eimeria*. Further details of the life cycle of *Cryptosporidium* spp. are contained in Chapter 8 and in a published review.[5]

5. *Caryospora* spp. — Heteroxenous Cycle with Asexual and Sexual Development in Both Hosts (Figure 7e)

It was once believed that the life cycle *Caryospora* spp. was completed in a single host species in much the same way as the life cycle of *Eimeria* spp.; however, we now know that at least some species involve both predator and prey hosts in which both asexual and sexual development occur. Within the intestine of snakes and raptors, the predator host, asexual stages (merogony), sexual stages, and oocyst formation occur. Oocysts are passed and then sporulate into infective forms that contain one sporocyst with eight sporozoites. If rodents are inoculated orally with oocysts of *C. simplex* or *C. bigenetica*, excystation occurs, sporozoites cross the intestinal wall and find their way to extraintestinal sites such as the tongue and the dermis and hypodermis of the facial area where asexual stages (merogony), sexual stages, oocyst formation, sporogony, and excystation occur. Sporozoites that excyst from thin-walled, extraintestinal oocysts then enter host cells and encyst as dormant sporozoites (caryocysts). These caryocysts are infective for the predator host when it ingests an infected prey host. Additional details of the life cycle of *Caryospora* spp. are contained in Chapter 9 and in a taxonomic review of the genus.[4]

6. *Sarcocystis* and *Frenkelia* spp. — Obligate Heteroxenous Cycle (Figure 7f)

Species of *Sarcocystis* develop in the subepithelium (lamina propria) of the intestine of carnivorous reptiles, birds or mammals; for *Frenkelia* spp. the only known definitive hosts are raptors. Within the definitive, carnivore host, only sexual stages of the life cycle occur, followed by oocyst formation and sporogony. Sporulated, *Isospora*-like oocysts are passed intact or as free sporocysts following disruption of the oocyst wall. Oocysts or sporocysts of each species will infect only a narrow range of intermediate hosts; for *Sarcocystis* spp. they include fishes, reptiles, birds, and mammals, and for *Frenkelia* spp. mammals are the only known intermediate hosts. Development in the intermediate host is confined to extraintestinal, asexual stages. For *Sarcocystis* spp., early merogonous development usually occurs in the vascular endothelium, followed by development of tachyzoites in intravascular mononuclear cells, and then formation of bradyzoites within cysts located in skeletal and cardiac muscle and in neural tissue. For *Frenkelia* spp., merogony takes place in the liver parenchyma and then bradyzoites form cysts in the brain. Bradyzoites of *Sarcocystis* and *Frenkelia* spp. are formed within tissue cysts from metrocytes, pre-merozoite mother cells that divide by endodyogeny. The life cycle is completed when a susceptible definitive host ingests mature tissue cysts containing bradyzoites. Additional aspects of the life cycles of *Sarcocystis* and *Frankelia* can be found in Chapter 6 and in published reviews.[7,10]

7. *Hammondia* and *Besnoitia* spp. — Obligate Heteroxenous Cycle (Figure 7g)

Development of *Besnoitia* and *Hammondia* spp. in the intestinal epithelium of carnivores includes asexual stages (merogony), sexual stages, and oocyst formation. Unsporulated oocysts are passed in the feces and then sporulate in the environment. These isosporan-like oocysts, often identical in shape to those of *Toxoplasma gondii*, infect only the intermediate hosts (reptiles and mammals). Initially, asexual development of tachyzoites occurs in the intestine, followed by bradyzoites forming tissue cysts. Characteristically, *Besnoitia* spp. form thick-walled cysts in subcutaneous or mesenteric tissues. The thinner-walled cysts of *Hammondia* spp. are found in skeletal muscle and neural tissue. Metrocytes are not formed. Ingestion of tissue cysts by the definitive host completes the life cycle. For additional information about the life cycles and biology of *Besnoitia* and *Hammondia* spp. the reader should refer to the works of Dubey[10] and Frenkel.[28]

8. *Toxoplasma gondii* — Heteroxenous Cycle with Optional Homoxenous Cycle (Figure 7h)

Within the intestine of felids (the definitive host) *T. gondii* develops through asexual stages (merogony and endodyogeny), sexual stages, and oocyst formation. Unsporulated oocysts are released and complete sporulation in the environment. Virtually all other vertebrates, including man, can serve as intermediate hosts. In the intermediate host sporozoites excyst from the oocysts, find their way into extraintestinal sites, enter host cells throughout the body where they proliferate as rapidly dividing tachyzoites, and then form numerous bradyzoites in tissue cysts. Metrocytes are not formed. Ingestion of tachyzoites or bradyzoites by other vertebrate hosts will initiate a similar tachyzoite, then bradyzoite cycle. Ingestion of either of these asexual forms by cats will lead to development of all stages in the gut and to tissue-dwelling zoites that are also found in the intermediate host.

Additional information on the live cycle of *T. gondii* can be found in Chapter 7 and in published reviews.[8,9]

REFERENCES

1. **Levine, N. D.,** Taxonomy and life cycles of coccidia, in *The Biology of The Coccidia,* Long, P. L., Ed., University Park Press, Baltimore, 1982, chap. 1.
2. **Current, W. L.,** Neonatal swine coccidiosis, *Anim. Health Nut.,* March 8, 1987.
3. **Garcia, L. S., Owen, R. L., and Current, W. L.,** Isosporiasis, in *Laboratory Diagnosis of Infectious Diseases: Principles and Practice,* Vol. 1, Ballows, A., Hausler, W. J., Jr., and Lennett, E. H., Eds., Springer-Verlag, New York, 1988, chap. 95.
4. **Upton, S. J., Current, W. L., and Barnard, S. M.,** A review of the genus *Caryospora* Leger, 1904 (Apicomplexa: Eimeriidae), *Syst. Parasitol.,* 8, 3, 1986.
5. **Current, W. L.,** *Cryptosporidium* spp., in *Parasitic Infections in the Compromised Host,* Walzer, P. D. and Genta, R. M., Eds., Marcel Dekker, New York, 1989, chap. 5.
6. **Current, W. L.,** Cryptosporidiosis, in *Diseases of Poultry,* 9th ed., Calnek, B. W., Ed., Iowa State University Press, Ames, IA, 1990, chap. 32.
7. **Dubey, J. P., Speer, C. A., and Fayer, R.,** *Sarcocystosis of Animals and Man,* CRC Press, Boca Raton, FL, 1988.
8. **Luft, B. J.,** *Toxoplasma gondii,* in *Parasitic Infections in the Compromised Host,* Walzer, P. D. and Genta, R. M., Eds., Marcel Dekker, New York, 1989, chap. 4.
9. **Dubey, J. P. and Beattie, C. P.,** *Toxoplasmosis of Animals and Man,* CRC Press, Boca Raton, FL., 1988.
10. **Dubey, J. P.,** *Toxoplasma, Hammondia, Besnoitia, Sarcocystis,* and other tissue cyst-forming coccidia of man and animals, in *Parasitic Protozoa,* Vol. 3, Krier, J. P., Ed., Academic Press, New York, 1977, 101.
11. **Dubey, J. P., Carpenter, J. L., Speer, C. A., Topper, M. J., and Uggla, A.,** Newly recognized fatal protozoan disease of dogs, *J. Am. Vet. Med. Assoc.,* 192, 1269, 1988.
12. **Levine, N. D.,** Introduction, history, and taxonomy, in *The Coccidia,* Hammond, D. M. and Long, P. L., Eds., University Park Press, Baltimore, 1973, chap. 1.
13. **Scholtyseck, E.,** Ultrastructure, in *The Coccidia,* Hammond, D. M. and Long, P. L., Eds., University Park Press, Baltimore, 1973, chap. 4.
14. **Chobotar, B. and Scholtyseck, E.,** Ultrastructure, in *The Biology of the Coccidia,* Long, P. L., Ed., University Park Press, Baltimore, 1982, chap. 4.
15. **Scholtyseck, E. and Piekarski, G.,** Elektronenmikroskopische Untersuchungen Uber die Merozoiten von *Eimeria stiedai* und *E. perforans* und *Toxoplasma gondii*: Sur systematischen Stellung von *T. gondii, Z. Parasitenkd.,* 26, 91, 1965.
16. **Senaud, J.,** Contribution a l'étude des Sarcosporidies et des Toxoplasmes, *Protistologica,* 5, 413, 1967.
17. **Hutchison, W. M.,** Experimental transmission of *Toxoplasma gondii, Nature (London),* 206, 961, 1965.
18. **Frenkel, J. K., Dubey, J. P., and Miller, N. L.,** *Toxoplasma gondii*: fecal forms separated from eggs of the nematode *Toxocara cati, Science,* 164, 432, 1969.
19. **Dubey, J. P., Miller, N. L., and Frenkel, J. K.,** The *Toxoplasma gondii* oocyst from cat feces, *J. Exp. Med.,* 132, 636, 1970.
20. **Frenkel, J. K., Dubey, J. P., and Miller, N. L.,** *Toxoplasma gondii*: fecal stages identified as coccidian oocysts, *Science,* 167, 893, 1970.
21. **Fayer, R.,** Gametogony of *Sarcocystis* sp. in cell culture, *Science,* 175, 65, 1972.
22. **Rommel, M., Haydorn, A. O., and Gruber, F.,** Beitrage zum Lebensxyklus der Sarkosporidien. I. Die Sporozyste von *S. tenella* in den Fazes der Katze, *Berl. Muench. Tieraerztl. Wochenschr.,* 85, 101, 1972.
23. **Current, W. L., Reese, N. C., Ernst, J. V., Bailey, W. S., Heyman, M. B., and Weinstein, W. M.,** Human cryptosporidiosis in immunocompetent and immunodeficient persons — studies of an outbreak and experimental transmission, *N. Engl. J. Med.,* 308, 1252, 1983.
24. **Levine, N. D.,** *The Phylum Apicomplexa,* Vol. I, CRC Press, Boca Raton, FL, 1988, chap. 10.
25. **Levine, N. D.,** *The Phylum Apicomplexa,* Vol. II, CRC Press, Boca Raton, FL, 1988, chap. 11.
26. **Barta, J. R.,** Phylogenetic analysis of the class sporozoea (Phylum Apicomplexa Levine, 1970): evidence for the independent evolution of heteroxenous life cycles, *J. Parasitol.,* 75, 195, 1989.
27. **Frenkel, J. K.,** *Besnoitia wallacei* of cats and rodents; with a reclassification of other cyst-forming isosporid coccidia, *J. Parasitol.,* 63, 611, 1977.
28. **Hammond, D. M.,** Life cycles and development of coccidia, in *The Coccidia,* Hammond, D. M. and Long, P. L., Eds., University Park Press, Baltimore, 1973, chap. 3.
29. **Wang, C. C.,** Biochemistry and physiology of coccidia, in *The Biology of the Coccidia,* Long, P. L., Ed., University Park Press, Baltimore, 1982, chap. 5.
30. **Box, E. D., Marchioldo, A. A., Duszynski, D. W., and Davis, C. P.,** Ultrastructure of *Sarcocystis* sporocysts from passerine birds and opossums: comments on classification of the genus *Isospora, J. Parasitol.,* 66, 68, 1980.
31. **Fayer, R. and Dubey, J. P.,** Comparative epidemiology of coccidia: clues to the etiology of equine protozoal myoencephalitis, *Int. J. Parasitol.,* 17, 615, 1987.

Chapter 2

STRUCTURE AND ULTRASTRUCTURE

S. J. Ball and R. M. Pittilo

TABLE OF CONTENTS

ABBREVIATIONS TO THE FIGURES

ACR — anterior preconoidal ring
AZ — adhesion zone
BS — beehive structure
C — conoid
CA — canaliculi·
CE — centriole
CW — cyst wall
CZ — cystozoite
E — endozoite
ER — granular endoplasmic reticulum
F — filopodium
FL — flagellum
G — Golgi body
HCM — host cell membrane
IF — intravacuolar fold
IM — inner membrane of pellicle
IMC — inner membrane complex
IT — intravacuolar tubule
L — lipid
La — lamellipodium
M — macrophage
M1—M5 — membranes of macrogamete pellicle and oocyst wall
MI — mitochondrion
MN — microneme
MP — micropore
MT — microtubule
MV — multimembranous vacuole
N — nucleus
NP — nuclear pore
NS — nuclear spindle
NU — nucleolus
OM — outer membrane of pellicle
OW — oocyst wall
P — plaque
PCR — posterior preconoidal ring
PG — polysaccharide granules
PL — pellicle
PR — polar ring
PT — protein granule
PV — parasitophorous vacuole
R — rhoptry
RB — residual body
RF — refractile body
SP — sporozoite
SW — sporocyst wall
V — vesicle
VA — vacuole
VT — vesicle with thick wall
W1 — wall-forming bodies type 1 (WF1)

W2 — wall-forming bodies type 2 (WF2)
W1G — W1 material with regular granular pattern
WG — granular W1 material
WL — W1-like material
W2L — W2-like material
X — granular part of outer layer of oocyst wall
Y — osmiophilic part of outer layer of oocyst wall
Z — inner layer of oocyst wall

I. INTRODUCTION

Most of the more fundamental aspects of the structure and ultrastructure of members of the families Eimeriidae and Sarcocystidae have been known for a number of years and presented in several excellent reviews.[1-3] Unfortunately, their functional organization is less well known. There is a remarkable constancy in the structure of some life cycle stages between different species and even between genera and much work during the past decade has been to record departures from the common organization, which may reflect adaptations to different microenvironments. So far there have been limited observations on sporogony due to the hard wall surrounding the zygote as it becomes an oocyst. Freeze fracture techniques should eventually help the study of this phase in the life cycle but these studies are in their infancy compared with transmission or even scanning electron microscopic investigations.

This chapter is intended as an update on the review of Chobotar and Scholtyseck[2] which together with the article by Scholtyseck[1] should be regarded as our basis.

II. THE OOCYST

A. PROBLEMS IN ULTRASTRUCTURAL EXAMINATION

Although the other stages in the life cycle of eimerian parasites have been well documented at the ultrastructural level the process of sporogony has received little attention. The reason for this stems from the fact that conventional techniques for preparation of tissues for transmission electron microscopy are virtually useless when applied to coccidian oocysts. This is because both the oocyst and sporocyst wall prevent the entry of plastic resin to the parasite. Birch-Andersen et al.[4] reported that both poor fixation and embedding of the cytoplasmic mass of *Eimeria brunetti* resulted from a variety of fixatives and resins when conventional techniques were employed. The problem is compounded in sporulated oocysts by the sporocyst wall which apparently has similar resistant properties to the oocyst wall.[5]

Despite these problems there have been a number of reports on the ultrastructure of oocysts or components of oocysts isolated from feces[6-18] (see also under oocyst wall formation). These reports have used mechanical techniques including sonication.[13] The main disadvantage of such techniques are that they result in disruption of the oocysts. However, they have been useful in revealing some aspects of oocyst ultrastructure and the process of excystation but have not aided studies on ultrastructure of sporogony.

Other methodologies have been used to examine coccidian oocysts. Jolley et al.[14] altered the oocyst walls of both *E. stiedai* and *E. tenella* chemically in order to make them porous to fixatives and resins. This method seems to result in the removal of the micropyle and in the case of *E. tenella* this structure is rendered visible after carbon dioxide treatment.[15] Wang and Stotish[19] produced electron micrograph autoradiographs of *E. tenella* after removing the outer walls using sodium hypochlorite. Dubremetz et al.[20] used the freeze-etch technique to study unsporulated oocysts of *E. tenella*. Sporulated oocyst of *E. stiedai* and *E. tenella* have also been studied by this method.[14] Probably the difficulties associated with the interpretation of freeze-etched replicas coupled with the expertise needed have prevented more widespread use of these methods.

The breakthrough for ultrastructural descriptions of sporogony resulted from a technique by which thin sections of coccidian oocysts could be reliably produced.[4] Essentially this technique involves embedding oocysts in bovine serum albumin (BSA) which is then frozen. Cryostat sections are then cut which are re-embedded in BSA. The BSA, containing sections of coccidian oocysts, is then cross-linked with glutaraldehyde and cut into small cubes which can be prepared for transmission electron microscopy using conventional techniques.[4] Using this technique ultrastructural details of sporogony have been described for *E. brunetti*,[5,21-23] *E. maxima*,[24,25] and *Toxoplasma gondii*.[26-28] Excystation was also described using this method for *T. gondii*.[29] Recently, Beesley and Latter[30] have used a freezing technique to fracture oocysts and reported ultrastructural aspects of sporulation of *E. tenella*.

Preparation of oocysts for scanning electron microscopy also presents problems in that distortion[15] and collapse[31] occur. It has been demonstrated that critical point drying results in loss of the outer layer of the oocyst wall in *Isospora lacazei*.[10] Despite these problems, scanning electron microscopy has been useful in revealing oocyst morphology both *in vitro* and *in vivo* and has even been used in a study on sporulation.[32] The surface of *E. stigmosa* oocysts from the goose has been described.[33] Strohlein and Prestwood[18] used both scanning and transmission electron microscopy to examine oocysts and sporocysts of *Sarcocystis suicanis* isolated from the mucosae of experimentally infected dogs. The sporocysts consisted of four plates which during excystation curled inwards and allowed the sporozoites to escape.[18] The sporocyst wall appeared similar to that of other coccidia when examined by transmission electron microscopy.[18] Oocysts of *Cryptosporidium parvum* have also been examined from the feces and the structure of the wall and the process of excystation reported.[34,35]

B. SPORULATION

Sporulation at the ultrastructural level for *E. brunetti* and *T. gondii* has been reviewed[2] (see Figures 1 through 4). There have only been two other reports on the ultrastructure of sporulation of *Eimeria* oocysts; sporulation of *E. maxima*[24] and sporulation of *E. tenella*.[30] There is also a report on observations made on the oocyst and sporocyst wall during sporogony of *S. alceslatrans*.[36] The limited ultrastructural studies made indicate that sporulation, like other developmental stages, show similarities between species. However, differences in detail exist. Electron translucent vacuoles are found at an early stage within the oocyst cytoplasm. There is evidence in *E. brunetti* that these are shed from the surface of the zygote and may function in waste disposal,[23] but this does not seem to occur with *E. maxima*[24] or *E. tenella*.[30] With *E. maxima* electron translucent vacuoles contained lipid prior to processing[24] and may therefore be analogous to the lipid globules found in *T. gondii* zygotes.[26] They may, along with the polysaccharide granules, form the food reserves for the oocyst. The formation of dense bodies which are closely associated with polysaccharide granules occurs in *E. maxima*[24] and *E. tenella*[30] and it seems likely that these structures are the precursors of the sporozoite refractile bodies.[23]

Scanning electron microscopy of sporulating *E. tenella* oocysts provided confirmatory observations to those seen using transmission electron microscopy.[32] The cytoplasm could be seen to contract during the early stages of sporulation and the formation of four sporoblasts was observed.[32]

Further studies using both scanning and transmission electron microscopy are required to consolidate the limited data available. There are a number of studies of sporulation in fish coccidia which may enhance our understanding. Preparation of fish coccidian oocysts for electron microscopy is less troublesome as they possess thinner walls and sporulation often occurs within the host.

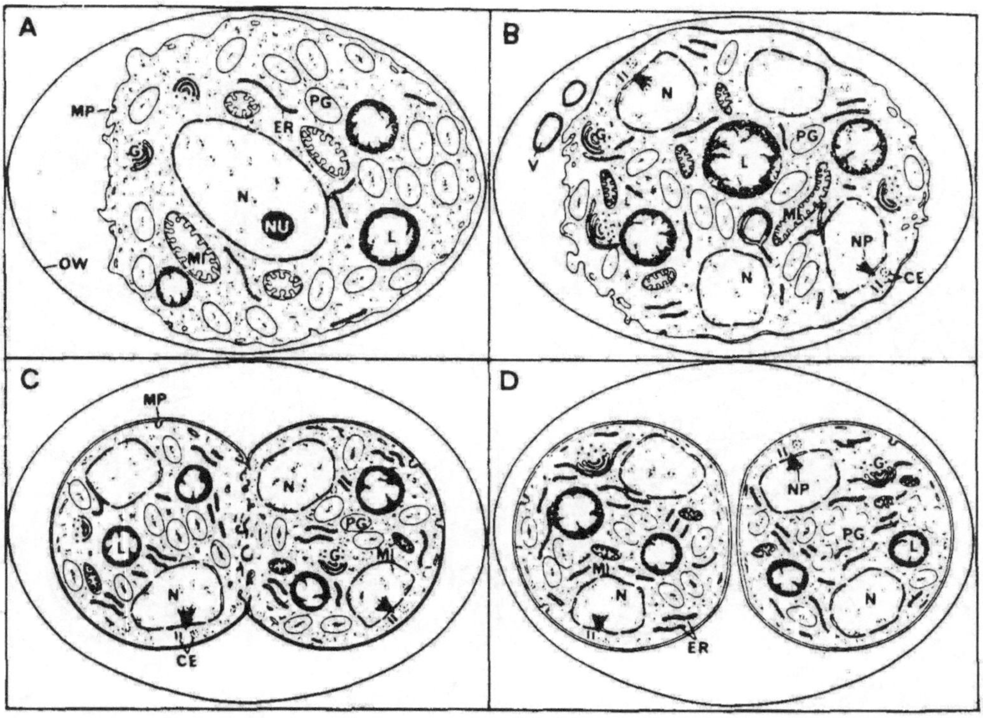

FIGURE 1. Diagrammatic representation of the changes observed during the development of the zygote and formation of the sporoblasts in *T. gondii.* (A) Unsporulated oocysts, (B) a sporulating oocyst, (C) formation of the sporoblasts, (D) early sporoblasts. (From Ferguson, D. J. P. et al., *Acta Pathol. Microbiol. Scand. Sect. B*, 87, 171, 1979. With permission.)

C. EXCYSTATION

There have been a number of studies on the ultrastructure of excystation of coccidian oocysts, including *T. gondii*,[12,29] *C. muris*,[37] and *C. parvum*.[34,35] Essentially there are two mechanisms of excystation. The sporozoites can escape through an aperture in the sporocyst wall that results from dissolution of the Stieda body. In species that lack a Stieda body, the sporocyst wall collapses and fragments releasing the sporozoites. While the former is known to be true for many eimerian species, ultrastructural studies show that the latter process is involved for *I. endocallimici*,[9] *T. gondii*,[12,29] and *S. suicanis*.[18] In the excystation of *C. parvum* a suture at the pole of the oocyst dissolves resulting in a slit-like opening through which the sporozoites escape.[34,35]

III. SPOROZOITE, MEROZOITE, AND ASEXUAL DEVELOPMENT

The structure of the sporozoite and the merozoite has been well documented in earlier reviews[1,2] and the process of merogony adequately explained.[2,38]

Although sporozoites and merozoites are very similar in appearance, the vast majority of sporozoites can be distinguished from merozoites by the presence of refractile bodies, usually one in the anterior and one in the posterior region. There have been a few exceptions where refractile bodies have been found in merozoites, e.g., from parasites grown in cell culture and from first generation merozoites of *E. falciformis* in the mouse.[39]

Sporozoites and merozoites have a smooth surface with a single microspore opening and they are covered by a single outer unit membrane (plasmalemma, plasma membrane).

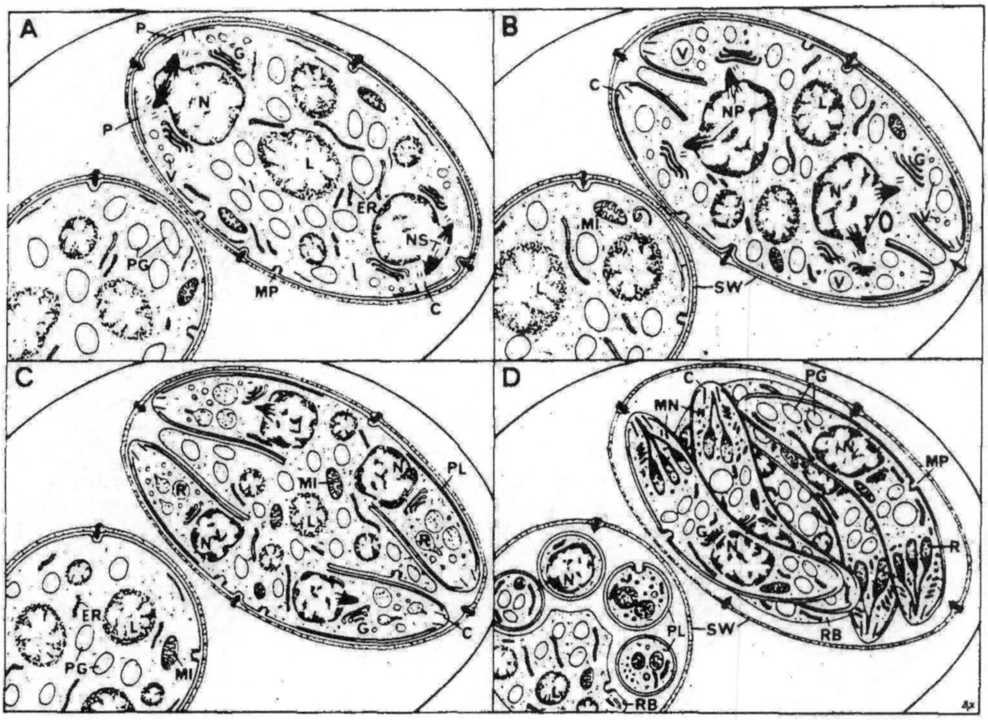

FIGURE 2. (A—C) Diagrammatic representation of the changes observed during the formation of the four sporozoites of *T. gondii*, (D) a fully sporulated sporocyst is shown. (From Ferguson, D. J. P., et al., *Acta Pathol. Microbiol. Scand. Sect. B*, 87, 253, 1979. With permission.)

Beneath this is the inner membrane consisting of two unit membranes which completely line the plasmalemma except at the gaps of the micropore and the anterior and posterior polar rings. There may be one or two anterior polar rings. The anterior ring(s) encircles a cylindrical conoid consisting of six to seven, and in some cases up to 18 to 20 diagonal subunits. From the single anterior ring or the inner of the double rings, 22 or more subpellicular microtubules run longitudinally nearly the whole length of the parasite. In the anterior region two or more rhoptries are present with ducts passing through the conoid and a few to many small rod-shaped micronemes. Also present in the cytoplasm are polysaccharide granules, mitochondria, lipid vesicles, Golgi body, cisternae of endoplasmic reticulum, ribosomes, and a nucleus (Figure 5).

Nichols and Chiappino[40] studied the conoid and anterior region of the endozoites (tachyzoites) of *T. gondii* in detail. The microtubules were anchored in the anterior polar ring and the insertions supported by blunt projections. It was postulated that the polar ring complex might move diagonally along the conoid subunits during conoid extension and retraction, rather than travel over several subunits in a straight anterior posterior direction. Movement along diagonal units could cause the observed twist to give body torsion.

Electron microscopy was used to examine *Sarcocystis* in the muscles of deer. *S. gracilis* was reported from the roe deer (*Capreolus capreolus*)[41] and *Sarcocystis* sp. from the fallow deer (*Cervus dama*).[42] Endopolygeny was demonstrated in the first and second meronts of *S. tenella* in lambs[43] and endodyogeny described during merogony in *Isospora felis*.[44] The ultrastructural features of the development and tissue cyst formation of *T. gondii* in the brains of mice have been studied[45] (Figure 6). Early cyst formation occurred predominantly in neurons. The tissue cyst wall developed as an adaptation of the lining of the parasitophorous vacuole. All cysts were retained in intact cells. There was no evidence of any degenerative changes in the cystozoites (bradyzoites) 22 months after infection.

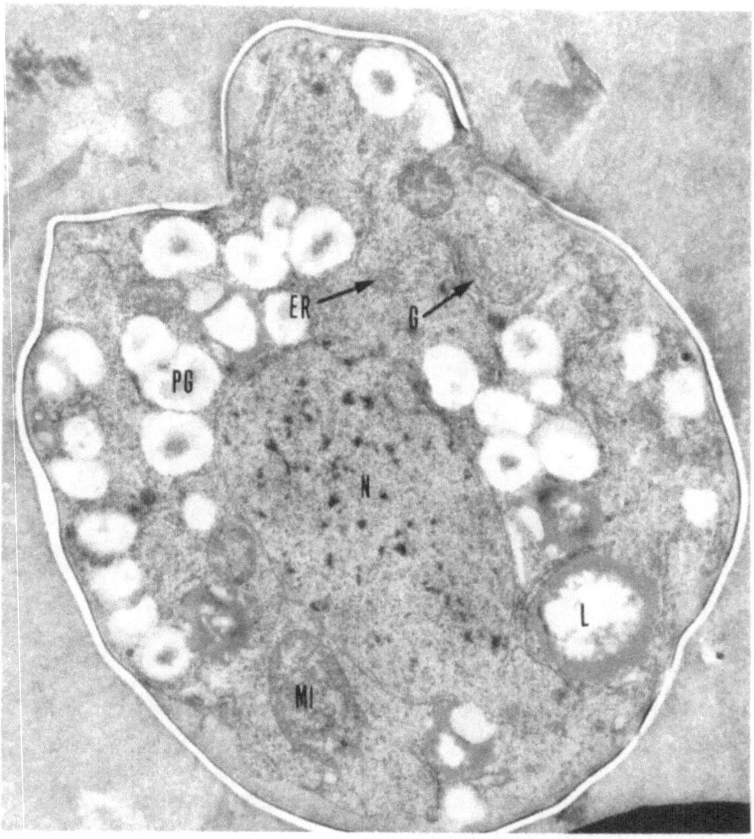

FIGURE 3. Transmission electron micrograph showing an unsporulated oocyst of *T. gondii*. The cytoplasm contains a nucleus, polysaccharide granules, lipid globules, Golgi bodies, mitochondria, and some rough endoplasmic reticulum. (Magnification × 15,000.) (From Ferguson, D. J. P. et al., *Acta Pathol. Microbiol. Scand. Sect. B*, 87, 171, 1979. With permission.)

IV. ENTRY INTO CELLS

The active and passive entry of sporozoites into cells *in vitro* was reviewed by Doran.[46] Chobotar and Scholtyseck[2] gave an account of host cell penetration and the parasitophorous vacuole and pointed out the discrepancies in the accounts of cell penetration. Despite additional work since then, especially on *T. gondii*, the entry of sporozoites and merozoites into host cells still needs clarification particularly because most work has employed cell cultures.

Penetration by these stages is preceded by parasite movements and extension of the conoid at the apical end which when applied to the host cell plasmalemma causes a large depression to form in it. The host cell membrane proliferates and the parasite enters as the invagination enlarges. The cell membrane fuses behind the invading parasite to enclose it in a parasitophorous vacuole. There are some reports describing the disruption of the host cell membrane at the point of entry leaving the parasite free for a short time in the host cell cytoplasm until the parasitophorous vacuole membrane is reformed. Others did not demonstrate the same disruption of the cell membrane, but instead, a breakdown of the invaginated host cell membrane around the parasite after entry and a rapid replacement by a newly formed membrane.[2]

The events involved in host cell invasion by the Apicomplexa may be considered as (1)

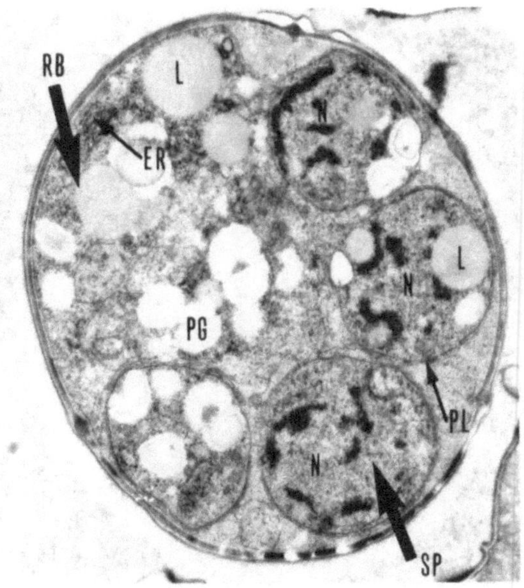

FIGURE 4. Transmission electron micrograph showing a mature sporocyst of *T. gondii*. Four sporozoites and the residual body can be seen. (Magnification × 15,000.) (From Ferguson, D. J. P. et al., *Acta Pathol. Microbiol. Scand. Sect. B*, 87, 171, 1979. With permission.)

initial recognition of cell surface; (2) attachment to cell surface; (3) creation of vacuole; (4) entry into vacuole and (5) sealing of the vacuole.

The conoid and apical complex of the anterior pole of the sporozoite and merozoite is of paramount importance in cell invasion and needs to make contact with the host cell to initiate active penetration. For *Eimeria* species and *T. gondii* the initial recognition of the cell surface does not appear to depend upon specific receptors because these parasites invade such a wide range of cultured cell types.[46] Also, *Toxoplasma* infects a wide range of animal species. Therefore, the attachment mechanism of the sporozoite would seem to have a general receptor recognition system. Cationized ferritin binding to anionic sites on surface membranes of primary turkey kidney cells inhibited the invasion of *E. meleagrimitis* sporozoites, as did the removal of *N*-acetylneuraminic acid from the cell surface.[47] It was considered that these data supported the hypothesis that there may be host cell membrane characteristics which were preferred sites for sporozoite invasion.

T. gondii is known to affect nonphagocytic as well as phagocytic cells. Active invasion into HeLa cells, for example, takes approximately 15 to 40 sec which is about one quarter the time for phagocytosis.[48] The conoid is protruded and contact made between the apical pole and host cell plasmalemma.

An interesting host cell reaction occasionally occurred during the entry of the parasite. The mouse peritoneal macrophages produced small finger-like filopodia around the tips of invading *Toxoplasma* endozoites and small continuous collars termed rudimentary lamellipodia[49] (Figure 7). Similar slender pseudopodia-like extensions from the surface of cultured canine kidney cells became attached to the extracellular parts of the invading merozoite of *S. muris*.[50]

It has been assumed for sometime that such unique and specialized organelles as the rhoptries and micronemes have an important function in coccidial parasites and it is generally agreed that their role is involvement in penetration. Studies on the invasion of endozoites of *T. gondii* into white cells occurring in the peritoneal cavity of mice indicate that rhoptries

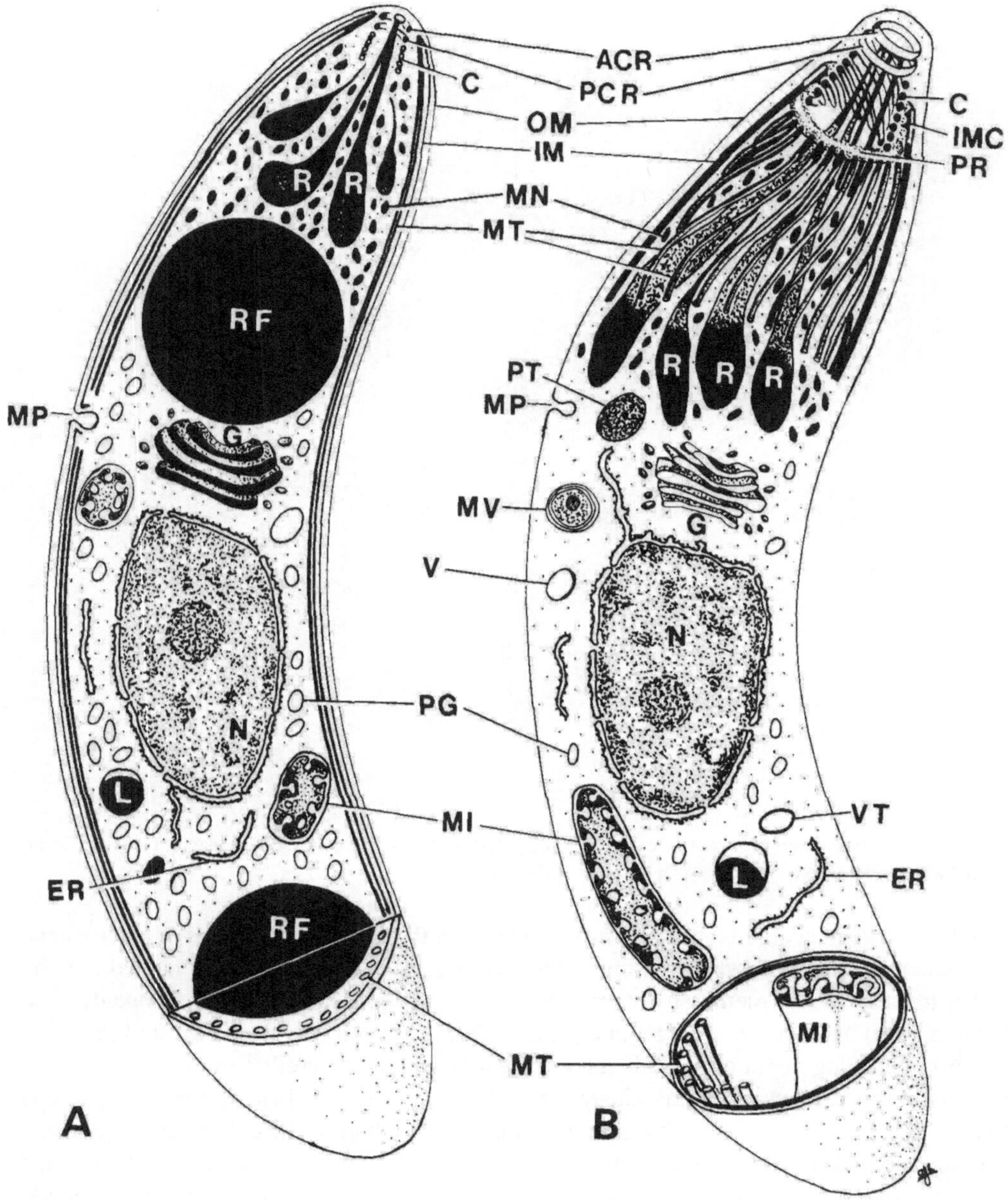

FIGURE 5. Diagrammatic representation of eimeriorine sporozoite (A) and merozoite/endozoite (B). (B adapted in part from Nichols, B. A. and Chiappino, M. L., *J. Protozool.*, 34, 217, 1987.)

shortened and emptied.[51] Membranes of the rhoptries fused with the limiting membrane at the anterior end of the parasite to form openings through which the products of the rhoptries could be released. During invasion by *S. muris* the rhoptries altered in appearance from electron dense to electron lucent.[50] In an ultrastructural study on *T. gondii* it was proposed that the rhoptries were true extrusive organelles.[52]

Usually sporozoites and merozoites are considerably constricted at the point of entry to the host cell, forming a ring-like junction of close continuity of host cell and parasite plasmalemma. As the anterior tip of *Toxoplasma* enters a macrophage the plasmalemma of the host cell may be interrupted leaving the parasite virtually free in the cytoplasm and not enclosed by a parasitophorous vacuole.[51,53] Clusters of vesicles bounded by typical unit

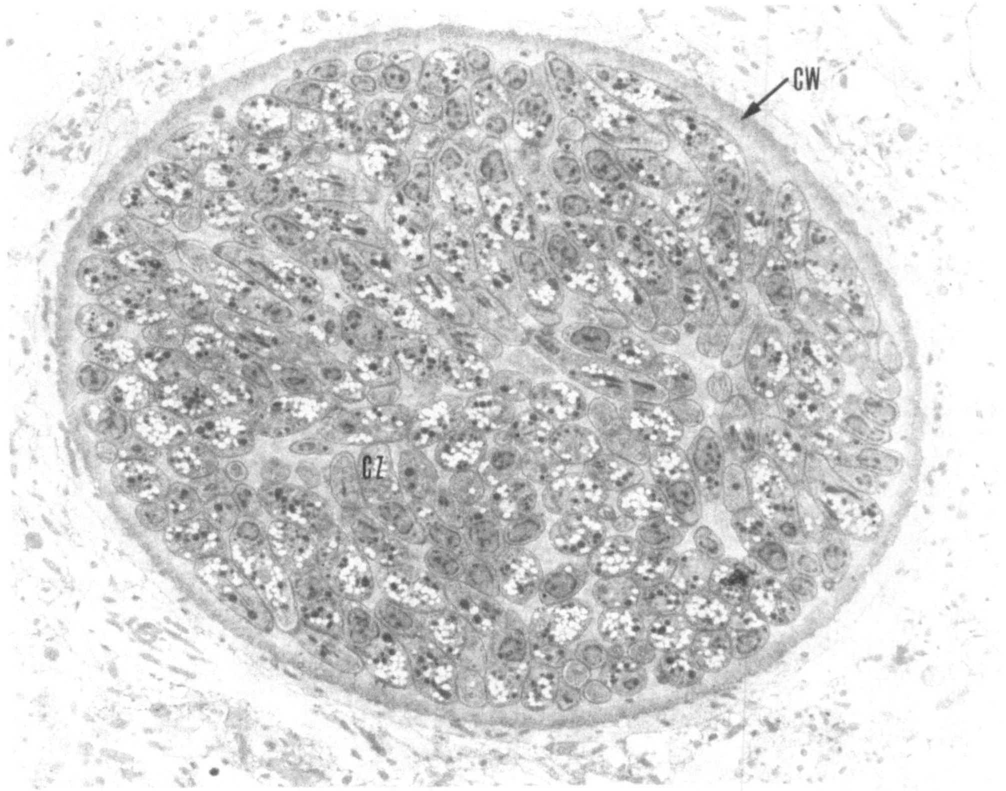

FIGURE 6. A tissue cyst of *T. gondii* containing many cystozoites within mouse brain, magnification × 3500. (Unpublished micrograph courtesy of Dr. D. J. P. Ferguson.)

membranes were recorded in the area around the tip of the parasite. Remnants of macrophage plasmalemma were observed adhering to the *Toxoplasma* endozoites as they moved into the cell cytoplasm.[53] The membrane around the parasite is reconstructed by components from both host cell and parasite, making the parasitophorous membrane a hybrid membrane.[48,51]

An interesting variation on cell entry was recorded by Entzeroth[50] studying merozoites of *S. muris* in cultures of canine kidney cells. Parasites vacated their first parasitophorous vacuoles to form second ones by enlarging the parasitophorous vacuole membrane at the anterior end and moving into the new space. This was brought about by a close membrane contact between the plasmalemma of the parasite and the limiting membrane of the second parasitophorous vacuole which causes constriction of the merozoite during movement. Simultaneously electron-dense granules accumulated in the anterior portion of the merozoite and the material of some of the granules as seen to be secreted into the lumen of the second parasitophorous vacuole. As Entzeroth[50] pointed out, although the function of the dense granules is unknown they are likely to play an important part in the biology of these parasites, possibly by stimulating adherence of the endoplasmic reticulum to the limiting membrane of the second parasitophorous vacuole or by causing an antigenic mimicry. In *S. muris* invasion, the invaginated host cell plasmalemma apparently becomes the limiting membrane of the primary parasitophorous vacuole. The second parasitophorous vacuole membranes were reinforced by a lamella of the host cell endoplasmic reticulum. Up to 2 min after invasion by *T. gondii* some rhoptries were found to contain dense material which appeared to re-organize into a mass of membrane-limited tubules.[51] These enlarged and tubules of identical appearance were demonstrated within the parasitophorous vacuole 6 to 13 min later. It was concluded that the intravacuolar tubules in contact with the parasitophorous membrane were derived from the rhoptry contents.[51]

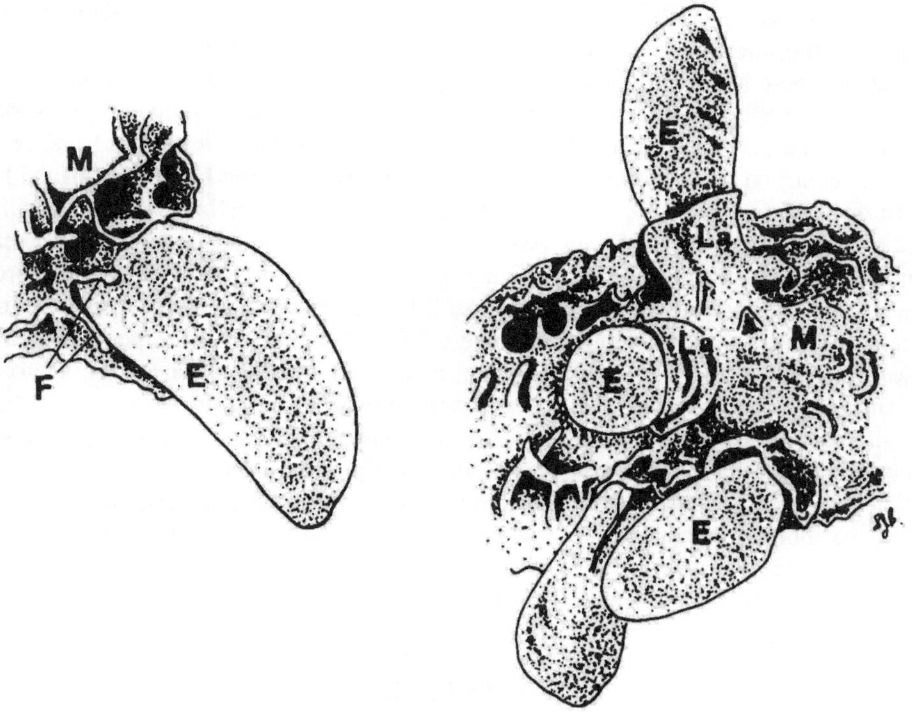

FIGURE 7. Diagrammatic representation of *Toxoplasma* endozoites (E) entering normal macrophages (M). Filopodia (F) and lamellipodia (La) are shown around the anterior end of the parasite. (Adapted from Chiappino, M. L., Nichols, B. A., and O'Connor, G. R., *J. Protozool.*, 31, 288, 1984.)

The constitution of the rhoptries is unknown. Endozoites of *T. gondii* from peritoneal exudate of infected mice were studied by cytochemical and electron-microscopical techniques and basic proteins were identified especially in the rhoptries[54] and two major proteins were demonstrated in the micronemes of the zoites of *S. tenella* from the esophagus of sheep.[55]

The apposition between host cell and parasite plasmalemmas is so tight that as the parasite moves into the cell the parasite surface is cleaned of absorbed molecules. This was concluded from observations of *Toxoplasma* endozoites with monoclonal antibody directed against major surface proteins invading HeLa cells.[56] Augustine and Danforth[57] examined the effects of hybridoma antibodies on the invasion of four species of poultry *Eimeria* into primary avian kidney cells. They depicted an invading sporozoite of *E. meleagrimitis* showing surface antigens sloughed off at the point of entry.

Penetration of host cells is usually associated with active movement of the parasite mainly by rotating.[48,49] Russell and Sinden[58] recognized three forms of locomotion by the sporozoites of *E. acervulina* and *E. tenella*; these were gliding, pivoting, and bending. During the gliding process of *Eimeria* sporozoites the body of the parasite is held in a fixed helical coil and movement is achieved only when the parasite is in contact with a substrate.[58]

The skeleton of subpellicular microtubules is necessary for the maintenance and adoption of the body shape. Microtubule inhibitors do not appear to affect these organelles ultrastructurally or prevent parasite movement.[58] On the other hand, cytochalasin B, an antimicrofilament agent, inhibits gliding and pivoting.[58] Experiments on *Eimeria* spp. using the ligand cationized ferritin as a probe for anionic sites demonstrated a rapid capping reaction located at the rear end.[58] In addition, latex beads were also transported backwards on the surface of *Eimeria* sporozoites.[59] It has been proposed that for the invasive parasite transmembrane components span the plasmalemma and act as a link between the external ligand (e.g., latex bead, ferritin) and a microfilamentous system (actin, myosin).[58-60] It was further proposed that a set of transmembrane components might link the actin filaments to the subpellicular microtubules thereby providing the necessary rigidity and also the tracks for a fast capping reaction.[59,60] King[59] hypothesized that in penetration, crosslinking between host cell surface components and parasite transmembrane components after contact at the front end of the sporozoite could be the first step in capping. Thus, if there is little breakdown of the host cell, as the components cap to the rear of the parasite, the host cell plasmalemma will be pulled over the parasite, e.g., *Cryptosporidium*. Alternatively, if the host cell cortex breaks down, capping to the rear results in invasion, e.g., *Eimeria*. In part, the differences recorded in the invasion processes between certain coccidial parasites could be due to the reaction of different host cells and to the ability of some cells also to phagocytose. Sporozoites of *E. acervulina* readily invaded chicken peripheral blood lymphocytes *in vitro*[61] and showed a close annular junction between host cell and parasite plasmalemma. Invading endozoites of *T. gondii* did not cause the same surface reaction in lymphocytes as in macrophages although occasionally a flocculent collar was seen around the tip of the parasite.

Toxoplasma can actively invade macrophages in which phagocytosis has been suppressed, but normally phagocytosis was considered to play a secondary role.[53] Entry by phagocytosis can be distinguished from active invasion in several ways including, the host cell membrane not being disrupted, the parasite being randomly orientated, and the parasite not changing shape on entry.[53]

V. GAMETOGONY

A. MACROGAMETOGENESIS

The process of macrogametogony is well documented for the coccidia and has been the subject of several reviews.[1,2,62] Recent studies, along with those that have not been covered by previous reviews, include reports on the macrogametes of *E. truncata* in the lesser snow goose,[63] *E. maxima*[64] and *E. acervulina*[65,66] in the chicken, *E. contorta* in the rat,[67] *E. stiedai* in the rabbit,[68] and *E. bakuensis* in sheep.[69] Macrogametes have also been reported for *Cryptosporidum* in the chicken[70,71] and in mice.[37,72] For *Sarcocystis* macrogametes there have been reports of *S. muris*[73] and *S. fusiformis*[74] in the small intestine of cats, *S. alceslatrans*[36] and *Sarcocystis* sp. in the small intestine of dogs.[75] Macrogametes of *Tyzzeria parvula* have been described in an intranuclear location in the goose[76] and macrogametes of *I. canis* have been reported from the dog.[77]

Although early changes in differentiation such as an increase in cell size are characteristic of gamonts destined to become macrogametes, it is the appearance of the wall-forming bodies which allow positive ultrastructural identification of macrogametes (Figure 8). All *Eimeria* species are thought to possess two types of wall-forming bodies designated type 1 (WF1) and type 2 (WF2), although this was difficult to confirm for *E. tuskegeenis*.[78] Both types of wall-forming bodies are found in *Isospora* species, but some species of *Sarcocystis* have only one type of wall forming body and this also seems to be true for *Cryptosporidium* sp. and *T. parvula*.[76] In *S. muris* there are dense bodies that are considered to be similar to the wall-forming bodies of *Eimeria* species.[73] In *Eimeria* species the wall-forming bodies of type 2 usually form before the wall-forming bodies of type 1 although in *E. stiedai*, it is the latter that appears first.[68]

Further confirmation that the Golgi bodies and the endoplasmic reticulum are involved in the formation of the wall-forming bodies has come from an examination of *E. truncata*[63] and *E. contorta*.[67] In the case of *E. acervulina*, Pittilo and Ball[66] were only able to confirm the involvement of endoplasmic reticulum in the formation of wall forming bodies of type 2. In this species wall-forming bodies of type 1 take on a number of distinct morphological forms.[66] Two types of wall-forming body of type 1 have been reported for *E. brunetti*[79] and

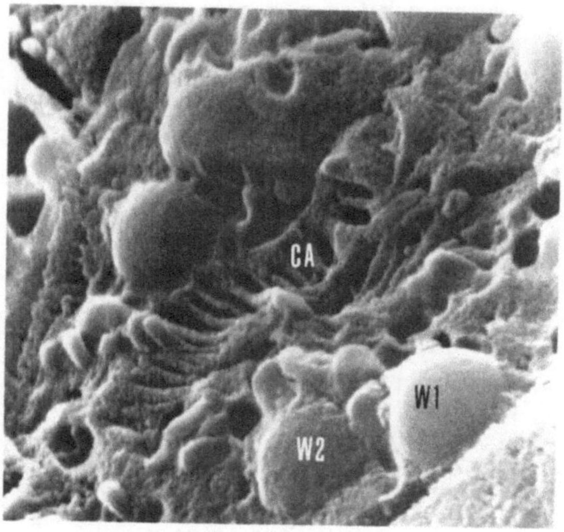

FIGURE 8. Scanning electron micrograph of a macrogamete that has been fractured within the tissue. Wall-forming bodies and canaliculi are easily resolved. (Magnification × 12,000.)

a range in osmiophilia has also been reported for the wall-forming bodies of type 1 in *E. maxima*[80] and *E. stiedai*.[68] Drug interactions are known to have an effect on wall-forming body structure (see below) and it may be that there are different pathways by which the wall-forming bodies can be formed which result in structures of varied morphology. Further studies are needed to consolidate our knowledge on the mechanisms the parasite uses to synthesize the wall-forming bodies. Fixatives do not seem to have much effect on the morphology of the wall forming bodies.[81] In *E. acervulina* the wall-forming bodies of type 2 have an additional surrounding membrane that lies internal to the granular endoplasmic reticulum.[66]

During development, macrogametes accumulate food reserves in the form of granules and lipid inclusions. Ultrastructurally, these granules have been demonstrated to be polysaccharide for *E. brunetti*.[82]

Intravacuolar folds and intravacuolar tubules along with membranous components are found within the parasitophorous vacuoles of many macrogametes. Intraparasitophorous vesicles have been reported to originate from the host cell cytoplasm and enter the macrogamete of *E. truncata* through the micropores by budding of the plasmalemma or by pinocytosis.[63] The vesicles were then broken down in Golgi associated vacuoles.[63] Intravacuolar tubules have been widely documented but although a nutritional hypothesis has been proposed for them this requires further confirmation. Clear evidence of structural connections between the intravacuolar tubules has been seen in *E. maxima*[80] and interconnections between intravacuolar tubules and the macrogamete cytoplasm have been observed for *E. contorta* where they also make contact with the macrogamete pellicle.[67] Scanning electron microscopy of fractured tissue containing *E. acervulina* macrogametes also shows interconnections and possible branching of intravacuolar tubules.[66] In *E. contorta* a new type of structure consisting of coils of microtubules interconnected by membranous bridges was found within the parasitophorous vacuole.[67] In some species, such as *E. stiedai*, there is close apposition between the parasite-limiting membranes and the host cell membrane of the parasitophorous vacuole.[68]

The number of membranes surrounding macrogametes differs between species and sometimes in reports for the same species when described by different workers. In *E. contorta* young gamonts have a single limiting membrane below which there are in places additional membranes, but the mature macrogametes have a pellicle consisting of three membranes.[67]

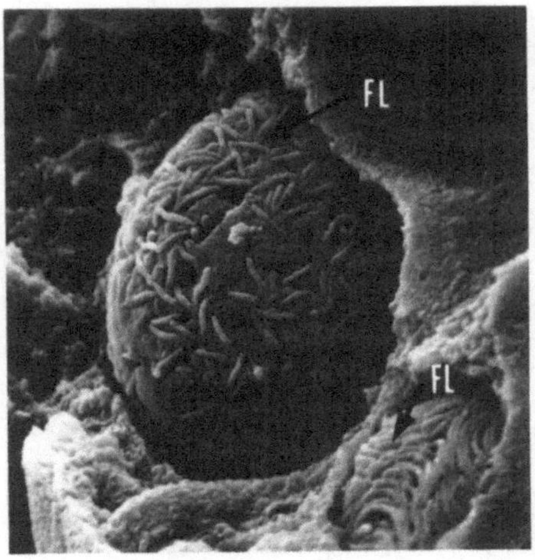

FIGURE 9. Scanning electron micrograph of *E. maxima* microgamonts. Flagella can be seen projecting out from the periphery of the parasites. (Magnification × 2500.)

A three-membraned pellicle is found in *Sarcocystis* sp. from the roe deer,[75] *S. muris*[73] and *S. fusiformis*.[74] A two-layered pellicle was seen in *Cryptosporidium* species from mouse small intestine.[72] Micropores are seen in many macrogametes.

In addition to the above organelles and inclusions, macrogametes typically have a large nucleus and a nucleolus is often present. Within the cytoplasm there are free ribosomes, mitochondria, and canaliculi. The canaliculi have been considered to be analogous to smooth endoplasmic reticulum within the coccidia. They are particularly prominent in *E. acervulina* macrogametes when observed by scanning electron microscopy.[66]

B. MICROGAMETOGENESIS

There is an overall uniformity in the development of microgamonts of species of *Eimeria* with a period of growth accompanied by nuclear division followed by a phase when microgametes are differentiated. In addition, two basic forms of microgamonts are recognized; those in which the microgamont retains a spherical shape with nuclei situated within the periphery of the limiting membrane around a single mass of cytoplasm and those with deep invaginations and fissures in the cytoplasm. Since the formation of microgametes is a surface occurrence the latter configuration increases surface area for the formation of a greater number of microgametes than could be produced from a surface of a single sphere.

Over the past 12 years, microgametogenesis has been studied in *E. perforans*[83] and *E. stiedai*[84] in rabbits; *E. necatrix*,[85] *E. acervulina*,[65] and *E. maxima*[86] in chickens; *E. truncata*[63] in geese; *E. bakuensis*[87] in sheep; and *E. bateri*[88,89] in quail. Development in these species followed the established eimerian pattern which can be summarized as follows. The young microgamont has nuclei, often with nucleoli, distributed throughout the cytoplasm which contains mitochondria and cisternae of granular endoplasmic reticulum. As the microgamont increases in size, nuclei with dispersed heterochromatin become located at the periphery. Paired centrioles are evident in the cytoplasm generally positioned between the nuclei and the microgamont surface. At an intermediate stage the chromatin condenses into about one half of the nucleus and paired flagella extend into the parasitophorous vacuole (Figure 9).

Polysaccharide granules are often present in the cytoplasm. Mitochondria are closely associated with the nuclei. The microgametes differentiate by the condensed osmiophilic portion of the nuclei each accompanied by a mitochondrion, protruding into the parasitophorous vacuole. As the microgamete matures its nucleus elongates to give a fusiform shape. The microgamete separates from the microgamont leaving behind the pale portion of the nucleus in the large mass of residual cytoplasm.

Within this overall pattern of development there are differences in detail that are not as yet well understood. The number of polysaccharide granules and mitochondria reported in microgamonts vary considerably and two or three flagella may be found in microgametes. Fissures and invaginations were observed in the microgamonts of *E. bakuensis*, *E. truncata*, *E. stiedai*, and *E. maxima*, but were absent in *E. bateri*, *E. necatrix*, and *E. perforans*.

In *T. gondii*,[90] *I. felis*,[91] and probably in other isosporan coccidia microgametogenesis follows a similar pattern as described for *Eimeria* spp. and produces microgametes of the same basic structure. The microgametes of *S. suihomonis*[92] are also similar but are formed by a different process. In contrast, typical nuclear binary fission has not been observed and hence a number of single nuclei are not present prior to microgamete formation. About 20 to 30 microgametes are produced simultaneously from the lobed extensions of the single nucleus of the microgamont.

The ultrastructural events of microgametogenesis in *Cryptosporidium* species have not been studied in the same detail as in the eimerian parasites. Some 19 species have been named[93] but with the present knowledge Levine[93] suggested that only four species should be considered valid; one species from each host class: mammals, birds, reptiles, and fishes. Recently, ultrastructural studies have been reported on *Cryptosporidium* infections in man,[94] chickens,[71] and mice.[37,72,95] After nuclear division 16 microgametes bud off from the periphery of the spherical microgamont. The microgametes are described as wedge — or bullet — shaped with a slightly expanded end covered with what appears to be a glycocalyx acting as an 'adhesive' membrane. The compact fusiform nucleus is surrounded by 11 microtubules. The microgametes lack flagella (Figure 10).

VI. FERTILIZATION

Mitosis occurs during merogony, microgametogenesis, and sporogony, and presumably, the single nucleus of the microgamont of *Sarcocystis* is polyploid until the microgametes bud off. Since in *Eimeria* both microgametes and macrogametes may be obtained from a single sporozoite or merozoite, there are no sex chromosomes.

The dispersal of microgametes for fertilization necessitates their escape from the host cell, requiring disruption of the host cell membrane, further possible penetration of host membranes, and penetration of the macrogamete. Little is known of these processes.

The zygote is the fertilized macrogamete and fertilization presumably takes place before oocyst wall formation, but may not be necessary for its initiation. Long[96] carried out a series of experiments to examine the effect of different temperatures on the production and sporulation of oocysts of *E. tenella* and *E. mitis* grown in chicken embryos. Temperature influenced sporulation rates. One cause suggested was the possible reduced viability of microgametes and macrogametes at lower temperatures resulting in the formation of the oocyst wall but no subsequent sporulation. Alternatively it was postulated that the results could equally be due to a slower development of microgametes and hence nonfertile macrogametes incapable of sporulation. This proposition supposes that the oocyst wall is formed without the presence of microgametes.

Attempts to record the process of fertilization by electron microscopy have had limited success. Ultrathin sections show only a minute portion of a macrogamete at one time and the process of fertilization could be of short duration.[2] Elwasilo[97] observed a microgamete of *E. maxima* adhering to the surface of a macrogamete. At the point of contact the membranes of both gametes appeared to have fused. Fertilization in *S. bovicanis* entailed the fusion of the microgamete and macrogamete membranes resulting in a connection through which there

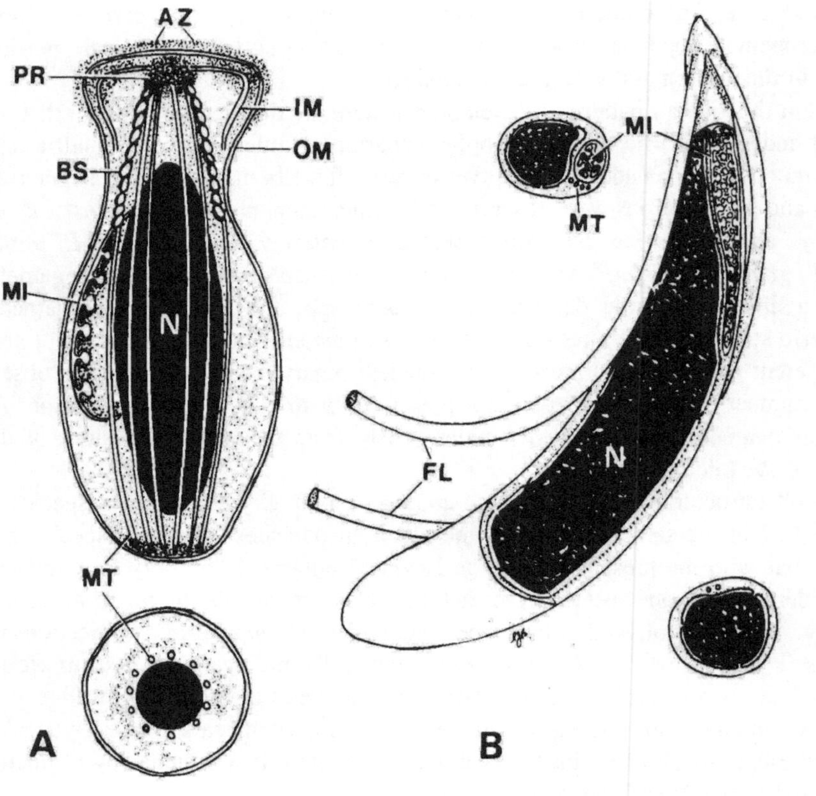

FIGURE 10. Diagrammatic representation of microgametes in longitudinal and transverse sections. (A) *Cryptosporidium*. (Adapted from Goebel, E. and Braendler, U., *Protistologica*, 18, 331, 1982.) (B) *Eimeria*.

was continuity of cytoplasm.[98] Through this junction microgamete nucleoplasm passed into the macrogamete cytoplasm. A similar finding was described in work on *Sarcocystis* sp. from the roe deer.[75]

Another pattern of fertilization has been observed in the few other eimerian coccidia studied in which the whole microgamete seems to be incorporated into the macrogamete. The process in *Goussia iroquoina* in fathead minnows (*Pimephales promelas*)[99] agrees in a general way with the earlier ultrastructural reports on *E. bovis*[100] and *E. tenella*.[101] Microgametes of *G. iroquoina* in young macrogametes were in vacuoles lined by the plasmalemma of the macrogamete plus two additional membranes. In mature macrogametes the cytoplasm of the gametes were separated only by the plasmalemma of each. In this and the two previous species the microgametes appear to enter the macrogamete by invagination of the plasmalemma of the macrogamete and fusion of the membranes of the gametes is considered to take place sometime after entry.

The surface association of *Cryptosporidium* spp. microgametes and macrogametes has been reported by light microscopy for *C. baileyi* in chickens[102] and *Cryptosporidium* sp., obtained from a naturally infected calf and infected into mice,[95] and by transmission electron microscopy for *Cryptosporidium* sp. in man[94] and in mice.[72] Microgametes were observed attached to the host cell membrane covering the macrogamete.[95,102] In addition a microgamete was recorded inside a macrogamete and attachments of a microgamete by its enlarged adhesive zone to the outer membrane of the parasitophorous vacuole (i.e., host membrane of the microvilli) surrounding a macrogamete were seen.[72] Penetration of the microgamete and fusion of nucleoplasm have not been observed. A possible, but questionable, micrograph of a microgamete nucleus touching the nucleus in a macrogamete has been published from infected human intestinal mucosa obtained by biopsy.[94]

VII. OOCYST WALL FORMATION

Oocyst wall formation has been extensively reviewed.[1,2,62] Recent reports not previously reviewed, exist for *E. maxima*[24,64] and *E. acervulina*,[65,66] in chickens (Figure 11), *E. contorta*[67] in rats, *E. truncata*[103] in the lesser snow goose, *E. bakuensis*[69] in sheep and *S. muris*[73] in cats.

It is widely accepted that where two types of wall-forming bodies are present, the wall-forming bodies of type 1 give rise to the outer layer of the oocyst wall while the wall-forming bodies of type 2 are responsible for formation of the inner layer of the oocyst wall. The wall-forming bodies disappear from the cytoplasm as the respective layers of the oocyst wall form. All the above studies support this. However, it is still not clear how the material of the wall-forming bodies is transported to the exterior of the oocyst. Understanding of this process is hindered by the difficulties associated with preparing oocysts for ultrastructural examinations (see above). Changes in the morphological appearance of the material of the wall-forming bodies during wall formation has been reported for some species. With *E. truncata*, the wall-forming bodies of type 1 become labyrinthine and degranulate to form the outer layer of the oocyst wall.[103] It may be that some biochemical alteration of the material of the wall forming bodies is necessary prior to wall formation. In *E. acervulina*[66] and *E. bakuensis*[69] the wall-forming bodies of type 2 are found in electron pale areas during formation of the oocyst wall. There are marked changes in the morphology of the wall-forming bodies of type 2 in *E. maxima* as formation of the outer layer nears completion.[24] The altered wall bodies of type 2 appear very similar to the material being deposited in formation of the inner wall layer.[24] It has been suggested that the longitudinally striated tubules seen in *E. contorta* (see above) may have a role in oocyst wall formation.[67] They are found in advanced stages parallel to the parasite's surface where they form an exterior layer.[67]

With *E. maxima*[24] and *E. acervulina*[66] formation of the oocyst wall commences by separation of the outer limiting membrane of the parasite to form a veil. In the case of *E. acervulina*, it appeared that the wall-forming bodies of type 1 were depositing their material between the limiting membrane of the parasite and the veil-forming membrane.[66] This is in agreement to that previously described for other species and a similar process is seen with *E. truncata*.[103]

Formation of the oocyst wall is associated in many species with the appearance of additional membranes at the periphery of the parasite. Pittilo and Ball[66] suggested for *E. acervulina* that these could be derived from the membranes surrounding the wall-forming bodies of type 1. Six sub-pellicular membranes were reported during oocyst wall formation in *E. truncata*.[103] Fully formed oocysts of *E. nieschulzi* have a membrane between the outer and inner layers of the oocyst wall.[10]

There are relatively few detailed reports on the process of formation of the inner layer of the oocyst wall and different processes have been reported by different workers for the same species. With *E. acervulina* it appears that the wall-forming bodies of type 2 discharge their contents between the cytoplasmic limiting membranes.[65,66] The material of the wall-forming bodies of type 2 is deposited between the third and fourth subpellicular membranes in *E. truncata*.[103]

Changes in the morphology of the oocyst wall during its formation and subsequent development have been reported. The outer layer of the oocyst wall in *E. maxima*,[24] *E. acervulina*,[66] and *E. bakuensis*[69] differentiates into two distinct morphological parts; an outer granular part and an inner osmiophilic part. The oocyst wall seen in thin sections of *E.*

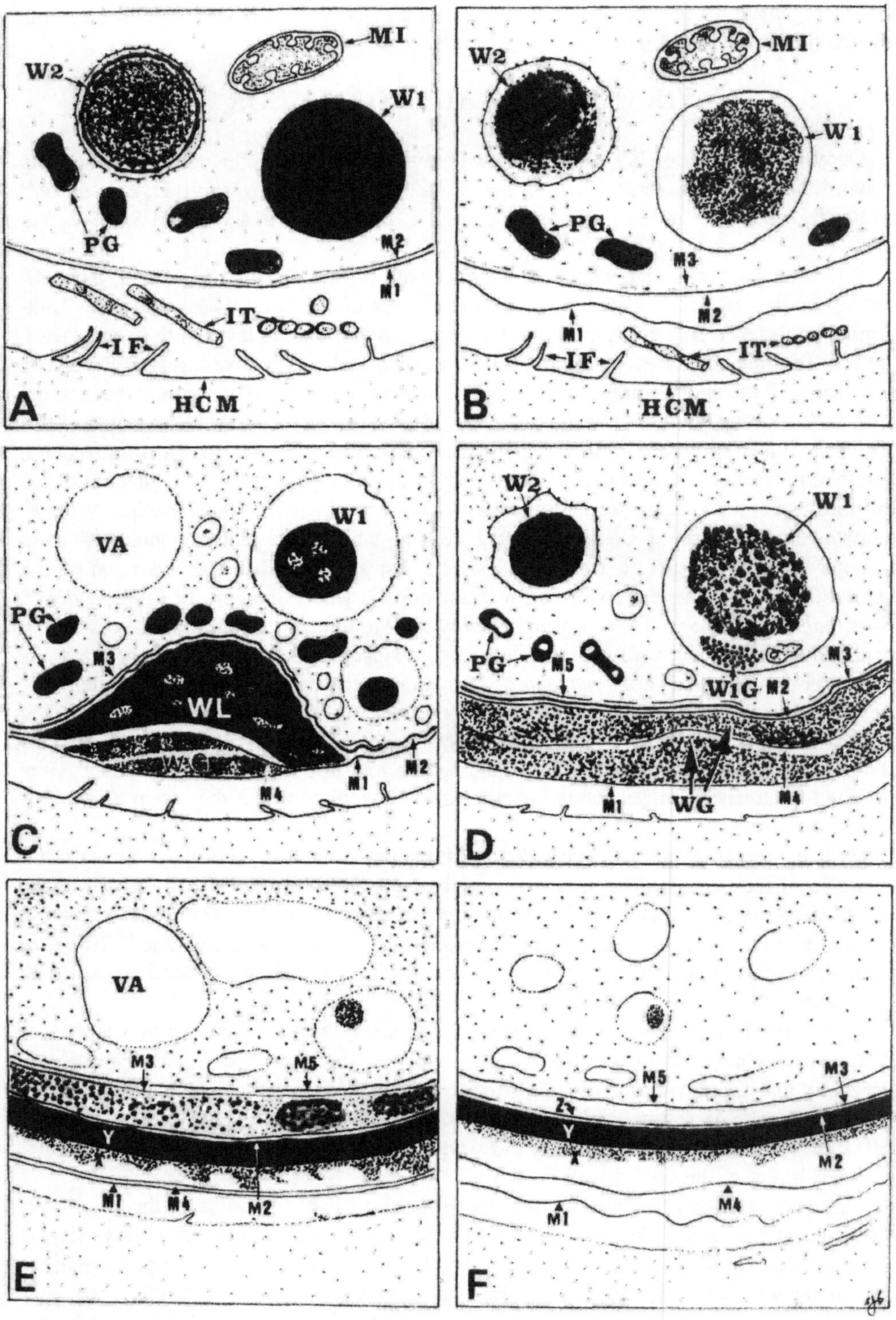

FIGURE 11. Diagrammatic representation of oocyst wall formation in *E. acervulina*. (Based on Pittilo, R. M. and Ball, S. J., *Parasitology*, 89, 1, 1984.)

maxima oocysts isolated from the feces is similar in morphology to the wall of oocysts of the same species examined in the tissues although the dimensions of the layers had altered and the inner part of the outer layer had increased in osmiophility.[24] In *S. muris* there are dense bodies in the cytoplasm that are similar to the wall-forming bodies of *Eimeria* species.[73] Wall formation resulted in the formation of an outer layer composed of a dense band apparently derived from the dense bodies.[73] The inner layer of the oocyst wall was composed of three unit membranes.[73]

The wall of developing oocysts of *C. muris* consists of three layers.[37] The outermost layer was considered to be a true oocyst wall and disintegrated as the oocyst matured, whereas the middle and innermost layers were assumed to develop into the sporocyst wall.[37] The oocyst wall of *C. parvum* examined in oocysts isolated from the feces has two layers, an outer irregular layer which is separated from the inner layer by a thin electron-lucent space.[34] The inner layer was considered to have an outer and inner zone.[34] An atypical oocyst of *Cryptosporidium* has been isolated from the feces of patients with clinically unremarkable cryptosporidiosis and shown to have a three-layered wall.[104]

VIII. ANTICOCCIDIAL DRUG ACTION

Comparative light microscope observations of treated and untreated infected intestines can provide information on the specific stages of a parasite affected by a drug and on the relative decrease in parasite population. In conjunction with experimental data a clear picture of drug action can be established. Electron microscope studies fail to give the same information because of sampling problems, but can elucidate further the action of drugs by detailed observations of structural changes in the parasite. One objection that can be raised against this type of study is that changes in dying or dead parasites could be interrupted as direct drug effects. Care must be taken to draw conclusions from only intact and not disintegrating stages.

Most anticoccidial drugs act on the asexual stages but some, such as arprinocid, amprolium, and dinitolmide adversely affect sporogony. Histological observations showed that one of these drugs, arprinocid, altered the wall-forming bodies in the macrogamete of *E. tenella*.[105] Subsequent ultrastructural observations confirmed this for all three drugs. Wall-forming bodies, especially these of type 2, were structurally abnormal and did not take part in oocyst wall formation[106-108] (Figure 12). One of the effects of sym. triazinones is similar.[109]

Another drug action revealed by electron microscopy is the dilation of endoplasmic reticulum to form large vacuoles or swollen perinuclear spaces in merozoites produced by arprinocid and its metabolites[107,110] or by ionophorous antibiotics.[111] In addition, the action of ionophores causing sporozoites to become swollen and develop an irregular surface shape dotted with pits is clearly demonstrated by transmission[112,113] and scanning electron microscopy.[113] Recently, the drug, diclazuril has been shown to affect different stages in the life cycle of *E. tenella*.[114] The organelles of the merozoites were produced within the cytoplasmic mass of the second generation meront where they remained because merozoites mainly failed to form. The endoplasmic reticulum in macrogametes was abnormal, and although the flagellar complexes formed within the main cytoplasm of the microgamont, microgametes were not differentiated.

IX. HOST-PARASITE INTERACTION

In addition to providing information on the fine structure of coccidian parasites, electron microscopy has been useful in examining the interactions that take place between the parasites and the host cells. The literature on this is vast and only a few more recent reports that have not been the subject of previous reviews will be outlined here.

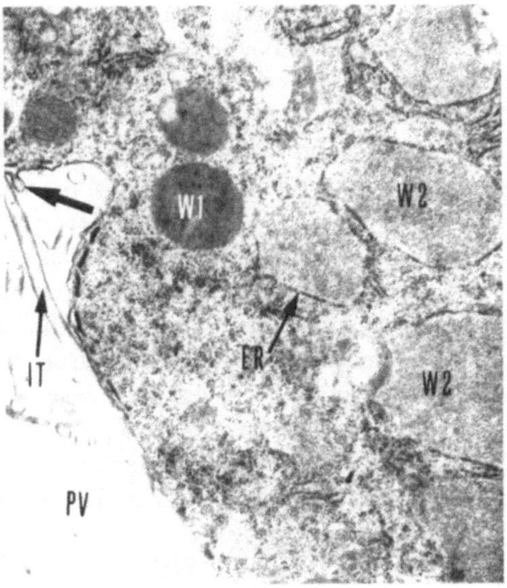

FIGURE 12. Transmission electron micrograph of *E. maxima* from a bird fed with the anticoccidial drug Amprolium at 125 ppm. Within the granular endoplasmic reticulum amorphous granular material, which represents developing wall-forming bodies of type 2, can be seen. A site of possible formation of intravacuolar tubules is shown (arrowed). (Magnification × 9000.)

Enterocytes in both turkeys and Bobwhite quail that are infected with second-generation schizonts and gamonts of *E. dispersa* develop spine-like projections that extend from the parasitophorous vacuole into the host cell cytoplasm.[115,116] This appears to be a parasite specific effect in that it has only been reported for this species and it is known to occur in two hosts.[116] It has been postulated that the spines may consist of skeletal filaments that are normally associated with the microvilli.[116] If the parasitophorous vacuole is formed by invagination of part of the enterocyte cell membrane on which microvilli are present, that is from the luminal surface, then this could explain the origin of the spines.[116] This hypothesis finds support in that the filaments of both the spines and the microvilli are 5 nm in diameter.[116] It has been suggested that other eimerian parasites may enter enterocytes by a different mechanism as spines do not form and the specific properties of the apical region of the enterocyte membrane are not retained in the parasitophorous vacuole.[116]

X. PRO-GAMONTS

In addition to light microscopy, scanning and transmission electron microscopy have been used to examine ovine coccidiosis in naturally acquired infections and the histopathological findings reported.[117] Three types of lesion are found in lambs infected with *E. bakuensis* (syn. *E. ovina*).[117] These are flat oocyst patches, raised oocyst patches, and polyps and all contain large numbers of gamonts and oocysts of *E. bakuensis* as well as a few meronts.[117] It appears that the polyps arise from oocyst patches as a result of progamonts[118] stimulating host cell replication in synchrony with their own division.[119] Light microscope observations suggest that both *E. crandallis* and *E. bakuensis* are able to stimulate host-cell mitosis and may be able to synchronize their division with that of the host cells.[119]

XI. FILAMENTOUS ORGANISMS

There have been many studies using scanning electron microscopy principally to examine the effects of coccidian parasites on the host intestinal wall. Recent studies include surface morphological studies on the effects of *T. gondii*[120,121] and *I. felis*[122] on the cat small intestine. The pathological effects observed are not within the scope of this chapter, but such studies have shed some information on the surface morphology of certain stages of the parasites. Recently, scanning electron microscopy has demonstrated filamentous organisms attached to the small intestinal mucosae in lambs infected with *E. bakuensis* and cats infected with *I. felis* but not *T. gondii*.[123] In the case of *E. bakuensis*, it seemed that the filamentous organisms had a preference for coccidia-infected cells.[123] Quantitative transmission electron microscopy had previously shown that there is a reduction in the microvillous length in the enterocytes of cats infected with *T. gondii*.[124]

REFERENCES

1. **Scholtyseck, E.,** Ultrastructure, in *The Coccidia: Eimeria, Isospora, Toxoplasma and Related Genera,* Hammond, D. M. and Long, P. L., Eds., University Park Press, Baltimore, 1973, 81.
2. **Chobotar, B. and Scholtyseck, E.,** Ultrastructure, in *The Biology of the Coccidia,* Long, P. L., Ed., University Park Press, Baltimore, 1982, 101.
3. **Tadros, W. and Laarman, J. J.,** Current concepts on the biology, evolution and taxonomy of tissue cyst-forming Eimeriid coccidia, *Advances in Parasitology,* Vol. 20, Baker, J. R. and Muller, R., Eds., Academic Press, London, 1982, 293.
4. **Birch-Andersen, A., Ferguson, D. J. P., and Pontefract, R. D.,** A technique for obtaining thin sections of coccidian oocysts, *Acta Pathol. Microbiol. Scand. Sect. B,* 84, 235, 1976.
5. **Ferguson, D. J. P., Birch-Andersen, A., Hutchison, W. M., and Siim, J. Chr.,** Light and electron microscopy on the sporulation of the oocysts of *Eimeria brunetti.* II. Development into the sporocyst and formation of the sporozoite, *Acta Pathol. Microbiol. Scand. Sect. B,* 86, 13, 1978.
6. **Nyberg, P. A. and Knapp, S. E.,** Scanning electron microscopy of *Eimeria tenella* oocysts, *Proc. Helminthol. Soc. Wash.,* 37, 29, 1970.
7. **Roberts, W. L., Speer, C. A., and Hammond, D. M.,** Electron and light microscope studies of the oocyst walls, sporocysts and excysting sporozoites of *Eimeria callospermophili* and *E. larimerensis, J. Parasitol.,* 56, 918, 1970.
8. **Speer, C. A., Hammond, D. M., Mahrt, J. L., and Roberts, W. L.,** Structure of the oocyst and sporocyst walls and excystation of sporozoites of *Isospora canis, J. Parasitol.,* 59, 35, 1973.
9. **Speer, C. A., Marchiondo, A. A., Duszynski, D. W., and File, S. K.,** Ultrastructure of the sporocyst wall during excystation of *Isospora endocallimici, J. Parasitol.,* 62, 984, 1976.
10. **Marchiondo, A. A., Duszynski, D. W., and Speer, C. A.,** Fine structure of the oocyst wall of *Eimeria nieschulzi, J. Protozool.,* 25, 434, 1978.
11. **Speer, C. A. and Duszynski, D. W.,** Fine structure of the oocyst walls of *Isospora serini* and *Isospora canaria* and excystation of *Isospora serini* from the canary *Serinius canarius* L., *J. Parasitol.,* 59, 35, 1975.
12. **Christie, E., Pappas, P. W., and Dubey, J. P.,** Ultrastructure of excystment of *Toxoplasma gondii* oocysts, *J. Protozool.,* 25, 438, 1978.
13. **Stotish, R. L., Wang, C. C., and Meyenhofer, M.,** Structure and composition of the oocyst wall of *Eimeria tenella, J. Parasitol.,* 64, 1074, 1978.
14. **Jolley, W. R., Allen, J. V., and Nyberg, P. A.,** Micropyle and oocyst wall changes associated with chemically mediated *in vitro* excystation of *Eimeria stiedai* and *Eimeria tenella, Int. J. Parasitol.,* 9, 199, 1979.
15. **Nyberg, P. A. and Knapp, S. E.,** Effect of sodium hypochlorite on the oocyst wall of *Eimeria tenella* as shown by electron microscopy, *Proc. Helminthol. Soc. Wash.,* 37, 32, 1970.
16. **Duszynski, D. W., Speer, C. A., Chobotar, B., and Marchiondo, A. A.,** Fine structure of the oocyst wall and excystation of *Eimeria procyonis* from the American raccoon *Procyonlotor, Z. Parasitenkd.,* 65, 131, 1981.

17. **Whitmire, W. M. and Speer, C. A.,** Ultrastructural localization of immunoglobin A and immunoglobin G receptors on oocysts, sporocysts, sporozoites and merozoites of *Eimeria falciformis, Can. J. Zool.*, 64, 778, 1986.
18. **Strohlein, D. A. and Prestwood, A. K.,** *In vitro* excystation and structure of *Sarcocystis suicanis* Erber, 1977 sporocysts, *J. Parasitol.*, 72, 711, 1986.
19. **Wang, C. C. and Stotish, R. L.,** Changes of nucleic acids and proteins in the oocysts of *Eimeria tenella* during sporulation, *J. Protozool.*, 22, 438, 1975.
20. **Dubremetz, J. F., Colwell, D., and Mahrt, J.,** Freeze etching of the oocyst of the coccidia *Eimeria nieschulzi, C.R. Hebd. Seances Acad. Sci. Ser. D*, 280, 2117, 1975.
21. **Ferguson, D. J. P., Birch-Andersen, A., Hutchison, W. M., and Siim, J. Chr.,** Observations on the ultrastructure of the late sporoblast and the initiation of sporozoite formation in *Eimeria brunetti, Acta Pathol. Microbiol. Scand. Sect. B*, 85, 110, 1977.
22. **Ferguson, D. J. P., Birch-Andersen, A., Hutchison, W. M., and Siim, J. Chr.,** Ultrastructural studies on the endogenous development of *Eimeria brunetti*. IV. Formation and structure of the oocyst wall, *Acta Pathol. Microbiol. Scand. Sect. B*, 85, 201, 1977.
23. **Ferguson, D. J. P., Birch-Andersen, A., Hutchison, W. M., and Siim, J. Chr.,** Light and electron microscopy on the sporulation of the oocysts of *Eimeria brunetti*. I. Development of the zygote and formation of the sporoblasts, *Acta Pathol. Microbiol. Scand. Sect. B*, 86, 1, 1978.
24. **Pittilo, R. M. and Ball, S. J.,** Ultrastructural observations on the sporogony of *Eimeria maxima, Int. J. Parasitol.*, 15, 617, 1985.
25. **Pittilo, R. M. and Ball, S. J.,** The ultrastructural development of the oocyst wall of *Eimeria maxima, Parasitology*, 81, 115, 1980.
26. **Ferguson, D. J. P., Birch-Andersen, A., Siim, J. Chr., and Hutchison, W. M.,** Ultrastructural studies on the sporulation of oocysts of *Toxoplasma gondii*. I. Development of the zygote and formation of the sporoblasts, *Acta Pathol. Microbiol. Scand. Sect. B*, 87, 171, 1979.
27. **Ferguson, D. J. P., Birch-Andersen, A., Siim, J. Chr., and Hutchison, W. M.,** Ultrastructural studies on the sporulation of oocysts of *Toxoplasma gondii*. II. Formation of the sporocyst wall, *Acta Pathol. Microbiol. Scand. Sect. B*, 87, 183, 1979.
28. **Ferguson, D. J. P., Birch-Andersen, A., Siim, J. Chr., and Hutchison, W. M.,** Ultrastructural studies on the sporulation of oocysts of *Toxoplasma gondii*. III. Formation of the sporozoites within the sporocysts, *Acta Pathol. Microbiol. Scand. Sect. B*, 87, 253, 1979.
29. **Ferguson, D. J. P., Birch-Andersen, A., Siim, J. Chr., and Hutchison, W. M.,** An ultrastructural study on the excystation of the sporozoites of *Toxoplasma gondii, Acta Pathol. Microbiol. Scand. Sect. B*, 87, 277, 1979.
30. **Beesley, J. E. and Latter, V. S.,** The sporulation of *Eimeria tenella* as revealed by a novel preparative method, *Z. Parasitenkd.*, 67, 255, 1982.
31. **Speer, C. A., Marchiondo, A. A., Mueller, B., and Duszynski, D. W.,** Scanning and transmission electron microscopy of the oocyst wall of *Isospora lacazei, Z. Parasitenkd.*, 59, 219, 1979.
32. **Long, M. S. and Strout, R. G.,** Sporulation of *Eimeria tenella* coccidia oocysts revealed by scanning electron microscopy, *Proc. Helminthol. Soc. Wash.*, 51, 320, 1984.
33. **Gajadhar, A. A., Rainnie, D. J., and Cawthorn, R. J.,** Description of the goose coccidium *Eimeria stigmosa* with evidence of intranuclear development, *J. Parasitol.*, 72, 588, 1986.
34. **Reduker, D. W., Speer, C. A., and Blixt, J. A.,** Ultrastructural changes in the oocyst wall during excystation of *Cryptosporidium parvum* (Apicomplexa, Eucoccidiorida), *Can. J. Zool.*, 63, 1892, 1985.
35. **Reduker, D. W., Speer, C. A., and Blixt, J. A.,** Ultrastructure of *Cryptosporidium parvum* oocysts and excysting sporozoite as revealed by high resolution scanning electron microscopy, *J. Protozool.*, 32, 708, 1985.
36. **Colwell, D. D. and Mahrt, J. L.,** Development of *Sarcocystis alceslatrans* in the small intestine of dogs, *Am. J. Vet. Res.*, 44, 1813, 1983.
37. **Uni, S., Iseki, M., Maekawa, T., Moriya, K., and Takada, S.,** Ultrastructure of *Cryptosporidium muris* (strain RN66) parasitizing the murine stomach, *Parasitol. Res.*, 74, 123, 1987.
38. **Hammond, D. M.,** Life cycles and development of coccidia, in *The Coccidia: Eimeria, Isospora, Toxoplasma, and Related Genera*, Hammond, D. M. and Long, P. L., Eds., University Park Press, Baltimore, 1973, 45.
39. **Scholtyseck, E. and Abdel Ghaffer, F.,** *Eimera falciformis* — merozoites with refractile bodies, *Z. Parasitenkd.*, 65, 117, 1981.
40. **Nichols, B. A. and Chiappino, M. L.,** Cytoskeleton of *Toxoplasma gondii, J. Protozool.*, 34, 217, 1987.
41. **Enzeroth, R.,** Light-, scanning-, and transmission electron microscope study of the cyst wall of *Sarcocystis gracilis* Ratz, 1909 (Sporozoa, Coccidia) from the roe deer *(Capreolus capreolus* L.), *Arch. Protistenk.*, 129, 183, 1985.
42. **Entzeroth, R., Chobotar, B., Scholtyseck, E., and Nemeseri, L.,** Light and electron microscope study of *Sarcocystis* sp. from the fallow deer *(Cervus dama), Z. Parasitenkd.*, 71, 33, 1985.

43. **Speer, C. A. and Dubey, J. P.**, An ultrastructural study of first — and second-generation merogony in the coccidian *Sarcocystis tenella*, *J. Protozool.*, 28, 424, 1981.
44. **Ferguson, D. J. P., Birch-Andersen, A., Hutchison, W. M., and Siim, J. Chr.**, Ultrastructural observations showing enteric multiplication of *Cystoisospora (Isospora) felis* by endodyogeny, *Z. Parasitenkd.*, 63, 289, 1980.
45. **Ferguson, D. J. P. and Hutchison, W. M.**, An ultrastructural study of the early development and tissue cyst formation of *Toxoplasma gondii* in the brains of mice, *Parasitol. Res.*, 73, 483, 1987.
46. **Doran, D. J.**, Behaviour of coccidia *in vitro*, in *The Biology of the Coccidia*, Long, P. L., Ed., University Park Press, Baltimore, 1982, 229.
47. **Augustine, P. C. and Danforth, H. D.**, Effects of cationized ferritin and neuraminidase on invasion of cultured cells by *Eimeria meleagrimitis* sporozoites, *J. Protozool.*, 31, 140, 1984.
48. **Werk, R.**, How does *Toxoplasma gondii* enter host cells, *Rev. Infect. Dis.*, 7, 449, 1985.
49. **Chiappino, M. L., Nichols, B. A., and O'Connor, G. R.**, Scanning electron microscopy of *Toxoplasma gondii*: parasite torsion and host-cell responses during invasion, *J. Protozool.*, 31, 288, 1984.
50. **Entzeroth, R.**, Invasion and early development of *Sarcocystis muris* (Apicomplexa, Sarcocystidae) in tissue cultures, *J. Protozool.*, 32, 446, 1985.
51. **Nichols, B. A., Chiappino, M. L., and O'Connor, G. R.**, Secretion from the rhoptries of *Toxoplasma gondii* during host-cell invasion, *J. Ultrastruct. Res.*, 83, 85, 1983.
52. **Porchet-Henneré, E. and Nicolas, G.**, Are rhoptries of coccidia really extrusomes?, *J. Ultrastruct. Res.*, 84, 194, 1983.
53. **Nichols, B. A. and O'Connor, G. R.**, Penetration of mouse peritoneal macrophages by the protozoan *Toxoplasma gondii*. New evidence for active invasion and phagocytosis, *Lab. Invest.*, 44, 324, 1981.
54. **de Souza, W. and Souto-Padrón, T.**, Ultrastructural localization of basic proteins on the conoid, rhoptries and micronemes of *Toxoplasma gondii*, *Z. Parasitenkd.*, 56, 123, 1978.
55. **Dubremetz, J. F. and Dissous, C.**, Characteristic proteins of micronemes and dense granules from *Sarcocystis tenella* zoites (Protozoa, Coccidia), *Mol. Biochem. Parasitol.*, 1, 279, 1980.
56. **Dubremetz, J. F., Rodriguez, C., and Ferreira, E.**, *Toxoplasma gondii*: redistribution of monoclonal antibodies on tachyzoites during host cell invasion, *Exp. Parasitol.*, 59, 24, 1985.
57. **Augustine, P. C. and Danforth, H. D.**, Effects of hybridoma antibodies on invasion of cultured cells by sporozoites of *Eimera*, *Avian Dis.*, 29, 1212, 1985.
58. **Russell, D. G. and Sinden, R. E.**, The role of the cytoskeleton in the motility of coccidian sporozoites, *J. Cell Sci.*, 50, 345, 1981.
59. **King, C. A.**, Cell motility of sporozoan protozoa, *Parasitol. Today*, 4, 315, 1988.
60. **Russell, D. G. and Sinden, R. E.**, Three-dimensional study of the intact cytoskeleton of coccidian sporozoites, *Int. J. Parasitol.*, 12, 221, 1982.
61. **Pittilo, R. M., Ball, S. J., Norton, C. C., and Gregory, M. W.**, Penetration of chicken peripheral blood lymphocytes by *Eimerian acervulina* sporozoites, *Ann. Trop. Med. Parasitol.*, 80, 569, 1986.
62. **Scholtyseck, E., Mehlhorn, H., and Hammond, D. M.**, Fine structure of macrogametes and oocysts of coccidia and related organisms, *Z. Parasitenkd.*, 37, 1, 1971.
63. **Gajadhar, A. A. and Stockdale, P. H. G.**, Ultrastructural studies of microgametogenesis and macrogametogenesis of *Eimeria truncata* of the lesser snow goose, *J. Protozool.*, 33, 345, 1986.
64. **Elwasila, M.**, Fine structure of the process of oocyst wall formation of *Eimeria maxima* (Apicomplexa; Eimeriina), *Acta Vet. Hung.*, 32, 3, 1984.
65. **Sénaud, J., Augustin, H., and Doens-Juteau, O.**, Observations ultrastructurales sur le développement sexué de la coccidie *Eimeria acervulina* (Tyzzer, 1929) dans l'epithelium intestinal du poulet: la microgametogenese et la macrogametogenese, *Protistologica*, 16, 241, 1980.
66. **Pittilo, R. M. and Ball, S. J.**, Electron microscopy of *Eimeria acervulina* macrogametogony and oocyst wall formation, *Parasitology*, 89, 1, 1984.
67. **Mueller, B. E. G., Desser, S. S., and Haberkorn, A.**, Ultrastructure of developing gamonts of *Eimeria contorta* (Protozoa, Sporozoa) with emphasis on the host parasite interface, *J. Parasitol.*, 67, 487, 1981.
68. **Pittilo, R. M., Ball, S. J., and Hutchison, W. M.**, Ultrastructural development of the macrogamete of *Eimeria stiedai*, *Protoplasma*, 104, 33, 1980.
69. **Pittilo, R. M., Ball, S. J., Catchpole, J., and Neal, C. R.**, Electron microscopy of *Eimeria bakuensis* (Syn. *E. ovina*) macrogametogony and oocyst wall formation, *Acta Vet. Hung.*, 36, 233, 1988.
70. **Nakamura, K. and Abe, F.**, Respiratory especially pulmonary and urinary infections of *Cryptosporidium* in layer chickens, *Avian Pathol.*, 17, 703, 1988.
71. **Itakura, C., Nakamura, H., Umemura, T., and Goryo, M.**, Ultrastructure of cryptosporidial life cycle in chicken host cells, *Avian Pathol.*, 14, 237, 1985.
72. **Goebel, E. and Braendler, U.**, Ultrastructure of microgametogenesis, microgametes and gametogamy of *Cryptosporidium* sp. in the small intestine of mice, *Protistologica*, 18, 331, 1982.
73. **Enzeroth, R., Chobotar, B., and Scholtyseck, E.**, Electron microscope study of gamogony of *Sarcocystis muris* (Protozoa, Apicomplexa) in the small intestine of cats *(Felis catus)*, *Protistologica*, 21, 399, 1985.

74. **Scholtyseck, E. and Hilali, M.,** Ultrastructural study of the sexual stages of *Sarcocystis fusiformis* in domestic cats, *Z. Parasitenkd.*, 56, 205, 1978.
75. **Entzeroth, R.,** Ultrastructure of gamonts and gametes and fertilization of *Sarcocystis* sp. from the Roe Deer (*Capreolus capreolus*) in dogs, *Z. Parasitenkd.*, 67, 147, 1982.
76. **Shibalova, T. A. and Morozova, T. I.,** Intranuclear development of macrogametes in the coccidium *Tyzzeria parvula*, *Tsitologiya*, 21, 969, 1979.
77. **Hilali, M., Ghaffar, F. A., and Scholtyseck, E.,** Ultrastructural study of the endogenous stages of *Isospora canis* in the small intestine of dogs, *Acta Vet. Acad. Sci. Hung.*, 27, 233, 1979.
78. **Current, W. L., Ernst, J. V., and Benz, G. W.,** Endogenous stages of *Eimeria tuskegeensis* (Protozoa, Eimeriidae), in the cotton rat *(Sigmodon hispidus)*, *J. Parasitol.*, 67, 204, 1981.
79. **Ferguson, D. J. P., Birch-Andersen, A., Hutchison, W. M., and Siim, J. Chr.,** Ultrastructural studies on the endogenous development of *Eimeria brunetti*. III. Macrogametogony and the macrogamete, *Acta Pathol. Microbiol. Scand. Sect. B*, 85, 78, 1977.
80. **Pittilo, R. M. and Ball, S. J.,** The fine structure of the developing macrogamete of *Eimeria maxima*, *Parasitology*, 79, 259, 1979.
81. **Knoebber, D., Entzeroth, R., Chobotar, B., and Scholtyseck, E.,** The effect of different fixatives on the fine structure of wall forming bodies of macrogametes of *Eimeria tenella*, *Eimeria maxima* and *Eimeria falciformis* (Apicomplexa, Coccidia, Eimeriidae), *Protistologica*, 16, 435, 1980.
82. **Ferguson, D. J. P., Birch-Andersen, A., Hutchison, W. M., and Siim, J. Chr.,** Cytochemical electron microscopy on polysaccharide granules in the endogenous forms of *Eimeria brunetti*, *Acta Pathol. Microbiol. Scand. Sect. B*, 85, 241, 1977.
83. **Streun, A., Coudert, P., and Rossi, G. L.,** Characterization of *Eimeria* species. II. Sequential morphologic study of the endogenous cycle of *Eimeria perforans* (Leuckart, 1979; Sluiter and Swellengrebel, 1912) in experimentally infected rabbits, *Z. Parasitenkd.*, 60, 37, 1979.
84. **Ball, S. J., Hutchison, W. M., and Pittilo, R. M.,** Ultrastructure of microgametogenesis of *Eimeria stiedai* in rabbits, *Acta Vet. Hung.*, 36, 229, 1988.
85. **Regal, D.-S.,** Licht-und Electronenoptische Untersuchungen an Entwicklungsstadien der Schizogonie und Gametogonie des Huhnercoccids *Eimeria necatrix*, *Zentralbl. Vet. Med. B.*, 24, 297, 1977.
86. **Ball, S. J., Pittilo, R. M., Joyner, L. P., and Norton, C. C.,** Scanning and transmission electron microscopy of *Eimeria maxima* microgametogenesis, *Parasitology*, 82, 131, 1981.
87. **Ball, S. J. and Pittilo, R. M.,** Ultrastructural observations of microgametogenesis in *Eimeria bakuensis* (Syn. *E. ovina*) of sheep, *Parasitol. Res.*, 74, 431, 1988.
88. **Hernandez-Rodriguez, S., Martinez-Gomez, F., and Navarrete, I.,** Estudio al microscopio electronico del ciclo evolutivo de *Eimeria bateri* Bathia et al, 1965, en *Coturnix coturnix japonica*. II. Microgametogenesis, *Rev. Iber. Parasitol.*, 44, 387, 1984.
89. **Martinez-Gomez, F., Hernandez-Rodriguez, S., and Navarrete, I.,** Estudio al microscopio electronico del ciclo evolutivo de *Eimeria bateri* Bathia et al, *1965*, *Coturnix coturnix japonica, III.* Microgametos, *Rev. Iber. Parasitol.*, 45, 1, 1985.
90. **Ferguson, D. J. P., Hutchison, W. M., Dunachie, J. F., and Siim, J. Chr.,** Ultrastructural study of early stages of asexual multiplication, and microgametogony of *Toxoplasma gondii* in the small intestine of the cat, *Acta Pathol. Microbiol. Scand. Sect. B*, 82, 167, 1974.
91. **Ferguson, D. J. P., Birch-Andersen, A., Hutchison, W. M., and Siim, J. Chr.,** Ultrastructural observations on microgametogenesis and the structure of the microgamete of *Isospora felis*, *Acta Pathol. Microbiol. Scand. Sect. B*, 88, 151, 1980.
92. **Mehlhorn, H. and Heydorn, A. O.,** Electron microscopical study on gamogony of *Sarcocystis suihominis* in human tissue cultures, *Z. Parasitenkd.*, 58, 97, 1979.
93. **Levine, N. D.,** Taxonomy and review of the coccidian genus (*Cryptosporidium*) (Protozoa, Apicomplexa), *J. Protozool.*, 31, 94, 1984.
94. **Bird, R. G. and Smith, M. D.,** Cryptosporidiosis in man: parasite life cycle and fine structural pathology, *J. Pathol.*, 132, 217, 1980.
95. **Current, W. L. and Reese, N. C.,** A comparison of endogenous development of three isolates of *Cryptosporidium* in suckling mice, *J. Protozool.*, 33, 98, 1986.
96. **Long, P. L.,** Observations on the oocyst production and viability of *Eimeria mivati* and *E. tenella* in the chorioallantois of chicken embryos incubated at different temperatures, *Z. Parasitenkd.*, 39, 27, 1972.
97. **Elwasila, M.,** The fine structure of an early stage in the process of fertilization of *Eimeria maxima* (Apicomplexa, Eimeriina), *Z. Parasitenkd.*, 69, 135, 1983.
98. **Sheffield, H. G. and Fayer, R.,** Fertilization in the coccidia: fusion of *Sarcocystis bovicanis* gametes, *Proc. Helminthol. Soc. Wash.*, 47, 118, 1980.
99. **Paterson, W. B. and Desser, S. S.,** Ultrastructural observations on fertilization and sporulation in *Goussia iroquoina* (Molnar and Fernando, 1974) in experimentally infected fathead minnows (*Pimephales promelas*, Cyprinidae), *J. Parasitol.*, 70, 703, 1984.

100. **Scholtyseck, E. and Hammond, D. M.,** Electron microscope studies of macrogametes and fertilization in *Eimeria bovis, Z. Parasitenkd.*, 34, 310, 1970.
101. **Madden, P. A. and Vetterling, J. M.,** Scanning electron microscopy of *Eimeria tenella* microgrametogenesis and fertilization, *J. Parasitol.*, 63, 607, 1977.
102. **Current, W. L., Upton, S. J., and Haynes, T. B.,** The life cycle of *Cryptosporidium baileyi* n.sp. (Apicomplexa, Cryptosporidiidae) infecting chickens, *J. Protozool.*, 33, 289, 1986.
103. **Gajadhar, A. A., Stockdale, P. H. G., and Cawthorn, R. J.,** Ultrastructural studies of the zygote and oocyst wall formation of *Eimeria tuncata* of the lesser snow goose, *J. Protozool.*, 33, 341, 1986.
104. **Baxby, D. and Blundell, N.,** Recognition and laboratory characteristics of an atypical oocyst of *Cryptosporidium, J. Infect. Dis.*, 158, 1038, 1988.
105. **McManus, E. C., Olson, G., and Pulliam, J. D.,** Effects of arprinocid on developmental stages of *Eimeria tenella, J. Parasitol.*, 66, 765, 1980.
106. **Pittilo, R. M., Ball, S. J., Joyner, L. P., and Norton, C. C.,** Ultrastructural changes in the macrogamete and early oocyst of *Eimeria maxima* resulting from drug treatment, *Parasitology*, 83, 285, 1981.
107. **Ball, S. J., Pittilo, R. M., Norton, C. C., and Joyner, L. P.,** Morphological effects of arprinocid on developmental stages of *Eimeria tenella* and *E. brunetti, Parasitology*, 91, 31, 1985.
108. **Ball, S. J., Pittilo, R. M., Norton, C. C., and Joyner, L. P.,** Ultrastructural studies of the effects of amprolium and dinitolmide on *Eimeria acervulina* macrogametes, *Parasitol. Res.*, 73, 293, 1987.
109. **Mehlhorn, H., Ortmann-Falkenstein, G., and Haberkorn, A.,** The effect of sym. triazinones on developmental stages of *Eimeria tenella, E. maxima* and *E. acervulina*: a light and electron microscopical study, *Z. Parasitenkd.*, 70, 173, 1984.
110. **Wang, C. C., Simashkevich, P. M., and Fan, S. S.,** The mechanism of anticoccidial action of arprinocid-1-N-oxide, *J. Parasitol.*, 67, 137, 1981.
111. **Mehlhorn, H., Pooch, H., and Raether, W.,** The action of polyether ionophorous antibiotics (monensin, salinomycin, lasalocid) on developmental stages of *Eimeria tenella* (Coccidia, Sporozoa) *in vivo* and *in vitro*: study by light and electron microscopy, *Z. Parasitenkd.*, 69, 457, 1983.
112. **Smith, C. K., II and Strout, R. G.,** *Eimeria tenella:* effect of narasin, a polyether antibiotic on the ultrastructure of intracellular sporozoites, *Exp. Parasitol.*, 50, 426, 1980.
113. **Smith, C. K., II, Galloway, R. B., and White, S. L.,** Effect of ionophores on survival, penetration and development of *Eimeria tenella* sporozoites *in vitro, J. Parasitol.*, 67, 511, 1981.
114. **Verheyen, A., Maes, L., Coussement, W., Vanparijs, O., Lauwers, F., Vlaminckx, E., Borgers, M., and Marsboon, R.,** *In vivo* action of the anticoccidial diclazuril (Clinacox) on the developmental stages of *Eimeria tenella*: an ultrastructural evaluation, *J. Parasitol.*, 74, 939, 1988.
115. **Long, P. L., Millard, B. J., and Lawn, A. L.,** An unusual reaction to an intracellular protozoon parasite *Eimeria dispersa, Z. Parasitenkd.*, 60, 193, 1979.
116. **Millard, B. J. and Lawn, A. L.,** Parasite-host relationships during the development of *Eimeria dispersa* Tyzzer 1929, in the turkey *(Meleagris gallopavo gallopavo)* with a description of intestinal intra-epithelial leukocytes, *Parasitology*, 84, 13, 1982.
117. **Gregory, M. W., Catchpole, J., Pittilo, R. M., and Norton, C. C.,** Ovine coccidiosis: observations on "oocyst patches" and polyps in naturally acquired infections, *Int. J. Parasitol.*, 17, 1113, 1987.
118. **Gregory, M. W., Catchpole, J., Norton, C. C., and Pittilo, R. M.,** A new coccidial stage, *Vet. Rec.*, 121, 383, 1987.
119. **Gregory, M. W., Catchpole, J., Norton, C. C., and Pittilo, R. M.,** Synchronised divisions of coccidia and their host cells in the ovine intestine, *Parasitol. Res.*, 73, 384, 1987.
120. **Hutchison, W. M., Pittilo, R. M., Ball, S. J., and Siim, J. Chr.,** *Toxoplasma gondii*: scanning electron microscope studies on the small intestine of infected and uninfected cats, *Acta Pathol. Microbiol. Scand. Sect. B.*, 87, 393, 1979.
121. **Hutchison, W. M., Pittilo, R. M., Ball, S. J., and Siim, J. Chr.,** Scanning electron microscopy of the cat small intestine and mucosal alteration observed during *Toxoplasma gondii* infection, *Ann. Trop. Med. Parasitol.*, 74, 427, 1980.
122. **Hutchison, W. M., Pittilo, R. M., Ball, S. J., and Siim, J. Chr.,** Scanning electron microscopy of the cat small intestine during *Isospora felis* infection, *Ann. Trop. Med. Parasitol.*, 75, 115, 1981.
123. **Gregory, M. W., Pittilo, R. M., Hutchison, W. M., and Ball, S. J.,** Scanning electron microscopy of filamentous organisms associated with coccidial infections in sheep, *Ann. Trop. Med. Parasitol.*, 79, 473, 1985.
124. **Ferguson, D. J. P., Hutchison, W. M., and Siim, J. Chr.,** The effect of endoenteric development of *Toxoplasma gondii* on the ultrastructure of epithelial cells of the small intestine of infected cats, *Acta Pathol. Microbiol. Scand. Sect. B,* 84, 189, 1976.

Chapter 3

HOST SPECIFICITY OF THE COCCIDIA

Michael H. Kogut

TABLE OF CONTENTS

I. INTRODUCTION

The term "host specificity" refers to the restriction of a species of parasite to one or more species of hosts.[1-4] However, little is known of the specific mechanisms that are involved in this resistance of a host to various parasites.

Historically, one of the dogmas of coccidial biology has been that these parasites have a rigid host specificity. However, some basic characteristics of coccidial biology (rigid host specificity and a direct life cycle) have undergone much scrutiny and are now being questioned.[5-7] Although members of the genus *Eimeria* are still considered monoxenous (a life cycle completed in only one host) and stenoxenous (each species parasitizes a single host species) these characteristics can no longer be regarded as absolute.[6-8] Even the belief that the *Eimeria* are exclusively parasites of the epithelium of the intestine has recently been invalidated by the finding of sites of parenteral development in other cells and organs of individual species of host.[9-15] Furthermore, members of the genera *Isospora, Sarcocystis, Toxoplasma*, and *Caryospora* were all, at one time or another, considered to be monoxenous. They now have all been shown to actually possess a heteroxenous life cycle which involves a direct fecal-oral route of infection in a definitive host, but also an optional indirect means of transmission in an intermediate host.[5,8,16-25] These genera of coccidia and their life cycles in different hosts will be discussed separately.

In most studies, host specificity of the coccidia has been judged by the completion of the endogenous stages of the parasite's life cycle culminating in oocyst production after the appropriate patent time. The majority of the studies reported have indicated that excystation is a nonspecific phenomenon that appears to occur readily in most foreign hosts.[8,26-28] Interestingly, the excystation of sporocysts of mammalian species of *Sarcocystis* appears to be somewhat dependent on the source of bile.[29] Thus, the chemical differences in the bile of a host can be the determining factor in the high degree of host specificity observed in the intermediate hosts of mammalian species of *Sarcocystis*. Similarly, Bejsovec found that oocysts of *Eimeria tetricus* from the black grouse (*Lyrurus tetrix*) passed unchanged through the alimentary tract of the chicken, the pheasant, and the Hungarian partridge.[30] No excystation appeared to have occurred when the intestinal material was microscopically examined. Therefore, it must be kept in mind that the failure to infect foreign hosts with some heterologous species of coccidia may be entirely due to the lack of excystation stimulus in the host.

II. SPECIFICITY OF COCCIDIAL GENERA: CROSS-TRANSMISSION IN FOREIGN HOSTS

A. *CARYOSPORA*

Parasites of the genus *Caryospora* possess a facultative heteroxenous life cycle with birds and reptiles serving as primary hosts and rodents as secondary hosts.[17-19,23,24,31,32] Both asexual and sexual development are found in the primary and secondary hosts. Development is limited to the intestinal epithelium in the primary host whereas development in the secondary host is extraintestinal.[18,19,23,24]

Species of *Caryospora* have been found to be transmitted via four routes including snake to snake, rodent to rodent, rodent to snake, and snake to rodent.[32] *Caryospora* spp. exhibit a specificity for their primary hosts,[31,32] but demonstrate a general lack of specificity for their secondary hosts.[18,19,32]

B. *ISOPORA*

Coccidia of the genus *Isospora*, which commonly shed oocysts in the feces of cats, dogs, and wild carnivores, have until recently been regarded as host specific and thought to develop only in the intestinal tract of the host. In fact, members of the genus have been considered to be very similar to the genus *Eimeria*, i.e., monoxenous, stenoxenous, and with a classical coccidial life cycle.[5] However, the life cycles of several feline and canine species of *Isospora* have been shown to be unlike that of *Eimeria* and exhibit a variation in their degree of host specificity. The genus *Isospora* contains two structurally and biologically distinct groups of parasites.[6] One group contains organisms with a direct life cycle of a single host. This group includes the two species of *Isospora* from the canary, *I. canaria* and *I. serini*. The second group contains organisms with a two-host life cycle. The parasites of this group encyst in the tissues of the intermediate hosts.

Rodents such as mice, rats and hamsters serve as intermediate hosts of the canine and feline coccidia, *I. felis* and *I. rivolta*.[33] Mahrt and Pellerdy were not able to transmit the dog *I. rivolta* to cats.[34,35] Dubey separated the canine and feline *I. rivolta* based on host specificity and found that they were actually two different species, *I. rivolta* (cat) and *I. ohioensis* (dog).[36] Dogs can also serve as an intermediate host for *I. felis* of cats and, likewise, cats and mice are also intermediate hosts for *I. canis* and *I. ohioensis* of dogs.[36,37] Finally, dogs and domestic chickens can be intermediate hosts for *I. rivolta* of cats.[37] Thus, it is possible for the *Isospora* of cats and dogs *(I. felis, I. rivolta, I. canis,* and *I. ohioensis)* to follow a monoxenous fecal-oral life cycle or a heteroxenous life cycle with an intermediate host.

The pathology, disease characteristics, and oocyst morphology of *Isospora* of sparrows and canaries are indistinguishable. However, Box has failed to transmit the parasite from sparrows to canaries.[39] The two isosporans of the canary, *I. canaria* and *I. serini*, appear to be highly monoxenous with direct life cycles,[40,41] with *I. serini* exhibiting a strong degree of host specificity.[42]

Isospora sp. apparently are more host specific for the definitive host than the intermediate host, with *I. arctophitheci* of marmosets being an exception.[43] Hendricks inoculated oocysts of *I. arctophitheci* to 35 primates, 12 carnivores, and 2 marsupials to determine the susceptibility of a wide range of hosts.[44] Oocysts were recovered from 12 of 14 species of animals investigated including 6 primate genera, 6 genera and species of carnivores, and a single species of marsupial. These parasites were also found to produce extraintestinal infections in rodents and chickens; i.e., intermediate hosts.

C. *SARCOCYSTIS*

Sarcocystis spp. have an obligatory two-host life cycle which is actually a predator-prey type life cycle.[6,45] Herbivores serve as the intermediate hosts which become infected by ingesting sporocysts or oocysts shed in feces of the definitive hosts. Asexual reproduction occurs in the vascular endothelium and muscle cells of the intermediate (prey) host.[6,16] Known intermediate hosts include birds, small rodents, reptiles, pigs, horses, cattle, sheep, and goats.[22,41,46-54] Carnivores, the definitive hosts of *Sarcocystis* spp., become infected by ingesting mature intramuscular cysts from the intermediate hosts.[16] Definitive hosts include dogs and cats, coyotes, foxes, wolves, owls, snakes, and raccoons.[22,46-60]

Mammalian species of *Sarcocystis* have exhibited a remarkable degree of host specificity in their intermediate hosts.[16,47,52,54,59] For example, *S. muris* is infectious for mice, but not rats, guinea pigs, or hamsters.[62] Similarly, various *Sarcocystis* spp. have not been found to be transmissible from one intermediate host species to another species using sporocysts. These include transmissions from sheep to cattle and vice versa, from sheep to goats and vice versa, from mule deer to sheep and cattle, and from cottontail rabbits to domestic rabbits.[61-68] However, Fayer et al.[51] found transmission of *Sarcocystis* spp. from bison, elk, moose, and cattle to cattle using sporocysts originally isolated from coyotes. Dubey et al.[69] has summarized in table form all of the reported attempts to cross-transmit various mammalian species of *Sarcocystis* from their known intermediate hosts to other mammalian intermediate hosts. It was found that not only are the *Sarcocystis* more specific for their intermediate hosts, but that those infecting large mammals (sheep, goats, ox) are more specific than those that infect smaller mammals (rats, mice, voles).[69]

The high degree of host specificity appears to be lacking in the definitive hosts of *Sarcocystis* spp. Several species of monkeys have been found to support the enteric stages of *S. suihominis* which normally follows a pig-man cycle.[70] Enteric stages of *S. bovicanis* (cattle-dog cycle) are supported by dogs, wolves, red fox, coyotes, and raccoons.[55,61,67,68,71] *Sarcocystis leporum* (cottontail rabbit-cat cycle) can also be supported by raccoons.[64] House cats and European ferrets also support the development of *S. muris*.[62]

In contrast to the mammalian species of *Sarcocystis*, the avian species of *Sarcocystis* have a wide variety of intermediate hosts which have been found to support the asexual stages of the parasites.[21,40] *Sarcocystis falcatula* obtained from muscle cysts from the family *Icteridae* (cowbirds and grackles) were fed to cats, opossums, rats and dogs with only the opossum found to be a suitable host.[6,21] Infections of the *S. falcatula* were found in sparrows and canaries, but not ducks fed sporocysts recovered from the definitive host, the opossum.[46] Canaries, zebra finches, budgerigars, pigeons, chickens, and guinea fowl with sporocysts were infected with *S. falcatula* from the opossum. Asexual development occurred in chickens and guinea fowl; pigeons were not susceptible to muscle stages but pre-muscle merogony occurred. The remaining hosts were completely susceptible to parasite development.[21,46,72-75] Thus, the avian *Sarcocystis* are not as rigidly host specific for intermediate hosts as the mammalian species.

D. *FRENKELIA*

The genus *Frenkelia* is made up of two known species with a life cycle involving an intermediate host (small mammals) and a definitive host (predatory birds).[61,69,76-78] *Frenkelia* as a genus are very similar morphologically to *Sarcocystis*,[69,76] although the cysts of *Frenkelia* are found only in the central nervous system.

Both species of *Frenkelia*, *F. glareoli* and *F. microti*, are specific for the same definitive host, *Buteo buteo* (buzzard),[69,76-78] but hawks are also suspected to be definitive hosts of *F. microti*.[69] Voles experimentally infected with *F. glareoli* were not infectious for owls, kestrels, or goshawks.[78] Additionally, *F. microti* has been found to be less specific for its intermediate host.[69,79-83]

E. *TOXOPLASMA*

There are only two species of *Toxoplasma* whose entire life cycles are known.[6,17,84] *Toxoplasma gondii* is best known with *T. hammondi* (synonymous with *Hammondia hammondi*) the second species. Oocysts from each species are indistinguishable from each other.

Toxoplasma gondii may follow a life cycle that is either monoxenous or heteroxenous.[6,17,84,85] Oocysts from feces of felids are infectious to the felids (definitive host) and numerous intermediate hosts. In felids that ingest sporulated oocysts, asexual stages and their gamonts develop in the intestinal epithelial cells. In intermediate hosts ingesting sporulated oocysts, sporozoites are released in the intestine. They then leave the intestine and may enter any cell in the body where they develop. Both definitive and intermediate hosts can become infected by ingesting infected animal tissues, by ingesting milk contaminated by the parasite, by transplacental transmission, by blood transfusion, or by organ transplantation.[6,17,86-91]

Toxoplasma gondii may be the most ubiquitous protozoan parasite. It has a very wide range of intermediate hosts but an extremely narrow range of final hosts.[75,85] In fact, all known final hosts are in the family Felidae, both domestic and wild. Therefore, the host specific nature of *Toxoplasma* is found in its definitive host. Thin walled tissue cysts may be found in cells of virtually any organ of a warm-blooded animal and some reptiles.[85] Fayer has estimated that there are at least 350 intermediate vertebrate hosts as well as invertebrates such as earthworms, cockroaches, and muscid flies which are exposed to oocysts.[92-95]

Toxoplasma usually parasitizes both definitive and intermediate hosts without producing clinical signs.[84] Natural resistance is an important consideration in explaining why members of a susceptible species do not develop clinical disease.[75,84] Strains of *T. gondii* vary in their pathogenicity and virulence in different host species.[6,8,75,84] Certain host species are genetically more resistant to effects of a primary infection with *T. gondii* than are others. Rats are highly resistant, genetically, while mice, cotton rats, hamsters, guinea pigs, and multimummate rats are genetically susceptible.[96-101] Similarly, chickens are more resistant than canaries.[99,102] The mechanisms for natural resistance and susceptibility are unknown, but are probably not due to defects in macrophage effector function.[98]

F. *CRYPTOSPORIDIUM*

Members of the genus *Cryptosporidium* are coccidial parasites which parasitize the microvillus border of host epithelial cells in a variety of host species including poultry, domestic animals, and man.[103-108] Infection of avian and mammalian hosts with *Cryptosporidium* is associated with self-limited diarrheal illness, but in the immunosuppressed host infection results in a severe, life-threatening, protracted diarrhea.[109-116]

Cryptosporidium spp. that have been collected from mammals exhibit very little host specificity with transmission occurring between a variety of mammalian host species including calves, goats, pigs, lambs, rodents, puppies, kittens, and man.[105,106,117-127] As a result, cryptosporidiosis associated with the documented cases of the disease in mammals is apparently caused by *C. parvum* or *C. muris* and is considered a zoonotic disease.

A review of the literature concerning avian spp. of *Cryptosporidium* leads to the conclusion that there is a limited host-range of these coccidia. Oral inoculation of *C. baileyi* produced no infections in neonatal mice or goats.[128] Similarly, an oral inoculation of a *Cryptosporidium* sp. (possibly *C. meleagridis*) resulted in no infections in adult athymic mice nor nursing gnotobiotic or conventional pigs, nor in six species of neonatal rodents including rats, gerbils, hamsters, and guinea pigs.[129] Therefore, unlike the mammalian species of *Cryptosporidium*, the avian species do not appear to be zoonotic. However, there is a certain lack of specificity of the avian species of *Cryptosporidium* when inoculated into other avian hosts. Cross transmission of both *C. baileyi* and *Cryptosporidium* sp. from the chicken has been demonstrated in turkeys, ducks, and geese, but not in quail.[128-130] However, the most interesting aspects of these cross transmission studies of *Cryptosporidium* into the turkey is that there appears to be a tissue tropism by the organism which is dependent upon the route of administration. For example, oral inoculation of *C. baileyi* into turkeys resulted in infections in the bursa of Fabricius, cloaca, and respiratory tract, but not in the small intestine.[128,131] Ocular inoculation of *C. baileyi* resulted in an infection of the cloaca but not the conjunctival epithelium in the turkey.[132] In contrast, intratracheal inoculation of *C. baileyi* into turkeys caused respiratory and conjunctival cryptosporidiosis while a nasal inoculation of *Cryptosproidium* sp. caused infection of the sinus and cloaca.[129,131] Natural cryptosporidiosis is found in the intestine and bursa.[133,134] The different species of avian *Cryptosporidium* exhibit a strong site (tissue) preference.

G. *EIMERIA*

The genus *Eimeria* is one of the most common and best known of the parasitic protozoa with over 900 species found in animals ranging from annelids to insects to reptiles and amphibians to birds to mammals.[3,4,135] Although the genus is well represented by species parasitizing a variety of hosts, the individual species are quite host specific.[3,8] As a result, some hosts can have more than one species of *Eimeria*, but one species of *Eimeria* seldom, if ever, occurs in more than one host.

When investigating the host specificity of *Eimeria*, cross transmission between genera seldom occurs although there are exceptions.[7,8,136-141] Levine and Ivens summarized 141 attempts to transmit ruminant coccidia from one host genus to another,[142] of these, only 10 were successful.[143-146] *E. ninakohlyakimovae* and *E. arloingi* were described to be infective to both sheep and goats. However, McDougald found that the former cannot be transmitted to sheep and he renamed the sheep parasite *E. ovinoidalis*.[146] Similarly, *E. arloingi* did not infect sheep and the sheep form was renamed *E. ovina*.[147] Additionally, Levine and Ivens have recently reviewed the literature dealing with the host specificity (cross-transmission) of the *Eimeria* spp. of rodents.[148] They found that the host specificity of 44 different species of *Eimeria* of the rodent was tested in 169 experiments by inoculation into one or more different species of host. Included in the 169 attempts were 161 rodent-to-rodent transmissions, 6 rodent-to-lagomorph, and 1 of rodent-to-carnivore and rodent-to-bird, respectively. Only those in the first category exhibited a successful transmission. Of these, only 14 of 112 attempts successfully transmitted a coccidium from one genera of rodent to another. Doran was able to produce patent infections of the turkey coccidium, *E. dispersa*, in Leghorn chickens, the Chukar partridge, ring-necked pheasant, and bobwhite quail.[149] The prepatent period was 6 h shorter in the quail and partridge than in chickens and pheasants. However, chickens and pheasants shed far fewer oocysts than did the partridge and the quail. The only similarity found among the four hosts was the size of the mature second and third generation schizonts. These findings were in basic agreement with some earlier work with *E. dispersa* in quail, turkeys, pheasants, and chickens, whereas[26,150,151] others have not been able to infect chickens with *E. dispersa*.[152] Norton also found that *E. colchici*,[153] an eimerian from the pheasant, produced an infection in turkeys that were inoculated with large numbers of oocysts. Finally, Mayberry and Marquardt were able to transmit *E. separata* from the rat to the mouse although reproduction was greatly reduced.[154]

Closely related species or subspecies may serve as adequate hosts for a species of *Eimeria*. Levine and Ivens found 39 of 49 attempts were successful in transmitting a coccidian from one species of rodent to another of the same genus.[148] Under natural conditions, only *Dipodomys paminthus mohavenis*, the kangaroo rat, can be infected with *E. mohavensis*. However, when comparing oocyst production in six species and eleven subspecies of *Dipodomys*, Doran found that all of these hosts produced oocysts and two of these hosts produced more oocysts than the natural host.[155] Similar results were described by Todd and Hammond where *E. callospermophili* from the ground squirrel *Spermophilus armatus* was transferrable to six different species of *Spermophilus* ground squirrels.[156] Finally, both *E. bateri* and *E. taldykurganica* from the Japanese quail, *Coturnix coturnix japonica*, will infect the common grey quail *Coturnix coturnix*.[139,140]

One last factor to be considered when investigating the host specificity of *Eimeria* is that cross transmission between families of hosts almost never occurs. However, a major exception occurs with *Eimeria chinchillae*, an eimerian parasite originally isolated from the chinchilla and described as a new species.[157] *E. chinchillae* was shown by de Vos to lack the normal host specificity characteristics associated with the genus *Eimeria*.[158] De Vos successfully transmitted this parasite to nine genera of rodents consisting of two families. Patent infections were also found in several wild rodents, laboratory white mice and laboratory white rats. Golden hamsters, guinea pigs, rabbits, and a shrew were not susceptible. In those animals which were susceptible, varying degrees of pathogenicity and oocyst production were found. Chinchillas could be infected with oocysts harvested from experimental infections of *P. natalensis, P. pumito*, white rats, and white mice.[158]

III. PROPOSED MECHANISMS OF HOST SPECIFICITY

The mechanisms by which development of coccidia in a foreign host is inhibited are unknown, but are undoubtedly complex and varied because the degree of specificity varies.

However, these factors responsible for mediating innate resistance to foreign species of coccidia appear to be highly effective. Experimentally, sporozoites from various species of *Eimeria* have been found to invade the intestinal mucosa of mature foreign hosts, but undergo limited or no further development.[7,8,26,159-164] Similar findings have been reported for the other genera of coccidia,[6,8,36] except for the cryptosporidia isolated from the chicken (*C. baileyi* and *C. meleagridis*). When inoculated into mammalian hosts, no developmental stages of these parasites were found in any tissues.

At present, the bases of the mechanisms responsible for the host specificity of the coccidia fall into three broad, but interconnected categories, including: (1) parasite biochemistry and nutrition, (2) host genetic background, and (3) host immune response. Since most of the research on the mechanisms of host specificity of the coccidia has utilized parasites of the genus *Eimeria*, I will concentrate this section on the *Eimeria* and add appropriate references to the other genera of coccidia where sufficient evidence exists.

A. PARASITE BIOCHEMISTRY AND NUTRITION

Desser postulated "that the progenitors of the eimeriid coccidia were inhabitants of the intestinal lumen of their vertebrate hosts, but they eventually invaded and became established in the intracellular niche.[165] Thus, they were able to find richer nutrient substrates and less competition." In agreement, Marquardt believed that the lack of development of *E. nieschulzi* in mice resulted from the lack of essential nutritional substances.[159,166] This was evident by the disappearance of early schizonts in the intestine of mice infected with *E. nieschulzi* that was not accompanied by any cellular reaction.[158] Pellerdy supported this hypothesis when he was not able to parenterally infect foreign hosts (rabbits and chickens) with *E. tenella* and *E. stiedae*, respectively.[12,13]

The metabolic requirements of a parasite generally vary with its developmental changes even within a single host. Since coccidia are obligate intracellular parasites, there are problems in isolating sufficient numbers of developing parasites to perform viable biochemical studies. The greatest assertion made from the accumulated data on coccidia metabolism is "that coccidia go through complicated life cycles without significantly changing the basic pattern of metabolic activities".[164] These activities include (1) consumption of polysaccharide storage prior to cell division,[168] (2) cell division,[169-171] (3) increase or preservation of polysaccharide storage,[168,172,173] and (4) breaking out to invade a new host. It is apparent that coccidial sporozoites are capable of excysting from oocysts and invading the intestinal epithelium of foreign hosts. Asexual development of the parasite appears to be most affected by the innate systems of the foreign host. Along these lines, glucose metabolism is apparently required by the parasite for asexual development.[167,173] Whether there is a direct correlation between glucose metabolism and asexual development in foreign hosts remains to be proven.

The cultivation of *E. tenella* in cell cultures has provided information delineating nutrients necessary for development. *p*-Aminobenzoic acid, thiamine, glutamine, pyridoxal, biotin, folic acid, ascorbic acid, riboflavin, nicotinamide, pyridoxone, vitamin A, menadione, choline chloride, vitamin B_{12}, calciferol, α-tocopherol, and inositol have been found to be necessary for development of *E. tenella* through second generation merogony.[174-178] The lack of host specificity *in vitro* indicates that in the *in vivo* situation any or all of these essential nutrients may be lacking, in an unavailable form to the parasite, or the parasite's requirements for a nutrient or nutrients may be greater than the requirements for host cells. Such studies emphasize the necessity of using biochemical tools to disclose the basic nature of host specificity.

B. GENETIC MAKEUP OF FOREIGN HOST

Marquardt has raised a strong argument for a genetic basis of host specificity of coccidia.[7] Citing the discoveries of a heteroxenous life cycle in most of the coccidial genera, Marquardt stated that "what we see now is that the ability to develop in the original (definitive) host has been lost. However, the genetic capacity to develop in the original host is still there." He believed that there is some sort of failure to turn on some part of the parasites' genome and not the loss of the genome which causes this characteristic of host specificity. Therefore, an *Eimeria* sp. from the rat could complete its life cycle in mice with certain genetic backgrounds accounting for the idea of genetic component of host specificity.[154,163] Vetterling hypothesized that "host specificity was relative to genetic proximity: close phylogenetic relatives would be more likely to support a common parasite"[160] following Doran's early studies with *E. mohavensis*.[155] In his studies, Vetterling found that *E. tenella* did not complete its life cycle in any bird other than its natural host. However, he was puzzled by the lack of endogenous stages in a bird (the pheasant) considered to be the closest phylogenetic relative of the chicken among those examined whereas a few macrogamonts were found in the chukar which is considered a more distant relative. Therefore, he concluded that the chukar and the chicken are either closer phylogenetic relatives than originally thought or that phylogenetic relationships are not as important in host specificity of *Eimeria* as his hypothesis stated. Mathis and McDougald produced hybrid birds by artificially inseminating Leghorn chickens with semen from guinea fowl or artificially inseminating Japanese quail with semen from Leghorn chickens.[179] The chicken/guinea fowl hybrids, as well as the parents of the intergeneric cross, were inoculated with *E. acervulina* or *E. tenella* from chickens or *E. grenieri* from guinea fowl. No oocysts were found in the feces of the hybrid birds. Oocysts of *E. acervulina* and *E. tenella* were found in the feces of chickens and oocysts of *E. grenieri* were found in the feces of the guinea fowl. Therefore, these hybrid birds were not susceptible to the coccidia specific for either parent. However, in contrast, the quail/chicken hybrids became infected with *Eimeria* from both parental birds including passing some oocysts during the normal patent period. The parental controls exhibited normal host specificity when infected with either chicken or quail coccidia, respectively. The results suggest "that absolute resistance to coccidiosis" is under the genetic control of the host. Mayberry et al. offered further proof that the transmission of *E. separata* from the rat to the mouse is dependent upon genotypic differences of the foreign host.[163] Several strains of mice, differing in coat-color genotypes, inoculated with *E. separata* which is normally found in the rat (*Rattus norvegicus*) passed oocysts in their feces. The location of the "genetic component" that dictates the susceptibility of the host to a parasitic infection was not determined.

Although many reports indicate that the susceptibility of a host to a nonspecific infection is under genetic control, most do not delineate a precise mechanism for the rejection of *Eimeria*, which may be a result of the host immune system.[161,179]

C. HOST IMMUNE RESPONSE

Innate resistance to any sort of foreign invader automatically suggests a role for the immune system of the host. Since the specificity of the immune system is genotypically determined, the basis of host specificity may ultimately be controlled by the genetic makeup of the host. Strain differences in the resistance of *Eimeria* infections in normal hosts have been suggested to have an immunological basis.[161,180,181]

Evidence that the immune response is involved in the prevention of coccidial development and their subsequent rejection by the foreign host falls into three categories: (1) the lack of specificity *in vitro*, (2) completion of the endogenous life cycle in immunologically compromised hosts, and (3) the role of macrophages as effector cells in host specificity.

1. Lack of Specificity *in vitro*

The general lack of host specificity of coccidia, especially the *Eimeria*, in cell cultures has been apparent since Patton first obtained development of *E. tenella* of chickens in bovine kidney cells, a mouse fibroblast cell line, and embryonic quail fibroblasts.[182] Doran cited nineteen different species of *Eimeria* that partially developed in cells from 30 foreign hosts, 4 species of *Isospora* that developed in cells from 9 different foreign hosts, 1 species of *Hammondia* that developed in cells from 3 hosts, 1 species of avian *Sarcocystis* developing in cells from 5 mammalian hosts, and *Toxoplasma gondii* developing in "a wide-variety of cells" from 11 different "warm-blooded animals".[183] There were no reports of the *in vitro* development of *Caryospora* nor *Cryptosporidium* at the time of Doran's review in 1982.

Since 1982, there have been fewer total reports on the cultivation of coccidia in cell culture but the majority report a lack of specificity by the parasites *in vitro*. For example, two eimerians, *E. debliecki* of swine and *E. tuskegeensis* of the cotton rat, partially developed (asexual cycles) in human fetal lung cells,[184,185] mouse kidney, rabbit kidney, or porcine testicle cells.[183] The asexual development of *Isospora suis* from pigs has been reported in cell cultures of bovine and human origin.[186,187] A human isolate of *Cryptosporidium* sp. has been shown to complete its development in porcine kidney and primary chicken kidney cell cultures;[188] whereas a calf isolate only completed its asexual development in a human rectal tumor cell line.[189] The development of *Caryospora bigenetica, C. duszynskii*, and *C. simplex* in cultured cell has been reported.[190-192] Interestingly, *C. duszynskii* did not multiply *in vitro* in human fetal lung cells, but instead formed caryocysts in those cells infected with sporozoites.[190] However, both *C. simplex* and *C. bigenetica* completed their life cycle *in vitro* in human fetal lung cells.[191,192] *In vitro* derived oocysts of *C. simplex* failed to sporulate and were thus noninfective; whereas oocysts of *C. bigenetica* sporulated *in vitro* and were infective to rodents.

In contrast to all of the information described above, the *in vitro* development of several species of *Sarcocystis* is extremely host specific.[193,194]

2. Completion of Endogenous Cycle in Immunocompromised Hosts

The use of different species of avian embryos has provided data that show that the coccidia are capable of development in a foreign host. These experiments also provide further proof of an immunological basis for host specificity since developing embryos are considered relatively immunologically incompetent.[195]

Long was the first to show that an eimerian parasite (*E. tenella*) would complete its entire life cycle in the chorioallantoic membranes (CAM) of the embryonating chicken embryo.[196] However, he was not able to find development of *E. tenella* in either quail or turkey embryos. Fitzgerald showed that *E. stiedae* of the rabbit developed in the CAM of the chicken embryo.[197] Long and Millard later found that a chicken-embryo attenuated strain of *E. tenella* would develop in duck and quail embryos, but *E. acervulina, E. mivati (E. mitis), E. brunetti*, and a pathogenic strain of *E. tenella* would not develop in these two foreign embryos.[198] Rollinson was also successful in cultivating *E. tenella* in quail embryos.[199] The development of the guinea fowl coccidia, *E. grenieri*, has been demonstrated in the CAM of embryonated chicken eggs while *E. dispersa*,[200] an eimeriian of the turkey, did not develop in either chicken or quail embryos.[150] Kogut et al.[201] have reported the serial passage of three strains of *Eimeria* from the chicken in embryonating turkey eggs. All three lines increased their production in turkey embryos; reproduction being equal to or better than the same lines maintained in chicken embryos. These same investigators also found that chicken embryo-adapted lines of *E. tenella* and *E. mitis* did not develop in duck embryos, but some oocysts were produced in Japanese quail embryos infected with *E. mitis*; these oocysts did not sporulate.[202] Finally, Long found no growth of *E. tenella* in goose embryos, but pretreatment of goose embryos with dexamethasone, an immunosuppressive drug, enabled *E. tenella* to achieve a partial development (second generation merogony).[203] The results from all of these studies show that the *Eimeria* are capable of growing and completing their life cycles in foreign embryonic host cells unhindered by a fully functional immune system and apparently with sufficient nutrients or biochemical precursors available to the parasite.

Doran's previous review identified no reports of the development of *Sarcocystis* spp., *Caryospora* spp., nor *Cryptosporidium* spp. in avian embryos.[183] He did find two reports on the complete development of *Besnoitia jellinosi* and a single report on the development of *B. besnoiti* in chicken embryos. Likewise, there were only two reports on the attempted development of a feline *Isospora* in avian embryos — one unsuccessful and one successful. Numerous authors reported the successful development of *Toxoplasma gondii* in avian embryos. However, the avian embryo apparently is the least sensitive means of cultivating the parasite when compared to cell cultures and mice.

Since 1982, there have been more attempts to cultivate various genera of coccidia in avian embryos although this reviewer has not found any reports in the literature to date on the development of *Caryospora* or *Sarcocystis* in embryos. There has been a single report on the successful complete development of *Isospora suis* of swine in chicken embryos.[204] However, the authors were not able to sporulate the oocysts produced in the embryo. Current and Long were the first to successfully cultivate *Cryptosporidium* in chicken embryos.[205] In this paper both human and calf isolates of *Cryptosporidium parvum* were cultured in chick embryos although too few oocysts were produced for serial passage. Naciri and colleagues did report the serial passage of *C. parvum* through chicken embryos (45 passes) before a severe reduction in oocyst output.[206] Interestingly, *C. baileyi* of the chicken exhibited no host specificity in avian embryos by producing oocysts in chicken, turkey, ring-necked pheasants, chukar or partridges, muscovy ducks, domestic ducks, Japanese and bobwhite quail, and guinea fowl embryos.[207]

McLoughlin was the first to attempt to alter the host specificity of *Eimeria* using another immunosuppressive drug, dexamethasone.[208] Patent infections of *E. meleagrimitis*, an eimeriian of turkeys, became established in dexamethasone-treated chickens. However, *E. tenella*, a parasite of chickens, was not transferable to dexamethasone-treated turkeys. Pellerdy and Durr were unable to infect hydrocortisone-treated mice and rats with *E. stiedae*, a coccidium of the rabbit.[209] In the same report, treatment with X-rays or hydrocortisone did not make rats or guinea pigs susceptible to *E. tenella*. The asexual development of *E. vermiformis*, a parasite of the mouse, was found in rats that had been treated with dexamethasone.[162] No parasites were found in any tissue other than the small intestine nor were any sexual stages observed in their histological specimens. However, one rat passed oocysts 11 days postinoculation, but the oocysts did not sporulate and a "positive identification could not be made". It was also noted that the asexual life cycle seen in the immunosuppressed rats was generally identical to that seen in the mouse. Thus, they concluded that "although host specificity could be altered, the organ specificity and appearance of the parasites in the abnormal host remained unchanged". Later, Todd and Lepp were able to demonstrate the completion of the life cycle of *E. vermiformis* in dexamethasone-treated rats.[210]

Rose and Millard have reported the successful completion of the life cycle of *E. vermiformis*, a coccidian of the mouse, in athymic (nu/nu) rats producing oocysts which sporulated and were infective for C57BL/6 mice.[161] The parasite did not complete its life cycle in cortisone acetate-treated or untreated euthymic (nu/+) littermates. Regardless of cortisone treatment, the parasites' life cycle in the euthymic mice was inhibited during asexual development. This study suggests a T lymphocyte role in the innate resistance of a host against a heterologous species of *Eimeria* although the precise mechanism is unknown. It also suggests that innate resistance may be directed especially against the sexual stages of life cycle.

We have further investigated the role of T lymphocytes in host specificity. We have used cyclosporin A (CsA), a fungal metabolite used as an immunosuppressive agent, as a probe in examining the mechanism of T cell-mediated rejection of *Eimeria* in abnormal hosts.[211] CsA directly affects the T helper cell subpopulation by inhibiting the expression of the gene for interleukin-2 and other lymphokines while sparing the immunocompetence of B cells and macrophages.[212] We have found that intramuscular injections of CsA to chicken given either singly 3 days *before* to 1 day *after* infection with *Eimeria* from the turkey or daily for 3 days prior to infection resulted in the production and release of small numbers of oocysts into the feces during the normal patent period. CsA injected singly on days 2 or 3 after infection with *Eimeria* from the turkey had no effect on the life cycle of the coccidia in a heterologous host. These results are suggestive of the role for lymphokines in the early response of heterologous hosts to foreign coccidia.

Further studies into the host specificity of immunosuppressed hosts were performed by Kogut and Long who injected chickens intravenously with crystalline silica particles,[213] which are selectively toxic to macrophages,[214-216] before oral inoculation with turkey *Eimeria* oocysts. Oocysts were produced by all chickens treated with silica. The greatest number of oocysts were discharged between 5 and 9 days postinoculation. This time period corresponded to the minimum prepatent period of *E. dispersa*. Untreated, uninfected control chickens did not shed oocysts. Interestingly, the effect was greatest in those chickens given silica on days 3 to 4 postinoculation. Thus, it appeared that the mechanisms involved in host specificity to *Eimeria* were effective against early gametocytes. This corresponds to the timing of the effector mechanism reported by Rose and Millard using the athymic rat model.[161] Because the parasites did not replicate to their full potential in any study using immunodeficient or immunosuppressed animals, the macrophage/T cell systems interacting together may be required for efficient rejection of *Eimeria* from foreign hosts.

3. Role of Macrophages

Few studies have reported the action of macrophages of foreign hosts on the growth of *Eimeria*. Hammond et al.[217] inoculated *E. bovis* of cattle into monolayers of mouse peritoneal macrophages. No development was noted, but the investigators did not determine whether the merozoites were affected in any way by the macrophages. Peritoneal macrophages obtained from foreign hosts were found to have profound effects on the viability (but not the morphology) of sporozoites of *E. tenella* or *E. grenieri* of the guinea fowl *in vitro*.[26] Their results showed that sporozoites of *E. tenella* and *E. grenieri* did not survive in macrophages from the guinea fowl or the chicken, respectively. Finally, survival of *E. tenella* was greatly reduced in macrophages from the turkey.[26] Survival and viability of sporozoites of *E. tenella* were determined by transferring the infected macrophages into chicken embryos while survival of *E. grenieri* was determined by the morphology of the parasite in the macrophages. *Toxoplasma gondii* zoites, if undamaged by antibodies or other factors, can invade and develop in macrophages in a wide range of hosts by being able to inhibit the fusion of lysosomes within phagosomes.[218-220] Sporozoites of *Eimeria* appear to be able to block lysosomal fusion in macrophages because they survive in these cells from a normal host.[221] However, it is possible that sporozoites are not capable of inhibiting lysosomal fusion in macrophages from foreign hosts.[26]

As described above, Kogut and Long provided further evidence for the role of macrophages in the host specificity of *Eimeria* by injecting a selective macrophage toxin into chickens before infection with heterologous species of *Eimeria*.[213] In subsequent studies, Kogut and Long (unpublished observations) injected a different macrophage toxin, carrageenan,[222,223] into chickens and found that the intraperitoneal injection of this substance does not "break" host specificity as well as the i.v. injection of silica. Birds given multiple small doses of carrageenan before inoculation were found to discharge small numbers of oocysts over a 10-day collection period. However, a large single dose of carrageenan given either 24 h before or 72 or 96 h after oocysts inoculation had no effect on the innate resistance of chicken to *Eimeria* of the turkey. The difference in these results and those found in silica-treated chickens can be explained by the heterogeneity of the monocyst-macrophage lineage. The i.v. injection of silica apparently affects the macrophage precursor cells, the monocytes or immature macrophages. The destruction of a large number of monocytes of immature macrophages will lead to fewer total macrophages at the local site at the time of infection and growth. However, i.p. injections of carrageenan would have a profound effect on the resident macrophages of the peritoneal cavity. It is doubtful that the resident peritoneal macrophages have a functional role in the innate resistance to an intracellular intestinal parasite which penetrates cells from the lumenal side of the intestine.

Further evidence for a monocyte/macrophage interaction in the rejection of heterologous species of *Eimeria* from a foreign host was provided by Kogut and colleagues.[224] In this study, the investigators studied the hematological changes that occur in chicken and turkey hosts following oral inoculation with homologous and heterologous species of *Eimeria*. *E. adenoeides* from the turkey caused consistently increased numbers of large mononuclear cells (monocytes) from 1 to 4 days after inoculation of chickens. This time period corresponds to the optimal time for the silica-treatment of chickens in order to get the completion of the life cycle of turkey *Eimeria*.[213] Similarly, when *Eimeria* of the chicken were inoculated into turkeys, a significant increase in the number of large mononuclear cells occurred; however, this monocytosis occurred much more rapidly (4 to 6 h postinoculation). This suggests a more efficient or functional effector mechanism for host specificity in turkeys.

Until recently the fate of sporozoites in foreign hosts was largely unknown. Of course, it was known that sporozoites from various species of *Eimeria* invade the intestinal mucosa of foreign hosts undergoing limited or no further development.[25,224-226] However, the fate of those sporozoites that did not develop was unknown. Long and Millard found that one could transfer the intestinal mucosa from guinea fowl inoculated 6 to 12 h before with *E. maxima* oocysts to susceptible chickens resulting in normal infections in these birds.[227] However, after 48 h there was no recovery. Interestingly, some of the sporozoites reached the liver of the guinea fowl where they survived for 48 h. How these sporozoites reached the liver was unknown. Kogut and Long have provided evidence that a proportion of sporozoites of *Eimeria* of the chicken after inoculation in normal and foreign hosts reach the peripheral blood and survive at least 4 days in the peripheral blood and possibly extraintestinal organs, including the liver, of chickens and turkeys.[228] Naciri and Yvore also found that chickens inoculated with homogenates of organs from mice infected with *E. tenella* produced patent infections in the birds.[229,230] The lungs of the *E. tenella*-inoculated mice were found to be particularly infective. A subsequent study reported that chickens shed oocysts after ingestion of lung homogenates of mice previously inoculated with *E. tenella*, *E. acervulina*, or *E. maxima*.[230] Even more interesting, was the discovery upon histological examination, of asexual development forms (trophozoites or schizonts) in the lungs of mice inoculated with *E. tenella*.[230]

In summary it can be seen that an enormous amount of literature has been accumulated concerning the host specificity of the coccidia. Underlying the advancement in our knowledge of host specificity are the discoveries involving the basic biology of the coccidia, particularly that of the life cycles of these parasites.

Further investigations into the host specificity of coccidia will not only enable us to understand more of the basic biology of the coccidia, but understanding those mechanisms responsible for host specificity can provide alternative methods for control of coccidiosis. The fact that evidence indicates that the immunology, genetics, and nutrition of the host play a role in the innate resistance to coccidia encourages research that may lead to alternative methods of coccidiosis control if chemotherapy can no longer remain a viable method.

ACKNOWLEDGMENTS

I wish to thank Mrs. Marilyn Schwartz for her excellent typing of this manuscript. New Jersey Agricultural Experiment Station Research Report Series Number F-06111-2-89, supported by Rutgers Research Council grant #2-02330.

REFERENCES

1. **Baer, J. G.,** *Animal Parasites,* McGraw-Hill, New York, 1971.
2. **Cheng, T. C.,** *General Parasitology,* Academic Press, New York, 1973.
3. **Levine, N. D.,** *Protozoan Parasites of Domestic Animals and of Man,* Burgess, Minneapolis, 1973.
4. **Noble, E. R. and Noble, G. A.,** *Parasitology: The Biology of Animal Parasites,* Lea and Febiger, Philadelphia, 1976.
5. **Duszynski, D. W.,** Who are the coccidia and do we really know where they live?, *J. Protozool.,* 28, 242, 1981.
6. **Fayer, R.,** Epidemiology of protozoan infections: the coccidia, *Vet. Parasitol.,* 6, 75, 1980.
7. **Marquardt, W. C.,** Host and site specificity in the coccidia: a perspective, *J. Protozool.,* 28, 243, 1981.
8. **Joyner, L. P.,** Host and site specificity, in *The Biology of the Coccidia,* Long, P. L., Ed., University Park Press, Baltimore, 1982, 35.
9. **Dresser, S. S.,** Extraintestinal development of eimeriid coccidia in pigs and chamois, *J. Parasitol.,* 64, 933, 1978.
10. **Lima, J. D.,** Development of *Eimeria* species in mesenteric lymph nodes of goats, *J. Parasitol.,* 65, 976, 1979.
11. **McCully, R. M., Basson, R. A., de Vos, V., and de Vos, A. J.,** Uterine coccidiosis of the impala caused by *Eimeria neitzi* sp. n., *Onderstepoort J. Vet. Res.,* 37, 45, 1970.
12. **Pellerdy, L.,** Attempts to alter the host specificity of eimeriae by parenteral infection experiments, *Acta Vet. (Brno),* 38, 43, 1969.
13. **Pellerdy, L.,** Parenteral infection experiments with *Eimeria stiedai, Acta Vet. Acad. Sci. Hung.,* 19, 171, 1969.
14. **Novilla, M. N., Carpenter, J. W., Spraker, T. R., and Jeffers, T. K.,** Parenteral development of eimerian coccidia in sandhill and whopping cranes, *J. Protozool.,* 28, 248, 1981.
15. **Overstreet, R. M.,** Species of *Eimeria* in non-epithelial sites, *J. Protozool.,* 28, 258, 1981.
16. **Dubey, J. P.,** *Toxoplasma Hammondia, Besnoitia, Sarcocystis,* and other tissue cyst-forming coccidia of man and animals, in *Parasitic Protozoa,* Vol. III, Krier, J. P., Ed., Academic Press, New York, 1979, 101.
17. **Cawthorn, R. J. and Stockdale, P. H. G.,** The developmental cycle of *Caryospora bubonis* Cawthorn and Stockdale, 1981 (Protozoa: Eimeriidae) in the great horned owl, *Bubo virginianus* (Gmelin), *Can. J. Zool.,* 60, 152, 1982.
18. **Upton, S. J., Lindsay, D. S., Current, W. L., and Barnard, S. M.,** Mouse-to-mouse transmission of *Caryospora simplex* (Apicomplexa: Eimeriidae), *J. Parasitol.,* 71, 395, 1985.
19. **Upton, S. J., Current, W. L., Ernst, J. V., and Barnard, S. M.,** Extraintestinal development of *Caryospora simplex* (Apicomplexa: Eimeridae) in experimentally infected mice, *Mus musculus, J. Protozool.,* 31, 392, 1984.
20. **Smith, D. D. and Frenkel, J. K.,** *Besnoitia darlingi* (Apicomplexa, Sarcocystidae, Toxoplasmatinae): transmission between opossums and cats, *J. Protozool.,* 31, 584, 1984.
21. **Box, E. D. and Smith, J. H.,** The intermediate host spectrum in a *Sarcocystis* species of birds, *J. Parasitol.,* 68, 668, 1982.
22. **Fayer, R. and Frenkel, J. K.,** Comparative infectivity for calves with oocysts of feline coccidia: *Besnoitia, Hammondia, Cystoisospora, Sarcocystis,* and *Toxoplasma, J. Parasitol.,* 65, 756, 1979.
23. **Wacha, R. S. and Christiansen, J. L.,** Development of *Caryospora bigenetica* n. sp. (Apicomplexa: Eimeriidae) in rattlesnakes and laboratory mice, *J. Protozool.,* 29, 272, 1982.
24. **Upton, S. J. and Barnard, S. M.,** Development of *Caryospora bigenetica* (Apicomplexa: Eimeriorina) in experimentally infected mice, *Int. J. Parasitol.,* 18, 15, 1988.
25. **Vetterling, J. M.,** *Eimeria tenella*: host specificity in gallinaceous birds, *J. Protozool.,* 23, 155, 1976.
26. **Long, P. L. and Millard, B. J.,** Rejection of *Eimeria* by foreign hosts, *Parasitology,* 78, 239, 1979.
27. **Sundermann, C. A., Lindsay, D. S., and Blagburn, B. L.,** *In vitro* excystation of *Cryptosporidium baileyi* from chickens, *J. Protozool.,* 34, 28, 1987.

28. **Fayer, R. and Leek, R. G.,** Excystation of *Sarcocystis fusiformis* sporocysts from dogs, *Proc. Helminthol. Soc. Wash.*, 40, 294, 1973.
29. **Woodmansee, D. B. and Powell, E. F.,** Cross-transmission and *in vitro* excystation experiments with *Sarcocystis muris*, *J. Parasitol.*, 70, 182, 1984.
30. **Bejsovec, J.,** The attempts on transfer of the coccidian *Eimeria tetricus* to nonspecific hosts *Phasianus colchicus*, *Perdix perdix*, and *Gallus gallus f. domestica*, *J. Protozool.*, (Suppl.), Abstr., 101, 1977.
31. **Upton, S. J. and Barnard, S. M.,** Experimental transmission of *Caryospora simplex* (Apicomplexa: Eimeridae) to Palestine ulpers, *Vipera xanthina plaestinae* (Serpentes: Viperidae), *J. Protozool.*, 33, 129, 1986.
32. **Upton, S. J., Current, W. L., and Barnard, S. M.,** A review of the genus *Caryospora* Leger, 1904 (Apicomplexa: Eimeriidae), *Syst. Parasitol.*, 8, 3, 1986.
33. **Frenkel, J. K. and Dubey, J. P.,** Rodents as vectors for feline coccidia, *Isospora felis* and *Isospora rivolta*, *J. Infect. Dis.*, 125, 69, 1972.
34. **Mahrt, J. L.,** Endogenous stages of the life cycle of *Isospora rivolta* in the dog, *J. Protozool.*, 14, 754, 1967.
35. **Pellerdy, L.,** Studies on the coccidia of the domestic cat, *Isospora novocati* sp. n., *Acta Vet. Hung.*, 24, 127, 1974.
36. **Dubey, J. P.,** Experimental *Isospora canis* and *Isospora felis* infection in mice, cats, and dogs, *J. Protozool.*, 22, 416, 1975.
37. **Dubey, J. P.,** *Isospora ohioensis* sp. n. proposed for *I. rivolta* of the dog, *J. Parasitol.*, 61, 462, 1975.
38. **Dubey, J. P. and Melhorn, H.,** Extraintestinal stages of *Isospora ohioensis* from dogs in mice, *J. Parasitol.*, 64, 689, 1989.
39. **Box, E. D.,** A *Toxoplasma* associated with an isosporan oocyst in canaries, *J. Protozool.*, 17, 391, 1970.
40. **Box, E. D.,** Exogenous stages of *Isospora serini* (Aragao) and *Isospora canaria* sp. n. in the canary *(Serinus canarius* L.), *J. Protozool.*, 22, 1975.
41. **Box, E. D.,** Life cycles of two *Isosopora* species in the canary, *Serinus canarius* Linnaeus, *J. Protozool.*, 24, 57, 1977.
42. **Novilla, M. N. and Box, E. D.,** Pathologic features of *Isospora serini* (Aragao) infection in the canary *(Serinus canarius* Linneaus), in *Research in Avian Coccidiosis, Proc. Georgia Coccidiosis Conf.*, McDougald, L. R., Joyner, L. P., and Long, P. L., Eds., University of Georgia, Athens, 1985, 217.
43. **Hendricks, L. D.,** A redescription of *Isospora arctopitheci* Rodhain 1933 (Protozoa: Eimeriidae) from primates of Panama, *Proc. Helminthol. Soc. Wash.*, 41, 229, 1974.
44. **Hendricks, L. D.,** Host range characteristics of the primate coccidian, *Isospora arctopitheci* Rodhain 1933 (Protozoa: Eimeriidea), *J. Parasitol.*, 63, 32, 1977.
45. **Ford, G. E.,** Prey-predator transmission in the epizootiology of ovine sarcosporidiosis, *Aust. Vet. J.*, 50, 38, 1974.
46. **Box, E. D. and Duszynski, D. W.,** Experimental transmission of *Sarcocystis* from icterid birds to sparrows and canaries by sporocysts from the opossum, *J. Parasitol.*, 64, 682, 1978.
47. **Bledsoe, B.,** Transmission studies with *Sarcocystis idahoensis* of deer mice *(Peromyscus maniculatus)* and gopher snakes *(Pituophis melanoleocus)*, *J. Wildlife Dis.*, 16, 195, 1980.
48. **Fayer, R. and Dubey, J. P.,** Development of *Sarcocystis fayeri* in the equine, *J. Parasitol.*, 68, 856, 1982.
49. **Dubey, J. P.,** A review of *Sarcocystis* of domestic animals and other animals of rats and dogs, *J. Am. Vet. Med. Assoc.*, 169, 1061, 1976.
50. **Dubey, J. P.,** Development of ox-coyote cycle of *Sarcocystis cruzi*, *J. Protozool.*, 29, 591, 1182.
51. **Fayer, R., Dubey, J. P., and Leek, R. G.,** Infectivity of *Sarcocystis* spp. from bison, elk, moose, and cattle for cattle via sporocysts from coyotes, *J. Parasitol.*, 68, 681, 1982.
52. **Fayer, R. and Dubey, J. P.,** Bovine sarcocystosis, *Compendium of Continuing Education for the Practicing Veterinarian*, 8, 130, 1986.
53. **O'Donoghue, P. J. and Ford, G. E.,** The asexual precyst development of *Sarcocystis tenella* in experimentally infected specific pathogen-free lambs, *Int. J. Parasitol.*, 14, 345, 1984.
54. **Tadros, W. and Laarman, J. J.,** Current concepts on the biology, evolution and taxonomy of tissue cyst-forming eimerid coccidia, in *Advances in Parasitology*, Vol. 20, Lumsden, W. H. R., Muller, R., and Baker, J. R., Eds., Academic Press, New York, 1982, 293.
55. **Crum, J. M., Fayer, R., and Prestwood, A. K.,** *Sarcocystis* spp. in white-tailed deer. I. Definitive and intermediate host spectrum with a description of *Sarcocystis odocoileocanis* n. sp., *J. Wildlife Dis.*, 17, 567, 1981.
56. **Levine, N. D. and Tadros, W.,** Named species and hosts of *Sarcocystis* (Protozoa: Apicomplexa: Sarcocystidae), *Syst. Parasitol.*, 2, 41, 1980.
57. **Adams, J. H., Levine, N. D., and Todd, K. S.,** *Eimeria* and *Sarcocystis* in racoons in Illinois, *J. Protozool.*, 28, 221, 1981.

58. **Zaman, V. and Colley, F. C.**, Light and electron microscopic observations of the life cycle of *Sarcocystis orientalis* n. sp. in the rat (*Rattus norvegicus*) and the Malaysian reticulated python *(Phyton reticulatus)*, *Z. Parasitenkd.*, 47, 169, 1975.
59. **Dubey, J. P. and Streitel, R. H.**, Shedding of *Sarcocystis* in feces of dogs and cats fed muscles of naturally infected food animals in the midwestern United States, *J. Parasitol.*, 62, 828, 1976.
60. **Rzepczyk, C. M.**, Evidence of a rat-snake life cycle for *Sarcocystis*, *Int. J. Parasitol.*, 4, 447, 1974.
61. **Frenkel, J. K., Heydorn, A. O., Melhorn, H., and Rommel, M.**, Sarcocystidae: *Nomina dubia* and available names, *Z. Parasitenkd.*, 58, 115, 1979.
62. **Ruiz, A. and Frenkel, J. K.**, Recognition of cyclic transmission of *Sarcocystis muris* by cats, *J. Infect. Dis.*, 133, 409, 1976.
63. **Fayer, R., Johnson, A. J., and Hildebrant, P. K.**, Oral infection of mammals with *Sarcocystis fusiformis* bradyzoites from cattle and sporocysts from dogs and coyotes, *J. Parasitol.*, 62, 10, 1976.
64. **Fayer, R. and Kradel, D.**, *Sarcocystis leporum* in cottontail rabbits and its transmission to carnivores, *J. Wildlife Dis.*, 13, 170, 1977.
65. **Hudkins, G. G. and Kistner, T. P.**, *Sarcocystis hemoionilatrantis* (sp. nov.), life cycle in mule deer and coyote, *J. Wildlife Dis.*, 13, 80, 1977.
66. **Collins, G. H. and Charleston, W. A. G.**, Studies on *Sarcocystis* species. IV. A species infecting dogs and goats; development in goats, *N.Z. Vet. J.*, 27, 260, 1980.
67. **Dubey, J. P.**, Coyote as a final host for *Sarcocystis* species of goats, sheep, cattle, elk, bison, and moose in Montana, *Am. J. Vet. Res.*, 41, 1227, 1980.
68. **Rickard, M. D. and Munday, B. L.**, Host specificity of *Sarcocystis* spp. in sheep and cattle, *Aust. Vet. J.*, 52, 48, 1970.
69. **Dubey, J. P., Speer, C. A., and Fayer, R.**, *Sarcocystis of Animals and Man*, CRC Press, Boca Raton, FL, 1989, 60.
70. **Fayer, R., Heydorn, A. O., Johnson, A. J., and Leek, R. G.**, Transmission of *Sarcocystis suihominis* from humans to swine to nonhuman primates *(Pan troglodytes, Macaca mulatta, Macaca irus)*, *Z. Parasitenkd.*, 59, 15, 1979.
71. **Fayer, R. and Johnson, A. J.**, *Sarcocystis fusiformis* infection in the coyote *(Canis latrans)*, *J. Infect. Dis.*, 131, 189, 1975.
72. **Duszynski, D. W. and Box, E. D.**, The opossum *(Didelphis virginiana)* as a host for *Sarcocystis deboner* from cowbirds *(Molothrus atec)* and grackles *(Cassidy mexicanus, Quiscalus quiscula)*, *J. Parasitol.*, 63, 326, 1978.
73. **Box, E. D. and Duszynski, D.**, *Sarcocystis* of passerine birds: sexual stages in the opossum *(Didelphis virginiana)*, *J. Wildlife Dis.*, 16, 209, 1980.
74. **Box, E. D., Meier, J. L., and Smith, J. H.**, Description of *Sarcocystis falcatula* Stiles, 1983, a parasite of birds and opossums, *J. Protozool.*, 31, 521, 1984.
75. **Smith, D. D.**, The Sarcocystidae: *Sarcocystis, Frenkelia, Toxoplasma, Besnoitia, Hammondia,* and *Cystoisospora*, *J. Protozool.*, 28, 262, 1981.
76. **Rommel, M., Krampitz, H. E., and Gersel, O.**, Beitrage zum Lebenszyklus der Frenkelien. III. Die sexulle Entwickling von *F. clethriomyobuteonis* in Mausebussard, *Z. Parasitenkd.*, 51, 139, 1976.
77. **Krampitz, H. E., Rommel, M., Geisel, O., and Kaiser, E.**, Beitrage zum Lebenszyklus, der Frenkelien. II. Die Ungeschlech Hiche Entwickling von *Frenkelia clethrionomyobuteonis* in der Rotelmaus, *Z. Parasitenkd.*, 51, 7, 1976.
78. **Rommel, M. and Krampitz, H. E.**, Beitrage zum Lebenszyklus der Frenkelien. I. Die Identitat von *Isospora buteonis* aus dem Mausebussard mit einer Frenkelienart (*F. clethrionomyobuteonis* spec. n.) aus der Rotelmaus, *Berl. Muench. Tierarztl. Wochenschr.*, 88, 338, 1978.
79. **Enemar, A.**, M-organisms in the brain of the Norway lemming, *Lemmus Lemmus*, *Ark. Zool.*, 18, 9, 1965.
80. **Hayden, D. W., King, N. W., and Murthy, A. S. K.**, Spontaneous *Frenkelia* infection in a laboratory-reared rat, *Vet. Pathol.*, 13, 337, 1976.
81. **Krampitz, H. E. and Rommel, M.**, Experimentelle Untersuchunger uber das Wirtsspektrum der Frenkelien der Erdmaus, *Berl. Muench. Tieraertztl. Wochenschr.*, 90, 17, 1977.
82. **Meingassner, J. G. and Burtscher, H.**, Doppelinfektion des Gehirns mit *Frenkelia* species and *Toxoplasma gondii* bei *Chinchilla laniger*, *Vet. Pathol.*, 14, 146, 1977.
83. **Rommel, M. and Krampitz, H. E.**, Weitere Untersuchunger uber das Zwischenwirtsspektrum and den Entwickl ungszyklus von *Frenkelia microti* aus der Erdmaus, *Zentralbl. Veterinaermed. Reihe B.*, 25, 273, 1978.
84. **Frenkel, J. K.**, Toxoplasmosis: Parasite life cycle, pathology, and immunology, in *The Coccidia: Eimeria, Isospora, Toxoplasma, and Related Genera*, Hammond, D. M. and Long, P. L., Eds., University Park Press, Baltimore, 1973, 343.
85. **Frenkel, J. K.**, Toxoplasmosis, *Microbiology 1984*, 212, 1984.
86. **Frenkel, J. K.**, Congenital toxoplasmosis-prevention or palliation, *Am. J. Obstet. Gynecol.*, 141, 359, 1981.

87. **Reynolds, E. S., Walls, W. W., and Pfeiffe, R. I.,** Generalized toxoplasmosis following renal transplantation, *Arch. Intern. Med.*, 118, 401, 1966.
88. **Roth, J. A., Siegel, S. E., Levine, A. S., and Berard, C. W.,** Fatal recurrent toxoplasmosis in a patient initially infected via a leukocyte transfusion, *Am. J. Clin. Pathol.*, 56, 601, 1971.
89. **Remington, J. S. and Desmonts, G.,** Toxoplasmosis, in *Infectious Diseases of the Fetus and Newborn Infant*, Remington, J. S. and Klein, J., Eds., W. B. Saunders, Philadelphia, 1983, 143.
90. **Remington, J. S., Jacobs, L., and Melton, M. L.,** Congenital transmission of toxoplasmosis from mother animals with acute and chronic infections, *J. Infect. Dis.*, 108, 163, 1961.
91. **Luft, B. J., Naot, Y., Araujo, F. G., Stinson, E. B., and Remington, J. S.,** Primary and reactivated *Toxoplasma* infection in patients with cardiac transplants, *Ann. Intern. Med.*, 99, 27, 1983.
92. **Fayer, R.,** Coccidian taxonomy and nomenclature, *J. Protozool.*, 28, 266, 1981.
93. **Dubey, J. P., Miller, N. L., and Frenkel, J. K.,** The *Toxoplasma gondii* oocyst from cat feces, *J. Exp. Med.*, 132, 636, 1970.
94. **Wallace, G. D.,** Experimental transmission of *Toxoplasma gondii* by filth flies, *Am. J. Trop. Med. Hyg.*, 20, 411, 1971.
95. **Wallace, G. D.,** Experimental transmission of *Toxoplasma gondii* by cockroaches, *J. Infect. Dis.*, 126, 545, 1972.
96. **Chinchilla, M., Guerrero, D. M., and Solano, E.,** Lack of multiplication of *Toxoplasma* in macrophages of rats *in vitro*, *J. Parasitol.*, 68, 952, 1982.
97. **Jacobs, L.,** Propagation, morphology, and biology of *Toxoplasma*, *Ann. N.Y. Acad. Sci.*, 64, 154, 1956.
98. **McCabe, R. E. and Remington, R. E.,** Mechanisms of killing of *Toxoplasma gondii* by rat peritoneal macrophages, *Infect. Immun.*, 52, 151, 1986.
99. **Lainson, R.,** Toxoplasmosis in England. II. Variation factors in the pathogenesis of *Toxoplasma* infections: the sudden increase of virulence of a strain after passage in multimammate rats and canaries, *Ann. Trop. Med. Parasitol.*, 49, 397, 1955.
100. **Sakurai, H., Igarashi, I., Omata, V., Saito, A., and Suzuki, N.,** Effects of neonatal spleen cell products on the multiplication of *Toxoplasma* in rat peritoneal macrophages, *J. Immunol.*, 131, 1527, 1983.
101. **Krahenbuhl, J. L. and Blazkovec, A. A.,** *Toxoplasma gondii*: immunotherapy of cutaneous hypersensitivity reactions in guinea pigs injected with living parasites, *Exp. Parasitol.*, 37, 83, 1975.
102. **Harboe, A. and Ericksen, S.,** Toxoplasmosis in chickens. 3. Attempts to provoke a systemic disease in chickens by infection with a chicken strain and a human strain of *Toxoplasma*, *Acta. Pathol. Microbiol. Scand.*, 35, 495, 1954.
103. **Blagburn, B. L., Lindsay, D. S., Giambrone, J. J., Sundermann, C. A., and Hoerr, F. J.,** Experimental cryptosporidiosis in broiler chickens, *Poult. Sci.*, 66, 442, 1987.
104. **Bermudez, A. J., Ley, D. H., Levy, M. G., Ficken, M. D., Guy, J. S., and Gerig, T. M.,** Intestinal and bursal cryptosporidiosis in turkeys following inoculation with *Cryptosporidium* sp. isolated from commercial poults, *Avian Dis.*, 32, 445, 1988.
105. **Tzipori, S.,** Cryptosporidiosis in animals and humans, *Microbiol. Rev.*, 47, 84, 1983.
106. **Fayer, R. and Ungar, B. L. P.,** *Cryptosporidium* spp. and cryptosporidiosis, *Microbiol. Rev.*, 50, 458, 1986.
107. **Wolfson, J. S., Richter, J. M., Waldron, M. A., Weber, D. J., McCarthy, D. M., and Hopkins, C. C.,** Cryptosporidiosis in immunocompetent patients, *N. Engl. J. Med.*, 312, 1278, 1985.
108. **Soave, R. and Armstrong, D.,** *Cryptosporidium* and cryptosporidiosis, *Rev. Infect. Dis.*, 8, 1012, 1986.
109. **Sallon, S., Deckelbaum, R. J., Schmid, I. I., Harlap, S., Baras, M., and Spira, D. T.,** *Cryptosporidium*, malnutrition, and chronic diarrhea in children, *Am. J. Dis. Child.*, 142, 312, 1988.
110. **Meisel, J. L., Perera, D. R., Meligro, C., and Rubin, C. E.,** Overwhelming watery diarrhea associated with a *Cryptosporidium* in an immunosuppressed patient, *Gastroenterology*, 70, 1156, 1976.
111. **Payne, P., Lancaster, L. A., Heinzman, M., and McCutchan, J. A.,** Identification of *Cryptosporidium* in patients with acquired immunodeficiency syndrome, *N. Engl. J. Med.*, 309, 613, 1984.
112. **Sloper, K. S., Dourmashkin, R. R., Bird, R. B., Slavin, G., and Webster, A. D.,** Chronic malabsorption due to cryptosporidiosis in a child with immunoglobulin deficiency, *Gut*, 23, 80, 1982.
113. **Kocoshis, S. A., Cibull, M. L., Davis, T. E., Hinton, J. T., Seip, M., and Banwell, J. G.,** Intestinal and pulmonary cryptosporidiosis in an infant with severe combined immune deficiency, *J. Pediatr. Gastroenterol. Nutr.*, 3, 49, 1984.
114. **Centers for Disease Control,** Cryptosporidiosis among children attending day-care centers — Georgia, Pennsylvania, Michigan, California, New Mexico, *MMWR*, 33, 599, 1984.
115. **Current, W. L., Reese, N. C., Ernst, J. V., Bailey, W. S., Heyman, M. B., and Weinstein, W. M.,** Human cryptosporiosis in immunocompetent and immunodeficient persons. Studies of an outbreak and experimental transmission, *N. Engl. J. Med.*, 308, 1252, 1983.
116. **Current, W. L.,** *Cryptosporidium* and cryptosporidiosis, in *Acquired Immune Deficiency Syndrome*, Gottlieb, M. S. and Groopman, J. E., Eds., Alan R. Liss, New York, 1984, 355.

117. **Upton, S. J. and Current, W. L.**, The species of *Cryptosporidium* (Apicomplexa: Cryptosporiidae) infecting mammals, *J. Parasitol.*, 71, 625, 1985.
118. **Anderson, B. C. and Bulgin, M. S.**, Enteritis caused by *Cryptosporidium* in calves, *Vet. Med. Sm. Anim. Clin.*, 76, 865, 1981.
119. **Reese, N. C., Current, W. L., Ernst, J. V., and Bailey, W. S.**, Human and calf cryptosporidiosis. A case report and results of comparative infections in mice and rats, *Am. J. Trop. Med. Hyg.*, 31, 226, 1982.
120. **Tzipori, S., Larson, J., Smith, M., and Lugft, R.**, Diarrhea in goat kids attributed to *Cryptosporidium* infection, *Vet. Rec.*, 111, 35, 1982.
121. **Moon, H. W. and Bemrick, W. J.**, Fecal transmission of calf cryptosporidiosis between calves and pigs, *Vet. Pathol.*, 18, 248, 1981.
122. **Moon, H. W., Schwartz, A., Welch, M. J., McCann, P. P., and Runnels, P. L.**, Experimental fecal transmission of human cryptosporidium to pigs, *Vet. Pathol.*, 19, 700, 1982.
123. **Links, I. J.**, Cryptosporidial infection of piglets, *Aust. Vet. J.*, 58, 60, 1982.
124. **Tzipori, S., McCartney, E., Lawson, G. H. K., Rowland, R. C., and Campbell, I.**, Experimental infection of piglets with *Cryptosporidium, Res. Vet. Sci.*, 31, 358, 1981.
125. **Tzipori, S., Angus, K. W., Campbell, I., and Clerihew, L. W.**, Diarrhea due to *Cryptosporidium* infection in artificially-reared lambs, *J. Clin. Microbiol.*, 14, 100, 1981.
126. **Tzipori, S., Angus, K. W., Gray, E. W., Campbell, I., and Allan, F.**, Diarrhea in lambs experimentally infected with *Cryptosporidium* isolated from calves, *Am. J. Vet. Res.*, 42, 1400, 1981.
127. **Angus, K. W., Tzipori, S., and Gray, E. W.**, Intestinal lesions in specific-pathogen-free lambs associated with a *Cryptosporidium* from calves with diarrhea, *Vet. Pathol.*, 19, 67, 1982.
128. **Current, W. L., Upton, S. J., and Haynes, T. B.**, The life cycle of *Cryptosporidium baileyi* n.sp. (Apicomplexa, Cryptosporidiidae) infecting chickens, *J. Protozool.*, 33, 289, 1986.
129. **Lindsay, D. S., Blagburn, B. L., and Sundermann, C. A.**, Host specificity of *Cryptosporidium* sp. isolated from chickens, *J. Parasitol.*, 72, 565, 1986.
130. **Lindsay, D. S., Blagburn, B. L., Sundermann, C. A., and Hoerr, F. J.**, Experimental infections in domestic ducks with *Cryptosporidium baileyi* isolated from chickens, *Avian Dis.*, 33, 69, 1989.
131. **Lindsay, D. S., Blagburn, B. L., and Hoerr, F. J.**, Experimentally induced infections in turkeys with *Cryptosporidium baileyi* isolated from chickens, *Am. J. Vet. Res.*, 48, 104, 1987.
132. **Lindsay, D. S., Blagburn, B. L., Hoerr, F. J., and Giambrone, J. J.**, Experimental *Cryptosporidium baileyi* infections in chickens and turkeys produced by ocular inoculation of oocysts, *Avian Dis.*, 31, 355, 1987.
133. **Goodwin, M. A., Steffens, W. L., Russel, I. D., and Brown, J.**, Diarrhea associated with intestinal cryptosporidiosis in turkeys, *Avian Dis.*, 32, 63, 1988.
134. **Bermudez, A. J., Ley, D. H., Levy, M. G., Fickens, M. D., Guy, J. S., and Gerig, T. M.**, Intestinal and bursal cryptosporidiosis in turkeys following inoculation with *Cryptosporidium* sp. isolated from commercial poults, *Avian Dis.*, 32, 445, 1988.
135. **Pellerdy, L.**, *Coccidia and Coccidiosis,* Paul Parey, Berlin, 1974.
136. **Bejosovec, J.**, Specificity of coccidians of wild galliform birds under condition of prolonged contact among various host species, *J. Protozool.*, 21, 454, 1974.
137. **Tsutsumi, Y.**, *Eimeria tsunodai* sp. nov. (Protozoa: Eimeriidae) a cecal coccidium of Japanese quails *(Coturnix coturnix japonica), Jpn. J. Vet. Sci.*, 34, 1, 1972.
138. **Tsunoda, K. and Muraki, Y.**, A new coccidium of Japanese quails: *Eimeria uzura* sp. nov., *Jpn. J. Vet. Sci.*, 33, 227, 1971.
139. **Ruff, M. D., Fagan, J. M., and Dick, J. W.**, Pathogenicity of coccidia in Japanese quail *(Coturnix coturnix japonica), Poult. Sci.*, 63, 55, 1984.
140. **Norton, C. C. and Pierce, M. A.**, The life cycle of *Eimeria bateri* (Protozoa: Eimeriidae) in the Japanese quail *Coturnix coturnix japonica, J. Protozool.*, 18, 57, 1971.
141. **Ruff, M. D.**, Life cycle and biology of *Eimeria lettyae* sp. n. (Protozoa: Eimeriidae) from the northern bobwhite, *Colinus virginianus, J. Wildlife Dis.*, 21, 361, 1985.
142. **Levine, N. D. and Ivens, V.**, The coccidian parasites (Protozoa, Sporozoa) or ruminants, *Ill. Biol. Monogr.*, 44, 1, 1970.
143. **Subramanian, G. and Jha, D.**, Cross-transmission of *Eimeria favrei* (Moussev and Marotel, 1902) Martin, 1909 and *E. ninakohlyakimovae* Yakimoff and Rastegaiff, 1980 in closely related hosts, *Ind. J. Microbiol.*, 6, 7, 1966.
144. **Sayin, F.**, The sporulated oocysts of *Eimeria ankarensis* n. sp. and of other species of *Eimeria* of buffalo in Turkey and transmission of four species of *Eimeria* from buffalo to cow calves, *Ankara Univ. Vet. Fakult. Deng.*, 15, 282, 1969.
145. **Fitzsimmons, W. M.**, Infection of goats with *Eimeria* spp. of sheep origin, *Vet. Rec.*, 76, 1099, 1964.
146. **McDougald, L. R.**, Attempted cross transmission of coccidia between sheep and goats and description of *Eimeria ovinoidalis* sp. n., *J. Protozool.*, 20, 109, 1979.

147. **Levine, N. D.,** Taxonomy and life cycles of coccidia, in *The Biology of the Coccidia*, Long, P. L., Ed., University Park Press, Baltimore, 1982, 1.
148. **Levine, N. D. and Ivens, V.,** Cross-transmission of *Eimeria* spp. (Protozoa, Apicomplexa) of rodents—a review, *J. Protozool.*, 35, 434, 1988.
149. **Doran, D. J.,** The life cycle of *Eimeria dispersa* Tyzzer, 1929 from the turkey in gallinaceous birds, *J. Parasitol.*, 64, 882, 1978.
150. **Tyzzer, E. E.,** Coccidiosis in gallinaceous birds, *Am. J. Hyg.*, 10, 269, 1929.
151. **Norton, C. C.,** Coccidiosis of pheasants and turkeys: recent experiments with pheasant coccidia at Weybridge, in *Proc. Int. Symp. Coccidia and Further Prospects for their Control*, Czechoslovakia Academy of Sciences, Prague, 1979, 174.
152. **Kogut, M. H. and Long, P. L.,** The effect of silica injections on the rejection of *Eimeria* from nonspecific hosts, *J. Parasitol.*, 67, 960, 1981.
153. **Norton, C. C.,** *Eimeria colchici* sp. n. (Protozoa: Eimeriidae) the cause of cecal coccidiosis in English covert pheasants, *J. Protozool.*, 14, 772, 1967.
154. **Mayberry, L. F. and Marquardt, W. C.,** Transmission of *Eimeria separata* from the normal host, *Rattus* to the mouse, *Mus musculus, J. Parasitol.*, 59, 198, 1973.
155. **Doran, D. J.,** Coccidiosis in the kangaroo rat of California, *Univ. Calif. Publ. Zool.*, 52, 31, 1953.
156. **Todd, K. S. and Hammond, D. M.,** Life cycle and host specificity of *Eimeria callospermophili* Henry, 1932 from the Uinta ground squirrel, *Spermophilus armatus, J. Protozool.*, 15, 1, 1968.
157. **de Vos, A. J. and Van der Westhuizen, I. B.,** The occurrence of *Eimeria chinchillae* n. sp. (Eimeriidae) in *Chinchilla laniger* (Molina, 1782) in South Africa, *J. S. Afr. Vet. Med. Assoc.*, 39, 81, 1968.
158. **de Vos, A. J.,** Studies on the host range of *Eimeria chinchillae* De Vos and Van der Westhuizen, 1968, *Onderst. J. Vet. Res.*, 37, 29, 1970.
159. **Marquardt, W. C.,** Attempted transmission of the rat coccidium *Eimeria nieschulzi* to mice, *J. Parasitol.*, 52, 691, 1966.
160. **Vetterling, J. M.,** *Eimeria tenella*: host specificity in gallinacecous birds, *J. Protozool.*, 23, 155, 1976.
161. **Rose, M. E. and Millard, B. J.,** Host specificity in eimerian coccidia: development of *Eimeria vermiformis* of the mouse, *Mus musculus*, in *Rattus norvegicus, Parasitology*, 90, 557, 1985.
162. **Todd, K. W., Lepp, D. L., and Trayser, G. V.,** Development of the asexual cycle of *Eimeria vermiformis* Ernst, Chobotar, and Hammond, 1971, from the mouse, *Mus musculus*, in dexamethasone-treated rats, *Rattus norvegicus, J. Parasitol.*, 57, 1137, 1971.
163. **Mayberry, L. F., Marquardt, W. C., Nash, D. J., and Plan, B.,** Genetic dependent transmission of *Eimeria separata* from *Rattus* to three strains of *Mus musculus*, an abnormal host, *J. Parasitol.*, 68, 1124, 1982.
164. **Naciri, M.,** Some observations on the development of *Eimeria tenella, E. acervulina,* and *E. Maxima* in a non-specific host, in *Research in Avian Coccidiosis, Proc. Georgia Coccidiosis Conf.*, McDougald, L. R., Joyner, L. P., and Long, P. L., Eds., University of Georgia, Athens, 1986, 46.
165. **Desser, S. S.,** Extraintestinal development of eimeriid coccidia in pigs and chamois, *J. Parasitol.*, 64, 933, 1978.
166. **Marquardt, W. C.,** Host and site specificity in the coccidia, in *The Coccidia: Eimeria, Isospora, Toxoplasma, and Related Genera*, Hammond, D. M. and Long, P. L., Eds., University Park Press, Baltimore, 1973, 23.
167. **Wang, C. C.,** Biochemistry and physiology of coccidia, in *The Biology of the Coccidia*, Long, P. L., Ed., University Park Press, Baltimore, 1982, 167.
168. **Wagner, W. H. and Foerster, O.,** Die PAS-AO methode, eine spezialfarburg fur coccidien in gewesbe, *Z. Parasitenkd.*, 25, 28, 1966.
169. **Hammond, D. M.,** Life cycles and development of coccidia, in *The Coccidia*, Hammond, D. M. and Long, P. L., Eds., University Park Press, Baltimore, 1973, 45.
170. **Ryley, J. F.,** Cytochemistry, physiology, and biochemistry, in *The Coccidia*, Hammond, D. M. and Long, P. L., Eds., University Park Press, Baltimore, 1973, 159.
171. **La Fon, S. W. and Nelson, D. J.,** Purine metabolism in the intact sporozoites and merozoites of *Eimeria tenella, Mol. Biochem. Parasitol.*, 14, 11, 1985.
172. **Fry, M., Hudson, A. T., Randall, A. W., and Williams, R. B.,** Potent and selective hydroxynaphthoquinone inhibitors of mitochondrial electron transport in *Eimeria tenella* (Apicomplexa: Coccidia), *Biochem. Pharmacol.*, 33, 2115, 1984.
173. **Smith, C. K. and Lee, D. E.,** Monosaccharide transport by *Eimeria tenella* sporozoites, *J. Parasitol.*, 72, 163, 1986.
174. **Warren, E. W.,** Vitamin requirements of the coccidia of the chicken, *Parasitology*, 58, 137, 1968.
175. **Sofield, W. L. and Strout, R. G.,** Amino acids essential for *in vitro* cultivation of *Eimeria tenella, J. Protozool.*, 21, 434, 1974.
176. **Ryley, J. F. and Wilson, R. G.,** Growth factor antagonism studies with coccidia in tissue culture, *Z. Parasitenkd.*, 40, 31, 1972.

177. **Latter, V. S. and Holmes, L. S.,** Identification of some nutrient requirements for the *in vitro* cultivation of *Eimeria tenella*, in *Proc. Int. Symp. Coccidia and Further Prospects of their Control*, Czechoslovak Academy of Sciences, Prague, 1979, 215.
178. **Doran, D. J. and Augustine, P. C.,** *Eimeria tenella*: vitamin requirements for development in primary cultures of chicken kidney cells, *J. Protozool.*, 25, 544, 1978.
179. **Mathis, G. F. and McDougald, L. R.,** Evaluation of interspecific hybrids of the chicken, guinea fowl, and Japanese quail for innate resistance to coccidia, *Avian Dis.*, 31, 740, 1987.
180. **Klesius, P. H. and Hings, S. E.,** Strain-dependent differences in murine susceptibility to coccidia, *Infect. Immun.*, 26, 1111, 1979.
181. **Rose, M. E., Owen, D. G., and Hesketh, P.,** Susceptibility to coccidiosis: effect of strain of mouse on reproduction of *Eimeria vermiformis*, *Parasitology*, 88, 45, 1984.
182. **Patton, W. H.,** *Eimeria tenella*: cultivation of the asexual stages in cultured animal cells, *Science*, 150, 767, 1965.
183. **Doran, D. J.,** Behavior of coccidia *in vitro*, in *The Biology of the Coccidia*, Long, P. L., Ed., University Park Press, Baltimore, 1982, 229.
184. **Lindsay, D. S., Current, W. L., and Upton, S. J.,** Development of *Eimeria tuskegeensis* (Protozoa: Eimeriidae) from the cotton rat, *Sigmodon hispidus*, in cell cultures, *J. Ala. Acad. Sci.*, 55, 248, 1984.
185. **Lindsay, D. S., Blagburn, B. L., Current, W. L., and Ernst, J. V.,** Development of the swine coccidium *Eimeria debliecki* Douwes, 1921 in mammalian cell cultures, *J. Protozool.*, 32, 669, 1985.
186. **Lindsay, D. S., Current, W. L., and Haynes, T. B.,** Comparative development of *Isospora suis* in piglets, chicken embryos, and cell culture, in *Proc. 4th Int. Symp. Neonatal Diarrhea*, 1983, 333.
187. **Lindsay, D. S. and Blagburn, B. L.,** Development of *Isospora suis* from pigs in primary porcine, and bovine cell cultures, *Vet. Parasitol.*, 24, 301, 1987.
188. **Current, W. L. and Haynes, T. B.,** Complete development of *Cryptosporidium* in cell culture, *Science*, 224, 603, 1984.
189. **Woodmansee, D. B. and Pohlenz, J. F.,** Development of *Cryptosporidium* sp. in a human rectal tumor cell line, in *Proc. 4th Int. Symp. Neonatal Diarrhea*, 1983, 306.
190. **Lindsay, D. S., Sundermann, C. A., and Blagburn, B. L.,** Caryocyst-like host cell formation by *Caryospora duszynskii* (Apicomplexa: Eimeridae) in human fetal lung cell cultures, *J. Protozool.*, 35, 32, 1988.
191. **Upton, S. J., Haynes, T. B., Current, W. L., and Barnard, S. M.,** Development of *Caryospora simplex* (Apicomplexa: Eimeriidae) from sporozoites to oocysts in human embryonic lung cell culture, *J. Protozool.*, 31, 398, 1984.
192. **Sundermann, C. A., Lindsay, D. S., Tibbs, R. E., and Bailey, M. A.,** Complete development of *Caryospora bigenetica* (Apicomplexa: Eimeriidae) *in vitro*, *J. Protozool.*, 35, 465, 1988.
193. **Speer, C. A., Whitmire, W. M., Reduker, D. W., and Dubey, J. P.,** *In vitro* cultivation of meronts of *Sarcocystis cruzi*, *J. Parasitol.*, 72, 677, 1986.
194. **Speer, C. A., Cawthorn, R. J., and Dubey, J. P.,** *In vitro* cultivation of the vascular phase of *Sarcocystis capracanis* and *Sarcocystis tenella*, *J. Protozool.*, 33, 486, 1986.
195. **Rose, M. E. and Orlans, E.,** Immunoglobulins in the egg, embryo, and young chicks, *Develop. Comp. Immunol.*, 5, 15, 1981.
196. **Long, P. L.,** Development of *Eimeria tenella* in avian embryos, *J. Parasitol.*, 56, 575, 1965.
197. **Fitzgerald, P. R.,** Development of *Eimeria stiedae* in avian embryos, *J. Parasitol.*, 56, 1252, 1970.
198. **Long, P. L. and Millard, B. J.,** Pathogenicity, immunogenicity and site specificity of an attenuated strain of *Eimeria tenella*, *J. Protozool.*, 52, 53A, 1975.
199. **Rollinson, D.,** Development of *Eimeria tenella* in quail embryos, *Trans. R. Soc. Trop. Med. Hyg.*, 70, 21, 1976.
200. **Long, P. L. and Millard, B. J.,** Studies on *Eimeria grenieri* in the guinea fowl *(Numida meleagris)*, *Parasitology*, 76, 1, 1978.
201. **Kogut, M. H., Gore, T. C., and Long, P. L.,** Serial passage of *Eimeria tenella* and *E. necatrix* in turkey embryos, *Parasitology*, 86, 199, 1983.
202. **Long, P. L., Johnson, J., and Gore, T. C.,** Attenuation of a strain of *Eimeria mivati* of U.S.A. origin by serial embryo passage, *Avian Dis.*, 26, 305, 1982.
203. **Long, P. L.,** Schizogony and gametogony of *Eimeria tenella* in the liver of chick embryos, *J. Protozool.*, 18, 17, 1971.
204. **Lindsay, D. S. and Current, W. L.,** Complete development of *Isospora suis* of swine in chicken embryos, *J. Protozool.*, 31, 152, 1984.
205. **Current, W. L. and Long, P. L.,** Development of human calf *Cryptosporidium* in chicken embryos, *J. Infect. Dis.*, 148, 1108, 1983.
206. **Naciri, M., Yvore, P., Boissieu, C., and Esnault, E.,** Multiplication de *Cryptosporidium muris* (Tyzzer, 1907) *in vitro* entretien d'une souche sur oeufs embryonnes, *Rec. Med. Vet.*, 162, 51, 1986.

207. **Lindsay, D. S., Sundermann, C. A., and Blagburn, B. L.,** Cultivation of *Cryptosporidium baileyi*: studies with cell cultures, avian embryos, and pathogenicity of chicken embryo-passaged oocysts, *J. Parasitol.*, 74, 288, 1988.
208. **McLoughlin, D. K.,** The influence of dexamethasone on attempts to transmit *Eimeria melegrimitis* to chickens and *E. tenella* to turkeys, *J. Protozool.*, 16, 145, 1969.
209. **Pellerdy, L. and Durr, V.,** Orale und parenterle Ubertragungsuersuche von Kokzidion auf nicht spizifische Wirte, *Acta Vet. Acad. Sci. Hung.*, 19, 253, 1969.
210. **Todd, K. S. and Lepp, D. L.,** Completion of the life cycle of *Eimeria vermiformis* Ernst, Chobotar, and Hammond, 1979, from the mouse, *Mus musculus* in dexamethasone-treated rats, *Rattus norvegicus*, *J. Parasitol.*, 58, 400, 1972.
211. **Eirmann, L. A. and Kogut, M. H.,** The role of the helper T cell in innate immunity against *Eimeria* infections in the chickens, *Exp. Parasitol.*, submitted.
212. **Kronke, M., Leonard, W. J., Depper, J. M., Arya, S. K., Wong-Stau, F., Gallo, R. C., Waldmann, T. A., and Green, W. C.,** Cyclosporine A inhibits T cell growth factor at the level of mRNA transcription, *Proc. Natl. Acad. Sci. U.S.A.*, 81, 5214, 1984.
213. **Kogut, M. H. and Long, P. L.,** The effect of silica injections on the rejection of *Eimeria* from nonspecific hosts, *J. Parasitol.*, 67, 960, 1981.
214. **Kessel, R. W. I., Monaco, L., and Marchisio, M. A.,** The specificity of the cytotoxic action of silica — a study *in vitro*, *Br. J. Exp. Pathol.*, 44, 351, 1963.
215. **Lotzovz, E. and Cudkowicz, G.,** Abrogation of resistance to bone marrow grafts by silica particles, *J. Immunol.*, 113, 343, 1974.
216. **Vigliani, E. C. and Pernic, B.,** Immunological aspects of silicosis, *Adv. Tuberculosis Res.*, 12, 230, 1963.
217. **Hammond, D. M., Fayer, R., and Minter, M. L.,** Further studies on *in vitro* development of *Eimeria bovis* and attempts to obtain second generation schizonts, *J. Protozool.*, 16, 298, 1969.
218. **Jones, T. C. and Hirsch, J. G.,** The interaction between *Toxoplasma gondii* and mammalian cells. II. The absence of lysosomal fusion within phagocytic vacuoles containing living parasites, *J. Exp. Med.*, 136, 1173, 1972.
219. **Jones, T. C., Len, L., and Hirsch, J. G.,** Assessment *in vitro* of immunity against *Toxoplasma gondii*, *J. Exp. Med.*, 141, 466, 1975.
220. **Khavkin, T. N. and Freidlin, L. S.,** A fluorescence phase contrast study of the interaction between *Toxoplasma gondii* and lysosomes in living cells, *Z. Parasitenkd.*, 52, 19, 1977.
221. **Long, P. L. and Rose, M. E.,** Growth of *Eimeria tenella in vitro* in macrophages from chicken peritoneal exudates, *Z. Parasitenkd.*, 48, 291, 1976.
222. **Allison, A. C., Harington, J. S., and Birbeck, M.,** An examination of the cytotoxic effects of silica on macrophages, *J. Exp. Med.*, 124, 141, 1966.
223. **Catanzaro, P. J., Schwartz, H. J., and Graham, R. C.,** Spectrum and possible mechanism of carrageenan cytotoxicity, *Am. J. Pathol.*, 64, 387, 1971.
224. **Lotze, J. C., Leek, R. G., Shalkop, W. T. and Behin, R.,** Coccidial parasites in the "wrong host" animal, *J. Parasitol.*, 47 (4 Sect. 2), 34, 1961.
225. **Levine, N. D. and Ivens, V.,** The coccidian parasites (Protozoa: Sporozoa) of rodents, *Ill. Biol. Monogr.*, 33, 1, 1965.
226. **Marquardt, W. C.,** Attempted transmission of the rat coccidium *Eimeria nieschulzi* to mice, *J. Parasitol.*, 52, 691, 1966.
227. **Long, P. L. and Millard, B. J.,** Studies on *Eimeria dispersa*, Tyzzer 1929 in turkeys, *Parasitology*, 78, 41, 1979.
228. **Kogut, M. H. and Long, P. L.,** Extraintestinal sporozoites of chicken *Eimeria* in chickens and turkeys, *Z. Parasitenkd.*, 70, 287, 1984.
229. **Naciri, M. and Yvore, P.,** Developpment d'*Eimeria tenella*, agent d'une coccidiose caecale du poulet chez un hote non specifique: existence d'une forme exointestinal infectante, *C.R. Acad. Sci. (Paris)*, 294, 219, 1982.
230. **Naciri, M.,** Some observations on the development of *Eimeria tenella*, *E. acervulina*, and *E. maxima* in a non-specific host, in *Research in Avian Coccidiosis, Proc. Georgia Coccidiosis Conf.*, McDougald, L. R., Joyner, L. P., and Long, P. L., Eds., University of Georgia, Athens, 1986, 46.

Chapter 4

EIMERIA: INFECTIONS OF THE INTESTINE

M. A. Fernando

TABLE OF CONTENTS

I. INTRODUCTION

Sporozoites of members of the genus *Eimeria* enter their respective hosts by penetrating epithelial cells of the intestinal mucosa. While most undergo endogenous development within the intestine, the life cycles of a few species occur at extraintestinal sites, e.g., *E. stiedai* in the biliary epithelium of the rabbit, *E. truncata* in the tubular epithelium of the goose kidney and *Eimeria* species of fish in epithelial and nonepithelial cells of the liver, kidney, and pancreas. Extraintestinal development has also been reported for species of *Eimeria* that normally occur within the intestine. Both asexual and sexual stages of *E. arloingi* and *E. christenseni* were found in the mesenteric lymph nodes,[1] asexual and sexual stages, including oocysts, of *E. rechenowi* and *E. gruis* in bronchial epithelial cells in the lungs of sandhill and whooping cranes,[2] and schizonts of *E. tenella* in the liver of chickens treated with corticosteroids.[3]

The present discussion, however, will be confined to infections of the intestine.

II. LIFE CYCLES AND DEVELOPMENT

The eimerian life cycle can be divided into sporogony, schizogony, and gametogony. Sporogony occurs outside the host and produces the infective stage while schizogony (asexual reproduction) and gametogony (sexual reproduction) occur intracellularly within a specific host animal.

A. SPOROGONY

Oocysts, which are the end products of infection in a host, are passed in the feces. Under appropriate environmental conditions, the oocysts undergo sporogony resulting in sporulated oocysts. Sporulated eimerian oocysts contain four sporocysts with two sporozoites within each sporocyst. Meiosis occurs during sporogony so that all stages in the life cycle, except the fertilized female gamete, contain a haploid amount of DNA.[4,5] This process requires the presence of oxygen and takes 24 h or longer depending on the ambient temperature, humidity, and the species of *Eimeria* in question. The optimum temperature has been noted as 29 and 30°C,[6] and exposure to either very high or very low temperatures prior to sporulation is detrimental to a high proportion of oocysts.[7] Conditions affecting sporogony and the viability and virulence of oocysts under varying environmental conditions are discussed in detail by Fernando.[8]

B. EXCYSTATION

When ingested by an appropriate host sporulated oocysts excyst within its intestine releasing the sporozoites. A variety of stimuli such as bile, trypsin, CO_2 and, in the chicken, the grinding action of the gizzard, help to break the oocyst wall and the Stieda and sub-Stieda bodies of the sporocyst. The processes of sporulation and excystation have been reviewed in detail by Hammond,[9] the physiological and biochemical aspects discussed by Ryley,[10] and Wang,[11] and the ultrastructure described by Chobotar and Scholtyseck.[12]

C. SITE FINDING BY SPOROZOITES

Sporozoites enter the host by penetrating villous or surface epithelial cells of the gut mucosa. Some species, such as *E. brunetti* and *E. praecox* of the domestic fowl, develop within epithelial cells at the site of entry while others travel within the mucosa to both epithelial and nonepithelial cells in which they normally undergo further development. For example, *E. bovis* of cattle develops in the endothelial cells of lacteals,[9] *E. zuernii* of cattle develop in cells within the lamina propria,[13] and *E. acervulina, E. maxima, E. necatrix,* and *E. tenella* of the domestic fowl develop in crypt epithelial cells.[14]

1. Transport of Sporozoites within the Intestinal Mucosa

Light and electron microscopic observations by Van Doorninck and Becker,[15] Challey and Burns,[16] Patillo,[17] Doran,[18] and Michael[19] demonstrated that sporozoites of *E. acervulina*, *E. necatrix*, and *E. tenella* of the chicken were transported to the crypts *via* the lamina propria by cells resembling macrophages. More recently, several workers have presented ultrastructural evidence to indicate that host cells transporting sporozoites of *E. tenella*,[20] *E. necatrix*,[21] *E. maxima*, and at least some of the cells transporting *E. acervulina*[22] were morphologically similar to those designated as granulated intraepithelial lymphocytes (IEL). *E. brunetti* and *E. praecox* develop in villous epithelial cells at their site of entry. The species that do not need a transport system to get to their final destination were found to be transported from the villous epithelial cells by host mononuclear cells only to return 18 to 24 h later to their site of entry to begin development.[22] There is now accumulated evidence to support the view that eimerian sporozoites are transported within host mononuclear cells from the surface epithelium of the gut to their preferred sites of development.

2. Transport of Sporozoites Extraintestinally

Eimerian sporozoites are also transported extraintestinally. We were able to produce patent infections in recipient chickens by feeding them blood, liver, and spleen of donors shortly after they were orally inoculated with large numbers of sporulated oocysts of *E. necatrix*,[21] *E. acervulina*, *E. brunetti*, *E. maxima*, and *E. praecox*,[22] demonstrating that sporozoites of these species travel outside the gut mucosa. Perry and Long[23] were able to produce patent infections in recipient chickens by feeding them livers of donors 12 h after infection with either *E. tenella* or *E. maxima*.

Whether the sporozoites are carried by host mononuclear cells, within which they are found in the lamina propria, is not known. We also do not know what proportion of the parasites are carried to extraintestinal sites and what proportion of those found there return to continue development in the intestine. This work discussed above indicates that extraintestinal travel within the normal host by eimerian sporozoites is perhaps more commonplace than hitherto acknowledged.

More evidence for extraintestinal migration of eimerian sporozoites has been provided by the work discussed earlier where extraintestinal development was found to occur in species of *Eimeria* that normally develop within the intestine. In fact, disseminated infections of *E. rechenowi* and *E. gruis* are almost the rule in sandhill and whooping cranes.[2]

D. SCHIZOGONY

Once within the appropriate cells, sporozoites undergo asexual reproduction by a process referred to as schizogony or merogony. Here, nuclear division is followed by cytoplasmic differentiation giving rise to merozoites. When mature, merozoites leave the cell in which they developed and enter other cells. Species of *Eimeria* usually undergo several generations of schizogony and was believed to be for the most part genetically predetermined for each species.[9] However, there is evidence to suggest that the number of cycles of asexual multiplication is not fixed. Long and Rose[24] were able to extend schizogony by treating chickens with betamethasone demonstrating that the host exerts some influence over the developmental cycles of these parasites. More recently, researchers have isolated what are termed "precocious" strains of several species of *Eimeria*. Jeffers[25] described a method of selection for precociousness. He used the earliest produced oocysts of *E. tenella* to provide the inoculum for each succeeding generation resulting in a reduction in the length of the life cycle. He found the strains obtained in this manner to be stable through at least 25 generations. Continued selection resulted in the complete absence of second generation schizogony within the selected strain.[26,27] This aspect of the life cycles of *Eimeria* species will be discussed later in this chapter.

Merozoites of most species reinfect cells that are close to those from which they emerge. This usually results in noninfected areas of the gut lying adjacent to heavily infected areas. In species such as *E. arloingi*, for example, an entire villus may be parasitized and next to a noninfected villus. This type of infection could help keep losses of merozoites down to a minimum. In other instances, merozoites need to travel long distances to continue their life cycle. The second generation merozoites of *E. necatrix* travel from the small intestine to the cecum of the chicken where third generation schizogony and gametogony occur. This may account for the low oocyst production in this species when compared to the large numbers of second generation merozoites that are usually found within the small intestine.

E. GAMETOGONY

It is not known with certainty when during the life cycle sexual differentiation occurs. Clones of *Eimeria* species of the chicken have been obtained by single sporozoite infections showing conclusively that this stage of the life cycle is not sexually differentiated.[28,29] Many species of *Eimeria* from the chicken have now been cloned, from single sporozoite infections. Canning[30] and Klimes et al.[31] found differences in nuclear structure and staining properties within groups of merozoites and suggested that these were sexually differentiated. Some of the precocious strains of *Eimeria* undergo gametogony after a single generation of schizogony, indicating the possibility that sexual differentiation occurs after the merozoites have entered cells to begin gametogony perhaps as a result of some sort of cytoplasmic stimulus exerted by the host cell.

Gametogony, or sexual reproduction, follows the last schizogonic cycle. Merozoites enter the appropriate cell and develop into male (micro) or female (macro) gamonts, also termed gametocytes. Microgamonts give rise to many microgametes by a process of division morphologically similar to schizogony while the macrogamonts do not divide but develop into a single macrogamete each. This is possible because, as stated earlier, all stages prior to the fertilized macrogamete contain a haploid amount of DNA.[5] The microgamete contains a nucleus, a mitochondrion, and two or three flagella each.[12] The microgametes leave the cell and fertilize the macrogamete while it is still within the host cell. It is not known at the present time how microgametes identify host cells harboring mature macrogametes ready to be fertilized. Perhaps the macrogamete either produces, or induces the host cell to produce, receptor molecules that are translocated onto the surface of the infected cell at the appropriate time.

Morphological and biochemical aspects of infection, schizogony, and gametogony are discussed in detail by Hammond,[9] Ryley,[10] and Wang.[11] The ultrastructure of these processes and of the varous stages in the life cycle has been reviewed by Chobotar and Scholtyseck.[12]

III. FACTORS AFFECTING LIFE CYCLES AND DEVELOPMENT

A. PARASITE FACTORS

In addition to the work on precocious strains discussed under schizogony, many other reports have appeared of successful attempts at obtaining precocious strains of all species of *Eimeria* of the domestic fowl. These have been reviewed in detail by Jeffers.[32,33] The shortening of the life cycles were found to be due to either fewer generations of schizogony and/or smaller and faster maturing schizonts.[34] These results question the validity of the prepatent period as a criterion for use in species identification.

Sporozoites of *E. tenella* inoculated into the allantoic cavity of developing chick embryos completed the entire life cycle in the chorioallantoic membrane.[35] After 42 passages, the second generation schizonts changed in size and the length of time they took to mature. When such strains were passaged through chickens the second generation schizonts remained small, and the crypt epithelial cells in which they developed did not enlarge or move into the lamina propria.[36] After 62 passages over a period of $2^1/_2$ years the characteristics of this strain became stabilized.[37] The work discussed above indicate that the life cycles and endogenous development of eimerian parasites are not fixed entities.

There is evidence for considerable variation or fluidity in many protozoan genomes giving rise to strain differences and altered behavior during all or part of their life cycles.[38-41] Corcoran et al.[41] found deletions with the loss of at least some coding sequences in cultured *Plasmodium falciparum*. These authors used pulsed field electrophoresis of chromosomal DNA to show chromosomal polymorphisms that resulted from the deletions. The isolation of precocious strains of *Eimeria* species and the changes in schizogony brought about by serial passages of *E. tenella* through chick embryos point to the fluidity of the genome of this group of parasites as well.

B. HOST FACTORS

The immune status of the host has been known to influence the development of parasites. A single infection stimulates at least partial immunity in the host, affecting the development of the parasite in subsequent infections. In *E. tenella*, Rose et al.[42] found a marked reduction in the numbers of developing parasites in immune chickens as compared to their nonimmune counterparts. The number of first generation schizonts was reduced by 87% and accounted for the greatest part in the reduction of infection. The effect was considered suppressive rather than a delaying one, confirming earlier observations quoted by these authors. Their experiments showed that this effect was due, at least partly, to failure of sporozoites transported by IEL to transfer to crypt enterocysts. Riley and Fernando[43] found that the transport of *E. maxima* sporozoites by IEL from surface to crypt epithelium was also affected by host immunity. In immune birds, as compared to nonimmune, a significantly greater number of IELs harboring sporozoites tended to remain in the lamina propria rather than migrate to the crypts. Long and Millard[44] had evidence to suggest the presence of occult endogenous stages of *Eimeria* in immune birds. These authors were able to induce low level oocyst production in immune chickens by corticosteroid treatment alone without any further inoculation of oocysts.

The genetic makeup of the host is perhaps the most important factor influencing both the susceptibility of that host to parasitic infection and the endogenous development of the parasite.[45] Rose et al.[46] found marked differences in the susceptibility of different strains of mice to primary infections with *E. vermiformis*. They found a greater reproduction of the parasite in the "susceptible" strains of mice and this was apparent not only as a higher peak output of oocysts but also as a prolongation of patency. These authors suggested the occurrences of additional cycles of schizogony. In another study Rose and Millard[47] compared the endogenous development of *E. vermiformis* in phenotypically normal BALB/c ("resistant") and C57BL/6 ("susceptible") mice in order to determine the basis for the differences in the pattern of oocyst production between the two strains. They found that development was similar up to, and including, 5 days postinfection. However, from day 9 onwards no parasites were seen in the "resistant" mice whereas large numbers of parasites were present in the intestinal epithelium of the "susceptible" mice up to, and including, day 14 postinfection. The parasites were also more widely distributed along the intestine in the "susceptible" mice at these later periods postinfection. This work clearly shows that the strain of host influences the endogenous development of intestinal *Eimeria* species.

C. OTHER FACTORS

Other factors such as anticoccidial drugs affect the development of eimerian parasites. Long and Millard[48] found eimerian sporozoites arrested in their development in quinolone treated chickens and showed that these lead to the relapse of infections seen after drug withdrawal. Drug-inhibited sporozoites of *E. tenella* were also found in the cecal tissue of chickens medicated with the anticoccidial drug, decoquinate, and given daily inoculations of a decoquinate sensitive strain of the parasite.[49]

Abnormal sporulation of *Eimeria* oocysts from chickens receiving various anticoccidial drugs has also been reported.[50,51] Norton and Joyner[51] were even able to select for abnormal bisporocystic oocysts of *E. maxima* which appeared during the slow development of resistance to increasing levels of a mixture of the anticoccidial drugs clopidol and methyl benzoate.

IV. HOST AND SITE SPECIFICITY

A. HOST SPECIFICITY

Species of *Eimeria* have historically been considered as having as high a degree of specificity as any other group of infectious agents.[6] However, they seem to be specific only as far as completion of their life cycles with the production of oocysts is concerned. Many species will excyst in, and parasitize cells in a variety of unrelated hosts. For example, Naciri[52] successfully infected mice with sporulated oocysts of *E. tenella*. She found stages resembling schizonts in the lungs of the infected mice and was able to transfer the infection to chickens by feeding them tissues of the infected mice. Kogut and Long[53] inoculated turkeys with a mixture of sporulated oocysts of *E. acervulina, E. brunetti, E. maxima, E. necatrix, E. praecox*, and *E. tenella*, and were able to infect chickens by feeding them the livers of turkeys. Host specificity in the coccidia is discussed in detail in Chapter 3.

B. SITE SPECIFICITY

Not only are eimerians host specific but they parasitize specific sites within the host and specific cell types within a tissue or an organ. In the chicken, species of *Eimeria* parasitize and develop in different regions of the gut with *E. acervulina* occupying the most proximal region and *E. tenella* and *E. brunetti* the most distal regions.[54] In addition, different stages of a single species can be specific to different regions of the intestine and to different cell types within that region. In *E. necatrix* of the chicken, for example, two generations of schizogony occur within crypt epithelial cells of the ileum but gametogony occurs in surface epithelial cells of the ceca. In *E. bovis* of cattle, the very large first generation schizonts are found in the posterior half of the small intestine within endothelial cells of the lacteals, second generation schizonts are within crypt epithelial cells of the large intestine and gametogony occurs in the surface epithelial cells of the cecum and colon.[9]

Site specificity has been studied in chickens by introducing the sporozoites of *E. tenella* parenterally. Injection *via* intravenous, intraperitoneal or intramuscular routes resulted in infection of the cecum where this species normally develops.[55-57] Perhaps the normal host plays a part in the site specificity exhibited by eimerian parasites even though, as discussed earlier, species of *Eimeria* have been found to develop at abnormal sites within the normal host.

Long[3] infected dexamethasone treated chickens orally with sporulated oocysts and intravenously with sporozoites of *E. tenella* and was able to demonstrate the presence of schizonts in the epithelial cells of the bile ducts. As stated earlier, sporozoites of *E. tenella* inoculated into the allantoic cavity of developing chick embryos completed the entire life cycle in the chorioallantoic membrane.[35] Of course, in the absence of host influences many species undergo schizogony and some are able to complete their life cycles *in vitro* in cultured cells.[18]

We do not know, however, with any degree of certainty how sporozoites find the specific sites at which they undergo normal development or why they must, while in the normal host, undergo development in specific cells at specific sites.

Species of *Eimeria* develop at specific locations within host cells. Most develop within the cytoplasm of epithelial cells, in a "parasitophorous vacuole" bounded by a unit membrane. A few, however, are known to undergo schizogony and others schizogony and gametogony within the nucleus of the host cell. *E. kotlani*, and *E. nocens* of geese undergo both schizogony and gametogony within the nuclei of intestinal epithelial cells.[58] All endogenous stages of *E. stigmosa* were found within the nucleoplasm of enterocysts of the goose intestine.[59] Merogony, but not gametogony, occurs within the nuclei of goose intestinal epithelial cells in *E. hermani* (Figure 1).[60,61] Species that develop within the nucleus were also found bounded by a unit membrane and inside a parasitophorous vacuole (Figure 2).[60] Shibalova[58] presented ultrastructural evidence to suggest that the sporozoites acquired the parasitophorous vacuole while still within the cytoplasm of the host cell. She suggested that the parasites travel into the nucleus through the nucleopore while still enveloped in the membrane of the parasitophorous vacuole.

V. HOST CELL CHANGES DURING PARASITE DEVELOPMENT

Most intracellular parasites induce changes within the host cell they parasitize and species of *Eimeria* are no exception.[62] Very soon after a cell is parasitized ultrastructural changes are evident. Mitochondria increase in number and come to lie adjacent to the parasitophorous vacuole, and there is an increase in the amount of rough endoplasmic reticulum.[63]

Often, the morphological changes that occur within the host cell are so drastic that it becomes difficult if not impossible to establish its original identity. For example, the intraepithelial lymphocytes of the gut within which first generation schizonts of *E. dispersa* of the turkey develop, change to the extent that the identity of the parasitized cell was not clarified till Lawn and Millard[64] described it in 1984. The same was true for the cells within which the large second generation schizonts of *E. necatrix* and *E. tenella* develop. They were identified as crypt cells that migrate out into the lamina propria during parasite development by Fernando et al. as recently as 1983.[65] Fernando et al.[66] found that within the lamina propria, these infected cells increased in size, their nuclei enlarged and their DNA content increased fivefold. Parasite induced modifications of host cells both at the morphological and at the biochemical level have been reviewed by Fernando.[62]

VI. INFECTIONS IN VARIOUS HOST SPECIES

A. INFECTIONS IN MAMMALS

1. Ruminants

The literature on coccidia and coccidiosis of both domesticated and free living ruminants has been reviewed extensively by Levine and Ivens[67] and by Pellerdy.[14] These authors have documented descriptions of *Eimeria* species that occur in ruminant hosts, their life cycles and endogenous development, geographic distribution and disease caused.

It is generally accepted that 19 to 20 species of *Eimeria* occur in cattle throughout the world. Of these, *E. zuerni* and *E. bovis* are usually associated with bovine coccidiosis but *E. alabamensis*, *E. auburnensis*, and *E. ellipsoidalis* are commonly seen and have been reported to cause diarrhea when present in large numbers.[68-70] Both *E. bovis* and *E. zuernii* develop in the lower small intestine, cecum and colon. *E. zuernii* is the causative agent of winter coccidiosis seen frequently in western Canada and in the midwestern U.S.[71,72] The mechanism(s) by which *E. zuernii* gives rise to coccidiosis in midwinter is not known with any certainty, although Stockdale and Niilo[73] were able to produce an acute form of coccidiosis by treating calves with dexamethasone 12, 15, and 16 days after inoculating them with *E. zuernii*.

Neurological signs are sometimes seen in association with intestinal coccidiosis in calves in northwestern U.S. and western Canada.[74-77] Both *E. bovis* and *E. zuernii* have been

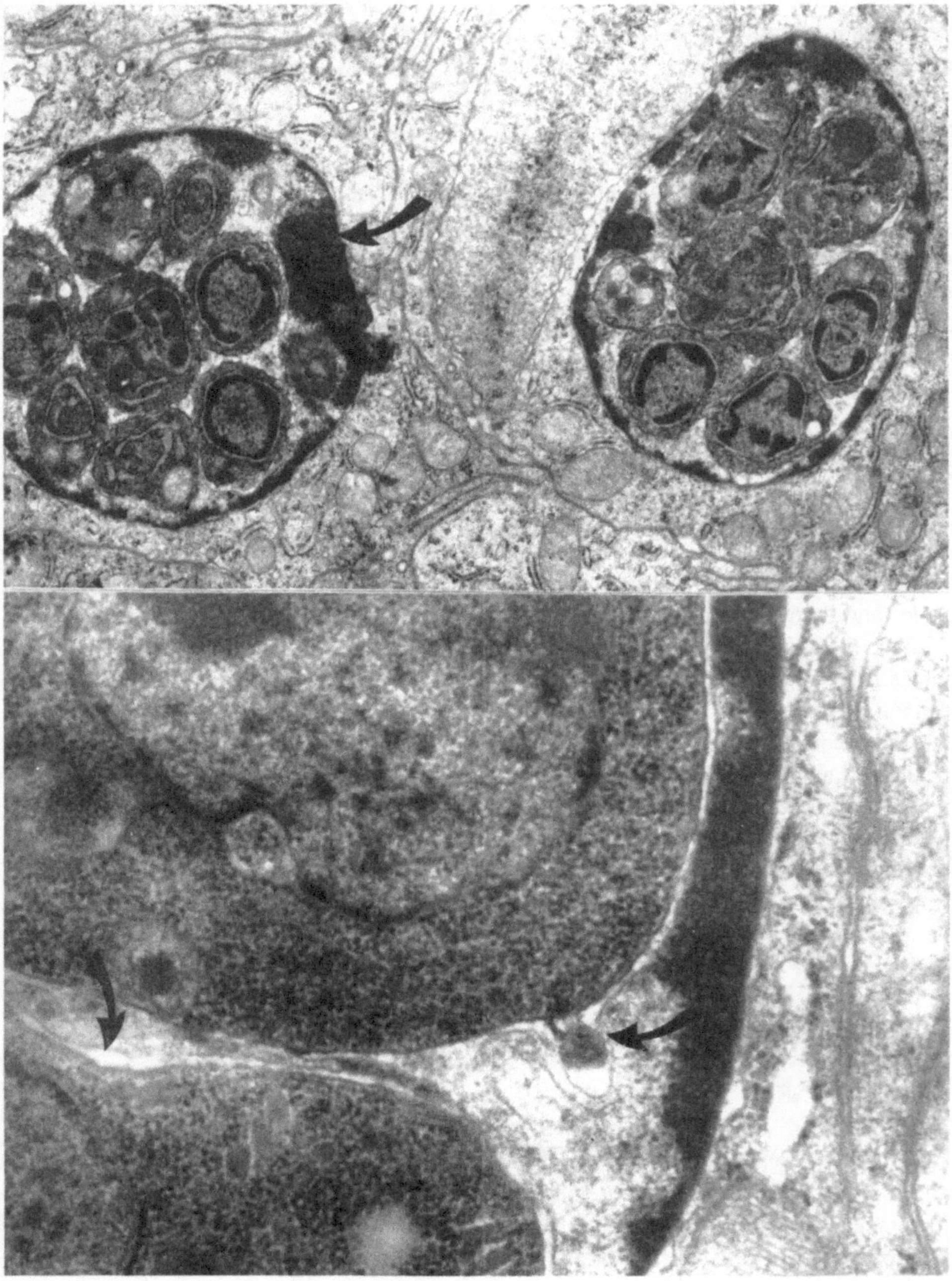

FIGURES 1 and 2. *Eimeria hermani* from experimentally infected domestic geese, 72 h postinfection, showing intestinal epithelial cell nuclei harboring schizonts. 1. (Top). Two nuclei each containing a mature schizont with merozoites in cross section. Arrow indicates a large mass of marginated host nuclear chromatin. (Magnification × 1300.) 2. Parts of two immature schizonts within one nucleus each lying within its own parasitophorous vacuole indicated by arrows. (Magnification × 43,700.)

associated with the disease and the death rate is reported to be high. The literature on this subject has been reviewed by Isler et al.[75]

2. Sheep and Goats

Although it was historically assumed that sheep and goats were infected with the same species of *Eimeria*, cross transmission experiments have proved that these two hosts have their own *Eimeria* species and that they do not cross infect.[78] Sheep and goats have at least 12 currently valid species of *Eimeria* each.

In sheep, coccidia can be found in animals of all ages but coccidiosis is mainly a disease of feedlot lambs which are usually infected with a mixture of species. *E. ovinoidalis, E. ahsata, E. ovina,* and *E. parva* are considered to be the pathogenic species.[14]

As opposed to sheep, coccidiosis is an important disease of goats. *E. ninakohlyakimovae* is perhaps the most pathogenic species even though *E. arloingi* is the most commonly seen.[79]

3. Swine

Isospora suis is the most important species of coccidia in swine and is the causative agent of neonatal coccidiosis. This species, however, will be discussed in Chapter 5. Nine species of *Eimeria* are found in swine and none of them are very pathogenic except under unusual circumstances such as in a debilitated host. Lindsay et al.[80] were not able to produce clinical coccidiosis in either nursing or weaned pigs with doses ranging from 8.0×10^5 to 1.0×10^7 oocysts of *E. debliecki*. In a survey of farms in southeastern United States by Lindsay et al.,[81] *E. debliecki, E. porci, E. scabra,* and *E. neodebliecki* were the species most often found.

4. Other Mammalian Hosts

The rabbit is the other domesticated mammalian host in which coccidiosis manifests itself as an important clinical disease. Except for the very pathogenic extraintestinal species, *E. stiedai*, which will be discussed in Chapter 6, all other species of rabbit *Eimeria* are intestinal parasites. At least five are known of which *E. perforans* is considered the commonest and is mildly pathogenic.[14]

B. INFECTIONS IN BIRDS

The genus *Eimeria* occurs in the domestic fowl, turkey, pheasant, and in many species of waterfowl. Passerine birds and birds of prey carry species of *Isospora*.

1. The Domestic Fowl

The group of eimerian parasites most extensively investigated is that of the domestic fowl. Seven species, *E. acervulina, E. brunetti, E. maxima, E. mitis, E. necatrix, E. praecox,* and *E. tenella* are currently considered valid species.[54] *E. tenella* and *E. necatrix* are the most pathogenic, followed by *E. brunetti* and *E. maxima*. *E. mivati* and *E. hagani* are controversial species and most investigators do not consider them to be valid. The endogenous development, biology, and pathogenicity of *Eimeria* species in the chicken have been reviewed by Long[54] and will not be discussed here.

2. Turkey and Pheasant

Seven species of *Eimeria* have been identified in the turkey. Of these, *E. adenoides* is considered the most pathogenic. *E. gallopavonis, E. meleagridis,* and *E. meleagrimitis* are also considered important in terms of clinical disease.[14]

3. Ducks and Geese

E. battakki, E. saitamae, and *E. schachdagica* are the commoner duck eimerians but *Tyzzeria perniciosa* is considered to be one of the most pathogenic species of coccidia.[14,60]

A very large number of species of *Eimeria* have been recorded from both domestic and free ranging geese.[14,60] Of these, *E. anseris, E. nocens,* and *E. kotlani* are considered pathogenic to domestic geese. Both schizogony and gametogony of three species, *E. kotlani, E. nocens,* and *E. stigmosa*, found primarily in the domestic goose, *Anser anser*, occur within the nuclei of gut epithelial cells.[58,59] Schizogony but not gametogony of *E. hermani* occurs within the nuclei of intestinal epithelial cells of *Branta canadensis* and experimentally infected domestic geese.[60,61]

VII. CONCLUDING REMARKS

It is not possible within the space of this chapter to discuss in detail the biology of *Eimeria* species that occur within the intestines of all animal species. I have, therefore, limited myself in the above discussion to areas that have not been adequately covered in previous reviews and have referred the readers to other reviews wherever they were available. In the section on infections within the host, I have purposely not dealt with chicken coccidiosis at any length as excellent reviews on this subject are readily available. For similar reasons, the *Eimeria* species of domestic and free ranging geese have received greater attention.

As noted in the above discussion, recent studies on the life cycles and endogenous development of *Eimeria* species of the chicken have demonstrated our lack of understanding of the developmental biology of *Eimeria* species, particularly at the molecular level. For example, we do not know whether "precocious" and embryo adapted strains of *Eimeria* species of the domestic fowl are the result of deletions of DNA sequences, as has been observed in cultured malaria parasites[41] or preexist in the population and are being selected for by our experimental procedures. The use of cloned strains obtained by single sporozoite infections to select for precociousness may at least partially answer these questions. In addition, pulsed field gel electrophoresis can be used to detect any chromosomal heterogeneity within these various clones and strains and may shed some light on the basis for the variations in their life cycles. These types of techniques have been used with success to study genetic polymorphisms in other protozoan parasites such as *Plasmodium* species,[41,42,82,83] *Leishmania*,[84,85] and the trypanosomes.[86]

At a time when the development of coccidial vaccines for use in the domestic fowl is receiving much attention, I would like to suggest that we may come out further ahead if we paid at least as much attention to understanding the developmental biology of these parasites.

REFERENCES

1. **Lotz, J. C., Shalkop, W. T., Leek, R. G., and Behin, R.**, Coccidial schizonts in mesenteric lymph nodes of sheep and goats, *J. Parasitol.*, 50, 205, 1964.
2. **Novilla, M. N., Carpenter, J. W., Spraker, T. R., and Jeffers, T. K.**, Parenteral development of eimerian coccidia in sandhill and whooping cranes, *J. Protozool.*, 28, 258, 1981.
3. **Long, P. L.**, Development (schizogony) of *Eimeria tenella* in the liver of chickens treated with corticosteroid, *Nature*, 225, 290, 1970.
4. **Canning, E. U. and Morgan, K.**, DNA synthesis, reduction, and elimination during life cycles of eimeriine coccidism, *Eimeria tenalla* and the haemogregarine, *Hepatozoon domerguei, Exp. Parasitol.*, 38, 217, 1975.
5. **Cornelissen, A. W. C. A., Overdulve, J. P., and Van der Ploeg, M.**, Determination of nuclear DNA of five eucoccidian parasites, *Isospora (Toxoplasma) gondii, Sarcocystis cruzi, Eimeria tenella, E. acervulina* and *Plasmodium berhgei*, with special reference to gametogenesis in *I. (T.) gondii, Parasitology*, 88, 531, 1984.
6. **Marquardt, W. C., Senger, C. M., and Seghetti, L.**, The effect of physical and chemical agents on the oocysts of *Eimeria zuernii* (Protozoa, Coccidia), *J. Protozool.*, 7, 186, 1969.
7. **Long, P. L.**, Pathology and pathogenicity of coccidial infections, in *The Coccidia*, Hammond, D. M. and Long, P. L., Eds., University Park Press, Baltimore, 1973, 253.
8. **Fernando, M. A.**, Pathology and pathogenesis, in *The Biology of the Coccidia*, Long, P. L., Ed., University Park Press, Baltimore, 1982, 287.
9. **Hammond, D. M.**, Life cycles and development of coccidia, in *The Coccidia*, Hammond, D. M. and Long, P. L., Eds., University Park Press, Baltimore, 1982, 45.

10. **Ryley, J. F.,** Cytochemistry, physiology and biochemistry, in *The Coccidia*, Hammond, D. M., and Long, P. L., Eds., University Park Press, Baltimore, 1973, 145.
11. **Wang, C. C.,** Biochemistry and physiology of coccidia, in *The Biology of the Coccidia*, Long, P. L., Ed., University Park Press, Baltimore, 1982, 167.
12. **Chobotar, B. and Scholtyseck, E.,** Ultrastructure, in *The Biology of the Coccidia*, Long, P. L., Ed., University Park Press, Baltimore, 1982, 101.
13. **Stockdale, P. H. G.,** Schizogony and gametogony of *Eimeria zuernii* (Rivolta, 1878) Martin, 1909, *Vet. Parasitol.*, 1, 367, 1976.
14. **Pellerdy, L. P.,** *Coccidia and Coccidiosis*, 2nd ed., Verlag Paul Parey, Berlin, 1974.
15. **Van Doorninck, W. M. and Becker, E. R.,** Transport of sporozoites of *Eimeria necatrix* in macrophages, *J. Parasitol.*, 43, 40, 1957.
16. **Challey, J. R. and Burns, W. C.,** The invasion of cecal mucosa by *Eimeria tenella* sporozoites and their transport by macrophages, *J. Protozool.*, 6, 238, 1959.
17. **Patillo, W. H.,** Invasion of cecal mucosa by sporozoites of *Eimeria tenella*, *J. Parasitol.*, 45, 253, 1959.
18. **Doran, D. J.,** The migration of *Eimeria acervulina* sporozoites to the duodenal glands of Liberkuhn, *J. Protozool.*, 13, 27, 1966.
19. **Michael, E.,** Sporozoites of *Eimeria acervulina* within intestinal macrophages in normal experimental infections. An ultrastructural study, *Z. Parasitenkd.*, 49, 33, 1976.
20. **Lawn, A. M. and Rose, M. E.,** Mucosal transport of *Eimeria tenella* in the cecum of the chicken, *J. Parasitol.*, 68, 1117, 1982.
21. **Al-Attar, M. A. and Fernando, M. A.,** Transport of *Eimeria necatrix* sporozoites in the chicken: effects of irritants injected intraperitoneally, *J. Parasitol.*, 73, 494, 1987.
22. **Fernando, M. A., Rose, M. E., and Millard, B. J.,** *Eimeria* spp. of domestic fowl: the migration of sporozoites intra- and extra-intestinally, *J. Parasitol.*, 73, 561, 1987.
23. **Perry, E. A. and Long, P. L.,** The extraintestinal stages of *Eimeria tenella* and *E. maxima* in the chicken, *Vet. Parasitol.*, 25, 9, 1987.
24. **Long, P. L. and Rose, M. E.,** Extended schizogony of *Eimeria mivati* in betamethazone-treated chickens, *Parasitology*, 60, 147, 1970.
25. **Jeffers, T. K.,** Alteration of *Eimeria tenella* through selection for precociousness, *J. Parasitol.*, 61, 1083, 1975.
26. **McDougald, L. R., and Jeffers, T. K.,** *Eimeria tenella*: (sporozoa; coccidia): gametogomy following a single asexual generation, *Science*, 192, 258, 1976.
27. **McDougald, L. R. and Jeffers, T. K.,** Comparative *in vitro* development of precocious and normal strains of *Eimeria tenella* coccidia, *J. Protozool.*, 23, 530, 1976.
28. **Shirley, M. W. and Millard, B. J.,** Some observations on the sexual differentiation of *Eimeria tenella* using single sporozoite infections in chicken embryos, *Parasitology*, 73, 337, 1976.
29. **Lee, E. H., Remmler, O. R., and Fernando, M. A.,** Sexual differentiation in *Eimeria tenella* (Sporozoa: coccidia), *J. Parasitol.*, 63, 155, 1977.
30. **Canning, E. U.,** Sexual differentiation in coccidia, in *Prog. in Protozoology. 4th Int. Congr. on Protozoology*, Clermont-Ferrand, 1973, 75.
31. **Klimes, B., Rootes, O. G., and Tanielian, Z.,** Sexual differentiation of merozoites of *Eimeria tenella*, *Parasitology*, 65, 131, 1972.
32. **Jeffers, T. K.,** Attenuation of coccidia — A review, in *Research in Coccidiosis*, McDougald, L. R., Joyner, L. P., and Long, P. L., Eds., University of Georgia, Athens, 1986, 482.
33. **Jeffers, T. K. and Shirley, M. W.,** Genetics, specific and infraspecific variation, in *The Biology of the Coccidia*, Long, P. L., Ed., University Park Press, Baltimore, 1982, 63.
34. **McDonald, V. and Shirley, M. W.,** *Eimeria mitis*: a comparison of the endogenous developmental stages of a line selected for early maturation and the parent strain, *Parasitology*, 88, 37, 1984.
35. **Long, P. L.,** Development of *Eimeria tenella* in avian embryos, *Nature*, 208, 509, 1965.
36. **Long, P. L.,** Endogenous stages of a "chick embryo-adapted" strain of *Eimeria tenella*, *Parasitology*, 66, 55, 1973.
37. **Long, P. L.,** Further studies on the pathogenicity and immunogenicity of an embryo-adapted strain of *Eimeria tenella*, *Avian Pathol.*, 3, 255, 1974.
38. **Dvorak, J.,** The natural heterogeneity of *Trypanosoma cruzi*: biological and medical implications, *J. Cell. Biochem.*, 24, 357, 1984.
39. **Van der Ploeg, L. H. T., Smits, M., Ponnudurai, T., Vermeulen, A., Meuwissen, J. H. E. Th., and Langsley, G.,** Chromosome-sized DNA molecules of *Plasmodium falciparum*, *Science*, 229, 658, 1985.
40. **Kemp, D. J., Corcoran, L. M., Coppel, R. L., Stahl, H. D., Bianco, A. E., Brown, G. V., and Anders, R. F.,** Size variation in chromosomes from independent cultured isolates of *Plasmodium falciparum*, *Nature*, 315, 347, 1985.
41. **Corcoran, L. M., Forsyth, K. P., Bianco, A. E., Brown, G. V., and Kemp, D. J.,** Chromosome size polymorphisms in *Plasmodium falciparum* can involve deletions and are frequent in natural parasite populations, *Cell*, 44, 87, 1986.

42. **Rose, M. E., Lawn, A. M., and Millard, B. J.,** The effects of immunity on the early events in the life-cycle of *Eimeria tenella* in the cecal mucosa of the chicken, *Parasitology,* 88, 199, 1984.
43. **Riley, D. and Fernando, M. A.,** *Eimeria maxima* (Apicomplexa): a comparison of sporozoite transport in naive and immune chickens, *J. Parasitol.,* 74, 103, 1988.
44. **Long, P. L. and Millard, B. J.,** The detection of occult coccidial infections by inoculating chickens with corticosteroid drugs, *Z. Parasitenkd.,* 48, 287, 1976.
45. **Wakelin, D.,** Genetic control of susceptibility and resistance to parasitic infection, *Adv. Parasitol.,* 16, 219, 1978.
46. **Rose, M. E., Owen, D. G., and Hesketh, P.,** Susceptibility to coccidiosis: effect of strain of mouse on reproduction of *Eimeria vermiformis, Parasitology,* 88, 45, 1984.
47. **Rose, M. E. and Millard, B. J.,** *Eimeria vermiformis*: host strains and the developmental cycle, *Exp. Parasitol.,* 60, 285, 1985.
48. **Long, P. L. and Millard, B. J.,** *Eimeria*: effect of meticlorpindol and methylbenzoate on endogenous stages in the chicken, *Exp. Parasitol.,* 23, 331, 1968.
49. **Jeffers, T. K. and Long, P. L.,** *Eimeria tenella*: immunogenicity of arrested sporozoites in chickens, *Exp. Parasitol.,* 60, 175, 1985.
50. **Joyner, L. P. and Norton, C. C.,** The anticoccidial effects of amprolium, dinitolmide and monensin against *Eimeria maxima, E. brunetti* and *E. acervulina, Parasitology,* 75, 155, 1977.
51. **Norton, C. C. and Joyner, L. P.,** The appearance of bisporocystic oocysts of *Eimeria maxima* in drug-treated chickens, *Parasitology,* 77, 243, 1978.
52. **Naciri, M.,** Some observations on the development of *Eimeria tenella, E. acervulina* and *E. maxima* in a non-specific host, in *Research in Coccidiosis,* McDougald, L. R., Joyner, L. P., and Long, P. L., Eds., University of Georgia, Athens, 1986, 46.
53. **Kogut, M. H. and Long, P. L.,** Extraintestinal sporozoites of chicken *Eimeria* in chickens and turkeys, *Z. Parasitenkd.,* 70, 287, 1984.
54. **Long, P. L.,** Coccodiosis in Poultry, *CRC Crit. Rev. Poult. Biol.,* 1, 25, 1987.
55. **Davis, S. F. M. and Joyner, L. P.,** Infection of the fowl by the parenteral inoculation of oocysts of *Eimeria, Nature,* 194, 996, 1962.
56. **Sharma, N. N. and Reid, W. M.,** Successful infection of chickens after parenteral inoculation of oocysts of *Eimeria* spp., *J. Parasitol.,* 48 (Suppl.), 33, 1962.
57. **Long, P. L. and Rose, M. E.,** Active and passive immunization of chickens against intravenously induced infections of *Eimeria tenella, Exp. Parasitol.,* 16, 1, 1965.
58. **Shibalova, T. A., Antukhaev, I. K., Ponizovski, A. K., and Morozova, T. I.,** Ultrastructure of life cycle stages in coccidia, in *4th Int. Congr. Parasitol.,* Short Communications, Section B, 1978, 73.
59. **Gajadhar, A. A., Rainnie, D. J., and Cawthorn, R. J.,** Description of the goose coccidium *Eimeria stigmosa* (Klimes, 1963), with evidence of intranuclear development, *J. Parasitol.,* 72, 588, 1986.
60. **Skene, R. C.,** Coccidia of Canada geese *(Branta canadensis)* at Kortright Waterfowl Park, Guelph, Ontario, M.Sc. thesis, University of Guelph, Guelph, Ontario, 1978.
61. **Skene, R. C. and Fernando, M. A.,** Intranuclear development of second and third generation schizonts of *Eimeria hermani* in geese, in *4th Int. Congr. Parasitol.,* Short Communications, Section B, 1978, 73.
62. **Fernando, M. A.,** *Eimeria*: parasite/cell interaction, in *Research in Avian Coccidiosis,* McDougald, L. R., Joyner, L. P., Long, P. L., Eds., University of Georgia, Athens, 1986, 36.
63. **Fernando, M. A. and Stockdale, P. H. G.,** Fine structural changes associated with schizogony in *Eimeria necatrix, Z. Parasitenkd.,* 43, 105, 1974.
64. **Millard, B. J. and Lawn, A. M.,** Parasite-host relationships during the development of *Eimeria dispersa* Tyzzer 1929, in the turkey *(Meleagris gallapava gallopavo)* with a description of intestinal intra-epithelial leucocytes, *Parasitology,* 84, 13, 1982.
65. **Fernando, M. A., Lawn, A. M., Rose, M. E., and Al-Attar, M. A.,** Invasion of chicken caecal and intestinal lamina propria by crypt epithelial cells infected with coccidia, *Parasitology,* 86, 391, 1983.
66. **Fernando, M. A., Pasternak, J., Barrell, R., and Stockdale, P. H. G.,** Induction of host nuclear DNA synthesis in coccidia-infected chicken intestinal cells, *Int. J. Parasitol.,* 4, 267, 1974.
67. **Levine, N. D. and Ivens, V.,** *The Coccidian Parasites (Protozoa, Sporozoa) of Ruminants,* University of Illinois Press, Urbana, 1970, 278.
68. **Radostits, O. M. and Stockdale, P. H. G.,** A brief review of bovine coccidiosis in Western Canada, *Can. Vet. J.,* 21, 227, 1980.
69. **Ernst, J. V. and Benz, G. W.,** Intestinal coccidiosis in cattle, *Vet. Clin. North Am. Food Anim. Pract.,* 2, 283, 1986.
70. **Kennedy, M. J. and Kralka, R. A.,** A survey of *Eimeria* spp. in cattle in central Alberta, *Can. Vet. J.,* 28, 124, 1987.
71. **Niilo, L.,** Bovine coccidiosis in Canada, *Can. Vet. J.,* 11, 91, 1970.
72. **Niilo, L.,** Experimental winter coccidiosis in sheltered and unsheltered calves, *Can. J. Comp. Med.,* 34, 20, 1970.

73. **Stockdale, P. H. G. and Niilo, L.,** Production of bovine coccidiosis with *Eimeria zuernii, Can. Vet. J.*, 17, 35, 1976.
74. **Isler, C. M., Bellamy, J. E. C., and Wobeser, G. A.,** Labile neurotoxin in serum of calves with "nervous" coccidiosis, *Can. J. Vet. Res.*, 51, 253, 1987.
75. **Isler, C. M., Bellamy, J. E. C., and Wobeser, G. A.,** Pathogenesis of neurological signs associated with bovine enteric coccidiosis: a prospective study and review, *Can. J. Vet. Res.*, 51, 261, 1987.
76. **Isler, C. M., Bellamy, J. E. C., and Wobeser, G. A.,** Characteristics of the labile neurotoxin associated with nervous coccidiosis, *Can. J. Vet. Res.*, 51, 271, 1987.
77. **Jubb, T. F.,** Nervous disease associated with coccidiosis in young cattle, *Aust. Vet. J.*, 65, 353, 1988.
78. **McDougald, L. R.,** Attempted cross-transmission of coccidia between sheep and goats and description of *Eimeria ovinoidalis* sp. n., *J. Parasitol.*, 26, 109, 1979.
79. **Craig, T. M.,** Epidemiology and control of coccidia in goats, *Vet. Clin. North Am. Food Anim. Pract.*, 2, 389, 1986.
80. **Lindsay, D. S., Blagburn, B. L., and Boosinger, T. R.,** Experimental *Eimeria debliecki* infections in nursing and weaned pigs, *Vet. Parasitol.*, 25, 39, 1987.
81. **Lindsay, D. S., Ernst, J. V., Current, W. L., Stuart, B. P., and Stewart, T. B.,** Prevalence of oocysts of *Isospora suis* and *Eimeria* spp. from sows on farms with and without a history of neonatal coccidiosis, *J. Am. Vet. Med. Assoc.*, 185, 419, 1984.
82. **Pologe, L. G. and Ravetch, J. V.,** A chromosomal rearrangement in a *P. falciparum* histidine-rich protein gene is associated with the knobless phenotype, *Nature*, 322, 474, 1986.
83. **Langsley, G. and Ponnudurai, T.,** *Plasmodium falciparum*: analysis of karyotype polymorphisms using chromosome-specific probes, *Exp. Parasitol.*, 65, 168, 1988.
84. **Bishop, R. P. and Miles, M. A.,** Chromosome size polymorphisms of *Leishmania donovani, Mol. Biochem. Parasitol.*, 24, 263, 1987.
85. **Giannin, S. H., Schittini, M., Keithly, J. S., Warburton, P. W., Cantor, C. R., and Van der Ploeg, L. H. T.,** Karyotype analysis of *Leishmania* species and its use in classification and clinical diagnosis, *Science*, 232, 762, 1986.
86. **Engman, D. M., Reddy, L. V., Donelson, J. E., and Kirchhoff, L. V.,** *Trypanosoma cruzi* exhibits inter and intra-strain heterogeneity in molecular karyotype and chromosomal gene location, *Mol. Biochem. Parasitol.*, 22, 115, 1987.

Chapter 5

ISOSPORA: INFECTIONS OF INTESTINE: BIOLOGY

David S. Lindsay

TABLE OF CONTENTS

I. INTRODUCTION

Oocysts of *Isospora* species are characterized by having two sporocysts each containing four sporozoites. The sporocyst may or may not have a Stieda body. Life cycle studies indicate that species with a Stieda body are generally monoxenous and confined to the intestines, whereas those species that lack a Stieda body often use paratenic hosts, may have latent stages in the host, and may be facultatively heteroxenous. All important (valid?) species of *Isospora* that infect man, nonhuman primates, dogs, cats, and domesticated mammals lack a Stieda body in their sporocysts. Generic names of *Levinia*[1] and *Cystoisospora*[2] have been proposed for the *Isospora* species which utilize paratenic hosts but these generic names have not gained widespread acceptance.

Levine listed 248 species of *Isospora* which had been described prior to 1986.[3] Most of these species are known only from oocysts found in the feces of the host animal. Until life cycle and cross transmission studies are conducted to determine more about the biology of these species, the validity of many of these coccidians is questionable.

Isospora species can cause serious disease in humans and pigs. Disease is seldom seen in nonhuman primates, dogs, or cats. *Isospora* species do not produce disease in horses, domestic ruminants, or domestic poultry and reports of isosporan oocysts in the feces of these hosts probably represent pseudoparasites that originated in feed contaminated with wild bird feces. The purpose of this chapter is to summarize what is known about the biology of the *Isospora* species of man, nonhuman primates, and domestic animals.

II. LIFE CYCLE

A. SPOROGONY

Sporogony, sporozoite formation is illustrated in Figure 1. This usually occurs outside the host and is the exogenous phase of the life cycle. Sporogony is dependent on moisture, temperature and adequate oxygen. Several controlled studies have been conducted on the sporogony of *Isospora* oocysts from dogs,[4,5] cats,[6] and pigs.[7] These studies indicate that temperatures greater than 40°C or less than 20°C inhibit sporogony of the oocysts. Rapid sporulation (<16 h) of oocysts occurs at 30 or 37°C. Structural events that occur during sporogony are similar for all species. Oocysts are passed in the feces and usually have a contracted sporont. A few oocysts will be passed that are in the sporoblast stage (two-celled stage). As the nucleus of the sporont divides a clear nuclear streak is formed, nuclear division occurs, and the sporont divides to form two uninucleate sporoblasts. Nuclear division occurs again and the nuclei are visible as clear areas at the poles of the sporoblast. Nuclear pyramids may be seen at the poles of the sporoblasts. The sporoblasts become elongate and form the sporocyst stage. Nuclear division occurs again and the outline of developing sporozoites soon becomes visible. When the sporozoites are fully visible the oocyst is considered to be sporulated. A small percentage (<2%) of oocysts are *Caryospora*-like and contain one sporocyst which encloses eight sporozoites.[7,8] It is not known if these *Caryospora*-like oocysts are infective.

B. EXCYSTATION

Excystation is the process by which sporozoites are released from the sporocysts/oocysts. The process is basically the same for all mammalian *Isospora* species studied to date[9-14] and is similar to what occurs in *Sarcocystis* spp.[15] and *Toxoplasma gondii*.[16] Studies have been conducted with intact oocysts or with sporocysts which have been mechanically freed from the oocyst wall. Pretreatment of oocysts with sodium hypochlorite solution[11,13] or cystine hydrochloride[14] under CO_2 enhances excystation of intact oocysts. Exposure of oocysts or sporocysts to sodium taurocholate solution (0.75%), bile (5%), or sodium taurocholate

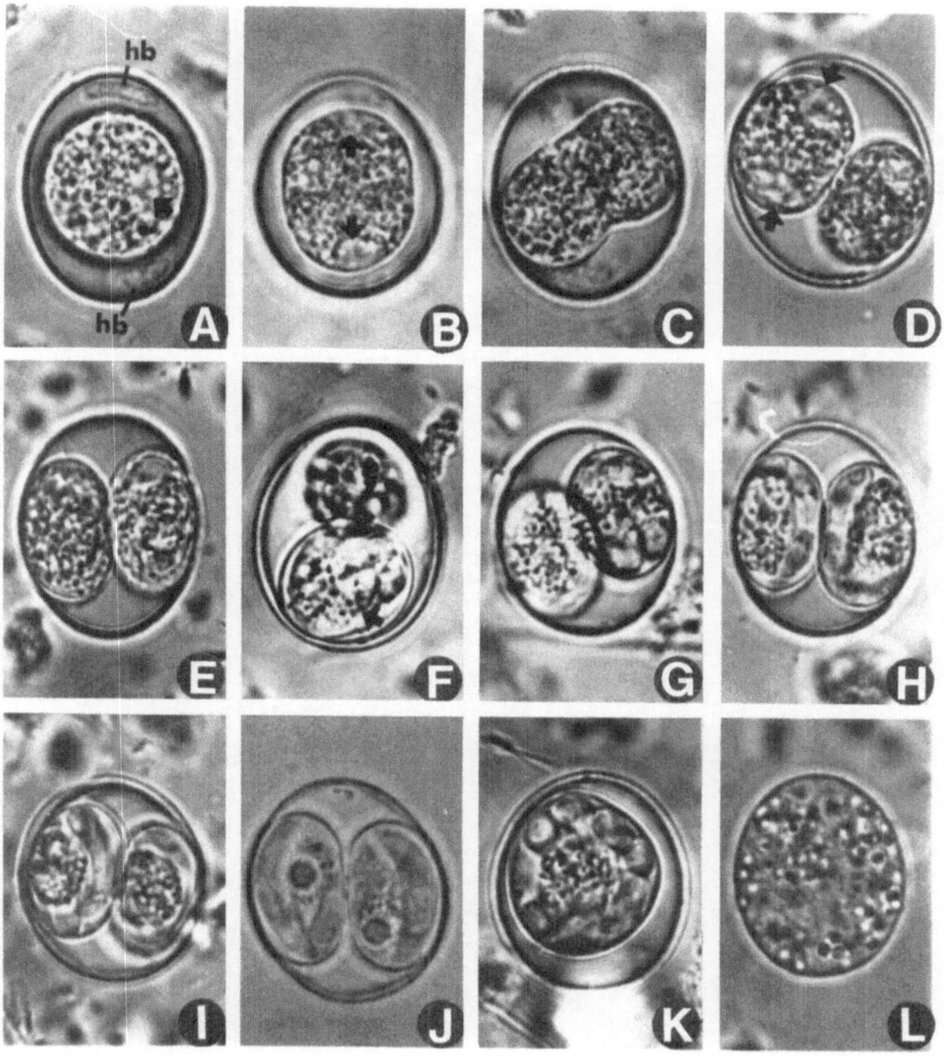

FIGURE 1. Photomicrographs of oocysts of *Isospora suis* during different stages of sporogony. Fecal flotations. All 1370×. (A) Unsporulated oocysts in sporont stage showing nucleus (arrow) and hazy bodies (hb). (B) Oocyst containing a binucleate sporont. Arrows point to nuclei. (C) Oocyst containing a sporont in the process of dividing. (D) Oocyst containing two binucleate sporoblasts with nuclei at poles (arrows). (E) Oocyst containing elongate sporocysts. (F) Oocyst in late sporocyst stage. Note developing sporozoite (arrows). (G) Early sporulated oocyst. Note the dispersed sporocyst residuum. (H) Sporulated oocyst in which the sporocyst residuum has condensed into an elongate form. (I) Sporulated oocyst in which the sporocyst residuum has contracted into a compact oval shape. (J) Sporulated oocyst in which the sporocyst residuum has contracted into a hyaline residuum with a few scattered granules. (K) *Caryospora*-like oocyst of *I. suis*. (L) Nonviable oocyst incapable of sporulating. (Modified from Lindsay, D. S., Current, W. L., and Ernst, J. V., *J. Parasitol.*, 68, 861, 1982.)

(0.75%) plus trypsin (0.25%) will cause activation of sporozoites. If bile is used, the host animal is of little importance.[14] Sporozoites become motile within the sporocysts and tumble or glide around one another. This movement is not continuous but is interrupted by periods of inactivity. Eventually the sporocyst wall opens along four plate-like junctions[11,12] and the sporozoites exit through the opening which is formed. Sporozoites exit oocysts through indentations or fractures which form in one or both ends of the oocyst wall.[11,14]

C. ENDOGENOUS DEVELOPMENT

The endogenous life cycle of mammalian *Isospora* species (Figures 2 and 3) is somewhat

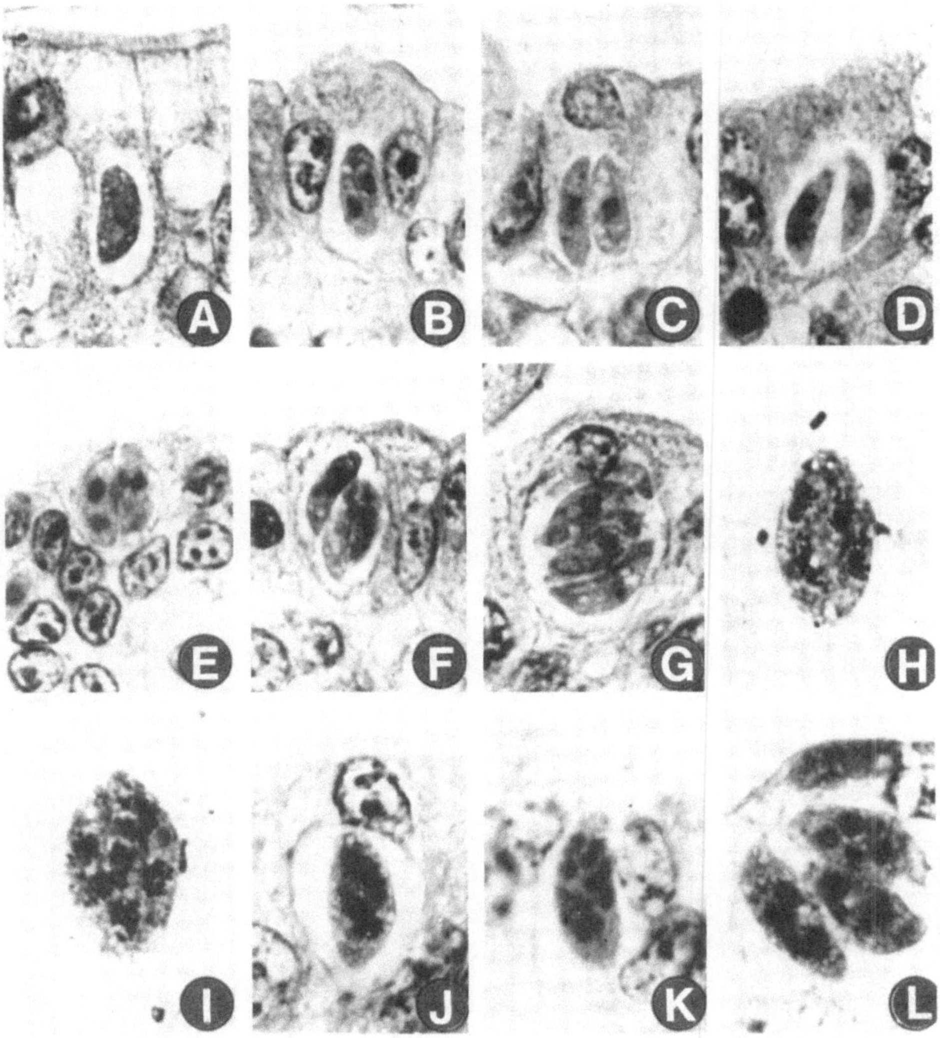

FIGURE 2. Endogenous asexual stages of *Isospora suis* in the small intestine of piglets. (A—G, J—L) Hematoxylin and eosin stained tissue sections, (H and I) Wright's-Giemsa stained mucosal smears. All 1400 ×. (A) Uninucleate zoite, 3 days PI. (B) Binucleate Type 1 meront, 3 days PI. (C) Two Type 1 merozoites, 3 days PI. (D) Binucleate Type 1 meront in same parasitophorous vacuole as a Type 1 merozoite, 3 days PI. (E) Two binucleate Type 1 meronts in the same parasitophorous vacuole, 3 days PI. (F) Group of 3 Type 1 merozoites, 3 days PI. (G) Group of 6 Type 1 merozoites, 3 days PI. (H) Tetranucleate Type 2 meront, 4 days PI. (I) Hexanucleate Type 2 meront, 4 days PI. (J) Single tetranucleate Type 2 meront, 4 days PI. (K) Single 12-nucleate Type 2 meront, 4 days PI. (L) Group of 3 multinucleate Type 2 meronts, 4 days PI. (Modified from Lindsay, D. S., Stuart, B. P., Wheat, B. E., and Ernst, J. V., *J. Parasitol.*, 66, 771, 1980.)

different from what is typical for *Eimeria* species. Sporozoites enter cells in the intestine but usually do not form rounded uninucleate trophozoites. Some sporozoites and/or merozoites leave the intestine and form dormant cyst stages (hypnozoites) in extraintestinal tissues.[17-19] Intestinal sporozoites retain their elongate sporozoite shape, become binucleate, and divide by endodyogeny to form two daughter merozoites. These daughter merozoites divide by endodyogeny an indefinite number of times. For this reason, the number of sequential asexual merogenous cycles cannot be determined and developmental stages are referred to as structural Types instead of generations. Eventually, multinucleate meronts are formed. These meronts are elongate and retain their merozoite shape. Several meronts may

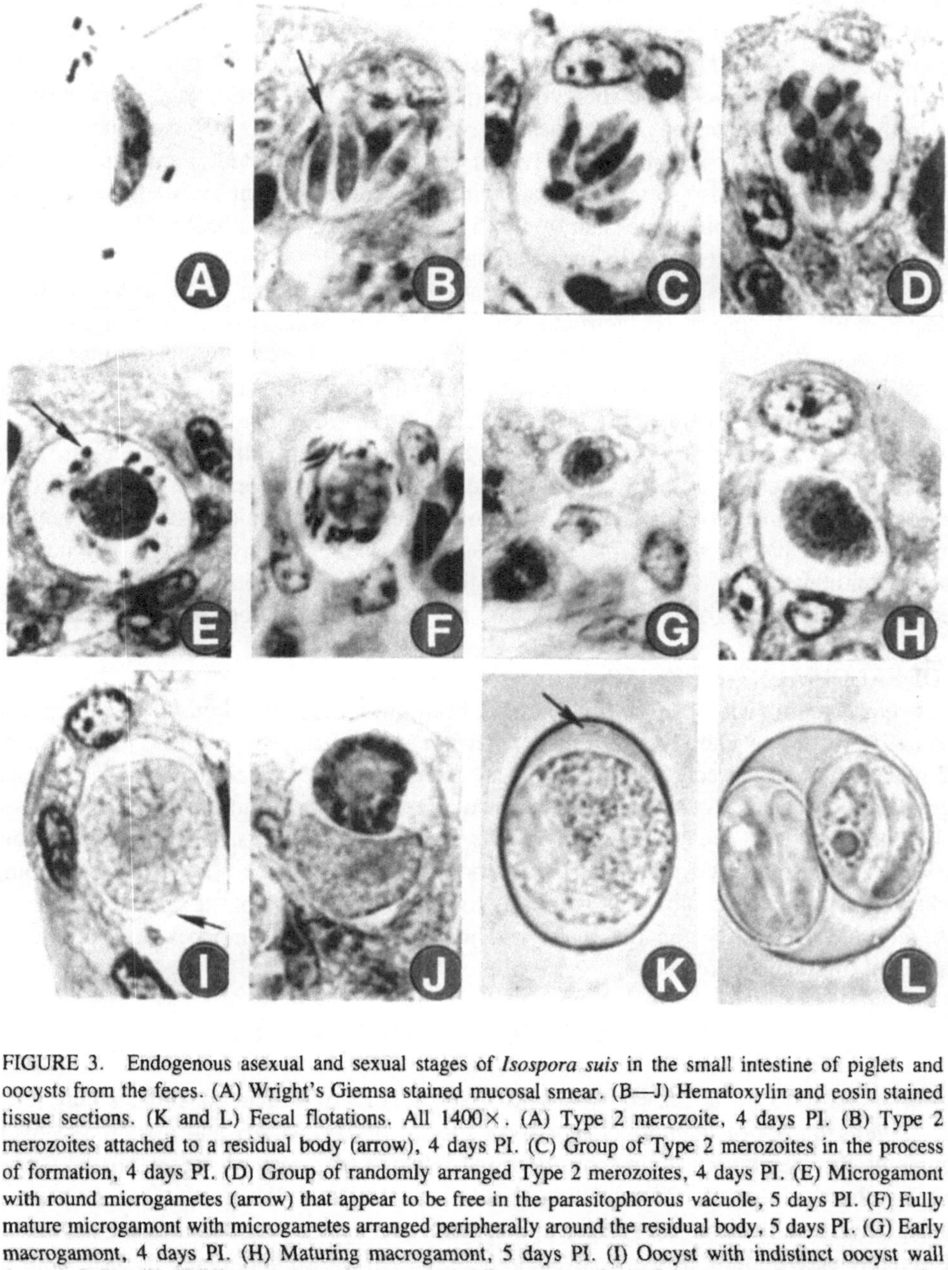

FIGURE 3. Endogenous asexual and sexual stages of *Isospora suis* in the small intestine of piglets and oocysts from the feces. (A) Wright's Giemsa stained mucosal smear. (B—J) Hematoxylin and eosin stained tissue sections. (K and L) Fecal flotations. All 1400×. (A) Type 2 merozoite, 4 days PI. (B) Type 2 merozoites attached to a residual body (arrow), 4 days PI. (C) Group of Type 2 merozoites in the process of formation, 4 days PI. (D) Group of randomly arranged Type 2 merozoites, 4 days PI. (E) Microgamont with round microgametes (arrow) that appear to be free in the parasitophorous vacuole, 5 days PI. (F) Fully mature microgamont with microgametes arranged peripherally around the residual body, 5 days PI. (G) Early macrogamont, 4 days PI. (H) Maturing macrogamont, 5 days PI. (I) Oocyst with indistinct oocyst wall (arrow), 5 days PI. (J) Microgamont and macrogamont that appear to be in the same parasitophorous vacuole, 5 days PI. (K) Unsporulated oocyst showing hazy body (arrow) between the sporont and oocyst wall. (L) Sporulated oocyst of *I. suis*. (Modified from Lindsay, D. S., Stuart, B. P., Wheat, B. E., and Ernst, J. V., *J. Parasitol.*, 66, 771, 1980.

occur in the same host cell. Eventually, sexual stages are formed. Microgamonts are multinucleate and produce biflagellated microgametes. Macrogamonts lack the highly eosinophilic wall-forming bodies found in most *Eimeria* species and the oocyst wall is usually inconspicuous. Microgamonts and macrogamonts may occur in the same host cell.

The endogenous life cycles in animals that ingest oocysts and in those that ingest paratenic hosts are similar.[20,21] The prepatent period may be shortened in these infections.[20-23]

D. EXTRAINTESTINAL STAGES IN DEFINITIVE AND PARATENIC HOSTS

Extraintestinal stages occur in the tissues of the definitive host in canine and feline *Isospora* species[17,18] and *I. belli* of humans.[19] Instead of undergoing the normal developmental cycle in the intestinal tract some sporozoites (merozoites?) leave and invade extraintestinal sites in the host. Mesenteric lymph nodes are most often infected but other tissues such as the liver, spleen, and tracheobronchial lymph nodes can be infected. Parasites are often found as single organisms resembling sporozoites, but some division by endodyogeny may occur in these extraintestinal sites and up to 15 parasites have been observed in an infected cell.[18] Infected host cells probably are macrophages.

Mice, rats, hamsters, dogs, cats, and cows have been shown to be paratenic hosts (transport hosts) for several *Isospora* species.[17,18,23-26] Sporozoites excyst from oocysts and invade extraintestinal tissues. Mesenteric lymph nodes are most often infected; other tissues such as the spleen, liver, and skeletal muscles are sometimes parasitized. Parasites are most often found as single organisms; division is rarely observed.[23] For this reason it is more accurate to refer to the host as a paratenic, rather than an intermediate host. Disease does not occur in the paratenic hosts.[21] Parasites remain viable for at least 23 months in their extraintestinal sites.[21] When the definitive host ingests a paratenic host the subsequent prepatent period may be shorter than when infections are initiated by oocysts. The number of oocysts produced by the definitive host and the patent period is similar to oocyst induced infections.[22] The tissues of paratenic hosts are not infectious for other paratenic hosts.[21]

E. DEVELOPMENT *IN VITRO*

Several mammalian *Isospora* species have been grown in cell cultures[27-31] with several divisions occurring by endodyogeny. Only *I. rivolta* and *I. suis* have produced multinucleate (>2 nuclei) meronts in cell cultures and these meronts did not reach maturity.[27,31] Sexual stages and oocysts have not been grown in cell cultures. Continuous cultivation of an *Isospora* species has not been achieved. *Isospora felis, I. rivolta,* and *I. suis* will develop from sporozoite to unsporulated oocyst in the chorioallantoic membrane (CAM) of developing chicken embryos.[32-34] Development is usually limited to the tissues of the CAM but meronts of *I. felis* have been seen in the livers and intestines of chick embryos that have been chemically immunosuppressed.[34]

III. *ISOSPORA* INFECTIONS OF MAN

Human isosporiasis is caused by *I. belli*. Early reports of human coccidiosis may refer to *I. hominis. Isospora hominis* is actually a species of *Sarcocystis* and a synonym of *S. hominis* or *S. suihominis* acquired by ingesting infected beef or pork, respectively. There is no structural means of differentiating these two species of *Sarcocystis* in human fecal samples or in intestinal biopsies. Human intestinal sarcocystosis can be a serious disease[35] and in many early reports it is impossible to determine if the authors really mean *I. belli* or a *Sarcocystis* species. *Isospora natalensis* has been reported in man.[36] Little is known about this parasite. Its oocysts resemble those of the *I. ohioensis* complex seen in dogs, *I. rivolta* of cats and *I. suis* of pigs. It may be a pseudoparasite originating from one of these animals.

Isospora belli infections are essentially cosmopolitan in distribution but are more common in tropical and subtropical regions.[37] Pigs, dogs, mice, rats, rabbits, guinea pigs, and rhesus monkeys are not suitable definitive hosts;[38] however, in one study patent infections were reported in two of three gibbons.[8] It is not known if these or other animals may serve as paratenic hosts for *I. belli*. The role of paratenic hosts in the transmission of *I. belli* needs to be investigated to establish if a possible mode of transmission other than by contaminated food or water exists. If paratenic hosts do exist it may help explain infections in areas where sanitation is adequate.

TABLE 1
Comparison of the Isosporan Oocysts from Nonhuman Primates[9,48,49,53]

	Oocysts[a]	Sporocysts	Original host
I. arctopitheci	21—30 by 21—25	13—21 by 10—16	*Callithrix jacchus*
I. callimico	13—21 by 12—17	10—13 by 7—9	*Callimico goeldii*
I. endocallimici	25—31 by 21—27	15—20 by 10—15	*Callimico goeldii*
I. scorzai[b]	23 by 20	14 by 9	*Cacajao rubicundus*

[a] All measurements are in micrometers.
[b] No range given.

Oocysts are passed in feces unsporulated, partially sporulated (sporoblast stage), and occasionally sporulated. They can sporulate in less than 24 h. Oocysts are elongate ellipsoidal with slightly tapered ends and measure 23—36 by 12—17 μm. Endogenous stages are located in enterocytes lining the villi of the small intestine and rarely the large intestine.[19,39,40] Stages are seldom found in enterocytes lining the crypts or in the lamina propria. All life cycle stages typical of *Isospora* species have been observed with light and transmission electron microscopy.[19,39,40] The number of asexual types present has not been determined.

Isospora belli causes serious and sometimes fatal disease in man. Symptoms of *I. belli* infection include diarrhea, steatorrhea, headache, fever, malaise, abdominal pain, vomiting, dehydration, and weight loss.[39-42] Blood is not usually discharged in the feces. Eosinophilia is seen in the blood of some patients. Disease is more severe in infants[42] and patients with AIDS[19,41,43-45] or other immune disorders.[46,47] Diarrhea in these patients is often very fluid, secretory-like, and leads to dehydration requiring hospitalization. The disease is often chronic with parasites present in the feces or biopsies for several months to years. Recurrences are common. The chronic nature of the disease may be related to an asexual stage that recycles or to reactivation of dormant extraintestinal stages which migrate back to the intestine and reinitiate the infection.

Main microscopic changes are villous atrophy and crypt hyperplasia.[19,39,40] Eosinophils may be present in the lamina propria in large numbers approaching those seen in eosinophilic enteritis. Plasma cells, lymphocytes, and polymorphonuclear leukocytes are present in increased numbers. Lymphatics may be dilated.

Treatment with trimethoprim-sulfamethoxazole[41,45] or pyrimethamine-sulfadiazine[39] is effective. Symptoms generally stop within 2 days; however, up to 47% of patients may have a recurrence within 2 to 20 weeks after treatment is terminated.[41] Recurrent infections are similar in nature to and respond to treatment as primary infections.[41,45]

IV. *ISOSPORA* INFECTIONS OF NONHUMAN PRIMATES

Little is known about the coccidial infections of nonhuman primates. Most of the *Isospora* species recorded are known only by oocyst structure (Table 1).

Isospora callimico[48] and *I. endocallimici*[9] were described from the feces of one and five Goeldi's marmosets (*Callimico goeldi*), respectively. No transmission or life cycle studies have been conducted with these species. *Isospora scorzai* was described from the feces of a Uakari monkey (*Cacajao rubicundus*) and was transmitted to a *Cebus nigrivittatus*.[49] The life cycle of *I. scorzai* is not known.

Isospora cebi was described from the feces of a *Cebus albifrons*.[50] The sporocysts of this species have Stieda bodies indicating a pseudoparasite of avian origin. *Isospora paponis* was described from Chacma baboons (*Papio ursinus*).[51] Oocysts sporulated endogenously

TABLE 2
Comparison of the *Isospora* Species Infecting Dogs[17,20,61,63-65]

	I. ohioensis	*I. neorivolta*	*I. burrowsi*	*I. canis*
Location[a]	SI, cecum, colon	Posterior SI	Posterior SI	Posterior SI
Site	Enterocytes	Laminia propria	Lamina propria, enterocytes	Lamina propria
Prepatent period (days)	5	6	6	9
Patent period (days)	N[c]	13—23	9—15	N
Oocysts[b]	19—27 by 18—23	N	17—22 by 16—19	34—40 by 28—32
Sporocysts	15—19 by 10—13	N	12—16 by 8—11	18—21 by 15—18
No. asexual types	2	4	2	3

[a] SI = small intestine.
[b] All measurements are in micrometers.
[c] N = No data available.

in the small intestines indicating that this is a *Sarcocystis* species. Additionally, sporulated oocysts of this species have been seen in the skeletal muscles of Chacma baboons.[52]

Isospora arctopitheci has been studied more than the other coccidia of nonhuman primates.[26,53-55] Hendricks conducted cross transmission studies with this parasite and claimed to have successfully transmitted it to members of 6 genera of New World nonhuman primates, four families of carnivores and one marsupial species.[54] This is an unusually large and diverse definitive host range and further experimental studies are needed to confirm or deny these initial findings.

The endogenous life cycle of *I. arctopitheci* occurs in the small intestine.[55] Developmental stages are located in enterocytes on the distal $^{2}/_{3}$ of the villi and parasite densities are greatest in the jejunum. Asexual multiplication was exclusively by endodyogeny. Eosinophilic bodies were present in gamonts. The prepatent period was about 7 days. Extraintestinal stages were not seen in the definitive host.

Experimental studies indicate that *I. arctopitheci* can be pathogenic.[55] Four of 13 titi marmosets (*Saguinus geoffroyi*) died after being inoculated with 1 to 2 × 10^5 oocysts. No clinical signs were seen in marmosets that died 3 and 5 days PI. Bloody diarrhea was seen in two marmosets that died 7 days PI. All nine other marmosets remained normal. Microscopic lesions were necrosis of apical enterocytes with subsequent exposure of the lamina propia.

V. *ISOSPORA* INFECTIONS OF DOGS

Several species of *Isospora* infect dogs (Table 2). Cats are not definitive hosts for *Isospora* species found in dogs.[56] Young dogs are more likely to be infected and surveys indicate that from 3 to 38% of dogs are positive for coccidial oocysts.[57] Stray dogs are more likely to be infected than are dogs with owners. This reflects greater exposure of stray dogs because of hunting for food and feeding on infected paratenic hosts.

It is unclear if coccidiosis is a serious problem in dogs. Diarrhea associated with the presence of coccidial oocysts in young dogs occurs but the clinical significance is not established because of the possibility of concurrent bacterial or viral infections. Published reports of naturally occurring canine coccidiosis are few[58-60] and further studies on natural cases are needed before firm conclusions can be made. Experimental infections have not usually been associated with disease.

Isospora canis has the largest oocysts of the canine *Isospora* species and is the only species that can be diagnosed by microscopical examination of oocysts. Disease was not produced in 25 6-week-old or 6 8-week-old pups inoculated with 1 to 1.5 × 10^5 *I. canis*

TABLE 3
Comparison of *Isospora rivolta* and *Isospora felis* of Cats[18,21,66-68]

	Isospora rivolta	*Isospora felis*
Location	Small intestine	Small intestine, cecum
Site	Enterocytes	Enterocytes
Prepatent period	4 to 7 days	7 to 11 days
Patent period	>14 days	10 to 11 days
Oocysts	18—28 by 16—23	38—51 by 27—39
Sporocysts	14—16 by 10—13	20—26 by 17—22
No. asexual types	3	3

oocysts.[61] Solid immunity follows primary infection and no oocysts are discharged after challenge. It has been suggested that the stress of weaning and shipping may enhance *I. canis* infections.[61] This needs further investigation since it may be related to a decrease in immunity and reactivation of extraintestinal stages.

Three *Isospora* species having smaller oocysts than *I. canis* can be found in dogs: *I. ohioensis, I. burrowsi*, and *I. neorivolta* (Table 2). Because they cannot be separated based on oocyst structure and because *I. ohioensis* was the first named, these oocysts are often referred to as *I. ohioensis*-like[59] or being in the *I. ohioensis* complex.

Isospora ohioensis can cause disease in experimentally infected 7-day-old pups but not weaned pups or young dogs.[62] Diarrhea was the major clinical sign seen in the 7-day-old pups. Microscopic changes were villous atrophy, necrosis of apical enterocytes, and cryptitis. Dogs developed an immunity which lasted about 2 months.

Little is known about the pathogenicity of *I. burrowsi* or *I. neorivolta* infection in dogs. Neither causes disease in experimentally infected dogs.[63-65]

Because the significance of diarrhea caused by coccidia in dogs is unclear, so is the treatment of the condition. Suspected clinical cases can be treated with a variety of drugs used alone or in combination. These include amprolium, sulfonamides, or trimethoprim plus sulfonamides.[57]

VI. *ISOSPORA* INFECTIONS OF CATS

Isospora rivolta and *I. felis* infect cats. Dogs do not serve as definitive hosts for these species.[66] Both feline *Isospora* species have extraintestinal stages in the cat definitive host and a variety of paratenic hosts. Life cycle data are presented in Table 3. From 3 to 36% of cats examined pass oocysts.[57] Stray cats are more likely to pass oocysts. Coccidiosis in cats is not thought to be a common problem and is usually seen only in naturally infected kittens in which other disease causing agents may be present. The drugs used to treat dogs are used to treat kittens.

Experimentally, *I. rivolta* can cause disease in newborn kittens.[21] Diarrhea occurs 3 to 4 days PI of 1×10^5 or 1×10^6 sporocysts. Microscopic changes consist of congestion, erosion of enterocytes, villous atrophy, and cryptitis. No disease was seen in 10- to 13-week-old kittens inoculated with up to 1×10^5 oocysts. Cats develop immunity to infection but it is not complete because some oocysts are shed after challenge.[21]

Experimental studies indicate that *I. felis* is not pathogenic for cats over 1 month of age.[67,68] Few signs of disease are seen in 6- to 13-week-old cats given 1 to 1.5×10^5 oocysts. Mild microscopic changes consisting of congestion, erosion of superficial enterocytes, and neutrophil infiltration may be seen. Four-week-old kittens are most susceptible and enteritis, emaciation, and death can occur after inoculation of 1×10^5 oocysts.[69]

VII. *ISOSPORA* INFECTIONS OF PIGS

Coccidiosis in pigs is a disease of nursing piglets.[70,71] *Isospora suis* is the cause of neonatal porcine coccidiosis. The disease is ubiquitous where pigs are farrowed in confinement and is responsible for 15 to 20% of the cases of piglet diarrhea seen at diagnostic laboratories in the U.S. and Canada.

Piglets develop diarrhea at 7 to 14 days of age. The diarrhea is initially pasty but becomes fluid after 2 to 3 days. The diarrhea is yellowish to gray in color; blood is never present. Piglets become covered with diarrhetic feces causing them to stay damp and smell like soured milk; they become lethargic but continue to nurse. Infections fail to respond to commonly used antibiotics and piglets within a litter, and litters are not equally affected. Morbidity is high and mortality is moderate. Microscopic changes consist of villous atrophy, villous fusion, necrotic enteritis, and crypt hyperplasia.[71-73] Experimental studies indicate that the development of clinical disease and microscopic lesions are dependent upon the number of oocysts inoculated and the age at which piglets are inoculated.[72,74] Doses of 5×10^4 oocysts or less generally produce diarrhea but no mortality in young (1- to 3-day-old) piglets; 7×10^4 to 3×10^5 oocysts cause low to moderate mortality, and doses of 4×10^5 or greater cause high mortality in young piglets. Weight gains of infected piglets are depressed.[75] Piglets that recover from infection show a strong degree of resistance to reinfection; no clinical signs develop after challenge and few or no oocysts are passed in the feces.[76]

Endogenous stages are found throughout the small intestine and occasionally the cecum and colon.[72,77,78] Parasite densities are highest in the jejunum and ileum. Developmental stages are found in enterocytes. Two types of meronts are produced.[78] Type 1 meronts are binucleate and divide by endodyogeny whereas Type 2 meronts are multinucleate. The prepatent period is 4 to 5 days and the patent period is 2 weeks or longer. Extraintestinal stages have not been found in infected piglets or in experimentally inoculated mice.[79]

The epidemiology of neonatal porcine isosporiasis is puzzling. The sow is a logical source of infection for newborn piglets, but experimental studies have failed to demonstrate *Isospora suis* oocysts in a significant number of sows.[71,80,81] *Isospora suis* oocysts were not found in the feces of 77 sows examined from seven farms with a problem of neonatal coccidiosis and only 1 of 172 sows examined from 27 farms without a history of neonatal coccidiosis was positive.[81] *Eimeria* spp. oocysts were found in 91% of these sows. Another study examined sows from two farms with neonatal coccidiosis in piglets on a daily basis for about 1 week prior to farrowing, at farrowing, and for about 2 weeks postfarrowing.[71] *Isospora suis* oocysts were not found in these sows; however, piglets nursing these sows developed coccidiosis and passed *I. suis* oocysts at 4 to 8 days of age. Microscopic examination of milk samples and placentas from these sows were negative for parasites. Once *I. suis* is established on a farm, it is probably maintained by piglets being infected from the contaminated farrowing crate.

Anticoccidial treatment of piglets has generally proven unrewarding.[70] Nursing piglets do not eat or drink enough to make in-feed or in-water treatment useful. Catching each piglet for dosing is time and labor intensive and probably only practical on small farms. Controlled studies indicate that amprolium, monensin, and furazolidone are not effective. Improved sanitation is the best means of controlling the disease.[80]

REFERENCES

1. **Dubey, J. P.,** *Toxoplasma, Hammondia, Besnoitia, Sarcocystis,* and other tissue cyst-forming coccidia of man and animals, in *Parasitic Protozoa,* Vol. 3, Krier, J. P., Ed., Academic Press, New York, 1977, 101.
2. **Frenkel, J. K.,** *Besnoitia wallacei* of cats and rodents: with a reclassification of other cyst-forming isosporid coccidia, *J. Parasitol.,* 63, 611, 1977.
3. **Levine, N. D.,** *The Protozoan Phylum Apicomplexa,* Vol. I, CRC Press, Boca Raton, FL, 1988, chap. 10.
4. **Lepp, D. L. and Todd, K. S., Jr.,** Sporogony of the oocysts of *Isospora canis, Trans. Am. Microsc. Soc.,* 95, 89, 1976.
5. **Mahrt, J. L.,** Sporogony of *Isospora rivolta* oocysts from the dog, *J. Protozool.,* 15, 308, 1968.
6. **Shah, H. L.,** Sporogony of the oocysts of *Isospora felis* Wenyon, 1923 from the cat, *J. Protozool.,* 17, 609, 1970.
7. **Lindsay, D. S., Current, W. L., and Ernst, J. V.,** Sporogony of *Isospora suis* Biester, 1934 of swine, *J. Parasitol.,* 68, 861, 1982.
8. **Zaman, V.,** Observations on human *Isospora, Trans. R. Soc. Trop. Med. Hyg.,* 62, 556, 1968.
9. **Duszynski, D. W. and File, S. K.,** Structure of the oocyst and excystation of *Isospora endocallimici* n. sp., from the marmoset *Callimico goeldii, Trans. Am. Microsc. Soc.,* 93, 403, 1974.
10. **Duszynski, D. W. and Speer, C. A.,** Excystation of *Isospora arctopitheci* Rodhain, 1933 with notes on a similar process in *Isospora bigemina* (Stiles 1891) Luhe 1906, *Z. Parasitenkd.,* 48, 191, 1976.
11. **Speer, C. A., Hammond, D. M., Mahrt, J. L., and Roberts, W. L.,** Structure of the oocysts and oocyst walls and excystation of sporozoites of *Isospora canis, J. Parasitol.,* 59, 35, 1973.
12. **Speer, C. A., Marchiondo, A. A., Duszynski, D. W., and File, S. K.,** Ultrastructure of the sporocyst wall during excystation of *Isospora endocallimici, J. Parasitol.,* 62, 984, 1976.
13. **Lindsay, D. S., Current, W. L., and Ernst, J. V.,** Excystation of *Isospora suis* Biester, 1934 of swine, *Z. Parasitenkd.,* 69, 27, 1983.
14. **McKenna, P. B. and Charleston, W. A. G.,** Activation and excystation of *Isospora felis* and *Isospora rivolta* sporozoites, *J. Parasitol.,* 68, 276, 1982.
15. **Box, E. D., Marchiondo, A. A., Duszynski, D. W., and Davis, C. P.,** Ultrastructure of *Sarcocystis* sporocysts from passerine birds and opossums: comments on classification of the genus *Isospora, J. Parasitol.,* 66, 68, 1980.
16. **Christie, E., Pappas, P. W., and Dubey, J. P.,** Ultrastructure of excystment of *Toxoplasma gondii* oocysts, *J. Protozool.,* 25, 438, 1978.
17. **Dubey, J. P.,** Experimental *Isospora canis* and *Isospora felis* infection in mice, cats, and dogs, *J. Protozool.,* 22, 416, 1975.
18. **Dubey, J. P. and Frenkel, J. K.,** Extra-intestinal stages of *Isospora felis* and *I. rivolta* (Protozoa: Eimeriidae) in cats, *J. Protozool.,* 19, 89, 1972.
19. **Restrepo, C., Macher, A. M., and Radany, E. H.,** Disseminated extraintestinal isosporiasis in a patient with acquired immune deficiency syndrome, *Am. J. Clin. Pathol.,* 87, 536, 1987.
20. **Dubey, J. P.,** Life cycle of *Isospora ohioensis* in dogs, *Parasitology,* 77, 1, 1978.
21. **Dubey, J. P.,** Life cycle of *Isospora rivolta* (Grassi 1879) in cats and mice, *J. Protozool.,* 26, 433, 1979.
22. **Dubey, J. P. and Streitel, R. H.,** *Isospora felis* and *I. rivolta* infections in cats induced by mouse tissue or oocysts, *Br. Vet. J.,* 132, 649, 1976.
23. **Frenkel, J. K. and Dubey, J. P.,** Rodents as vectors for the feline coccidia, *Isospora felis* and *Isospora rivolta, J. Infect. Dis.,* 125, 69, 1972.
24. **Dubey, J. P. and Mehlhorn, H.,** Extraintestinal stages of *Isospora ohioensis* from dogs in mice, *J. Parasitol.,* 64, 689, 1978.
25. **Fayer, R. and Frenkel, J. K.,** Comparative infectivity for calves of oocysts of feline coccidia: *Besnoitia, Hammondia, Cystoisospora, Sarcocystis,* and *Toxoplasma, J. Parasitol.,* 65, 756, 1979.
26. **Hendricks, L. D. and Walton, B. C.,** Vertebrate intermediate hosts in the life cycle of an isosporan from a non-human primate, *Proc. Int. Congr. Parasitol.,* 1, 96, 1974.
27. **Fayer, R.,** Cultivation of feline *Isospora rivolta* in mammalian cells, *J. Parasitol.,* 58, 1207, 1972.
28. **Fayer, R. and Mahrt, J. L.,** Development of *Isospora canis* (Protozoa: Sporozoa) in cell cultures, *Z. Parasitenkd.,* 38, 313, 1972.
29. **Fayer, R. and Thompson, D. E.,** *Isospora felis*: development in cultured cells with some cytological observations, *J. Parasitol.,* 60, 160, 1974.
30. **Fayer, R., Gamble, H. R., and Ernst, J. V.,** *Isospora suis*: development in cultured cells with some cytological observations, *Proc. Helminthol. Soc. Wash.,* 51, 154, 1984.
31. **Lindsay, D. S. and Blagburn, B. L.,** Development of *Isospora suis* from pigs in primary porcine and bovine cell cultures, *Vet. Parasitol.,* 24, 301, 1987.
32. **Arcay, L.,** Nuevo coccidia de gato: *Cystoisospora frenkeli* sp nova (Toxoplasmatinae) y su desarrollo en la membrana corioalantoidea de embrion de pollo, *Acta Cient. Venez.,* 32, 401, 1981.

33. **Lindsay, D. S. and Current, W. L.,** Complete development of *Isospora suis* of swine in chicken embryos, *J. Protozool.*, 31, 152, 1984.
34. **Gutierrez, F. and Arcay, L.,** Cultivo de *Cystoisospora felis* Frenkel, 1977 (*Isospora felis* Wasielewski, 1904, Wenyon, 1923) en la membrana corioalantoidea de embrion de pollo, *Acta Cient. Venez.*, 38, 474, 1987.
35. **Bunyaratvej, S., Bunyawongwiroj, P., and Nitiyanant, P.,** Human intestinal sarcosporidiosis: report of six cases, *Am. J. Trop. Med. Hyg.*, 31, 36, 1982.
36. **Elsdon-Dew, R.,** *Isospora natalensis* (sp. nov.) in man, *J. Trop. Med. Hyg.*, 56, 149, 1953.
37. **Faust, E. C., Giraldo, L. E., Giraldo, G., and Bonfante, R.,** Human coccidiosis in the western hemisphere, *Am. J. Trop. Med. Hyg.*, 10, 343, 1961.
38. **Jeffery, G. M.,** Human coccidiosis in South Carolina, *J. Parasitol.*, 42, 491, 1956.
39. **Trier, J. S., Moxey, P. C., Schimmel, E. M., and Robles, E.,** Chronic intestinal coccidiosis in man: intestinal morphology and response to treatment, *Gastroenterology*, 66, 923, 1974.
40. **Brandborg, L. L., Goldberg, S. B., and Breidenbach, W. C.,** Human coccidiosis — a possible cause of malabsorption: the life cycle in small-bowel mucosal biopsies as a diagnostic feature, *N. Engl. J. Med.*, 283, 1306, 1970.
41. **DeHovitz, J. A., Pape, J., Boncy, M., and Johnson, W. D., Jr.,** Clinical manifestations and therapy of *Isospora belli* infection in patients with the acquired immunodeficiency syndrome, *N. Engl. J. Med.*, 315, 87, 1986.
42. **Liebman, W. M., Thaler, M. M., DeLorimier, A., Brandborg, L. L., and Goodman, J.,** Intractable diarrhea of infancy due to intestinal coccidiosis, *Gastroenterology*, 78, 579, 1980.
43. **Forthal, D. and Guest, S. S.,** *Isospora belli* enteritis in three homosexual men, *Am. J. Trop. Med. Hyg.*, 33, 1060, 1984.
44. **Gaska, J. A., Tietze, K. J., and Cosgrove, E. M.,** Unsuccessful treatment of enteritis due to *Isospora belli* with spiramycin: a case report, *J. Infect. Dis.*, 152, 1336, 1985.
45. **Whiteside, M. E., Barkin, J. S., May, R. G., Weiss, S. D., Fischl, M. A., and MacLeod, C.,** Enteric coccidiosis among patients with the acquired immunodeficiency syndrome, *Am. J. Trop. Med. Hyg.*, 33, 1065, 1984.
46. **Henry, K., Bird, R. G., and Doe, W. F.,** Intestinal coccidiosis in a patient with alpha-chain disease, *Br. Med. J.*, 1, 542, 1974.
47. **Greenberg, S. J., Davey, M. P., Zierdt, W. S., and Waldmann, T. A.,** *Isospora belli* infection in patients with human T-cell leukemia virus type I-associated adult T-cell leukemia, *Am. J. Med.*, 85, 435, 1988.
48. **Hsu, C. and Melby, E. C., Jr.,** *Isospora callimico* n. sp., (Coccidia: Eimeriidae) from Goeldi's marmoset *(Callimico goeldii), Lab. Anim. Sci.*, 24, 476, 1974.
49. **Arcay, L.,** Coccidiosis en monos y su comparacion con la isosporosis humana, con descripcion de una nueva especie de *Isospora* en *Cacajao rubicundus* (Uakari monkey o mono chucuto), *Acta Biol. Venez.*, 5, 203, 1967.
50. **Marinkelle, C. J.,** *Isospora cebi* sp. n. aislada de un mico de Colombia *(Cebus albifronis), Rev. Brasil Biol.*, 29, 35, 1969.
51. **McConnell, E. E., De Vos, A. J., Basson, P. A., and De Vos, V.,** *Isospora papionis* n. sp. (Eimeriidae) of the Chacma baboon *Papio ursinus, J. Protozool.*, 18, 28, 1971.
52. **McConnell, E. E., Basson, P. A., Thomas, S. E., and De Vos, V.,** Oocysts of *Isospora papionis* in the skeletal muscles of Chacma baboons, *Onderst. J. Vet. Res.*, 39, 113, 1972.
53. **Hendricks, L. D.,** A redescription of *Isospora arctopitheci* Rhodhain, 1933 (Protozoa: Eimeriidae) from primates of Panama, *Proc. Helminthol. Soc. Wash.*, 41, 229, 1974.
54. **Hendricks, L. D.,** Host range characteristics of the primate coccidian, *Isospora arctopitheci* Rodhain 1933 (Protozoa: Eimeriidae), *J. Parasitol.*, 63, 32, 1977.
55. **Olcott, A. T., Speer, C. A., and Hendricks, L. D.,** Endogenous development of *Isospora arctopitheci* Rodhain, 1933 in the marmoset *Saguinus geoffroyi, Proc. Helminthol. Soc. Wash.*, 49, 118, 1982.
56. **Dubey, J. P.,** *Isospora ohioensis* sp. n. proposed for *I. rivolta* of the dog, *J. Parasitol.*, 61, 462, 1975.
57. **Kirkpatrick, C. E. and Dubey, J. P.,** Enteric coccidial infections *Isospora, Sarcocystis, Cryptosporidium, Besnoitia,* and *Hammondia, Vet. Clin. North Am. Small Anim. Pract.*, 17, 1405, 1987.
58. **Correa, W. M., Correa, C. N. M., Langoni, H., Volpato, O. A., and Tsunoda, K.,** Canine isoporosis, *Canine Pract.*, 10, 44, 1983.
59. **Dubey, J. P., Weisbrode, S. E., and Rogers, W. A.,** Canine coccidiosis attributed to an *Isospora ohioensis*-like organism: a case report, *J. Am. Vet. Med. Assoc.*, 173, 185, 1978.
60. **Olson, M. E.,** Coccidiosis caused by *Isospora ohioensis*-like organisms in three dogs, *Can. Vet. J.*, 26, 112, 1985.
61. **Lepp, D. L. and Todd, K. S., Jr.,** Life cycle of *Isospora canis* Nemeseri, 1959 in the dog, *J. Protozool.*, 21, 199, 1974.
62. **Dubey, J. P.,** Pathogenicity of *Isospora ohioensis* infection in dogs, *J. Am. Vet. Med. Assoc.*, 173, 192, 1978.

63. **Mahrt, J. L.,** Endogenous stages of the life cycle of *Isospora rivolta* in the dog, *J. Protozool.*, 14, 754, 1967.
64. **Dubey, J. P. and Mahrt, J. L.,** *Isospora neorivolta* sp. n. from the domestic dog, *J. Parasitol.*, 64, 1067, 1978.
65. **Trayser, C. V. and Todd, K. S., Jr.,** Life cycle of *Isospora burrowsi* n. sp. (Protozoa: Eimeriidae) from the dog *Canis familiaris*, *Am. J. Vet. Res.*, 39, 95, 1978.
66. **Shah, H. L.,** *Isospora* species of the cat and attempted transmission of *I. felis* Wenyon, 1923 from the cat to the dog, *J. Protozool.*, 17, 603, 1970.
67. **Hitchcock, D. J.,** The life cycle of *Isosopora felis* in the kitten, *J. Parasitol.*, 41, 383, 1955.
68. **Shah, H. L.,** The life cycle of *Isospora felis* Wenyon, 1923, a coccidium of the cat, *J. Protozool.*, 18, 3, 1971.
69. **Andrews, J. M.,** Coccidiosis in mammals, *Am. J. Hyg.*, 6, 784, 1926.
70. **Lindsay, D. S.,** Diagnosing and controlling *Isospora suis* in nursing pigs, *Vet. Med.*, 83, 443, 1989.
71. **Stuart, B. P. and Lindsay, D. S.,** Coccidiosis in swine, *Vet. Clin. North Am. Food Anim. Pract.*, 2, 455, 1988.
72. **Stuart, B. P., Lindsay, D. S., Ernst, J. V., and Gosser, H. S.,** *Isospora suis* enteritis in piglets, *Vet. Pathol.*, 17, 84, 1980.
73. **Eustis, S. L. and Nelson, D. T.,** Lesions associated with coccidiosis in nursing piglets, *Vet. Pathol.*, 18, 21, 1981.
74. **Stuart, B. P., Gosser, H. S., Allen, C. B., and Bedell, D. M.,** Coccidiosis in swine: dose and age response to *Isospora suis*, *Can. J. Comp. Med.*, 46, 317, 1982.
75. **Lindsay, D. S., Current, W. L., and Taylor, J. R.,** Effects of experimental *Isospora suis* infection on morbidity, mortality, and weight gains of nursing pigs, *Am. J. Vet. Res.*, 46, 1511, 1985.
76. **Stuart, B. P., Sisk, D. B., Bedell, D. M., and Gosser, H. S.,** Demonstration of immunity against *Isospora suis* in swine, *Vet. Parasitol.*, 9, 185, 1982.
77. **Biester, H. E. and Murray, C.,** Studies in infectious enteritis of swine. VII. *Isospora suis* n. sp. in swine, *J. Am. Vet. Med. Assoc.*, 85, 207, 1934.
78. **Lindsay, D. S., Stuart, B. P., Wheat, B. E., and Ernst, J. V.,** Endogenous development of the swine coccidium, *Isospora suis* Biester 1934, *J. Parasitol.*, 66, 771, 1980.
79. **Stuart, B. P., Bedell, D. M., and Lindsay, D. S.,** Coccidiosis in swine: a search for extraintestinal stages of *Isospora suis*, *Vet. Rec.*, 110, 82, 1982.
80. **Ernst, J. V., Lindsay, D. S., and Current, W. L.,** Control of *Isospora suis*-induced coccidiosis on a swine farm, *Am. J. Vet. Res.*, 46, 643, 1985.
81. **Lindsay, D. S., Ernst, J. V., Current, W. L., Stuart, B. P., and Stewart, T. B.,** Prevalence of oocysts of *Isospora suis* and *Eimeria* spp. from sows on farms with and without a history of neonatal coccidiosis, *J. Am. Vet. Med. Assoc.*, 185, 419, 1984.

Chapter 6

SARCOCYSTIS: INFECTION AND DISEASE OF HUMANS, LIVESTOCK, WILDLIFE, AND OTHER HOSTS

R. J. Cawthorn and C. A. Speer

TABLE OF CONTENTS

I. INTRODUCTION AND HISTORICAL BACKGROUND

Sarcocystosis causes significant economic losses to the livestock industry; and this group of parasites can also be an important cause of human illness and could also impact negatively on wildlife populations. Although species of *Sarcocystis* were first recognized in the mid-1800s, critical information on systematics, developmental cycles, pathogenesis, immunity, and epidemiology has become available relatively recently (i.e., beginning in the early 1970s). This review is designed to critically examine and overview information on structure and developmental cycles, systematics and diagnosis, techniques, pathogenesis of infection, immunological responses, epidemiology, control, and directions for future research on species of *Sarcocystis*.

Species of the genus *Sarcocystis* are apicomplexan parasites of the order Eucoccidiorida, closely related to the genera *Besnoitia, Caryospora, Cystoisospora (Isospora), Frenkelia, Hammondia, Neospora,* and *Toxoplasma*. The systematic status of species of *Sarcocystis* is given in Chapter 1. The parasite was originally described as "milky white threads" (=Miescher's tubules) in the skeletal musculature of a house mouse by Miescher;[1] the genus name *Sarcocystis* (sarco = muscle) was proposed by Labbe.[2] However, the correct type species of the genus is *Sarcocystis miescheriana*[2,3] (from pigs) as the parasite *S. muris* (in mice) was unnamed until 1886 by Railliet.[4]

Unfortunately, systematics of species of *Sarcocystis* was confusing prior to demonstration of an obligatorily heteroxenous life cycle (asexual multiplication [merogony, sarcocyst formation] in intermediate hosts; sexual reproduction [gametogony and sporogony] in definitive hosts) utilizing predator-prey relationships.[5,6] Previous to this, numerous species of *Sarcocystis* were named based on the presence of sarcocysts in intermediate hosts; however, these species descriptions were inadequate and type specimens were not available. Additionally, one species of *Sarcocystis* can utilize one or more species of intermediate host or one or more species of definitive host. Also, one species of intermediate host or definitive host can support development of two or more species of *Sarcocystis* simultaneously. Sarcocysts have species-specific sarcocyst walls; however, two or more species of *Sarcocystis* can have ultrastructurally similar sarcocyst walls. Consequently, earlier named species of *Sarcocystis* needed to be redescribed.

One group, led by Heydorn and colleagues,[7] suggested new names by combining names (i.e., generic or specific) of intermediate and definitive hosts, for species of *Sarcocystis* already described from cattle and sheep. Similarly, Frenkel et al.[8] suggested that *S. muris* (rather than *S. miescheriana*) would be the type species of the genus *Sarcocystis*. Another group, led by Levine[9] supported the retention of old names, even if a given name was demonstrated to originally apply to more than one taxonomic unit; the name simply requires redescription.[10] The view of the latter group has been supported by the International Commission of Zoological Nomenclature.[11,12] Further, new species of *Sarcocystis* are readily proposed while adhering to the International Code of Zoological Nomenclature (Third Edition — 1985).

The life cycles of numerous species of *Sarcocystis in vivo* have been recently reviewed.[13] These parasites are pathogens of intermediate hosts, causing both acute and chronic disease. Intermediate hosts can be variously treated and can develop protective immune responses. Species of *Sarcocystis* are an important cause of reproductive failure in livestock.[14] Additionally, *in vitro* studies are becoming increasingly important. Fayer[15] demonstrated gametogony of *S. falcatula in vitro*. Perhaps one of the most important steps to understanding host-parasite relationships of *Sarcocystis* spp. was the successful *in vitro* cultivation of the merogonous phase of *S. cruzi* by Speer and Dubey.[16]

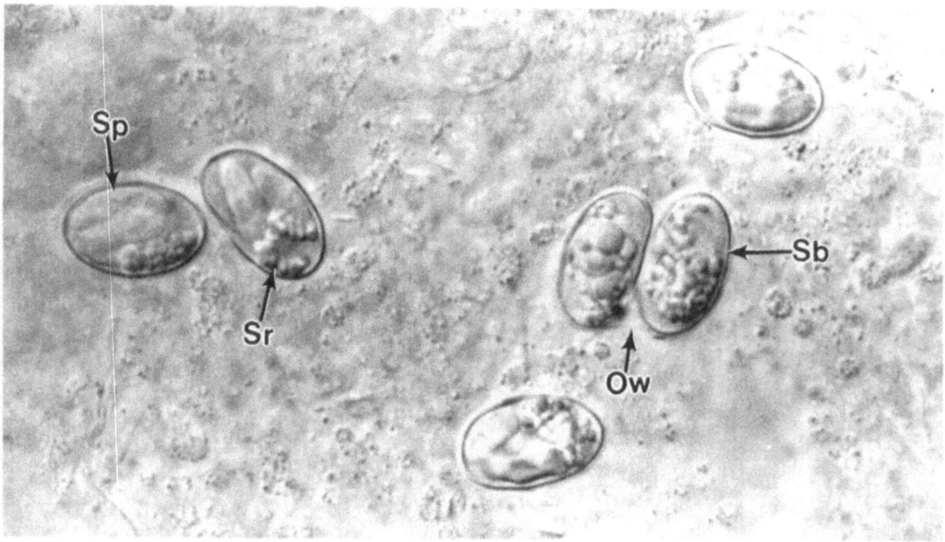

FIGURE 1. Interference-contrast photomicrograph showing oocysts and sporocysts of *Sarcocystis cruzi*. The oocyst contains two sporoblasts held together by a thin oocyst wall (Ow); the sporocysts contain sporozoites (Sp) and a sporocyst residuum (Sr). (Magnification ×1500.)

II. STRUCTURE AND DEVELOPMENTAL CYCLES

Species of *Sarcocystis* undergo the typical apicomplexan life cycle with three developmental stages: gametogony, sporogony and schizogony (= merogony);[17] gametogony and sporogony occur within the definitive host while schizogony occurs in the intermediate host. Life cycles of species of *Sarcocystis* are obligatorily heteroxenous and transmission is facilitated by predator-prey or scavenger-carrion relationships. Schizogony (asexual multiplication) leading to cyst formation (containing metrocytes and bradyzoites [the latter infectious to definitive hosts]) occurs in striated or cardiac muscle of intermediate hosts (prey or carrion). Bradyzoites undergo gametogony (sexual reproduction) in the small intestine of definitive hosts (predator or scavenger). Subsequently, sporulated oocysts (or sporocysts), infectious to intermediate hosts, are passed in feces (Figure 1). Intermediate hosts are infected primarily via ingestion of oocyst-contaminated food or water (see Chapter 1, Figure 7).

A significant variation to the above generalized life cycle of *Sarcocystis* spp. is that single (or different) individuals of a given species of host can act as both intermediate and definitive hosts. Pak[18] demonstrated that Corsican foxes, *Vulpes corsac*, act as intermediate and definitive hosts of *S. corsaci*. Recently Matuschka and colleagues[19,20] suggested that omnivorous island-dwelling lizards, because of cannibalism, scavenging, autonomous shedding of tails and high intraspecific aggression, are well-suited to cycling of *Sarcocystis* spp. within one species of host. Three species of *Sarcocystis* have been described from such lizards; sarcocysts are concentrated in tail musculature.[19,20] Consequently, cannibalism or scavenging may be ecologically important (and common) in transmission of *Sarcocystis* spp. among other carnivorous or omnivorous hosts.

The colorless, sporulated oocysts of *Sarcocystis* spp. are passed in feces of definitive hosts; there is no oocyst residuum or micropyle (Figure 1). Usually only single, elongate sporocysts are found in feces as the relatively thin (<1 μm thick) oocyst wall ruptures during movement (transport?) of the oocyst from within the lamina propria into the lumen of the small intestine. Box and Duszynski[21] suggested fully sporulated oocysts are periodically forced out at villar tips by intestinal contractions with simultaneous rupture of oocyst walls. This movement of oocysts (sporocysts) could also prolong the patent period of *Sarcocystis* infections.

Each *Sarcocystis* sporocyst contains four sporozoites. The sporocyst residuum is granular and often located terminally (Figure 1); the appearance (compact or dispersed) of the residuum varies within the patent period and with length of storage.[22] Sporocyst dimensions (i.e., length, width) are also affected either by the patent period or storage medium.[22] Sporocysts, which are readily transported (dispersed) by arthropods, retain their viability in the environment for extended periods.[23,24] Sporocysts can resist freezing and disinfectants.[25-27]

Sporocysts of *Sarcocystis* spp. have a similar appearance when viewed with the light microscope. The surface structural features of sporocysts of *S. cruzi, S. falcatula, S. hirsuta*, and *S. miescheriana* are similar; each sporocyst wall consists of four plates joined at sutures which appear as ridges on the surface of the sporocyst.[28,29] The arrangement of plates of sporocyst walls is similar to that in *Toxoplasma gondii* and may be similar to those species of *Isospora* which lack Stieda bodies.[30-32] Detailed scanning electron microscopical examination of sporocysts of *Isospora* and *Cystoisospora* spp. is required to confirm the above. However, sporocyst structure is unlikely to be useful for systematic purposes at the generic or species level, as suggested by Box et al.[28] and later refuted by Ferguson et al.[30] The surface topography of sporocysts of *S. cruzi, S. falcatula, S. hirsuta* and *S. miescheriana* appear similar (there are differences in length and width although the range of dimensions overlaps). Sporocyst walls consist of a thin continuous outer layer and a thick inner layer, composed of four plates joined at their lateral margins (Figures 2 and 3).[28]

Excystation of sporozoites of *Sarcocystis* spp. occurs in the small intestine of intermediate hosts. *In vitro*, excystation requires pretreatment with reducing agents, i.e., NaOCl, and treatment with excysting fluid (trypsin and bile or bile salts) for *S. capracanis, S. cruzi, S. hirsuta*, and *S. tenella*.[33] Another reducing agent, cysteine hydrochloride, also enhances excystation of *S. cruzi* and *S. miescheriana*.[29,34] The ultrastructural effect of reducing agents on sporocysts of *Sarcocystis* spp. is unknown. Apparently excysting fluid causes plate separation, beginning with the equatorial sutures and progressing towards the poles of the sporocysts. As the plates separate and curl inward, the sporocysts collapse, releasing the sporozoites[29] (Figures 2 and 3).

In vivo, some sporocysts of *Sarcocystis* spp. pass intact through the gastrointestinal tract of intermediate hosts. A high proportion of these sporocysts retain their viability, complicating protocols for experimental studies and perhaps enhancing dispersal of sporocysts in the environment.[35,36]

Sporozoites of *Sarcocystis* spp. move, via an unknown mechanism(s) to the site of first-generation schizogony in intermediate hosts. The banana-shaped sporozoites are ultrastructurally similar to merozoites and bradyzoites, possessing the organelles typical of a member of the Apicomplexa. However, only sporozoites have a crystalloid body comprising electron-lucent and -dense bodies, usually located in the posterior portion of the sporozoite (Figures 3 and 4). This organelle may be similar to the refractile bodies of Eimeria spp., serving as an energy or amino acid reserve. Sporozoites and bradyzoites, but not merozoites, possess rhoptries.[13]

Schizogony occurs in various locations and cell types in intermediate hosts, depending on the species of *Sarcocystis*. Schizonts develop free in the host cell cytoplasm (Figures 4 and 5). Merozoites are formed by a type of asexual reproduction identified as endopolygeny; many merozoites develop simultaneously, first internally and then at the surface of the schizont in a budding process (Figures 4 and 5). The banana-shaped merozoites are similar ultrastructurally to sporozoites and bradyzoites, although they lack rhoptries, and do not have the crystalloid body of sporozoites (Figure 6). Species of *Sarcocystis* utilizing ruminants as intermediate hosts have two or more generations of schizogony in vascular endothelial cells.[13] First-generation schizonts develop in larger circulatory vessels than second-generation schizonts. For example, first-generation schizogony of *S. capracanis* occurs in gut-associated arteries of goats at 15 to 17 days postinoculation (DPI) with second-generation schizogony

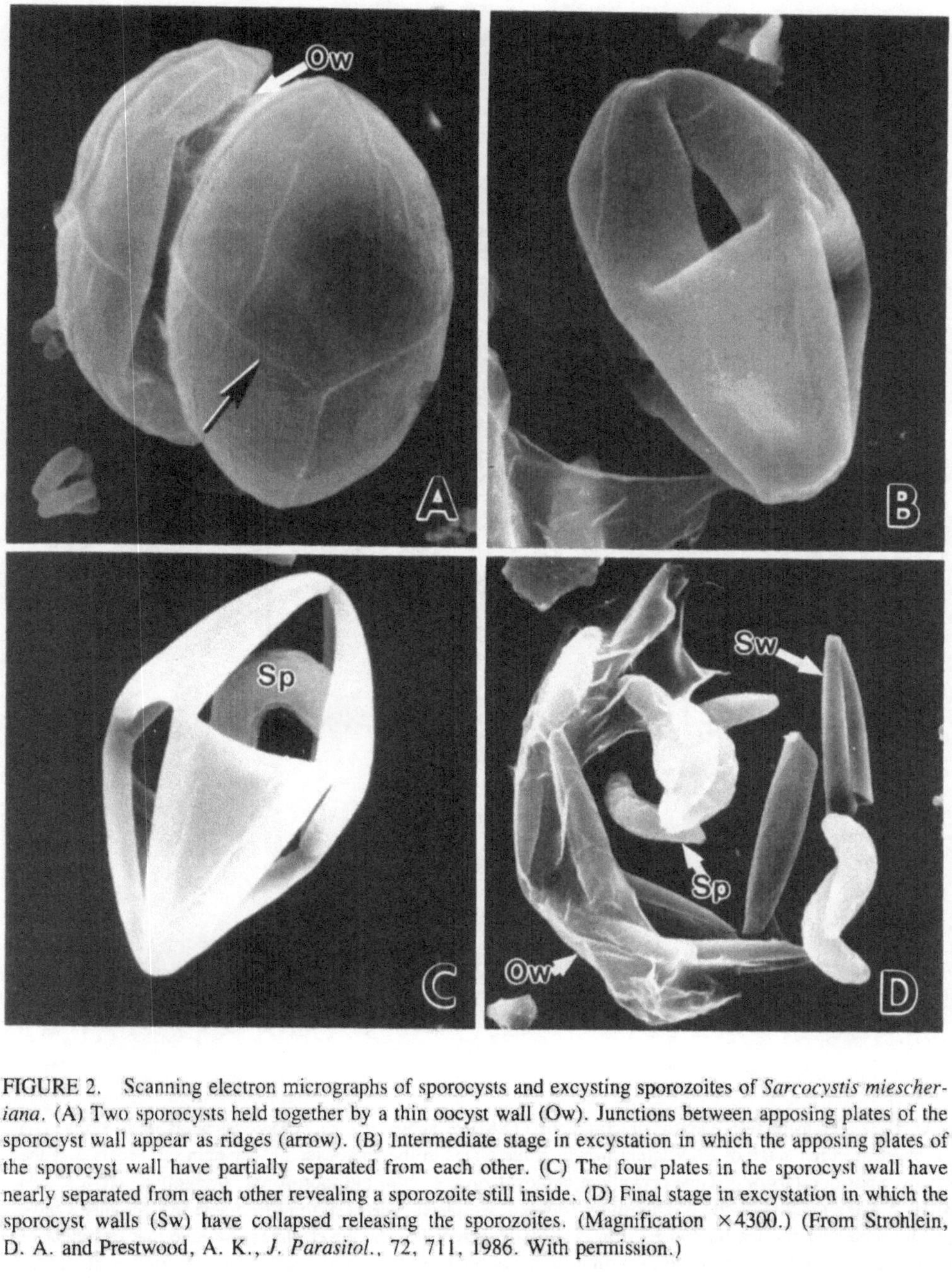

FIGURE 2. Scanning electron micrographs of sporocysts and excysting sporozoites of *Sarcocystis miescheriana*. (A) Two sporocysts held together by a thin oocyst wall (Ow). Junctions between apposing plates of the sporocyst wall appear as ridges (arrow). (B) Intermediate stage in excystation in which the apposing plates of the sporocyst wall have partially separated from each other. (C) The four plates in the sporocyst wall have nearly separated from each other revealing a sporozoite still inside. (D) Final stage in excystation in which the sporocyst walls (Sw) have collapsed releasing the sporozoites. (Magnification ×4300.) (From Strohlein, D. A. and Prestwood, A. K., *J. Parasitol.*, 72, 711, 1986. With permission.)

occurring in capillary endothelial cells of all body organs at 14 to 24 DPI.[37] Consequently, sarcocyst formation is evident beginning 30 DPI. In contrast, many species of *Sarcocystis* utilizing rodents as intermediate hosts have only one generation of schizogony which occurs within hepatocytes. First-generation schizonts of *S. rauschorum* develop in hepatocytes of varying lemmings at 4 to 6 DPI; sarcocysts are formed beginning 9 DPI.[22] Even more unusual, some schizogony of *S. falcatula* occurs in modified endothelial cells of birds (Kuppfer cells of liver, sinusoidal lining cells of spleen).[38] However, because of variability among species of *Sarcocystis*, there may be exceptions to the above broad generalizations.

Often there is a detectable parasitemia associated with *Sarcocystis*-schizont rupture. Merozoites are free in the bloodstream or in phagocytic cells.[39-41] The extracellular merozoites undergo pellicular damage and eventually lyse;[42] containment of merozoites (or sporozoites

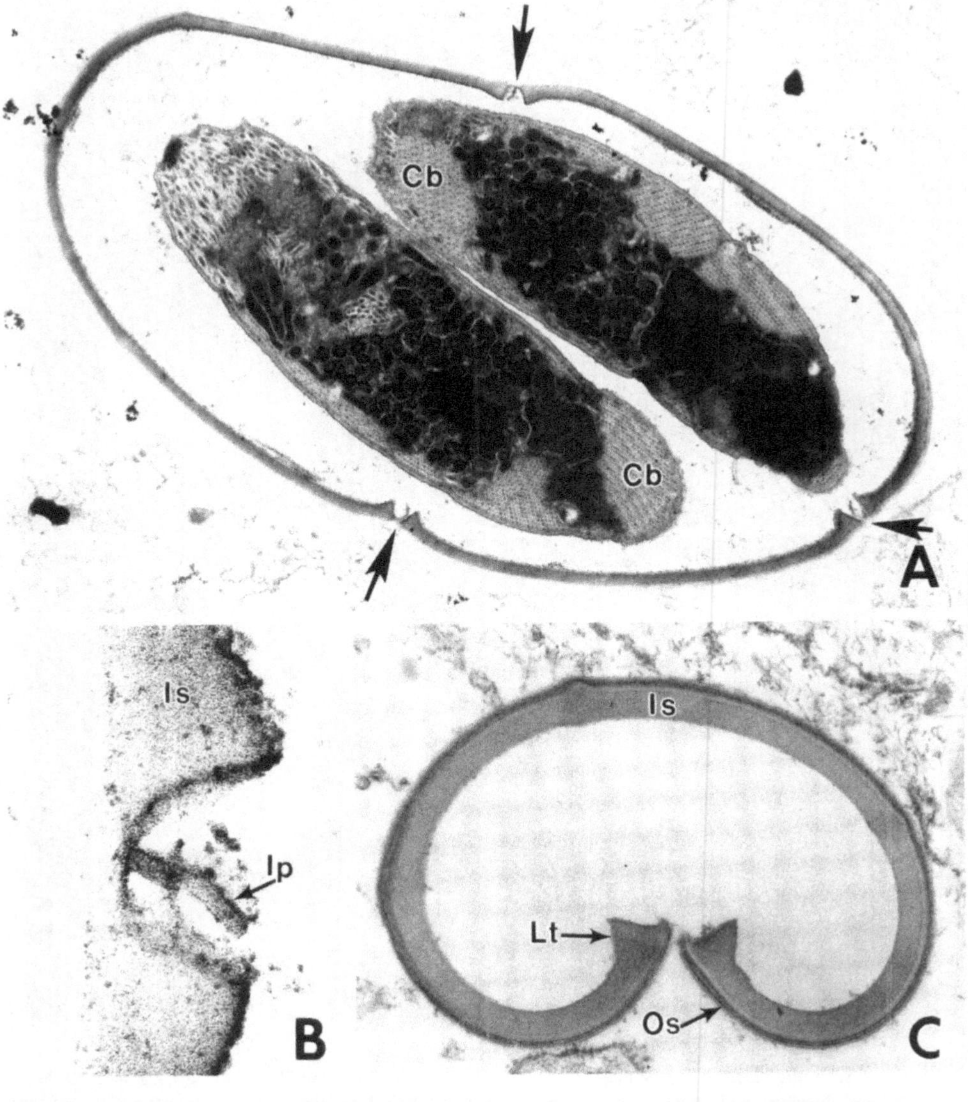

FIGURE 3. Transmission electron micrographs (TEM) of sporocysts and sporocyst walls of *S. cruzi*. (A) Sporocyst in initial stage of excystation in which the plates in the sporocyst are beginning to separate at the sites of apposition (arrows). Note crystalloid bodies (Cb) in each of the sporozoites. (Magnification ×9500.) (B) High magnification of site of apposition between two plates located at top of A; note that the plates have partially separated and the interposed strip is partially attached to the lower plate; also note that treatment of the sporocyst with NaOCl prior to excysting fluid has removed the outer layer of the sporocyst wall. (Magnification ×99,000.) (C) Collapsed sporocyst wall showing lip-like thickenings (Lt) at margin of plate and outer (Os) and inner (Is) layers of the sporocyst wall. (Magnification ×30,000.)

or bradyzoites) within white blood cells ensures protection to the parasite against circulating antibody and movement of zoites to further sites of development. Because the merozoites are free in the cytoplasm, they avoid destruction within phagolysosomes.[42] Whether zoites are phagocytosed or actively penetrate into leucocytes is unknown.

There is some light microscopical (but not ultrastructural) evidence that reproduction (by endodyogeny) of merozoites of several species of *Sarcocystis* occurs in leukocytes.[39,43-45] These observations require further study.

Sarcocyst formation, primarily in striated or cardiac musculature, or in neural tissue of intermediate hosts, begins with merozoites from end-stage schizogony (Figure 7). In most

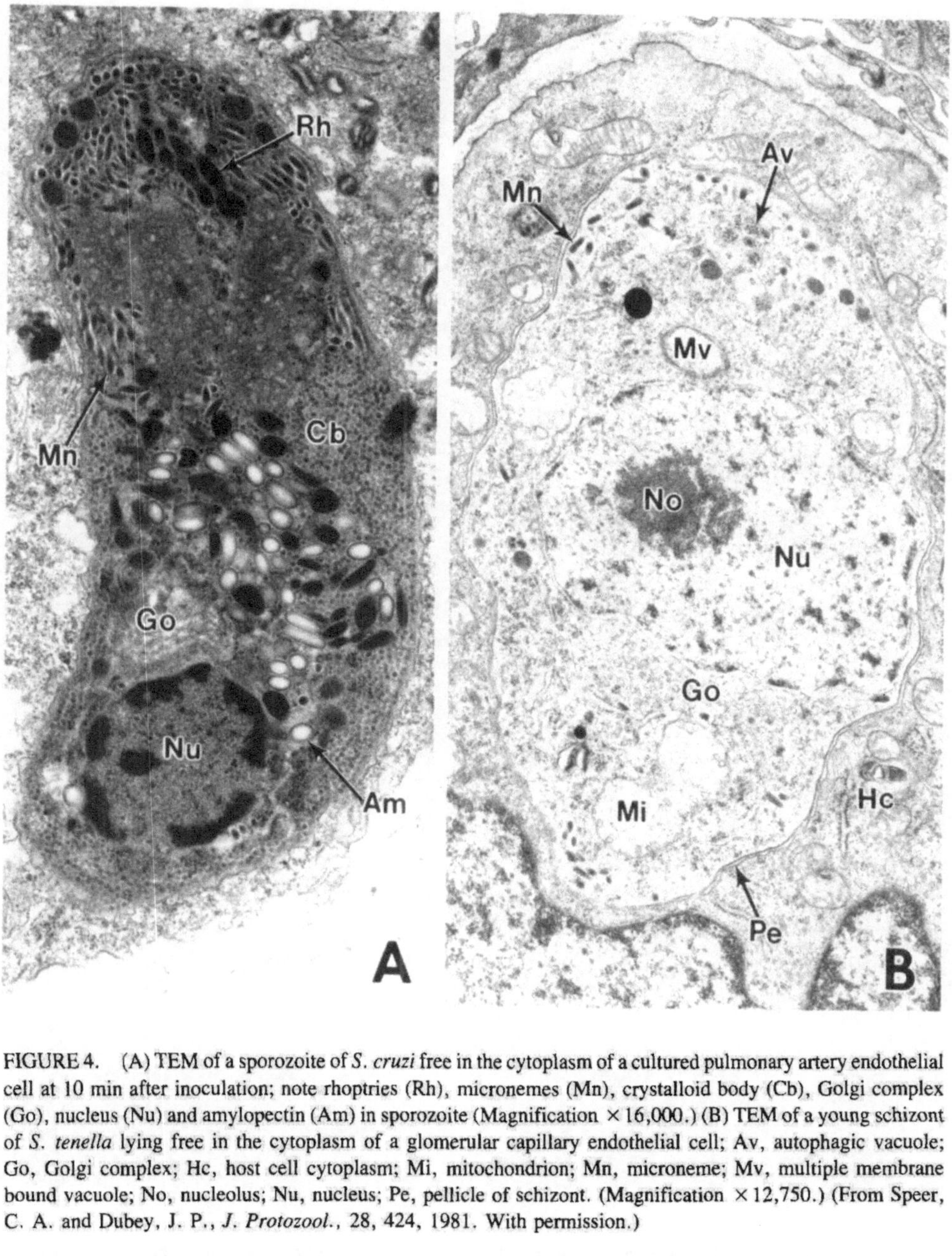

FIGURE 4. (A) TEM of a sporozoite of *S. cruzi* free in the cytoplasm of a cultured pulmonary artery endothelial cell at 10 min after inoculation; note rhoptries (Rh), micronemes (Mn), crystalloid body (Cb), Golgi complex (Go), nucleus (Nu) and amylopectin (Am) in sporozoite (Magnification × 16,000.) (B) TEM of a young schizont of *S. tenella* lying free in the cytoplasm of a glomerular capillary endothelial cell; Av, autophagic vacuole; Go, Golgi complex; Hc, host cell cytoplasm; Mi, mitochondrion; Mn, microneme; Mv, multiple membrane bound vacuole; No, nucleolus; Nu, nucleus; Pe, pellicle of schizont. (Magnification × 12,750.) (From Speer, C. A. and Dubey, J. P., *J. Protozool.*, 28, 424, 1981. With permission.)

Sarcocystis spp. there is little temporal overlap between schizogony and sarcocyst development. However, the process of schizogony in *S. falcatula* (avian intermediate hosts) persists extensively after sarcocyst formation begins.[38] The number of generations of schizogony, prior to sarcocyst formation, appears to be species specific.

Sarcocysts form when a merozoite enters a muscule or neural cell and becomes enclosed in a parasitophorous vacuolar membrane (Pm) (within a parasitophorous vacuole).[13] The Pm differentiates into a primary sarcocyst wall (Pw) consisting of the Pm and adjacent underlying electron-dense layer (Figure 8). There is early development of a granular layer immediately beneath the Pw, containing tiny granules, and vesicles arise from invaginations in the Pm. In many *Sarcocystis* spp., septa develop from the granular layer, extending into the interior of a sarcocyst and divide it into variably sized units containing metrocytes and/or bradyzoites

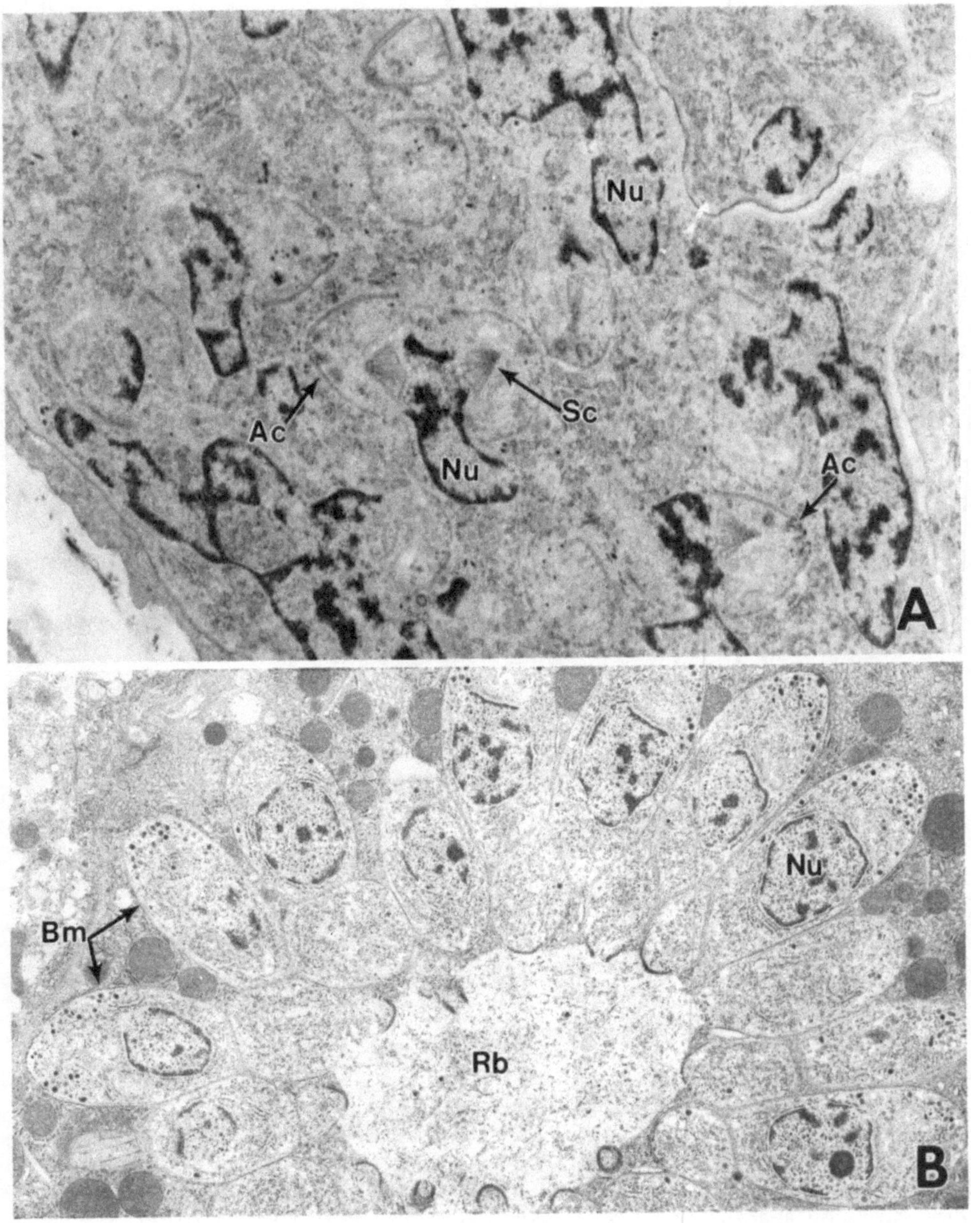

FIGURE 5. (A) TEM of first-generation schizont of *S. tenella* in intermediate stage of development showing formation of the apical complexes (Ac) of merozoites in association with each lobe of an irregularly shaped nucleus (Nu); Sc, spindle polar cone. (Magnification ×12,400.) (From Speer, C. A. and Dubey, J. P., *J. Protozool.*, 28, 424, 1981. With permission.) (B) Nearly mature schizont of *S. rauschorum* with merozoites (Bm) budding from a central residual body (Rb); Nu, nucleus of merozoite. (Magnification ×8000.)

(Figure 8). The structure (form) of septa is highly variable. Minute undulations are present in the Pw of mature sarcocysts of all species of *Sarcocystis*. However, the form of the Pw is highly variable (simple to complex) although for each species of *Sarcocystis* there is an ultrastructurally unique sarcocyst wall. Dubey et al.[13] recognized 24 distinct types of sarcocyst walls. The Pw can undergo significant chronological changes depending on species (see Friesen et al.[46]) and these changes are important in systematics and diagnosis of *Sarcocystis* spp. In *S. rauschorum* of varying lemmings the Pw is relatively simple[46] while the

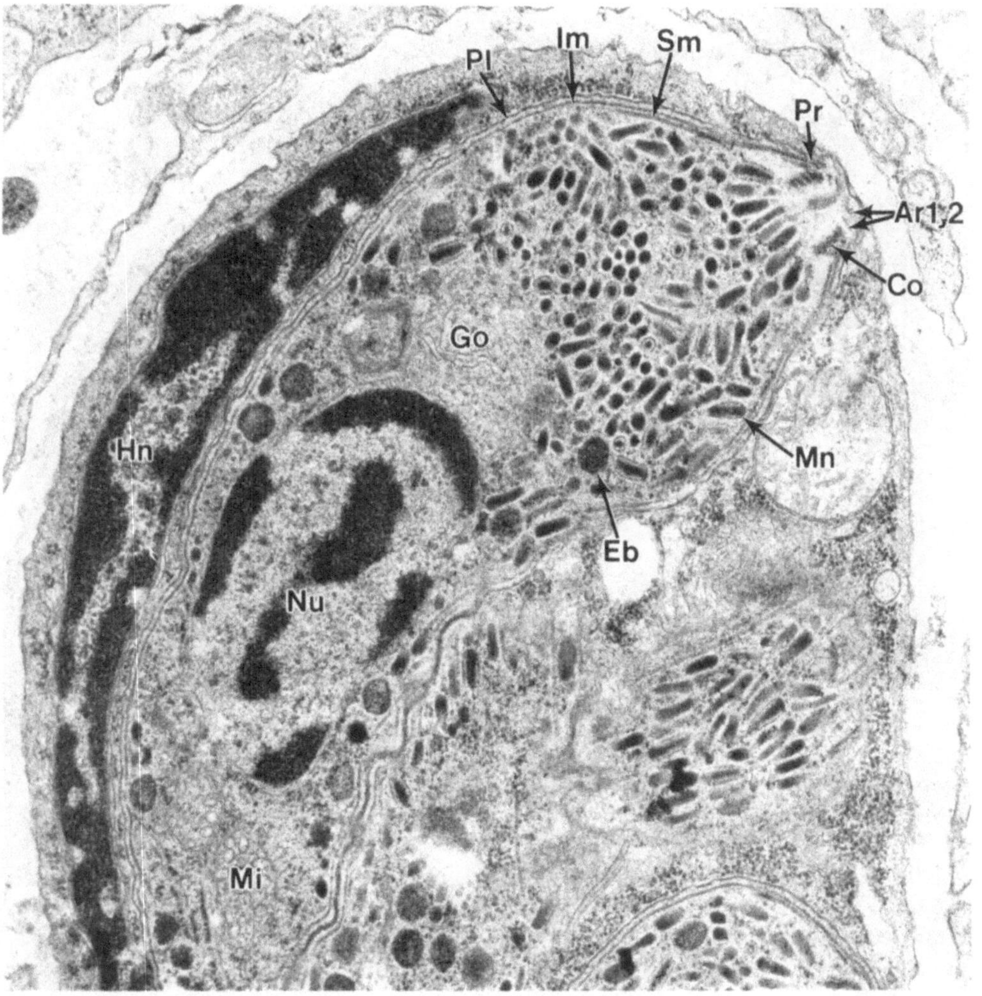

FIGURE 6. TEM of merozoites of *S. tenella*; Arl, 2, apical rings 1 and 2; Co, conoid; Eb, electron-dense body; Go, Golgi complex; Hn, host cell nucleus; Im, inner membrane complex of merozoite pellicle; Mi, mitochondrion; Mn, microneme; Nu, nucleus; Pl, plasmalemma of pellicle; Pr, polar ring; Sm, subpellicular microtubule. (Magnification ×26,600.) (From Speer, C. A. and Dubey, J. P., *J. Protozool.*, 28, 424, 1981. With permission.)

Pw of *S. singaporensis* of rats is complex (Figure 8). The latter Pw undergoes marked changes in the Pw including minute undulations as a young sarcocyst through an increasing lengthening of protrusions (with shape changes) to a noticeable shortening (and increasing irregularity of shape) of protrusions as the sarcocysts age. The presence of vesicles and membranous whorls within the ground substance or protrusions of sarcocyst walls could reflect the status of metabolic activity within the sarcocyst; as sarcocysts become infective (i.e., contain bradyzoites) and later became senescent, the number of these inclusions decreases.

Within a developing sarcocyst, the end-stage merozoite transforms into a round-to-ovoid metrocyte. Usually metrocytes reproduce rapidly, asexually, via the process of endodyogeny, producing two daughter cells[13] (Figure 9). Additionally, in at least one species of *Sarcocystis, S. dirumpens* of rodents, metrocytes are formed both by endodyogeny and by endopolygeny.[47] After several generations of metrocytes producing metrocytes, bradyzoites (infectious to definitive hosts) develop. Metrocytes, merozoites, and bradyzoites can all undergo endo-

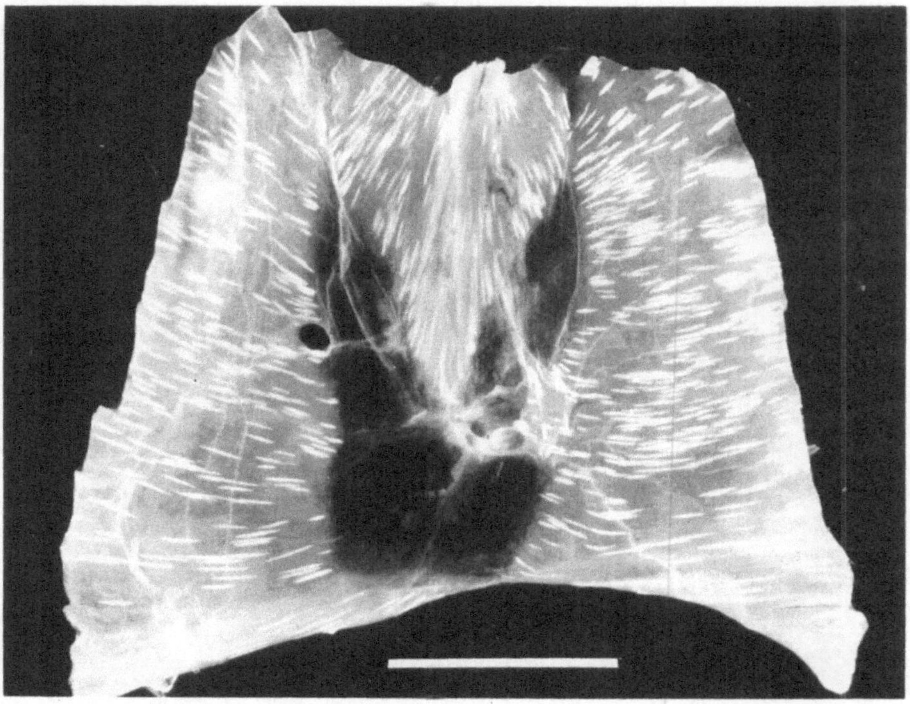

FIGURE 7. Sarcocysts of *S. campestris* in diaphragm of a Richardson's ground squirrel (*Spermophilus richardsonii*). (Bar = 1 cm.) (From Cawthorn, R. J., Wobeser, G. A., and Gajadhar, A. A., *Can. J. Zool.*, 61, 370, 1983. With permission.)

dyogeny. Bradyzoites have an exocytosis pore through which electron-dense granules are discharged into the Pv. It has been proposed that this material also stimulates adhesion of the host cell endoplasmic reticulum to Pv with formation of a three-membrane complex. It may also be involved in antigenic mimicry, preventing immune or inflammatory responses.[48,49] As sarcocysts age, fewer metrocytes and more bradyzoites are present. In old sarcocysts, the center of the cysts is fluid-filled as bradyzoites degenerate; bradyzoites may be physiologically deprived the greater distance they are from the periphery of the cyst.

Depending on species of *Sarcocystis*, the time required for bradyzoites to develop and become infectious to definitive hosts can be short or extended. The number of schizogonous generations may be an important factor. For example, *S. rauschorum* has one generation of schizogony, within hepatocytes of varying lemmings, and cysts are evident in striated musculature beginning 9 DPI. Bradyzoites were present at least as early as 28 DPI (likely as early as 21 DPI)[22] (Figure 10). Such rapid development of infectious stages could ensure transmission of the parasite among rodents which are short-lived and relatively mobile prey. However, *S. gigantea* of sheep which has one generation of schizogony, produces bradyzoites as early as 119 DPI which are not infective until 230 to 265 DPI.[50,51] The ecological significance of delayed development to infectivity of bradyzoites is unknown. In contrast, *S. hemionilatrantis* of mule deer with three generations of schizogony, produces bradyzoites as early as 90 DPI.[44] Thus, there is great variability among *Sarcocystis* spp. for the time required for development of (infectious) bradyzoites.

Sarcocysts can be short-lived or persist throughout the life of the intermediate host. In *S. miescheriana* of swine, sarcocysts resolve as early as 40 DPI although they mature only at 27 to 80 DPI; sarcocysts are rare at 180 DPI.[52,53] In contrast, sarcocysts of *S. gigantea* persist more than 4 years in sheep.[50,51] The mechanism of sarcocyst degeneration is unknown. Perhaps there is spontaneous rupture resulting in an inflammatory response or perhaps inflammatory cells surrounding "normal" sarcocysts are able to remove the cysts.[53]

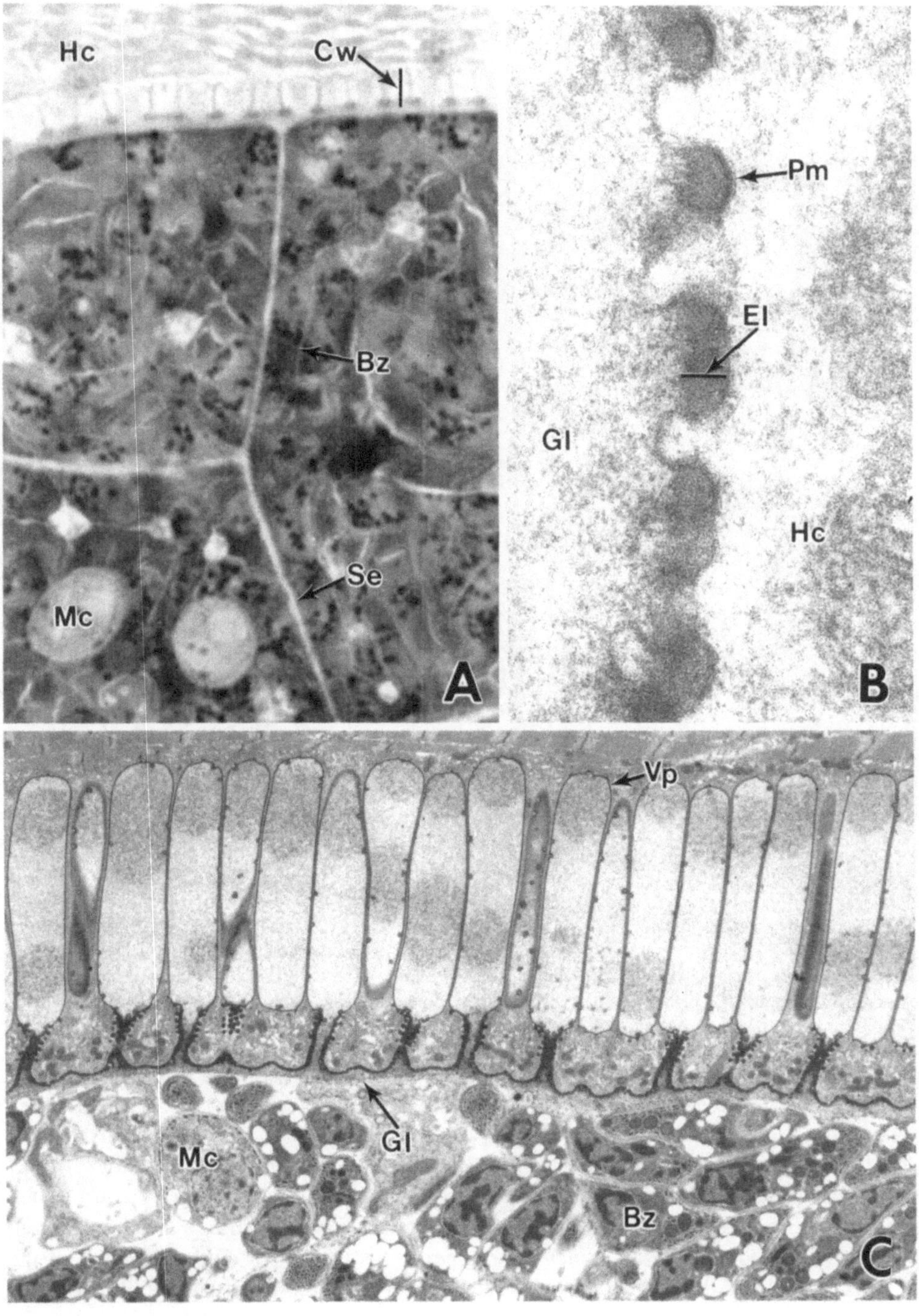

FIGURE 8. (A) Phase-contrast photomicrograph of a portion of a sarcocyst of *S. hemionilatrantis* in the diaphragm of a mule deer. (Magnification ×2000.) (From Dubey, J. P. and Speer, C. A., *J. Wildl. Dis.*, 21, 219, 1985. With permission.) (B) High magnification TEM of primary cyst wall of *S. rauschorum* which consists of the parasitophorous vacuolar membrane (Pm) and the electron-dense layer (El); Gl, granular layer of cyst wall; Hc, host cell cytoplasm. (Magnification ×99,000.) (C) TEM of portion of the sarcocyst wall of *S. singaporensis* showing villar protrusions (Vp), bradyzoites (Bz), metrocytes (Mc), and the granular layer (Gl). (Magnification ×5000.)

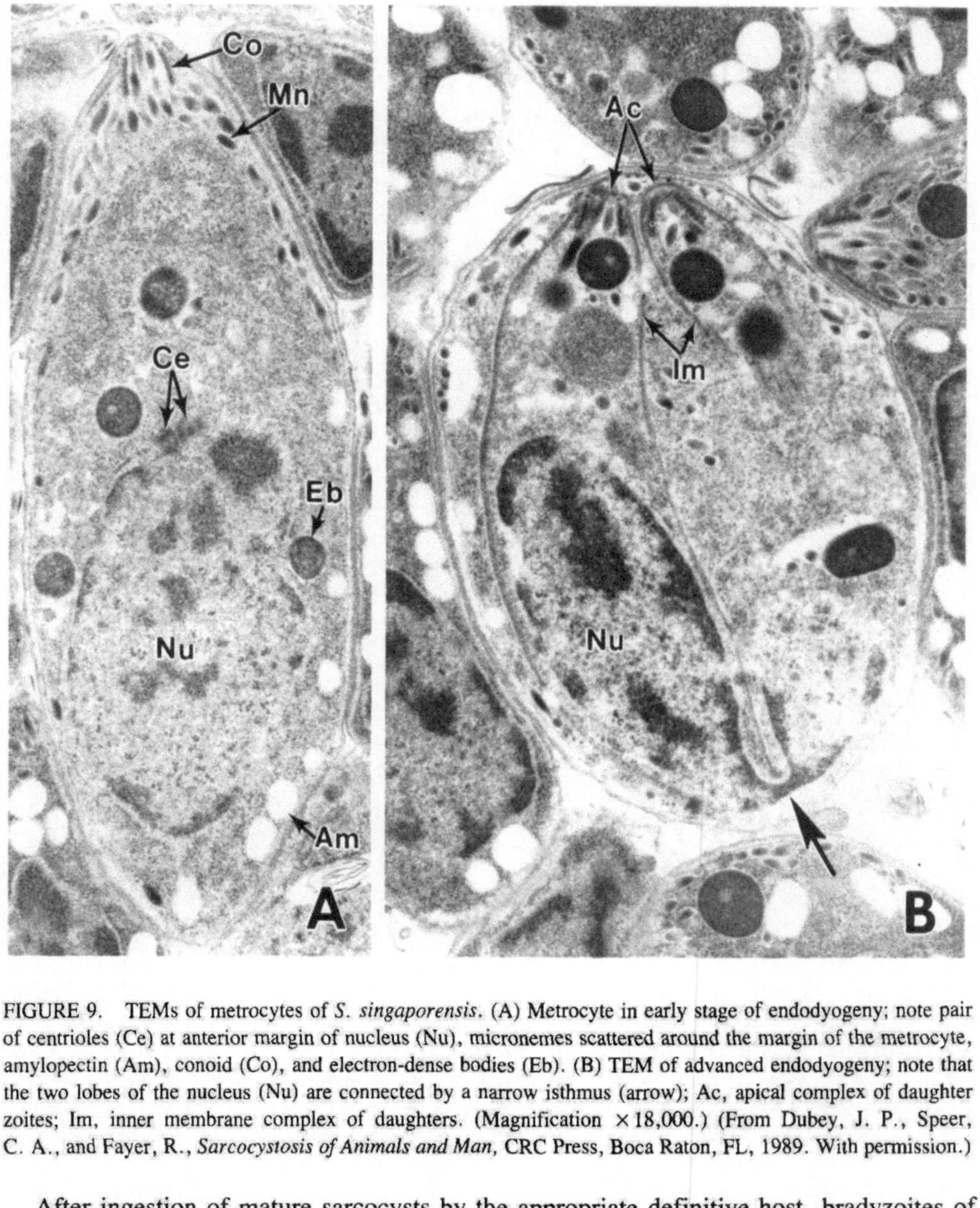

FIGURE 9. TEMs of metrocytes of *S. singaporensis*. (A) Metrocyte in early stage of endodyogeny; note pair of centrioles (Ce) at anterior margin of nucleus (Nu), micronemes scattered around the margin of the metrocyte, amylopectin (Am), conoid (Co), and electron-dense bodies (Eb). (B) TEM of advanced endodyogeny; note that the two lobes of the nucleus (Nu) are connected by a narrow isthmus (arrow); Ac, apical complex of daughter zoites; Im, inner membrane complex of daughters. (Magnification ×18,000.) (From Dubey, J. P., Speer, C. A., and Fayer, R., *Sarcocystosis of Animals and Man*, CRC Press, Boca Raton, FL, 1989. With permission.)

After ingestion of mature sarcocysts by the appropriate definitive host, bradyzoites of most species of *Sarcocystis* rapidly (i.e., within 18 h) complete gametogony and oogony (Figures 11 to 14). However, gamonts of *S. idahoensis* of gopher snakes are not apparent until 5 DPI.[54] Gamonts develop primarily in enterocytes (goblet or epithelial cells) of the small intestine. Although the mechanism of sexual determination (or regulation) is unknown, it is likely predetermined in bradyzoites as the bradyzoites instantaneously differentiate as either micro- or macrogamonts. There is no apparent sexual dimorphism of bradyzoites of *S. cruzi*.[55]

As bradyzoites of *Sarcocystis* spp. undergo gametogony, they develop an oval-spherical shape and there is dissolution of the apical complex[13] (Figures 11 and 12). Macrogamonts are distinguished from microgamonts by having electron-dense granules and an exocytosis pore, which is ultrastructurally distinct from a micropore.[13] Electron-dense granules are discharged from the exocytosis pore: some granules fuse in the Pv as a thin electron-dense layer (oocyst wall); other granules could be enzymes causing lysis of the host cell. The

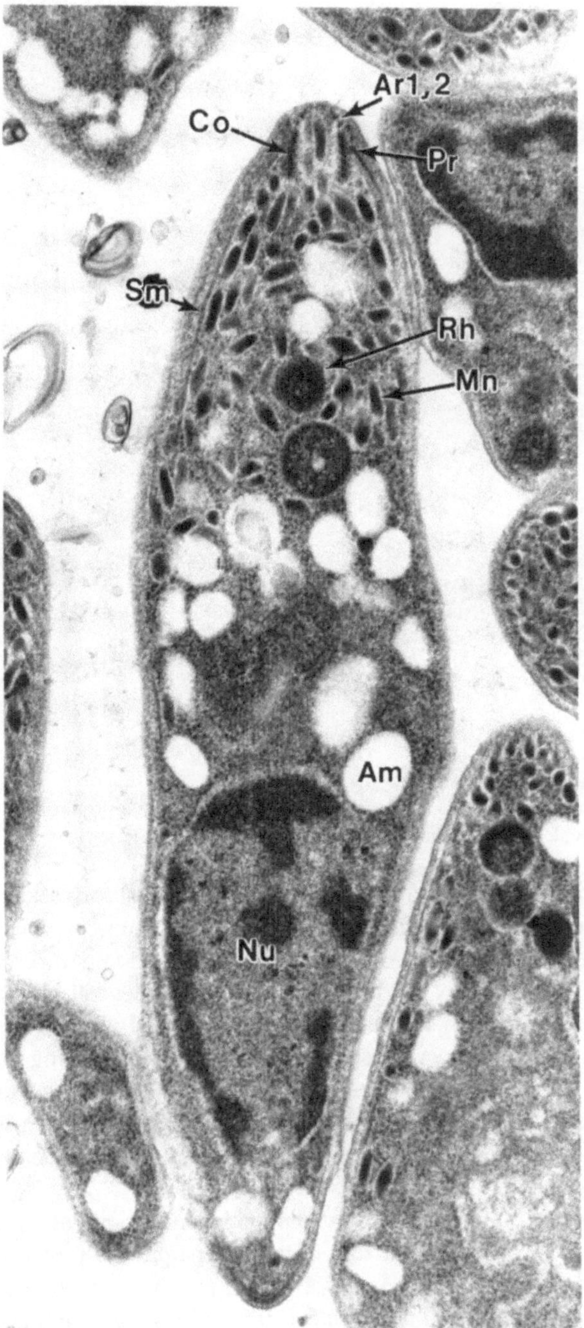

FIGURE 10. TEM of bradyzoite of *S. singaporensis* showing amylopectin (Am), apical rings 1 and 2 (Ar1,2), conoid (Co), micronemes (Mn), nucleus (Nu), polar ring (Pr), rhoptry (Rh) and subpellicular microtubule (Sm). (Magnification ×25,000.)

electron-dense bodies of *Sarcocystis* macrogamonts correspond to the wall-forming bodies (of the oocyst wall) in other coccidia such as *Eimeria* spp. The oocyst wall consists of two layers, an outer electron-dense layer and an inner moderately electron-dense layer composed of two to three membranes (Figure 14). Macrogamonts start development in intestinal epithelium; however, as gametogony proceeds, the parasitized host cell degenerates and lyses.

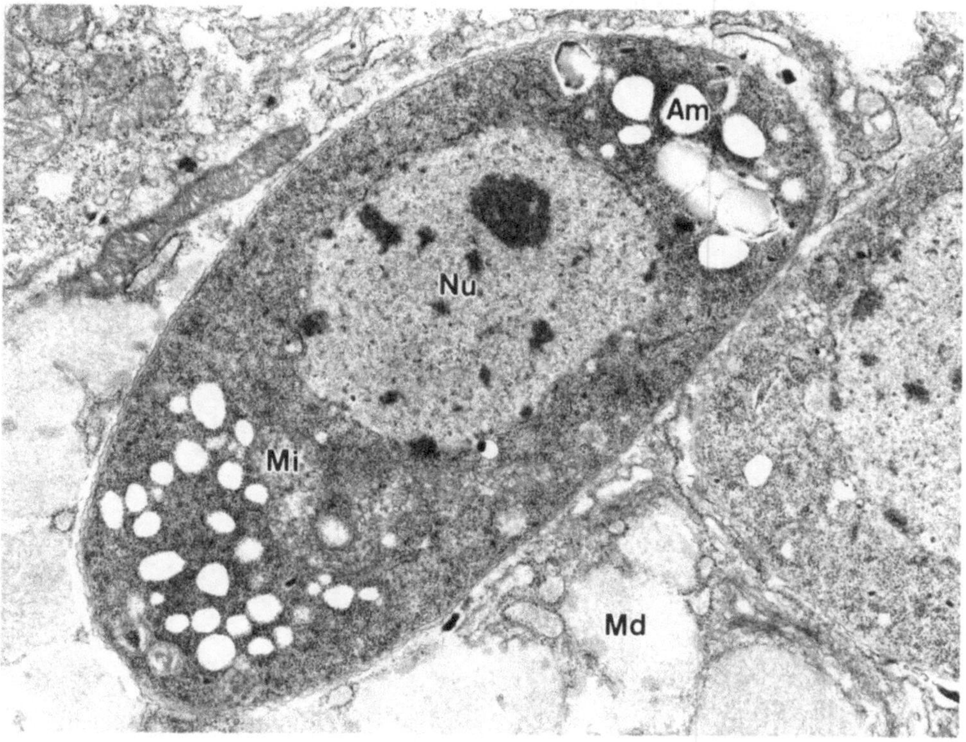

FIGURE 11. TEM of bradyzoite of *S. singaporensis* in early stage of transformation to a gamont; note mucin droplets (Md) of host goblet cell; Am, amylopectin; Mi, mitochondrion; Nu, nucleus. (Magnification ×13,000.)

The macrogamont moves into the lamina propria where oocyst formation and sporogony are completed. Whether the macrogamont is in a new host cell, and if so, what type, is unknown.

Microgamonts of *S. cruzi* have no exocytosis pores and do not discharge electron-dense granules into the Pv. Gametogony is usually completed in goblet cells of intestinal epithelium; up to 15 biflagellated microgametes develop in a host cell.[13]

Fertilization, for most *Sarcocystis* spp., occurs within 18 h of ingestion of bradyzoites by definitive hosts[56] (Figure 13). Oocyst wall formation, induced by fertilization, begins prior to fusion of gamete nuclei[57] (Figure 13). The nucleus of the microgamete traverses a cytoplasmic bridge to enter the macrogamont; fusion of the gamete nuclei results in a zygote (young oocyst) (Figure 14).

Sporogony of species of *Sarcocystis* occurs in the lamina propria of the intestine. Bledsoe[58] suggested that the highly vascular structure of the lamina propria provides adequate oxygen supply for the aerobic process of sporogony. By light microscopy, sporogony of *Sarcocystis* spp. is similar.[13,59,60] Initially, primary nuclear division results in an oocyst with two polar nuclei. After the primary cytokinesis and simultaneous second nuclear division, there are two sporoblasts each containing two polar nuclei. Terminally, a third nuclear division and second cytokinesis result in a sporulated oocyst containing two sporocysts, each of which contains four sporozoites. There is some variability among species of *Sarcocystis*. Occasionally, sporulated oocysts are found in mesenteric lymph nodes of definitive hosts; the origin and significance of these oocysts or sporocysts is unknown.[61] The process of sporulation, which among *Sarcocystis* spp. is asynchronous, has not been examined ultrastructurally. Subsequently, sporulated oocysts and sporocysts move into the intestinal lumen, and are passed in the feces of the definitive host, completing the life cycle of *Sarcocystis* spp.

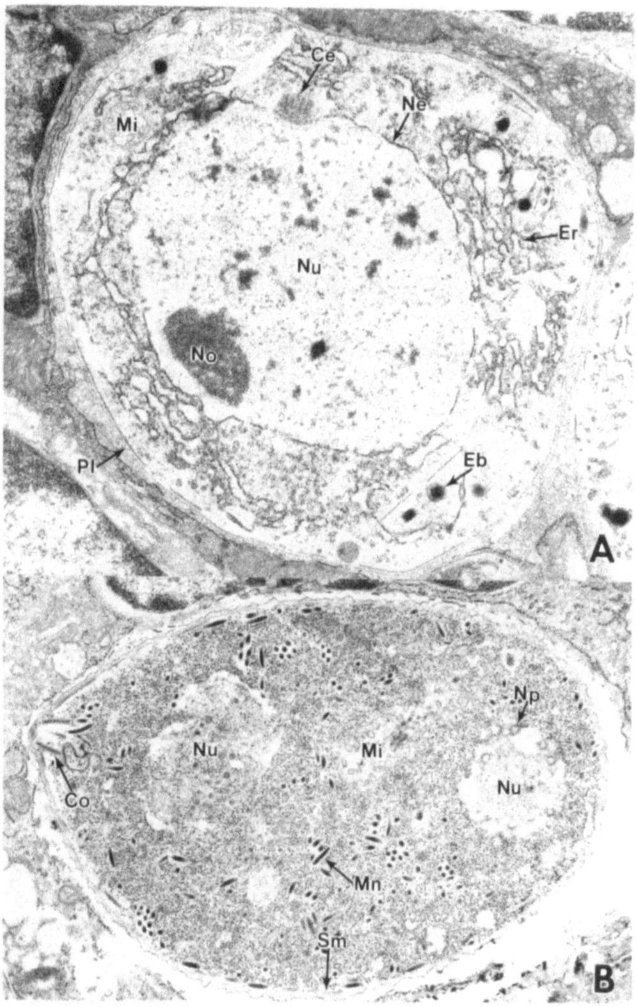

FIGURE 12. (A) TEM of macrogamont of *S. cruzi*; Ce, centrioles; Eb, electron-dense body; Er, endoplasmic reticulum; Mi, mitochondrion; Ne, nuclear envelope; No, nucleolus; Nu, nucleus, Pl, plasmalemma of macrogamont. (Magnification × 15,840.) (B) TEM of early microgamont of *S. cruzi* showing two lobes of nucleus (Nu); Co, conoid; Mi, mitochondrion; Mn, micronemes; Np, nuclear pore; Sm, subpellicular microtubule. (Magnification × 15,180.) (From Dubey, J. P., Speer, C. A., and Fayer, R., *Sarcocystosis of Animals and Man*, CRC Press, Boca Raton, FL, 1989. With permission.) (Figures reduced by $^1/_3$.)

III. SYSTEMATICS AND DIAGNOSIS

Cyst-forming coccidia comprise eight genera of apicomplexan parasites including species of *Besnoitia, Caryospora, Cystoisospora (Isospora), Frenkelia, Hammondia, Neospora, Sarcocystis,* and *Toxoplasma*. Apparently, heteroxenous life cycles of coccidia evolved independently several times, reflecting changes in feeding behavior of ancestral definitive hosts.[62] Classification of members of the apicomplexan class Sporozoea, including *Sarcocystis* spp., should be based on biologically valid characters, including biological and ultrastructural features.[62] Among species of *Sarcocystis*, intermediate hosts are only of ecological, not phylogenetic, significance.[19,62,63]

In the genus *Sarcocystis* there are approximately 130 named species, of which both

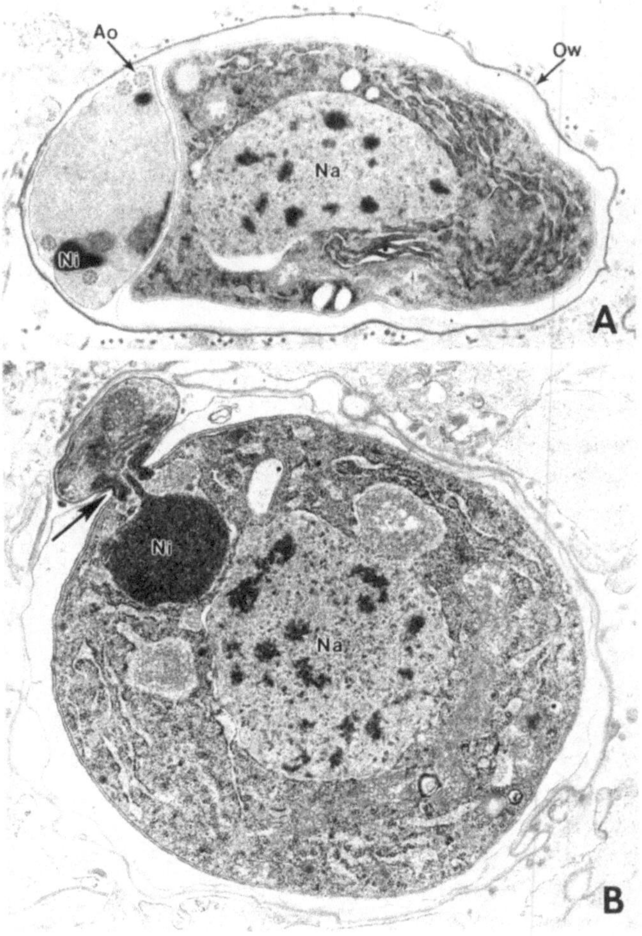

FIGURE 13. TEMs of fertilization in *S. cruzi*. (A) Early stage in fertilization of macrogamont by microgamete; note that the oocyst wall (Ow) has already formed; Ao, axoneme of microgamete flagellum; Na, nucleus of macrogamont; Ni, nucleus of microgamete. (Magnification ×13,000.) (From Dubey, J. P., Speer, C. A., and Fayer, R., *Sarcocystosis of Animals and Man*, CRC Press, Boca Raton, FL, 1989. With permission.) (B) Intermediate stage in fertilization in which the microgamete nucleus (Ni) is entering the macrogamont through a cytoplasmic bridge (arrow) created by fusion of the plasmalemmae of the microgamete and macrogamont. (Magnification ×15,770.) (From Sheffield, H. G. and Fayer, R., *Proc. Helminthol. Soc. Wash.*, 47, 118, 1980. With permission.) (Figures reduced by $^1/_3$.)

intermediate and definitive hosts are recognized for 64 species.[10] Historically, naming of new species of *Sarcocystis* was based on structure of sarcocysts and host species. However, recent transmission electron microscope studies with ultrastructural examination of sarcocysts demonstrate that some earlier descriptions refer to more than one species in a given host and consequently redescriptions are necessary.[10] Relatively few biological and ultrastructural features, including sarcocyst wall ultrastructure, transmission studies and isoenzyme characterization, are useful for adequate descriptions of *Sarcocystis* spp.[10,13,64] Comparison of small subunit ribosomal RNA is a newly emerging, useful tool for systematic studies.[65]

Presently, there are several difficulties in speciating members of the genus *Sarcocystis*.[13] Structure of the sarcocyst is affected by its age, physiological state, type of host cell

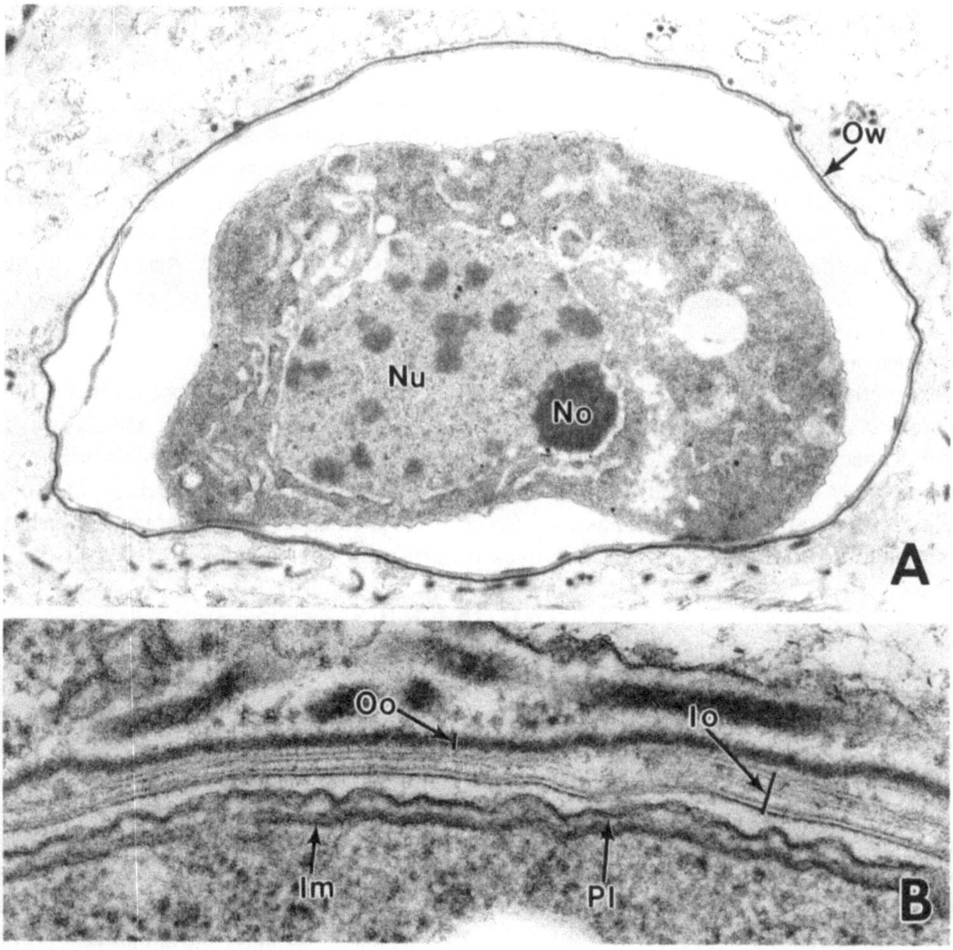

FIGURE 14. TEMs of oocysts of *S. cruzi*. (A) Oocysts; No, nucleolus; Nu, nucleus; Ow, oocyst wall. (Magnification ×11,000.) (B) High magnification of oocyst wall showing inner (Io) and outer (Oo) layers; note that the inner layer is moderately electron-dense and consists of two to three membranes and that the outer layer is electron-dense; Im, inner membrane complex of zygote; Pl, plasmalemma of zygote. (Magnification ×80,000.)

parasitized, and fixative used. Light microscopy is inadequate to determine species characters of sarcocysts. Villar protrusions (i.e., shape and size) vary with preparation. Sporulated oocysts and sporocysts of *Sarcocystis* spp. are structurally similar and size dimensions overlap among many species. Host specificity is not a singularly adequate criterion for speciation as species of *Sarcocystis* exhibit highly variable specificity for both intermediate and definitive hosts.[26] However, species of *Sarcocystis* utilizing small mammals as intermediate hosts are more host specific than those utilizing large mammals.[13]

Diagnosis to the generic level is relatively easy in definitive hosts. Only species of *Sarcocystis* and *Frenkelia*, of the genera of cyst-forming coccidia, undergo endogenous sporulation in definitive hosts. Species of *Frenkelia* develop only in the intestinal tract of birds; consequently generic diagnoses can be confirmed only by transmission studies.[66-68] Sporocysts, detected in fresh feces by flotation-concentration techniques, of many *Sarcocystis* spp., are similar in size and shape with overlapping dimensions.[13] No serological tests are available for detection of prepatent or patent infections of *Sarcocystis* spp. in definitive hosts.

Diagnosis to the generic level is becoming easier in intermediate hosts of cyst-forming coccidia. Serologic tests are useful, although none is available commercially and standardization is lacking for detection of *Sarcocystis* spp.[13] Biopsy specimens can aid histological detection of schizonts and sarcocysts. Detection of parasitemia is impractical on a routine basis. Histological techniques, including immunohistochemistry based on immunoperoxidase staining, are becoming more precise, allowing differentiation of schizonts (and cysts) of *Sarcocystis* spp. from other coccidia.[69] Unfortunately, in *Sarcocystis*-induced abortions, the parasites are often not present in fetal or placental tissues.[14] Sarcocysts can be detected in the musculature of intermediate hosts by gross examination, unstained squash preparations, digestion methods and histological studies. Tissue digestion methods are sensitive although speciation of bradyzoites is not possible; ultrastructural examination is precise although tedious, time-consuming, and expensive. A thorough clinical history is useful to confirm sarcocystosis as a disease entity in a specific outbreak.

Recently, *Neospora caninum* has been described as a cyst-forming coccidium.[70] There are several structural differences in schizonts and cysts between *N. caninum* and other cyst-forming coccidia, including species of *Sarcocystis*. This new parasite, although its host range is unknown, can cause fatal myelitis, meningoencephalitis, and myositis in dogs, cats, and cattle.[70-72] Overall, generic diagnosis of *Sarcocystis* is relatively easy; however, species differentiation within the genus is a complex undertaking.

IV. TECHNIQUES

For experimental and diagnostic purposes, sporulated oocysts or sporocysts of *Sarcocystis* spp. can be detected and recovered from feces of definitive hosts via flotation-concentration methods. As detailed in Dubey et al.,[13] we prefer Sheather's sugar solution as the flotation medium; phase-contrast illumination aids detection of sporocysts. Sporocysts are recovered from feces by utilizing 2 *M* Sheather's sugar solution, with recovery of sporocysts at the water-sugar solution interface after concentration by centrifugation. *Sarcocystis* sporocysts can be directly recovered from patent infections in definitive hosts by scraping and homogenizing pieces of intestinal epithelium; sporocysts are concentrated by centrifugation with water.[13] Recovered sporocysts are stored at 4°C in Hanks' balanced salt solution (HBSS) with antibiotics and fungistats.[73] Although sporocysts recovered from feces are usually cleaner than those recovered from the lamina propria, we have observed no difference in infectivity between the two groups of sporocysts.

Sporozoites of *Sarcocystis* spp. are excysted utilizing the methodology of Cawthorn et al.[74] with the modification of 5.3% (v/v) NaOCl for the 30 min incubation at 4°C. We now also utilize this method to obtain sporozoites of *S. hirsuta*. A Percoll gradient will remove debris from the excysted sporozoites; this provides a much cleaner inoculum for inoculation into tissue cultures and for inoculation into hosts for immunological studies. The effective concentration of the gradient varies depending on the species of *Sarcocystis*. Viability of sporozoites is tested by inoculation into tissue cultures. Sporozoites of *S. capracanis*, *S. cruzi*, *S. hirsuta*, and *S. tenella* develop into schizonts *in vitro*, depending on cell type utilized.[42,75] For species-specific purposes, sporocysts of *Sarcocystis* spp. should be obtained only from experimentally infected, laboratory-reared definitive hosts.

Bradyzoites of *Sarcocystis* spp. can be recovered from sarcocysts by digestion with artificial gastric fluid (e.g., pepsin). The muscle of intermediate hosts is homogenized, incubated at 37°C in a pepsin-HCl solution for up to 2 h.[76] Bradyzoites are concentrated by centrifugation, subsequently cleaned of host tissue by concentration with Percoll and washed with HBSS.[77] A useful modification is centrifugation for 20 min at each step. This method is highly sensitive for detection of bradyzoites in prevalence studies; however, species of *Sarcocystis* are indistinguishable. Immature sarcocysts will not be detected as metrocytes do not survive digestion. Phase-contrast illumination markedly aids observation of bradyzoites in the digest mixture. For species-specific studies, bradyzoites of *Sarcocystis* spp. should be harvested from experimentally infected, laboratory-reared intermediate hosts.

Recently we have developed techniques for cryopreservation of zoites of *S. cruzi* and *S. hirsuta*. The method used provides effective, long-term storage of sporozoites and merozoites, and maintenance of reference strains thus reducing the need of recycling of parasites *in vivo*.[78,79] Sporozoites of *S. cruzi* and *S. hirsuta* and merozoites of *S. cruzi*, retain high rates of viability when frozen in liquid nitrogen at −196°C for periods of at least 180 days. Ten percent dimethylsulfoxide provided better protection than 10% glycerol.

Presently, serological testing for *Sarcocystis* spp. using IHA and ELISA is based on bradyzoite antigen.[80] Bradyzoites are readily attained as above. However, immunological testing for acute disease will be more sensitive and specific by utilizing sporozoites or merozoites as the antigen. Examination of serum from naturally infected cattle by Western blotting techniques revealed that there are multiple antigens from the zoites of *S. cruzi* that are recognized by these cattle. We recently found that some antigens appear to be common to all three parasite stages while some appear to be specific to each of the sporozoites, merozoites, or bradyzoites. Sporozoites can be obtained as above; merozoites have recently become readily obtainable from *in vitro* cultivation.[16]

Relatively few species of *Sarcocystis* have been cultured *in vitro*. Although sexual stages of *Sarcocystis* spp. were grown in tissue culture beginning in 1970,[81] only recently have the disease-causing stages (schizogony) of *Sarcocystis* spp. in livestock been cultivated[16,75,81,82] (Figures 15 and 16). In addition to *S. capracanis* of goats, *S. cruzi* of cattle, and *S. tenella* of sheep, merozoites of *S. hirsuta* of cattle can now be produced *in vitro*.

Speer et al.[75] suggested that successful *in vitro* cultivation of schizonts of *Sarcocystis* spp. depends on host cell type becasue *S. cruzi* sporozoites developed to schizonts in bovine pulmonary artery endothelial cells or bovine monocytes, but did not develop in Madin-Darby bovine kidney cells nor mouse macrophages. We recently observed that sporozoites or merozoites of *S. cruzi* also develop well in rat cardiac muscle and rat striated muscle cells. Neither sporozoites nor merozoites of *S. cruzi* formed sarcocysts in these muscle cells. In fact, to date no one has succeeded in culturing merozoites to sarcocysts *in vitro* which will likely require the appropriate biochemical trigger or appropriate cell type. Elucidating mechanisms of immunity and pathogenesis, development of serological testing and testing of prophylactic agents should be facilitated by the ability to cultivate the most pathogenic phase of *Sarcocystis* spp. *in vitro*.

Gametogony and sporogony of four *Sarcocystis* spp. have been completed *in vitro*.[83] Sexual development of species of *Sarcocystis in vitro* is similar to that occurring *in vivo*.[13] For successful cultivation of sexual stages of *Sarcocystis* spp., the source of cell (i.e., cell line must be derived from the appropriate natural definitive host) is important; in contrast, the type of host cell (regardless of source) appears essential for the *in vitro* cultivation of schizonts of *Sarcocystis* spp. In our recent studies bradyzoites of *S. cruzi* did not develop in canine kidney cells; predictably these bradyzoites did not develop in either feline lung or feline kidney cells.

V. PATHOGENESIS OF INFECTION

Most species of *Sarcocystis* are not pathogenic to definitive hosts. Perhaps clinical disease results from immediate-type hypersensitivity reactions.[84,85] Although *S. hominis* in Europe and Asia and *S. suihominis* in Europe can cause diarrhea in humans, the clinical syndrome of *S. suihominis* is usually more severe, including nausea, gastric pain, dry heaves, vomiting, dyspnea and rapid pulse.[86] However, *S. hominis* can, in association with bacterial infections, cause acute segmental eosinophilic enteritis or segmental necrotizing enteritis.[87] The prevalence and severity (of disease) of sarcocystosis in humans reflects cultural habits of food

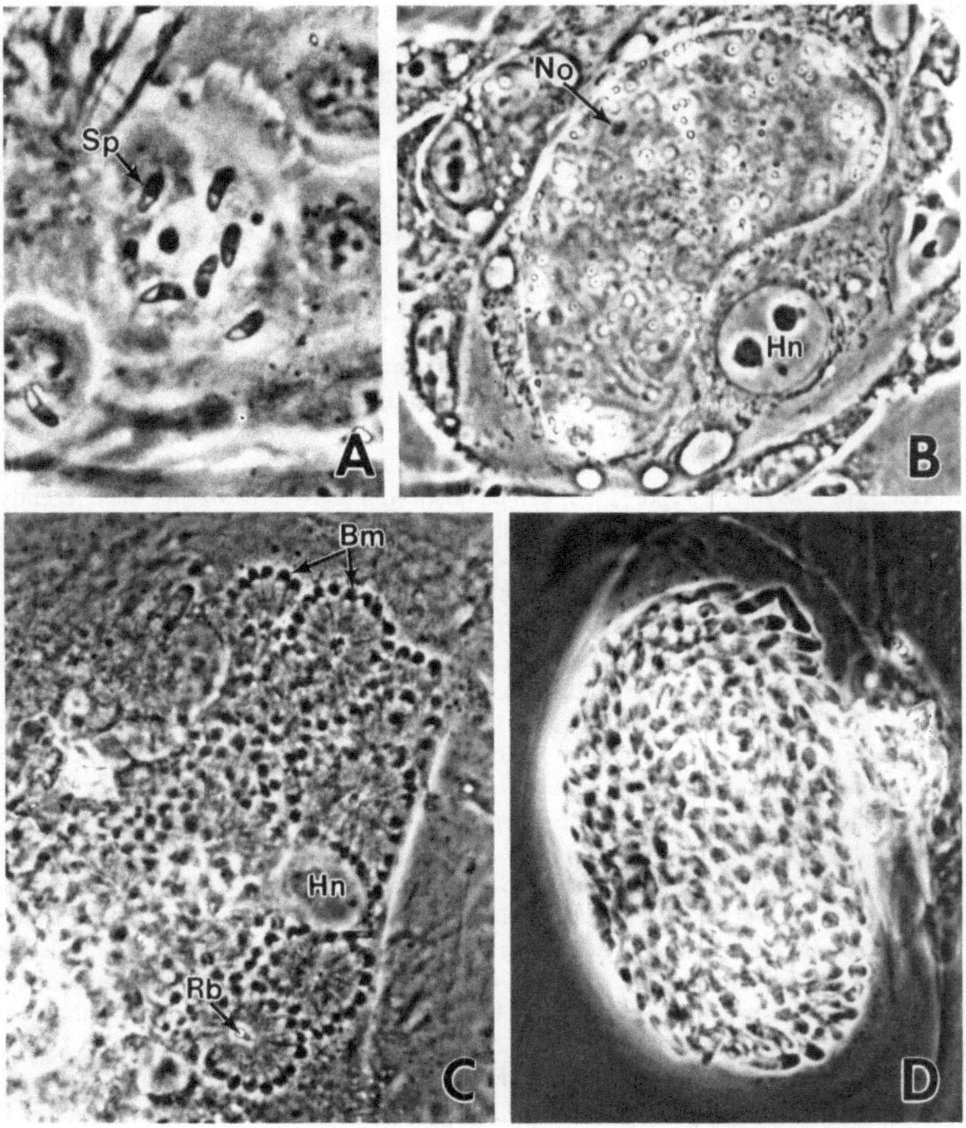

FIGURE 15. Phase-contrast photomicrographs of first-generation schizonts of *S. cruzi* (A, B and D) and *S. tenella* (C) in cultured cells. (A) Intracellular sporozoites (Sp) one hour after inoculation. (Magnification ×680.) (B) Large intermediate schizont 42 days after inoculation of sporozoites; Hn, host cell nucleus; No, nucleolus of parasite. (Magnification ×840.) (From Speer, C. A., et al., *J. Parasitol.*, 72, 677, 1986.) (C) Large schizont with budding merozoites (Bm); Hn, host cell nucleus; Rb, residual body (Rb). (Magnification ×840.) (From Speer, C. A. et al., *J. Protozool.*, 33, 486, 1986. With permission.) (D) Mature schizont with completely formed merozoites. (Magnification ×840.)

preparation and consumption.[88,89] Although there are several reports of *Sarcocystis*-intestinal infection as an important mortality factor in raptorial birds, the data, derived from necropsies of wild, naturally infected birds, are difficult to interpret.[22] Dubey et al.[13] suggested that intestinal disturbances observed in mammalian carnivores (i.e., dogs, coyotes) could reflect dietary changes (i.e., dry dog food to uncooked meat) associated with experimental protocols. Overall, sarcocystosis of definitive hosts is presently of little veterinary or medical significance.

Several species of *Sarcocystis* are pathogenic to intermediate hosts. Severity of acute

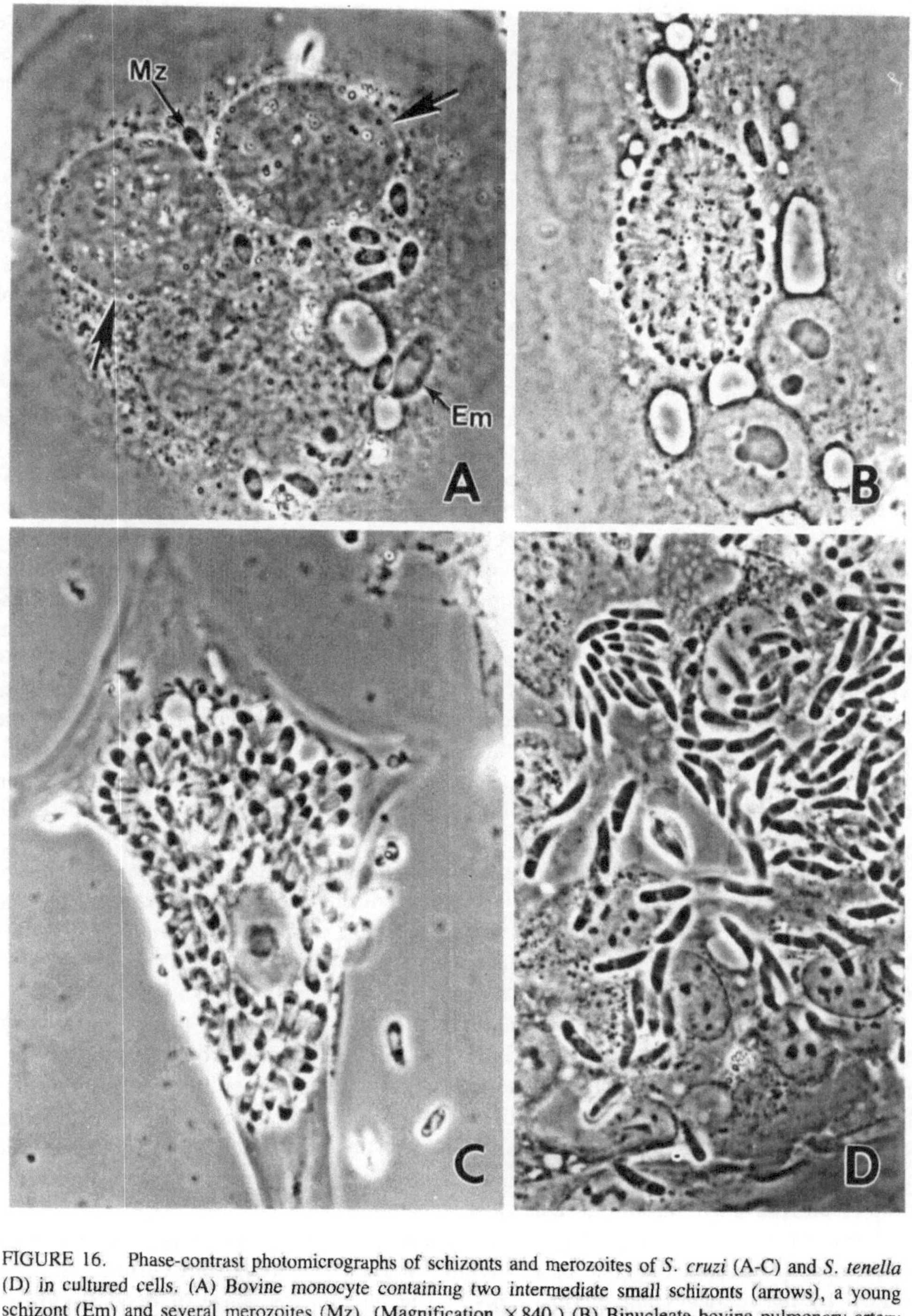

FIGURE 16. Phase-contrast photomicrographs of schizonts and merozoites of *S. cruzi* (A-C) and *S. tenella* (D) in cultured cells. (A) Bovine monocyte containing two intermediate small schizonts (arrows), a young schizont (Em) and several merozoites (Mz). (Magnification ×840.) (B) Binucleate bovine pulmonary artery endothelial (CPA) cell infected with small schizont with budding merozoites. (Magnification ×840.) (C) Mature small schizont with merozoites in cultured CPA cell. (Magnification ×840.) (D) Merozoites in process of escaping from host cells. (Magnification ×840.) (Figure D from Speer, C. A. et al., *J. Protozool.*, 33, 486, 1986. With permission.)

and chronic disease is strongly proportional to the number of sporocysts ingested. Age or size of the intermediate host does not affect the severity of disease. The magnitude of number of sporocysts which is pathogenic is species specific. Overall, the smaller the host species, the fewer sporocysts required for disease to develop. Various stressors (e.g., severe weather, inadequate nutrition) can increase susceptibility to infection and severity of disease. Acute sarcocystosis is caused by schizogonous phases in vascular endothelium or other cells of various organs while chronic disease results from sarcocyst formation in striated or cardiac musculature of intermediate hosts.

Acute sarcocystosis can cause fever, anorexia, weight loss, muscle twitching, prostration, and death in intermediate hosts.[13,14] Other signs include reproductive failure (i.e., abortion, stillbirth, premature birth), decrease or cessation of milk production, wool breakage, anemia, tissue damage, and clotting dysfunction. At gross necropsy there is widespread petechial to ecchymotic hemorrhage, serous atrophy of fat, excess straw-colored fluid in body cavities, icterus, and lymphoid hyperplasia.[14] Histologically, schizonts of *Sarcocystis* spp. may or may not be present, and if they are present, schizonts may or may not be associated with lesions.[13] Microscopically, hemorrhage with associated mononuclear cell infiltration is widespread. There is degeneration and necrosis of various organs, including heart, skeletal muscle, and kidneys. Localized tissue necrosis can be severe, and is lethal depending on the species of *Sarcocystis*. Hepatic insufficiency caused by extensive necrosis of hepatocytes is likely to be lethal to varying lemmings infected with *S. rauschorum*.[90] In birds infected with *S. falcatula*, extensive, prolonged schizogony in pulmonary endothelium causes acute, fatal interstitial pneumonitis; the concomitant hepatitis could also be fatal.[38,91,92] The inflammatory response is caused by antigens or metabolic products derived from sporozoites or merozoites or from disrupted host cells.[13,14] Myositis reflects penetration of muscle cells by merozoites. Perhaps endothelial cells process and present antigens to T lymphocytes, which, after sensitization, cause destructive immune responses.[100] Edema and ascites fluid is caused by hypoproteinemia. Fever could be caused by schizont rupture and parasitemia, with release of pyrogens from schizonts or from indirect stimulation of prostaglandin release.[13] The mechanism of anemia is unknown; a combination of hemorrhage, sequestration of erythrocytes in the spleen, toxin production (i.e., parasite metabolites) or capillary extravasation could be involved. Acute sarcocystosis may cause immunosuppression, leading to increased susceptibility to other infectious agents.[93]

Acute sarcocystosis occurring during pregnancy of intermediate hosts can cause reproductive failure. Clinically ill hosts abort.[13] There is a discrepancy between experimental and natural infections which may reflect differences in strains of species of *Sarcocystis*. Unfortunately, the mechanism of *Sarcocystis*-induced abortion is unclear; many hypotheses have been proposed although none has been confirmed.

Chronic sarcocystosis can cause inappetence, slow growth (i.e., poor weight gain), hyperexcitability, hypersalivation, nasal discharge, hair loss, muscle atrophy, central nervous system signs, and death of intermediate hosts.[14] Grossly, the only significant lesion is serous atrophy of fat with mineralization.[13] Clinical chemistry and hematological parameters are normal. The pathogenesis of chronic disease is not well understood; there is markedly reduced inflammatory response with numerous sarcocysts in muscle and neural tissues. Hair loss could result from microthrombi or inadequate host nutrition.[13]

The mechanism of sarcocyst degeneration is unknown; perhaps sarcocysts spontaneously rupture and inflammation results, or apparently "normal" sarcocysts are surrounded, then invaded by macrophages and plasma cells.[94] The possible sequence of degeneration of sarcocysts through granuloma formation in naturally infected waterfowl was described by Wobeser and Cawthorn[95] who suggested cell-mediated immunity caused the granulomatous myositis. Other causes for granuloma formation include spontaneous lysis of old cysts, reinfection of an immune host, or drug-induced cyst destruction.[94] There is not likely to be a direct relationship between eosinophilic granulomatous myositis and recent *Sarcocystis*-infection of naive hosts.[94] The syndrome has not yet been reproduced in livestock experimentally infected with several species of *Sarcocystis*.[13] Recently Gajadhar et al.[96] demonstrated a previously undescribed species of *Sarcocystis* associated with eosinophilic myositis in cattle. Perhaps this parasite normally infects a wildlife host. We have observed degenerated sarcocysts of *S. hirsuta* in eosinophilic granulomas in cattle. Perhaps eosinophilic myositis in cattle is an abnormal response to sarcocyst degeneration.[97] *Sarcocystis*-specific IgE may be concentrated in degenerating sarcocysts in cattle with eosinophilic myositis, causing a Type I hypersensitivity response. Consequently, the bradyzoites must produce a stage-specific allergen.[97] Overall, sarcocystosis of intermediate hosts can be a highly significant disease.

VI. IMMUNOLOGICAL RESPONSES

The immunological responses of definitive hosts infected with *Sarcocystis* spp. have not been well studied. Apparently a Type-I immediate hypersensitivity response can develop.[84,85] Although definitive hosts can be repeatedly reinfected, the prepatent period is lengthened, the patent period is shortened and sporocyst output is reduced in dogs infected with *S. levinei*.[98] Whether definitive hosts can be immunized and sporocyst production inhibited is unknown.

In intermediate hosts, immunological studies have primarily been based on responses against antigens obtained from *Sarcocystis cruzi* bradyzoites.[13] We are now able to utilize antigens derived from sporozoites and merozoites of *S. cruzi*. Analysis of proteins and antigens of sporozoites, merozoites, and bradyzoites of *S. cruzi* indicated that they have similar and distinct proteins.[99,100] Interestingly, merozoites obtained *in vitro* have temporal variation in appearance of several proteins.[82] Potent (i.e., highly immunogenic) antigens may be present. Based on antisporozoite and antimerozoite monoclonal antibodies, there are epitopes on the surface of live merozoites which could be constitutent molecules within sporozoites and some sporozoite molecules may be precursors to surface or internal antigens of bradyzoites.[99,100] Antigen processing and presentation may occur either directly by schizont-infected endothelial cells or indirectly by adjacent endothelial cells.[100]

There are several studies which demonstrate a humoral response against several *Sarcocystis* spp. in cattle, mice, sheep, and swine which is detectable with several serological tests.[13] Elevation of IgM occurs first but its level decreases to low levels as sarcocysts mature. In contrast, levels of IgG peak as sarcocysts mature and remain at relatively high levels. There is no anamnestic response to challenge infections.[80] There is much cross reactivity (shared antigens) among species of *Sarcocystis*; antibody titers are much higher with antigen from homologous species of *Sarcocystis*.[80]

Cell-mediated immunity to *Sarcocystis*-infection in intermediate hosts develops rapidly. There is a persistent lymphocytic and macrophage response which moderates with increasing time after inoculation of sporocysts.[43] Acute sarcocystosis induces immunosuppression which can exacerbate concomitant infection with other disease agents.[93]

Protective immunity can be induced in intermediate hosts by immunizing with live or irradiated sporocysts of the homologous species of *Sarcocystis*.[101-103] The immunizing stage is not known although it is likely to be the sporozoite or first generation schizont.[13] Colostrum is not protective.[104] Protective immunity can persist with *S. capracanis* in the absence of sarcocysts.[93] We have also observed a persistent immune response to bradyzoites of *S. cruzi* in the absence of sarcocysts.

VII. EPIDEMIOLOGY

A given species of *Sarcocystis* tends to have a cosmopolitan distribution with relatively high prevalence, limited only by the geographical distributions of the appropriate intermediate and definitive hosts. Species of *Sarcocystis* have variable specificity for hosts: one species of *Sarcocystis* can develop in several species of intermediate host or in several species of definitive host. A given species of host can support development of several *Sarcocystis* spp.

Relatively low numbers of sporocysts are required to infect and consequently maintain infection within the populations of intermediate and definitive hosts.[26] Sporocysts, although susceptible to heat and drying, are relatively resistant to freezing and disinfectants.[22,25,105] Environmental stressors do not affect sporulation of *Sarcocystis* spp. as sporocysts are fully infectious when passed in feces. Transport hosts, including various invertebrates, can protect and disseminate sporocysts in the environment; this phenomenon can enhance transmission ecologically or can interfere with experimental protocols by contaminating the laboratory environment.[23,24] Finally, recent demonstration that species of *Sarcocystis* can utilize one host species as both intermediate and definitive hosts will ensure reevaluation of various host-parasite interactions.[19]

VIII. CONTROL

To prevent human disease associated with *S. suihominis* and *S. hominis*, education regarding proper meat preparation (i.e., thorough cooking or freezing) and handling is essential. Promiscuous defecation and proper disposal of human feces to avoid sporocyst contamination of livestock feed and water supplies will reduce transmission of the above parasites. No prophylactic or treatment regimes have been tested against human sarcocystosis.

To control sarcocystosis in intermediate hosts by management is impractical in wildlife populations. However, in livestock (including game ranching) operations, management can be highly effective in reducing the impact of sarcocystosis. Good hygiene is necessary to reduce sporocyst contamination of feed, bedding, and water supplies. Complete elimination of infection is probably impossible; the aim is to reduce epidemics (both acute and chronic) of sarcocystosis. Carnivores should be fed only dry food or cooked or frozen meat or meat by-products; they should not be allowed to scavenge on livestock carcasses. Carcasses and offal of livestock should be buried or incinerated. Although these management techniques are simple they may be difficult to implement.

Several anticoccidial drugs have been tested for preventive use or treatment against sarcocystosis in intermediate hosts; none has been tested in definitive hosts.[13] All these drugs are active against the schizogonous phase of *Sarcocystis*-infection and thus could be used to reduce the effects of acute sarcocystosis. Effective prophylactic drugs include amprolium, Bay® G1783, halofuginone, primaquine diphosphate, salinomycin, sulfaquinoxalin plus pyrimethamine, and Zoalene®.[106-112] The only effective chemotherapeutic agents are halofuginone and oxytetracycline; the latter may be prohibitively expensive.[108,110,112] Only sulfaquinoxaline plus pyrimethamine is effective against sarcocysts.[111] Extrapolation on drug use cannot be made from tests on one species of *Sarcocystis* to other species.[111] No farm management schemes have been devised or tested for the above drugs. Little research has been conducted on prophylactic or therapeutic drugs against acute or chronic sarcocystosis.

IX. PROSPECTS FOR THE FUTURE

The development of an *in vitro* system for growing numerous merozoites and for studying schizogony of various species of *Sarcocystis* facilitates investigation of immune and pathogenic mechanisms in sarcocystosis.[100] The system also aids development of diagnostic testing, chemoprophylaxis and chemotherapy, and immunoprophylaxis and immunotherapy. Additionally, basic questions on physiological and biochemical aspects of host-parasite interactions of *Sarcocystis* spp. can be examined in depth.

Overall, there is great need to develop better methods to detect *Sarcocystis*-induced disease and to facilitate chemo- or immunotherapy and prevention of *Sarcocystis* infections in various intermediate hosts. Accurate diagnosis of sarcocystosis during acute infection associated with schizogony may permit intervention with therapeutic drugs or help confirm etiologic involvement of *Sarcocystis* spp. in various disease states, including reproductive failure. The demonstration that cattle respond to antigens which are common to all three zoite forms and which are specific to each of the sporozoite, merozoite, and bradyzoite forms of *S. cruzi*, may allow the development of immunological tests to evaluate the immune response of cattle throughout infection or reinfection. The use of acute stage-specific antigens and measurement of specific IgM responses could maximize early serological detection of sarcocystosis. Immunoperoxidase stains are exceptionally sensitive for detecting specific antigens in tissue specimens[113] and may be useful for detecting *Sarcocystis* spp. in fetal and placental tissues, which would aid etiological diagnosis of abortion.

It is likely that DNA probes will be developed which will be specific for *Sarcocystis* spp. Because DNA probes are specific for parasite DNA, they have the potential of providing a more accurate assessment of a *Sarcocystis* infection.

Some anticoccidial drugs are effective against acute sarcocystosis, as indicated earlier. However, during parasitemia most viable merozoites are within phagocytic cells and could be protected from the efficacy of therapeutic drugs.[39,42] Liposomes (phospholipid vesicles) have tremendous potential as carriers of drugs to reticuloendothelial tissues and blood.[120,121] New antiparasitic drugs must be developed with known modes of action, especially those which exploit metabolic differences between parasite and host.[122]

Although intermediate hosts can develop protective immunity against *Sarcocystis* infection (or disease), the mechanism of induction of immunity is unknown. Immune therapeutic reagents could be developed with identification of antibodies and surface antigens of zoites of *Sarcocystis* spp. which prevent penetration of parasites into host cells.[123]

Presently, there are major efforts to expedite development of vaccines and devise implementation of vaccination programs against various infectious agents.[124] Ideally, a vaccine should be directed against the invasive stages (i.e., sporozoites); however, it could be difficult to achieve a persistent sterile immunity in intermediate hosts of apicomplexan parasites, including *Sarcocystis* spp.[125] Perhaps *Sarcocystis* zoites are susceptible to cell-mediated immune responses, especially as merozoites are within phagocytic cells for extended periods.[114]

Basic knowledge of host-parasite interactions is inadequate and there is a need to understand immune responses which are involved in the protection of the host.[115] In effective vaccines, immune responses are boosted to particular epitopes on parasite antigens.[116] Antigen mixtures must be purified to avoid immunologically-induced lesions. Vaccine efficacy, especially in chronic parasitic infections such as sarcocystosis, can be difficult to demonstrate, where there are no (or few) disease-specific diagnostic criteria.[117] There are many difficulties with antiparasitic drugs including residues in animal products used for human consumption and secondary toxicities to other organisms in the environment.[118] Anti-*Sarcocystis* livestock vaccines should incorporate antigens which are cheap to produce, easy to store and have long shelf-lives.[118] The vaccines should be effective in very young animals (i.e., less than 4 weeks old), require only one or two doses to induce lifelong immunity by induction of strong immunological memory and produce no side effects. Overall, antiparasitic vaccines may be potentially safer, cheaper and more efficacious than drugs; however, vaccines should be only one component of integrated control programs.[118,119] There is a need for a major research effort to develop suitable prophylactic and therapeutic immune reagents against species of *Sarcocystis*, primarily those which are important to the livestock industry.

REFERENCES

1. **Miescher, F.,** Uber eigenthumliche Schlauche in den Muskelen einer Hausmaus, *Berl. Verh. Naturforsch Ges.*, 5, 198, 1943.
2. **Labbe, A.,** Sporozoa, *Das Tierreich*, 5, 116, 1899.
3. **Kuhn, J.,** Untersuchungen uber die Trichinenkrankheit der Schweine, *Mittheilungen des Landwirtschaftlichen Institutes der Universitat du Halle*, 1865, 1.
4. **Railliet, A.,** *[Miescheria tenella,] Bull. Mem. Soc. Cent. Med. Vet.*, 40, 130, 1886.
5. **Rommel, M. and Heydorn, A. O.,** Beitrage zum Lebenszyklus der Sarkosporidien. III. *Isosopora hominis* (Railliet and Lucet, 1891) Wenyon, 1923, eine Dauerform der Sarkosporidien des Rindes und des Schweins, *Berl. Muench. Tieraertzl. Wochenschr.*, 85, 143, 1972.
6. **Rommel, M., Heydorn, A. O., and Gruber, F.,** Beitrage zum Lebenszyklus der Sarkosporidien. I. Die Sporozyste von *S. tenella* in den Fazes der Katzke, *Berl. Muench. Tieraertzl. Wochenschr.*, 85, 101, 1972.
7. **Heydorn, A. O., Gestrich, R., Mehlhorn, H., and Rommel, M.,** Proposal for a new nomenclature of *Sarcosporidia, Z. Parasitenkd.*, 48, 73, 1975.
8. **Frenkel, J. K., Heydorn A. O., Melhorn, H., and Rommel M.** Sarcocystinae *Nomina dubia* and available names, *Z. Parasitenkd.*, 58, 115, 1979.
9. **Levine, N. D. and Tadros, W.,** Named species and hosts of *Sarcocystis* (Protozoa: Apicomplexa: Sarcocystidae), *Syst. Parasitol.*, 2, 41, 1980.
10. **Levine, N. D.,** The taxonomy of *Sarcocystis* (Protozoa, Apicomplexa) species, *J. Parasitol.*, 72, 372, 1986.
11. **Melville, R. V.,** *Nomina dubia* and available names, *Z. Parasitenkd.*, 62, 105, 1980.
12. **Melville, R. V.,** Reply to Frenkel, Mehlhorn and Heydorn on protozoan *Nomina dubia, J. Parasitol.*, 70, 815, 1980.
13. **Dubey, J. P., Speer, C. A., and Fayer, R.,** *Sarcocystosis of Animals and Man*, CRC Press, Boca Raton, FL, 1989.
14. **Fayer, R. and Dubey, J. P.,** Bovine sarcocystosis, *Comp. Cont. Educ. Pract. Vet.*, 8, F130, 1986.
15. **Fayer, R.,** Gametogony of *Sarcocystis* sp. in cell culture, *Science*, 175, 65, 1972.
16. **Speer, C. A. and Dubey, J. P.,** Vascular phase of *Sarcocystis cruzi* cultured *in vitro, Can. J. Zool.*, 64, 209, 1986.
17. **Corliss, J. O.,** Classification and phylogeny of the Protista, *Acta Protozoolog.*, 1, 251, 1974.
18. **Pak, S. P.,** Occurrence of *Sarcocystis* in *Vulpes corsac* (in Russian), in X vsesoyuznaya Koniferentsiya po prirodenoi ochagovosti bolezei 9 — 1 oktyabrya 1979, Dushanbe Tezisy dokladov 1, Alma-Ata, U.S.S.R., *Nauka Kaz. S.S.R.*, 133, 1979.
19. **Matuschka, F.-R. and Bannert, B.,** Cannibalism and autotomy as predator-prey relationship for monoxenous Sarcosporidia, *Parasitol. Res.*, 74, 88, 1987.
20. **Matuschka, F.-R. and Bannert, B.,** Recognition of cyclic transmission of *Sarcocystis stehlinii* n. sp. in the Gran canarian giant lizard, *J. Parasitol.*, 75, 383, 1989.
21. **Box, E. D. and Duszynski, D. W.,** *Sarcocystis* of passerine birds: sexual stages in the opossum *(Didelphis virginiana), J. Wildl. Dis.*, 16, 209, 1980.
22. **Cawthorn, R. J., Gajadhar, A. A., and Brooks, R. J.,** Description of *Sarcocystis rauschorum* sp. n. (Protozoa: Sarcocystidae) with experimental cyclic transmission between varying lemmings (*Dicrostonyx richardsoni)* and snowy owls *(Nyctea scandiaca), Can. J. Zool.*, 62, 217, 1984.
23. **Markus, M. B.,** Flies as natural transport hosts of *Sarcocystis* and other coccidia, *J. Parasitol.*, 66, 361, 1980.
24. **Smith, D. D. and Frenkel, J. K.,** Cockroaches as vectors of *Sarcocystis muris* and of other coccidia in the laboratory, *J. Parasitol.*, 64, 315, 1978.
25. **Barutzki, D., Erber, M., and Boch, J.,** Moglichkeiten der Desinfektion bei Kokzidiose (Eimeria, Isospora, Toxoplasma, Sarcocystis), *Berl. Muench. Tieraerztl.* Wochenschr., 94, 451, 1981.
26. **Cawthorn, R. J., Wobeser, G. A., and Gajadhar, A. A.,** Description of *Sarcocystis campestris* sp. n. (Protozoa: Sarcocystidae): a parasite of the badger *Taxidea taxus* with experimental transmission to the Richardson's ground squirrel, *Spermophilus richardsonii, Can. J. Zool.*, 61, 370, 1983.
27. **Leek, R. G.,** Infection of sheep with frozen sporocysts of *Sarcocystis ovicanis, Proc. Helminthol. Soc. Wash.*, 53, 297, 1986.
28. **Box, E. D., Marchiondo, A. A., Duszynski, D. W., and Davis, C. P.,** Ultrastructure of *Sarcocystis* sporocysts from passerine birds and opossums: comments on classification of the genus *Isospora, J. Parasitol.*, 66, 68, 1980.
29. **Strohlein, D. A. and Prestwood, A. K.,** *In vitro* excystation and structure of *Sarcocystis suicanis* Erber, 1977, sporocysts, *J. Parasitol.*, 72, 711, 1986.
30. **Ferguson, D. J. P., Birch-Andersen, A., Siim, J. C., and Hutchison, W. M.,** Scanning electron microscopy of the oocyst and sporocyst of *Toxoplasma gondii, Acta Pathol. Microbiol. Immunol. Scand. Sect. B*, 90, 269, 1982.

31. **Speer, C. A., Hammond, D. M., Mahrt, J. L., and Roberts, W. L.,** Structure of the oocyst and sporocyst walls and excystation of sporozoites of *Isospora canis, J. Parasitol.*, 59, 35, 1973.
32. **Speer, C. A., Marchiondo, A. A., Duszynski, D. W., and File, S. K.,** Ultrastructure of the sporocyst wall during excystation of *Isospora endocallimici, J. Parasitol.*, 62, 894, 1976.
33. **Cawthorn, R. J., Reduker, D. W., Speer, C. A., and Dubey, J. P.,** *In vitro* excystation of *Sarcocystis capracanis, Sarcocystis cruzi* and *Sarcocystis tenella* (Apicomplexa), *J. Parasitol.*, 72, 880, 1986.
34. **Fayer, R. and Leek, R. G.,** Excystation of *Sarcocystis fusiformis* sporocysts from dogs, *Proc. Helminthol. Soc. Wash.*, 40, 294, 1973.
35. **Box, E. D.,** Recovery of *Sarcocystis* sporocysts from feces after oral administration, *Proc. Helminthol. Soc. Wash.*, 50, 348, 1983.
36. **Munday, B. L.,** Demonstration of viable *Sarcocystis* sporocysts in the faeces of a lamb dosed orally, *Vet. Parasitol.*, 17, 355, 1985.
37. **Dubey, J. P., Speer, C. A., Epling, G. P., and Blixt, J. A.,** *Sarcocystis capracanis*: development in goats, dogs, and coyotes, *Int. Goat Sheep Res.*, 2, 252, 1984.
38. **Smith, J. H., Neill, P. J. G., and Box, E. D.,** Pathogenesis of *Sarcocystis falcatula* (Apicomplexa: Sarcocystidae) in the budgerigar (*Melopsittacus undulatus*). III. Pathologic and quantitative parasitologic analysis of extrapulmonary disease, *J. Parasitol.*, 75, 270, 1989.
39. **Dubey, J. P.,** Development of ox-coyote cycle of *Sarcocystis cruzi, J. Protozool.*, 29, 591, 1982.
40. **Dubey, J. P.,** Quantitative parasitemia in calves fed *Sarcocystis cruzi* sporocysts from coyotes, *Am. J. Vet. Res.*, 43, 1085, 1982.
41. **Cawthorn, R. J. and Brooks, R. J.,** Histological observations on precystic merogony and metrocyte formation of *Sarcocystis rauschorum* (Apicomplexa: Sarcocystidae) in varying lemmings, *Dicrostonyx richardsoni, Can. J. Zool.*, 63, 2907, 1985.
42. **Speer, C. A. and Dubey, J. P.,** Ultrastructure of *in vivo* lysis of *Sarcocystis cruzi* merozoites, *J. Parasitol.*, 67, 961, 1981.
43. **Dubey, J. P., Speer, C. A., and Epling, G.,** Sarcocystosis in newborn calves fed *Sarcocystis cruzi* sporocysts from coyotes, *Am. J. Vet. Res.*, 43, 2147, 1982.
44. **Dubey, J. P., Kistner, T. P., and Callis, G.,** Development of *Sarcocystis* in mule deer transmitted through dogs and coyotes, *Can. J. Zool.*, 61, 2904, 1983.
45. **Fayer, R.,** Multiplication of *Sarcocystis bovicanis* in the bovine bloodstream, *J. Parasitol.*, 65, 980, 1979.
46. **Friesen, D. L., Cawthorn, R. J., Speer, C. A., and Brooks, R. J.,** Ultrastructural development of the sarcocyst of *Sarcocystis rauschorum* (Apicomplexa: Sarcocystidae) in the varying lemming *Dicrostonyx richardsoni, J. Parasitol.*, 75, 422, 1989.
47. **Hafner, U. and Frank, W.,** Morphological studies on the muscle cysts of *Sarcocystis dirumpens* (Hoare 1933) Hafner and Matuschka 1984 in several host species revealing endopolygeny in metrocytes, *Parasitol. Res.*, 72, 453, 1986.
48. **Entzeroth, R.,** Electron microscope study of host-parasite interactions of *Sarcocystis muris* (Protozoa, Coccidia) in tissue culture and *in vivo, Parasitol. Res.*, 70, 131, 1984.
49. **Entzeroth, R.,** Invasion and early development of *Sarcocystis muris* (Apicomplexa, Sarcocystidae) in tissue cultures, *J. Protozool.*, 32, 446, 1985.
50. **Dubey, J. P., Leek, R. G., and Fayer, R.,** Prevalence, transmission, and pathogenicity of *Sarcocystis gigantea* of sheep, *J. Am. Vet. Assoc.*, 188, 151, 1986.
51. **Munday, B. L. and Obendorf, D. L.,** Development and growth of *Sarcocystis gigantea* in experimentally-infected sheep, *Vet. Parasitol.*, 15, 203, 1984.
52. **Barrows, P. L., Prestwood, A. K., Adams, D. D., and Dykstra, M. J.,** Development of *Sarcocystis suicanis* Erber, 1977 in the pig, *J. Parasitol.*, 68, 674, 1982.
53. **Barrows, P. L., Prestwood, A. K., and Green, C. E.,** Experimental *Sarcocystis suicanis* infections: disease in growing pigs, *Am. J. Vet. Res.*, 43, 1409, 1982.
54. **Bledsoe, B.,** *Sarcocystis idahoensis* sp. n. in deer mice *Peromyscus maniculatus* (Wagner) and gopher snakes *Pituophis melanoleucus* (Daudin), *J. Protozool.*, 27, 93, 1980.
55. **Cornelissen, A. W. C. A., Overdulve, J. P., and Van Der Ploeg, M.,** Cytochemical studies on nuclear DNA of four eucoccidian parasites, *Isospora (Toxoplasma) gondii, Eimeria tenella, Sarcocystis cruzi* and *Plasmodium berghei, Parasitology*, 88, 25, 1984.
56. **Mehlhorn, H. and Heydorn, A. O.,** Electron microscopical study on gamogony of *Sarcocystis suihominis* in human tissue cultures, *Parasitol. Res.*, 58, 97, 1979.
57. **Sheffield, H. G. and Fayer, R.,** Fertilization in the coccidia: fusion of *Sarcocystis bovicanis* gametes, *Proc. Helminthol. Soc. Wash.*, 47, 118, 1980.
58. **Bledsoe, B.,** Sporogony of *Sarcocystis idahoensis* in the gopher snake, *Pituophis melanoleucus* (Daudin), *J. Parasitol.*, 65, 875, 1979.
59. **Cawthorn, R. J. and Brooks, R. J.,** Light microscopical observations on sporogony of *Sarcocystis rauschorum* (Protozoa: Sarcocystidae) in snowy owls *(Nyctea scandiaca), Can. J. Zool.*, 63, 1455, 1985.
60. **Fayer, R.,** Development of *Sarcocystis fusiformis* in the small intestine of the dog, *J. Parasitol.*, 60, 660, 1974.

61. **Shimura, K., Ito, S., and Tsunoda, K.,** Sporocysts of *Sarcocystis cruzi* in mesenteric lymph nodes of dogs, *Natl. Inst. Anim. Health Q. (Jpn.),* 21, 186, 1981.
62. **Barta, J.,** Phylogenetic analysis of the class Sporozoea (Phylum Apicomplexa Levine, 1970): evidence for the independent evolution of heteroxenous life cycles, *J. Parasitol.*, 75, 195, 1989.
63. **Tadros, W. and Laarman, J. J.,** Current concepts on the biology, evolution and taxonomy of tissue cyst-forming eimeriid coccidia, *Adv. Parasitol.*, 20, 293, 1982.
64. **Ford, G. E., Fayer, R., Adams, M., O'Donoghue, P. J., Dubey, J. P., and Baverstock, P. R.,** Genetic characterisation by isoenzyme markers of North American and Australian isolates of species of *Sarcocystis* (Protozoa: Apicomplexa) from mice, sheep, goats and cattle, *Syst. Parasitol.*, 9, 163, 1987.
65. **Johnson, A. M., Illana, S., Hakendorf, P., and Baverstock, P. R.,** Phylogenetic relationships of the apicomplexan protist *Sarcocystis* as determined by small subunit ribosomal RNA comparison, *J. Parasitol.*, 74, 847, 1988.
66. **Cerna, Z. and Kvasnovska, Z.,** Life cycle involving bird-bird relation in *Sarcocystis* coccidia with the description of *Sarcocystis accipitris* sp. n., *Folia Parasitol. (Prague),* 33, 305, 1986.
67. **Lindsay, D. S., Ambrus, S. I., and Blagburn, B. L.,** *Frenkelia* sp.-like infection in the small intestine of a red-tailed hawk, *J. Wildl. Dis.*, 23, 677, 1987.
68. **Tuggle, B. N. and Schmeling, S. K.,** Parasites of the bald eagle (*Haliaeetus leucocephalus*) of North America, *J. Wildl. Dis.*, 18, 501, 1982.
69. **Smith, T. S. and Herbert, I. V.,** Experimental microcyst *Sarcocystis* infection in lambs: serology and immunohistochemistry, *Vet. Rec.*, 119, 547, 1986.
70. **Dubey, J. P., Carpenter, J. L., Speer, C. A., Topper, M. J., and Uggla, A.,** Newly recognized fatal protozoan disease of dogs, *J. Am. Vet. Assoc.*, 192, 1269, 1988.
71. **Dubey, J. P. and Lindsay, D. S.,** Fatal *Neospora caninum* infection in kittens, *J. Parasitol.*, 75, 148, 1989.
72. **Dubey, J. P., Leathers, C. W., and Lindsay, D. S.,** *Neospora caninum*-like protozoon associated with fatal myelitis in newborn calves, *J. Parasitol.*, 75, 146, 1989.
73. **Leek, R. G. and Fayer, R.,** Survival of sporocysts of *Sarcocystis* in various media, *Proc. Helminthol. Soc. Wash.*, 46, 151, 1979.
74. **Cawthorn, R. J., Reduker, D. W., Speer, C. A., and Dubey, J. P.,** *In vitro* excystation of *Sarcocystis capracanis, Sarcocystis cruzi* and *Sarcocystis tenella* (Apicomplexa), *J. Parasitol.*, 72, 880, 1986.
75. **Speer, C. A., Cawthorn, R. J., and Dubey, J. P.,** *In vitro* cultivation of the vascular phase of *Sarcocystis capracanis* and *Sarcocystis tenella, J. Protozool.*, 33, 486, 1986.
76. **Collins, G. H., Charleston, W. A. G., and Wiens, B. G.,** Studies on *Sarcocystis* species VI: a comparison of three methods for the detection of *Sarcocystis* species in muscle, *N.Z. Vet. J.*, 28, 173, 1980.
77. **Farooqui, A. A., Adams, D. D., Hanson, W. L., and Prestwood, A. K.,** Studies on the enzymes of *Sarcocystis suicanis*: purification and characterization of an acid phosphatase, *J. Parasitol.*, 73, 681, 1987.
78. **Mutetwa, S. M. and James, E. R.,** Low temperature preservation of *Plasmodium* spp., *Parasitology,* 90, 589, 1985.
79. **Morii, T., Matsui, T., Iijima, T., Abe, N., Yoshizawa, A., Chikatsune, M., and Watanabe, M.,** Cryopreservation of *Leucocytozoon caulleryi* sporozoites, *J. Protozool.*, 35, 356, 1988.
80. **Gasbarre, L. C., Suter, P., Fayer, R.,** Humoral and cellular immune responses in cattle and sheep inoculated with *Sarcocystis, Am. J. Vet. Res.*, 45, 1592, 1984.
81. **Fayer, R.,** *Sarcocystis*: development in cultured avian and mammalian cells, *Science,* 168, 1104, 1970.
82. **Speer, C. A., Whitmire, W. M., Reduker, D. W., and Dubey, J. P.,** *In vitro* cultivation of meronts of *Sarcocystis cruzi, J. Parasitol.*, 72, 677, 1986.
83. **Becker, B., Mehlhorn, H., and Heydorn, A. O.,** Light and electron microscopic studies on gamogony and sporogony of 5 *Sarcocystis* species *in vivo* and in tissue cultures, *Zbl. Bakt. Hyg., I. Abt. Orig. A,* 244, 394, 1979.
84. **Fayer, R., Heydorn, A. O., Johnson, A. J., and Leek, R. G.,** Transmission of *Sarcocystis suihominis* from humans to swine to nonhuman primates *(Pan troglodytes, Macaca mulatta, Macaca irus), Parasitol. Res.*, 59, 15, 1979.
85. **Speer, C. A., Pond, D. B., and Ernst, J. V.,** Development of *Sarcocystis hemionilantrantis* Hudkins and Kistner, 1977 in the small intestine of coyotes, *Proc. Helminthol. Soc. Wash.*, 47, 106, 1980.
86. **Heydorn, A. O.,** Sarcosporidia-infected meat as a possible source of human disease, *Arch. Lebensmittelhyg.*, 28, 27, 1977.
87. **Bunyaratvej, S., Bunyawongwiroj, P., and Nitiyanant, P.,** Human intestinal sarcosporidiosis: report of six cases, *Am. J. Trop. Med. Hyg.*, 31, 36, 1982.
88. **Jain, P. C. and Shah, H. L.,** *Sarcocystis hominis* in cattle in Madhya Pradesh and its public health importance, *Indian Vet. J.*, 6, 650, 1987.
89. **Murrell, K. D., Fayer, R., and Dubey, J. P.,** Parasitic organisms, *Adv. Meat Res.*, 2, 311, 1986.
90. **Stackhouse, L. L., Cawthorn, R. J., and Brooks, R. J.,** Pathogenesis of infection with *Sarcocystis rauschorum* (Apicomplexa) in experimentally infected varying lemmings *(Dicrostonyx richardsoni), J. Wildl. Dis.*, 23, 566, 1987.

91. **Smith, J. H., Meier, J. L., Neill, P. J. G., and Box, E. D.,** Pathogenesis of *Sarcocystis falcatula* in the budgerigar. I. Early pulmonary schizogony, *Lab. Invest.*, 56, 60, 1987.
92. **Smith, J. H., Meier, J. L., Neill, P. J. G., and Box, E. D.,** Pathogenesis of *Sarcocystis falcatula* in the budgerigar. II. Pulmonary pathology, *Lab. Invest.*, 56, 72, 1987.
93. **Dubey, J. P.,** Immunity to sarcocystosis: modification of intestinal coccidiosis, and disappearance of sarcocysts in dairy goats, *Vet. Parasitol.*, 13, 23, 1983.
94. **O'Toole, D.,** Experimental ovine sarcocystosis: sequential ultrastructural pathology in skeletal muscle, *J. Comp. Pathol.*, 97, 51, 1987.
95. **Wobeser, G. and Cawthorn, R. J.,** Granulomatous myositis in association with *Sarcocystis* sp. infection in wild ducks, *Avian Dis.*, 26, 412, 1981.
96. **Gajadhar, A. A., Yates, W. D. G., and Allen, J. R.,** Association of eosinophilic myositis with an unusual species of *Sarcocystis* in a beef cow, *Can. J. Vet. Res.*, 51, 373, 1987.
97. **Granstrom, D. E., Ridley, R. K., Baoan, Y., Gershwin, L. J., Nesbitt, P. M., and Wempe, L. A.,** Type-I hypersensitivity as a component of eosinophilic myositis (muscular sarcocystosis) in cattle, *Am. J. Vet. Res.*, 50, 571, 1989.
98. **Srivastava, P. S., Sinha, S. R. P., Juyal, P. D., and Saha, A. K.,** Host resistance and faecal sporocyst excretion in dogs exposed to repeated infection with *Sarcocystis levinei, Vet. Res. Commun.*, 11, 185, 1987.
99. **Burgess, D. E., Speer, C. A., and Reduker, D. W.,** Identification of antigens of *Sarcocystis cruzi* sporozoites, merozoites, and bradyzoites with monoclonal antibodies, *J. Parasitol.*, 74, 828, 1988.
100. **Speer, C. A. and Burgess, D. E.,** *In vitro* development and antigen analysis of *Sarcocystis, Parasitol. Today*, 4, 46, 1988.
101. **Dubey, J. P.,** Development of immunity to sarcocystosis in dairy goats, *Am. J. Vet. Res.*, 42, 800, 1981.
102. **Fayer, R. and Dubey, J. P.,** Protective immunity against clinical sarcocystosis in cattle, *Vet. Parasitol.*, 15, 187, 1984.
103. **Leier, H., Boch, J., and Erber, M.,** Investigations on immunizing mice against *Sarcocystis muris* with irradiated sporocysts, *Berl. Muench. Tierarztl. Wschr.*, 95, 231, 1982.
104. **Fayer, R., Johnson, A. J., and Lunde, M.,** Abortion and other signs of disease in cows experimentally infected with *Sarcocystis fusiformis* from dogs, *J. Infect. Dis.*, 134, 624, 1976.
105. **Heydorn, A. O.,** Zur Widerstandfahigkeit von *Sarcocystis bovicanis* Sporozysten, *Berl. Muench. Tieraerztl. Wochenschr.*, 93, 267, 1980.
106. **Fayer, R. and Dubey, J. P.,** Protective immunity against clinical sarcocystosis in cattle, *Vet. Parasitol.*, 15, 187, 1984.
107. **Fayer, R. and Johnson, A. J.,** Effects of amprolium on acute sarcocystosis in experimentally infected calves, *J. Parasitol.*, 61, 932, 1975.
108. **Heydorn, A. O., Haralambidis, S., and Matuschka, F. R.,** Zur Chemoprophylaxe und Therapie der aukten Sarkosporidiose, *Berl. Muench. Tieraerztl. Wochenschr.*, 94, 229, 1981.
109. **Leek, R. G. and Fayer, R.,** Amprolium for prophylaxis of ovine *Sarcocystis, J. Parasitol.*, 66, 100, 1980.
110. **Leek, R. G. and Fayer, R.,** Experimental *Sarcocystis ovicanis* infection in lambs: salinomycin chemoprophylaxis and protective immunity, *J. Parasitol.*, 69, 271, 1983.
111. **Rommel, M. R., Schwerdtfeger, A., and Blewaska, S.,** The *Sarcocystis muris*-infection as a model for research on the chemotherapy of acute sarcocystosis of domestic animals, *Zbl. Bakt. Hyg. I. Abt. Orig. A*, 250, 268, 1981.
112. **Voigt, W. P. and Heydorn, A. O.,** Chemotherapy of sarcosporidiosis and theileriosis in domestic animals, *Zentralbl. Bakteriol. Parasitenkd. Infektionstr. Hyg. I. Abt. Orig. Reihe A*, 250, 256, 1986.
113. **Larsson, L.-I.,** *Immunocytochemistry: Theory and Practice*, CRC Press, Boca Raton, FL, 1988, 272.
114. **James, S. L. and Scott, P.,** Induction of cell-mediated immunity as a strategy for vaccination against parasites, in *The Biology of Parasitism*, Englund, P. T. and Sher, A., Eds., Alan R. Liss, New York, 1988, 249.
115. **Schild, G. C.,** Strategies for viral vaccine: scientific challenge and regulatory aspects, in *Symp. Br. Soc. Parasitol.*, McLaren, D. J., Ed., Cambridge University Press, Cambridge, 1989, S13.
116. **Roitt, I. M.,** Basic concepts and new aspects of vaccine development, *Symp. Br. Soc. Parasitol.*, McLaren, D. J., Ed., Cambridge University Press, Cambridge, 1989, S7.
117. **Mitchell, G. F.,** The Wellcome Trust Lecture. Problems specific to parasite vaccines, in *Symp. Br. Soc. Parasitol.*, McLaren, D. J., Ed., Cambridge University Press, Cambridge, 1989, S19.
118. **Morrison, W. I.,** Immunological control of ticks and tick-borne diseases, in *Symp. Br. Soc. Parasitol.*, McLaren, D. J., Ed., Cambridge University Press, Cambridge, 1989, S19.
119. **Gutteridge, W. E.,** Parasite vaccines versus anti-parasite drugs; rivals or running mates, in *Symp. Br. Soc. Parasitol.*, McLaren, D. J., Ed., Cambridge University Press, Cambridge, 1989, S87.
120. **Coune, A.,** Liposomes as drug delivery system in the treatment of infectious diseases. Potential applications and clinical experience, *Infection*, 16, 141, 1988.

121. **Swenson, C. E., Popescu, M. C., and Ginsberg, R. S.,** Preparation and use of liposomes in the treatment of microbial infections, *Crit. Rev. Microbiol.*, 15, Suppl. 1, S1, 1988.
122. **Hart, D., Langridge, A., Barlow, D., and Sutton, B.,** Antiparasitic drug design, *Parasitol. Today*, 5, 117, 1989.
123. **Whitmire, W. M., Kyle, J. E., Speer, C. A., and Burgess, D. E.,** Inhibition of penetration of cultured cells by *Eimeria bovis* sporozoites by monoclonal immunoglobulin G antibodies against the parasite surface protein P20, *Infect. Immun.*, 56, 2538, 1988.
124. **McLaren, D. J.,** Vaccines and vaccination strategies, in *Symp. Br. Soc. Parasitol.*, McLaren, D. J., Ed., Cambridge University Press, Cambridge, 1989, S1.
125. **Sher, A.,** Vaccination against parasites: special problems imposed by the adaptation of parasitic organisms to the host immune response, in *The Biology of Parasitism*, Englund, P. T. and Sher, A., Eds., Alan R. Liss, New York, 1988, 169.

Chapter 7

TOXOPLASMA: BIOLOGY, PATHOLOGY, IMMUNOLOGY, AND TREATMENT

Alan M. Johnson

TABLE OF CONTENTS

I. INTRODUCTION

Toxoplasma gondii is a ubiquitous protistan parasite of animals. Work on this parasite, and toxoplasmosis, the disease it causes, has been the subject of thousands of publications. A two-volume bibliography published in 1983[1] contained 12,476 references. Therefore, the number of publications on the biology, pathology, and immunology of *T. gondii* published at the time of writing this chapter must be in the tens of thousands. It is impossible to distill all available information into this chapter, accordingly this chapter concentrates on work in the major areas of interest of *T. gondii* research published during the last 5 years and thus remains an enormous task. Research carried out prior to 1982 on pathology and immunology was well covered earlier.[2]

Other older reference works and several recent books, book chapters, and reviews on specific areas of *T. gondii* research will be referenced where appropriate. In addition, some of the material published well before 1982 will be covered here in order to place the work discussed in this chapter in perspective. Earlier work is used as a basis from which to suggest the directions in which future research on *T. gondii* and toxoplasmosis is likely to proceed.

II. LIFE CYCLE

Although the parasite was first identified and named in 1908,[3] it was not until 1970 that its full life cycle was completed.[4-8] The main stages and routes of transmission in the life cycle of *T. gondii* are outlined in Figure 1. The life cycle can be divided into five main stages: schizogony, gametogony, oocyst formation, the tachyzoite, and the bradyzoite (tissue cyst stage).

Schizogony is an asexual multiplication stage which occurs only in Felidae.[9,10] Ingestion of tissue cysts initiates the process which occurs in the intestinal epithelium.[4-8]

Bradyzoites, the resting stage of the parasite present in tissue cysts, are relatively resistant to digestive juices although the cyst wall is destroyed immediately.[11] After bradyzoites enter the feline intestinal cells they undergo several development stages designated "Types" A—E. Knowledge on the development of merozoite types occurring during schizogony was elucidated by Dubey and Frenkel[12] and Dubey.[13]

Gametocytes occur in the surface epithelium of the entire length of the small intestine, but more commonly in the ileum. They can be seen from 3 to 15 days after tissue cyst ingestion. Fusion of a male and female gamete leads to oocyst formation which is then detached from the intestinal epithelium and discharged. Oocyst formation occurs from 3 to 5 days after tissue cyst ingestion.[14]

The length of time required for sporulation of the oocyst is from 1 to 5 days depending on several environmental factors such as temperature and oxygen tension.[14,15] The sporulated oocyst contains two sporocysts each containing four sporozoites. The oocyst can survive in an inimical environment under a wide range of adverse conditions. They remain infective for at least 410 days at 4°C but lose their viability after 30 days at 37°C. Unprotected oocysts, kept outside and exposed to direct sunlight, remained infective for 46 days. If the oocysts were covered when kept outdoors they remained infectious for at least 410 days.[16] The longevity of the infectious oocyst under conditions similar to those in which cats bury their feces is of major importance in the transmission of toxoplasmosis. Herbivores such as sheep and goats can ingest the oocyst directly and humans may contact the infectious stage while cleaning out cats' litter bins or when playing in sand pits.[17]

The tachyzoite is the proliferative stage of the parasite. Within a few hours of ingestion of tissue cysts tachyzoites are present in the lamina propria of the small intestine. Later they can be found in mesenteric lymph nodes, liver, lungs, and brain, but they slowly disappear from these tissues by the 14th day. In nonfeline hosts tachyzoites constitute the first stage of infection after ingestion of an oocyst or tissue cyst.

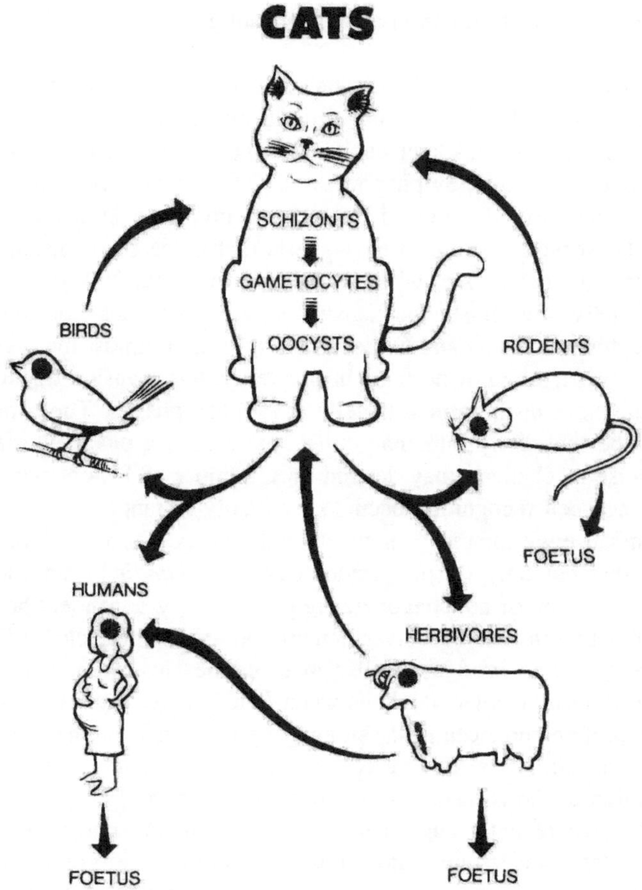

FIGURE 1. Major pathways of the life cycle of *Toxoplasma gondii*.

Unlike other protistan parasites such as *Leishmania* and *Trypanosoma*,[18,19] the surface of *T. gondii* tachyzoites does not contain a significant amount of glycoconjugates. Fluorescence staining[20] failed to find binding of the lectins concanavalin A, wheat germ agglutinin, or soybean agglutinin, but more sensitive techniques have found that concanavalin A does bind to the insoluble tachyzoite fraction,[21] and that although the number of binding sites for concanavalin A was low,[22] the affinity constant of the binding was of the same order of magnitude as that determined for mammalian cells.

It has been proposed that a wheat germ agglutinin-like lectin on the *Trypanosoma cruzi* trypomastigote surface recognizes and attaches to an *N*-acetylglucosamine-containing receptor on the host cell surface prior to infection by the parasite.[23] However, because of the lack of lectins on the tachyzoite surface, it is likely that an alternative mechanism of recognition of host cells and *T. gondii* is involved. Blackwell et al.[24] concluded that the presence of the type 3 complement receptor (CR3) and the lectin-like mannosyl/fucosyl receptor on macrophages was necessary for recognition with *Leishmania donovani*. Mosser and Edelson[25] also confirmed that the receptor for C3bi was involved in the recognition of *Leishmania major* promastigotes with murine macrophages. Therefore, we attempted to see whether the CR3 receptor was also involved in the recognition of *T. gondii* and a murine macrophage cell line.[26] A range of MAb which bind to the murine C3 receptor, a MAb that binds to the Fc receptor, and a MAb that binds to the Ia antigen, were added to tissue culture before the addition of RH strain tachyzoites. Inhibition of *T. gondii* uptake was measured by a ^{3}H-uracil assay.[27] The results suggested that *T. gondii* binds to the C3 receptor or the structurally related lymphocyte-function-associated antigen on the surface of the murine macrophages.

Whether tachyzoites enter host cells by phagocytosis, active invasion, or both, has been controversial. Early work using light[28] and electron microscopy[29,30] concluded that active penetration was involved, whereas later chemical[31] and electron microscopy studies[32] suggested that tachyzoites enter cells by phagocytosis. Others[33] have concluded that *T. gondii* actively enters host cells by complicated interactions involving both the parasite and the host cell, and the latest work[34] in this area suggests that at least two mechanisms are used by *T. gondii* to penetrate cells. One, which accounted for about 20% of the infected cells, appeared to require active invasion by the tachyzoites, the other was a process of phagocytosis or induced phagocytosis. Nichols and O'Connor[35] had found almost the reverse. Although a few tachyzoites were engulfed by normal phagocytes, active invasion was the predominant means by which the parasites became intracellular in macrophages. The consensus of these studies appears to be that tachyzoites can enter host cells via phagocytosis and/or active invasion. The precise mechanism may depend on a number of factors such as the type of host cell, whether cell-cell recognition occurs specifically or randomly, the age of the host cells or their chemical environment,[36] or whether the tachyzoite is still viable. Kimata et al.[37] have shown that not only do phagocytes engulf *T. gondii* by an energy-dependent process, but active invasion of nonphagocytes requires energy-dependent host cell activity.

The process of active invasion involves penetration enhancing factor (PEF)[38] which has been described as an acidic protein, similar in properties to an enzyme. Lycke et al.[39] extracted a protein with a molecular weight between 70 to 150 kDa from RH strain tachyzoites and found that the protein enhanced host cell entry by the parasite. They suggested that the PEF may reside in the rhoptries. During cell-cell contact the parasite secretes PEF which aids in the penetration of the parasite into the host cell (Section III.C).

Saffer et al.[40] have recently suggested a role for parasite phospholipase A2 in the penetration of RH strain tachyzoites into human fibroblast monolayers. Locksley et al.[41] found that the generation of a substance they termed compound X by murine peritoneal macrophages correlated with the extent of phagocytosis of RH strain tachyzoites and with intracellular survival of the parasites. Prior antibody coating of the parasite or activation of macrophages, either of which stopped the intracellular tachyzoites from multiplying, were associated with the production of leukotriene D4 and not compound X. They suggested that the lack of leukotriene production by macrophages in this situation may explain the relative lack of neutrophil inflammatory response in toxoplasmosis.

Hughes[42] has recently reviewed the literature on oxidative killing of intracellular *T. gondii* mediated by macrophages, and suggests that more than one pathway is involved in parasite destruction. There is also evidence to suggest that oxygen-independent processes may be involved in parasite killing.[43,44] Of course, not all parasites are usually killed. Some appear to be able to switch off the oxidative burst,[43] and not only survive in the phagocyte, but reproduce and multiply.

Tachyzoites occur within vacuoles in their host cells with a definite space between the parasite and vacuole wall.[45] The parasites can develop within a multitude of nucleated cell types and can even multiply in enucleate cells.[46-48] However, tachyzoites have not been found to multiply in cell-free media and cannot survive in cell-free tissue culture medium at 37°C for more than 6 h.[49] It is this obligately intracellular nature of *T. gondii* that has most retarded research on the parasite.

From the site of primary infection in the gut tachyzoites spread from cell to cell and are disseminated by the lymphatics and bloodstream. Multiplication of tachyzoites within cells is by a process of internal budding called endodyogeny[50] or repeated endodyogeny[51] although it has been reported that the virulent RH strain undergoes atypical extracellular fission possibly as a result of population pressures.[52] During normal multiplication as many as eight or more parasites may accumulate in the host cell before it disintegrates and new cells are infected. This is the usual mode of spread of the parasite in the acute stages of infection.

The fluidity in the cell membrane lipid layers of *T. gondii* tachyzoites is different to that of a range of other cell systems, due to differences in the lipid composition.[53] *T. gondii* has a low cholesterol/phospholipid ratio, many unsaturated fatty acid chains, and high amounts of phosphatidyl choline and low amounts of sphingomyelin.

Tissue cyst formation is characterized by a slowing of the multiplication of the tachyzoite leading to an accumulation of parasites within the large intracellular cyst-like structures.[54] Multiplication of the encysted organisms is by endodyogeny.[55,56] These quiescent parasites are called bradyzoites. The structure of the bradyzoite differs from that of the tachyzoite in that there is a large accumulation of amylopectin granules in the cytoplasm of the former,[57] and the nucleus is subterminal.

The factors influencing tissue cyst formation are still unresolved. Although tissue cysts usually appear at the same time as the humoral response becomes detectable[58,59] it has been thought unlikely that immune mechanisms are involved. Cyst production has been demonstrated in immunosuppressed mice and humans, and in tissue cultures where immune mechanisms could not be operative.[60-62] However, Brinkmann et al.[63] have recently compared parasite-specific antibody titer, antigen-specific T-cell response, and the number of tissue cysts developing in the brains of immunocompetent mice, with the identical parameters developing in B-cell-deficient mice bearing the $H\text{-}2^k$ haplotype. The results suggested that elevated humoral antibody responses favored tissue cyst formation in the brain, and that cell-mediated immunity may also control tissue cyst formation in the brain. Clearly, more in-depth studies need to be carried out on this very important transition stage in the parasite's life cycle. Tachyzoites are susceptible to antibodies and treatment with a range of drugs (see Section VIII), whereas the bradyzoites inside tissue cysts are much more resistant to these effects. Therefore, if we could prevent the tachyzoite transition to bradyzoite stage, we would be able to eliminate the parasite from the host. In fact, the transition from tachyzoite to bradyzoite is not cyclically innate in the parasite and it is possible to maintain the infection in the tachyzoite stage indefinitely by rapid, continuous passage.[64]

The size of individual tissue cysts varies widely. They may contain as few as two or as many as several hundred bradyzoites depending upon the duration of the infection.[54] It has been considered that the tissue cyst wall contains parasitic material and is independent of the host-cell membrane.[54,55,65] Very recently, Sims et al.[66] used reduced silver salts and protargol silver staining in a light and electron microscopy study of tissue cysts. The results suggested that the tissue cyst wall was made up, at least partly, of neuronal cytoskeleton components. A layer of neurofibrillae enclosed within the host cell membrane separated the tissue cysts from the extracellular compartment. The most likely hypothesis regarding the formation of the tissue cyst, is that of Matsubayashi and Akao[67] who suggest that it is produced by the accumulation of *T. gondii* substances on the limiting membrane of the vacuole in which the parasites are located. Although antibody to the parasite enters the host cell, it does not pass through the limiting membrane of the vacuole and hence encysted bradyzoites are unaffected.[67,68]

III. ANTIGENIC STRUCTURE

Since the last major reviews of the antigenic structure of *T. gondii*,[69,70] dozens of papers have been published on this topic. This section reviews the major findings on the antigens of *T. gondii*. The headings used are more for convenience rather than strict scientific definition, as we still have little definitive information on the cellular localization of many of the antigens of the parasite.

TABLE 1
Apparent Molecular Weights of the Major Surface Antigens of *Toxoplasma gondii* RH Strain[a] Tachyzoites

Apparent molecular weights (kDa)	Ref.
Reduced	
43, 35, 27, 14	72
43, 35, 22, 14	73
41,[b] 35,[b] 30, 22[b]	74
41, 35, 32, 23, 21	75
43, 35, 30, 22	76
38, 35, 28, 26	77
Nonreduced	
33,[b] 30,[b] 27, 18[b]	74
38, 31, 23, 21	75

[a] The RH strain is the most commonly used *T. gondii* strain. It was isolated in 1941 from a fatal case of human encephalitis,[71] and has been passaged in mice every 2 to 3 days, almost continuously since.
[b] These weights not stated in the paper, just read from the autoradiographs in the paper.

A. SURFACE ANTIGENS

Despite, or perhaps because of, the burgeoning literature on the surface antigens of the tachyzoite of *T. gondii*, there is still confusion over this aspect of the parasite's biology. This is caused to some extent by an over confidence in the ability of SDS-PAGE to consistently identify apparent molecular weights to within 1 kDa. In addition, the increasing likelihood that *T. gondii* has strain-specific antigenic differences (see Section III.E), and that at least some of its major antigenic proteins are posttranslationally modified (see this section), means that studies done in different laboratories are unlikely to obtain exactly the same apparent molecular weights for these surface antigens. These facts are often overlooked in discussing differences in results. Despite these limitations a summary of the apparent molecular weights of the major surface antigens of RH strain tachyzoites are given in Table 1.

For the reasons given above it is not appropriate to generalize on the apparent molecular weights of the surface antigens of *T. gondii* because it has led to much confusion regarding the relationship between a 35 and a 30 kDa antigen.[78-81] However, the most recent generalizations[75] (P43, P35, P30, P23, P22) will be used here for the sake of brevity.

Most research has been carried out on the antigen designated P30. This antigen makes up about 5% of the total protein of the RH strain tachyzoite,[74] and has been considered to be not significantly glycosylated,[82] and to contain an immunodominant region with repetitive epitopes.[83,84] However, the gene that encodes P30 has recently been cloned and sequenced[85] (see Section IV.A), and it does not contain repetition in its linear primary sequence. Therefore, aggregation of P30, and not repeat epitopes, appears to be the explanation for the previous results.[83,84] The sequence also reveals a potential N-linked glycosylation site, so previous studies not detecting glycosylation of the protein may not have been sensitive enough.

The P30 gene has a coding capacity for 34.7 kDa of primary translation product, and Burg et al.[85] suggest that this is consistent with the apparent molecular weight of 30 kDa, including a presumptive hydrophobic signal sequence. This of course offers a solution to the problem of why some papers report bands of 35 kDa, some 30 kDa, and some both, and shows why discussions comparing results[75] should take into account posttranslational modifications. The P30 antigen is apparently anchored to the surface plasma membrane via a glycophosphatidylinositol anchor.[86]

Purified P30 is able to induce proliferation in the T cells of naturally infected humans but not those of seronegative individuals. The anti-parasite activity induced by culture supernatant fluids from peripheral blood mononuclear cells stimulated by the addition of P30 was apparently due to the production of γ interferon.[87]

In the presence of complement, MAb to P30 lyse the parasite,[74] but immune complexes formed between MAb to P30 and live RH strain tachyzoites are shed from the parasite's surface at the level of the moving junction governing the entry of the parasite into a host cell.[80] Sibley et al.[88] found that RH strain tachyzoites secrete membranous vesicles that form a reticulate network within macrophage phagosomes, less than 5 min after entry. These modified vacuoles could then resist normal endocytic processing and digestion, allowing the tachyzoites to divide and multiply. Not surprisingly, heat-killed or specific antibody-coated tachyzoites did not secrete the intraphagosomal membrane network, and were engulfed and digested. On purification, the intraphagosomal membrane was found to consist primarily of P30. In addition to the secretion of P30, tachyzoites were found to also secrete a 22 kDa calcium-binding protein upon entry into the host cell, but its point of formation is not known. Radioimmunoprecipitation and silver staining of SDS-PAGE treated *T. gondii*-derived network vesicles[89] showed a pattern of proteins (41, 35, 29, and 22 kDa) similar to that found on the parasite's cell surface (see Section III.A). However, a major 32 kDa protein found in the purified intravacuolar networks was not found on the surface of intact tachyzoites. MAb binding suggested that this protein was identical to P30 and that the apparent molecular weight difference was due to posttranslational modifications mentioned previously.

B. EXCRETORY/SECRETORY ANTIGENS

Although excretory/secretory (ES) antigens constitute as many as 90% of the *T. gondii* antigens circulating in the serum of infected hosts,[90] they have received little attention compared with the amount of research carried out on surface antigens. Most of the earlier work on ES antigens was done by Hughes.[90-92] More recently Roques et al.,[93] Chumpitazi et al.,[94] and Decoster et al.[76] have investigated ES antigens.

This last group found that sera from naturally infected humans suffering from acute toxoplasmosis immunoprecipitated ^{35}S methionine-labeled ES antigens, with apparent molecular weights of 97 and 67 kDa. However, the sera of patients with chronic toxoplasmosis recognized about 20 different ES antigens. The major ones had apparent molecular weights of 108, 97, 86, 69, 60, 57, 42, 39, and 28.5 kDa. By contrast, the same sera immunoprecipitated only four antigens (43, 35, 30, and 22 kDa) on the surface of tachyzoites. It appears that the parasite contains many intracellular antigens that are excreted/secreted during infection.

Although the toxoplasmin skin test developed over 40 years ago[95] is now little used because of problems with safety and lack of specificity, a recent modification of the test using ES antigens shows 100% specificity and of least 94.8% sensitivity in comparison with the indirect immunofluorescence test for the measurement of serum antibody to *T. gondii*.[96] The use of ES antigens in diagnostic tests for cell-mediated immunity to *T. gondii* is discussed in Section V.A.

C. INTRACELLULAR ANTIGENS

Schwartzman et al.[97] used seven different MAb to mammalian β-tubulin to characterize this protein in RH strain tachyzoites. All seven of the MAb immunoblotted a protein of 53 to 54 kDa, consistent with β-tubulin, on SDS-PAGE, and bound to protein in the subpellicular microtubules, but not the polar ring or conoid, in an indirect immunofluorescence assay.

The *T. gondii* genes coding for α- and β-tubulin have recently been cloned and sequenced,[98] and this is discussed in Section IV.A.

Two groups[99,100] have investigated the antigens associated with *T. gondii* rhoptries. As discussed in Section I, it has been tentatively believed that the PEF of *T. gondii* is located in the rhoptries, and recent work using an assay of penetration enhancement and MAb to rodlike organelles resembling rhoptries, adds weight to this belief. When MAb raised against a rhoptry enriched subcellular fraction of *T. gondii* were used to immunoblot the proteins contained in RH strain tachyzoites and sporozoites and bradyzoites of the 76K strain of *T. gondii* after SDS-PAGE, two major proteins of 60 and 55 kDa were identified.[100] Pulse-chase analysis of radiolabeled tachyzoites demonstrated that the 60 and 55 kDa rhoptry proteins were derived from a 68 to 66 kDa doublet which was processed about 30 min after biosynthesis. Interestingly, Kimata and Tanabe[101] produced a MAb that concentrated around the tips of parasites penetrating into the host cell cytoplasm, and they speculated that the antigen fraction involved had something to do with the PEF. The MAb immunoblotted a 66 kDa protein after SDS-PAGE. Thus, it is possible that this protein is similar to one of the rhoptry proteins found by Sadak et al.[100]

Sibley and Sharma[102] used a MAb termed F3G3 in immunoblotting and immunoelectron microscopy experiments on RH strain tachyzoites. They found that the MAb, the antigen of which had previously been shown to immunize mice against toxoplasmosis,[103] reacted with an epitope within plasma membrane-derived vesicles, located beneath the tachyzoite cell surface membrane. After SDS-PAGE the MAb immunoblotted a 58 and a 28 kDa protein which, although located primarily beneath the tachyzoite surface membrane, could become exposed extracellularly, perhaps after cell membrane disruption or during developmentally regulated release or secretion.

D. NUCLEOSIDE TRIPHOSPHATE HYDROLASE

Unlike almost all other *T. gondii* antigens, the precise biological roles of which are not known, the nucleoside triphosphate hydrolase (NTPase) is well characterized biologically, at least *in vitro*. This is due mainly to the work of Asai.[104-106]

The NTPase is localized in the cytosol of RH strain tachyzoites. It has an apparent molecular weight of 240 to 260 kDa made up of four equivalent subunits of 63 kDa and appears to constitute about 8% of the total tachyzoite protein. This means that unless the NTPase and P30 are somehow one in the same protein, then these two proteins account for about 10 to 15% of all tachyzoite proteins. Dithiol compounds are essential activators of the NTPase, and magnesium is the most effective activating metal ion, although manganese and calcium were also active. The NTPase has a wide substrate specificity for both ribo- and deoxyribonucleoside triphosphates, and its higher adenosine triphosphate hydrolase activity, at least, is unique to *T. gondii*. However, the adenosine triphosphate hydrolase activity in the absence of dithiothreitol was similar in *T. gondii* (22.8 μmol/mg protein/h). The major difference between *T. gondii* and the other nine protists tested (*Giardia lamblia, Leishmania brasiliensis, Trichomonas vaginalis, Trypanosoma cruzi, Amoeba proteus, Entamoeba histolytica, Plasmodium berghei, Paramecium caudatum*, and *Tetrahymena thermophila*), was that in the presence of dithiothreitol, the hydrolytic activity of the other protists increased by a maximum of 1.3 fold, whereas that of *T. gondii* increased 513 fold.

Akao[107] found that the ATPase activity of RH strain tachyzoites was most conspicuous between two layers of the tachyzoite surface membrane, although mitochondria, the toxonemes, Golgi complex, cytoplasmic vacuoles, and so-called dense body were also found to be positive for ATPase activity.

E. STAGE-SPECIFIC AND STRAIN-SPECIFIC ANTIGENS

Early work[15] had suggested antigenic similarity among all five stages (schizont, gametocyte, sporozoite, tachyzoite, and bradyzoite) of the life cycle of the parasite, and strain specific antigens had not been identified. At least five studies had been unable to find qualitative antigenic differences among RH strain, C37 strain, and C56 strain tachyzoites,[72,108] among RH strain, C strain, and P strain tachyzoites,[109] between RH strain and ME49 tachyzoites,[110] or between RH strain and C strain tachyzoites[111] although minor quantitative differences were noted. However, it has now been found that *T. gondii* possesses stage-specific antigens, and probably also has strain-specific antigens.

Fluorescein-labeled antiserum against bradyzoites of the ME49 strain has been found not to react against RH strain tachyzoites, whereas antiserum raised against the tachyzoite does show partial reaction against the bradyzoite.[110] Strain variation did not appear to influence these results, as no difference was found in the reaction of tachyzoites of both the RH and ME49 strains with fluorescein-labeled antisera raised against RH strain tachyzoites or ME49 strain bradyzoites. Sporozoites of the C strain of *T. gondii* have been found to possess two surface membrane proteins of 65 and 25 kDa that are not present on the tachyzoite stage of either the C strain or RH strain.[111] These proteins can be radioiodinated and are antigenic, as murine antiserum raised against sporozoites reacted with them on immunoblotting. In addition, the tachyzoite membrane proteins P30 and P22 were not detectable, and P43 was markedly reduced on the surface of radioiodinated sporozoites. This was later confirmed using antisera from humans naturally infected via an oocyst-transmitted infection. The antisera, but not normal control sera, also recognized prominent sporozoite proteins of 190, 67, and 25 kDa. Although the antisera did react with proteins of >100, 62, and 30 kDa on tachyzoites, they did not identify these antigens on sporozoites. Major antigens of 97, 66, 44, and 22 kDa were common to both sporozoites and tachyzoites.[109] Antigens specific to tachyzoites and not found on bradyzoites may be important in the diagnosis of toxoplasmosis. In the acute stage of human toxoplasmosis, the IgG antibody response appears to be directed against antigens on the tachyzoite which are different from those on the bradyzoite and which stimulate the IgG antibody response during the chronic stage of the infection.[112]

As stated above, strain-specific antigens of *T. gondii* have only been found very recently. Contrary to their previous findings,[109,111] Kasper et al.[113] have now reported qualitative antigenic differences among the RH strain, C strain, and P strain tachyzoites. Specifically, C strain and P strain tachyzoites possessed antigens of 62, 37, 23, 8, and 7 kDa, while RH strain tachyzoites did not. In addition, each strain had unique antigens of 39 and 26 kDa (RH strain), 54 kDa (P strain), and 88 and 25 kDa (C strain). The results appeared to vary depending on which type of test was used to detect the differences. More recently, Weiss et al.[77] found antigenic differences among tachyzoites of the RH strain, C56 strain and T100 strain after immunoblotting SDS-PAGE gels with antisera from patients with HIV infection and cerebral toxoplasmosis. All strains had antigens of 105, 80, 65, 48, 38, 35, 28, and 26 kDa, but there were unique antigens at 130, 71, and 55 kDa (RH strain) and at 24 kDa (C56 strain). The RH and T100 strains contained antigens at 91 and 21 kDa, that were not possessed by the C56 strain.

F. OTHER POTENTIALLY DIAGNOSTIC ANTIGENS

Much of the research discussed in Section III had been carried out in an attempt to identify antigens that can be used to increase the reliability of the diagnosis of toxoplasmosis and/or provide a vaccine against toxoplasmosis, as well as to investigate the biology of the parasite. The work discussed in this sub-section also fits into this category, but is not specific enough to belong in one of the sub-sections previously discussed. Much of the recent work has been carried out by Potasman et al.[114-116] Work carried out before 1986 has been reviewed previously.[69]

Most surprisingly, it appears that the sera of adults and children who had no previous exposure to *T. gondii* can react with immunoblots of RH strain tachyzoites after SDS-

PAGE.[114] Of 15 adults, the IgG in the sera of 6 reacted with a 125 to 120 kDa antigen, 10 reacted with a 92 kDa antigen, 8 reacted with a 68 kDa antigen, 5 reacted with a 50 to 48 kDa antigen, 7 to a 35 kDa antigen, 10 to a 30 kDa antigen, 5 to a 27 kDa antigen, and 6 to a 22 kDa antigen. However, there was no antigen which was consistently identified by the sera from all 15 patients. The IgM in the patients sera reacted with antigens of 115 kDa (13 patients), 105 kDa (all 15), 85 to 80 kDa (all 15), 66 kDa (9), 60 to 56 kDa (9), 50 to 48 kDa (12), 30 kDa (10), and 14 to 10 kDa (6). The immunoblots of the childrens' sera showed considerable variability in the number and distribution of bands. These results highlight the need for appropriate negative controls in immunoblotting experiments, and question the value for diagnosis[79] of kits[78,117] that use only the P30 fraction as antigen.

Identical immunoblotting was performed with sera from 17 patients who showed rising total antibody titers to *T. gondii*,[115] and from 12 congenitally infected infants and 7 of the mothers.[116] All sera from the 17 patients with acute toxoplasmosis intensely stained a 4 kDa antigen (IgM immunoblot) and a 35 kDa antigen (IgG immunoblot), but sera from chronically infected patients reacted with these antigens weakly (IgG) or not at all (IgM). Similarly, the sera of the congenitally infected infants and their mothers strongly recognized antigens of 115 kDa and 35 kDa on IgG immunoblotting. Although the sera of all the mothers recognized an antigen of 4 kDa on IgM immunoblotting, the sera of only two of the 12 infected infants recognized this antigen. These results suggested that the 35 kDa antigen provoked the strongest and most consistent antibody response following infection with *T. gondii*.

Verhofstede et al.[118] also found that a 35 kDa antigen elicited the first IgG response soon after infection. Over time as the disease become chronic, 13 to 15 major antigenic bands appeared on immunoblotting. The immunoblots (IgG) of 28 patients showed the following antigens: 60 to 59 kDa (in all 28), 54 to 53 kDa (all 28), a doublet at 40 to 39 kDa (all 28), 35 kDa (all 28), 32 kDa (all 28), 29 kDa (all 28), 28 kDa (26), and 27 kDa (all 28). Natural antibodies were also found in 14 of 16 sera from patients with no previous exposure to *T. gondii*. These antibodies reacted against antigens of 98 kDa (in 1 patient), 95 kDa (2), 85 kDa (2), 78 kDa (7), 68 kDa (1), 65 kDa (1), 60 kDa (3), 55 kDa (3), 52 kDa (2), 50 kDa (4), 40 kDa (1), 33 kDa (1), 30 kDa (3), and 25 kDa (2).

IV. MOLECULAR BIOLOGY

A. NUCLEIC ACID CHARACTERIZATION

Unlike the numerous studies on other parasitic protists, the new techniques in molecular biology/gene cloning have been applied to *T. gondii* over only the last 3 to 4 years. As early as 1967, Neimark and Blaker[119] had determined that RH strain tachyzoite DNA had a density of 1.712 g/cm^3 and a guanine plus cytosine (G + C) content of 53%. Then, in 1971, Perrotto et al.[120] found a G + C content of 57% and a total of 100×10^{-15} g of DNA per tachyzoite. In addition, they found that *T. gondii* tachyzoites could synthesize DNA independently of the host cell and could incorporate preformed pyrimidines as well as pyrimidine precursors, unlike other obligately intracellular parasites. Pyrimidine precursors were preferentially utilized over preformed pyrimidines, and there was little detectable utilization of purine precursors.

Then after a gap of almost 13 years, work on the molecular biology of *T. gondii* began to gain momentum. Cornelissen et al.[121] found absolute DNA values of 96×10^{-15} g/sporozoite cell, and 164×10^{-15} g/tachyzoite cell. Therefore, the tachyzoite appears to have a total genome size of about 8×10^7 base pairs. Borst et al.[122] isolated a closed circular duplex DNA fraction from tachyzoites. A major part of this circular DNA consisted of 12-μm circles containing a cruciform with 0.5-μm tails which were part of a 1.7-μm palindrome. Borst et al.[122] believed that the cruciforms were not present *in vivo* and that the 12-μm circles represented mitochondrial DNA. Johnson et al.[123] found that the tachyzoite DNA was not significantly methylated.

As mentioned in Section III.A and C, the genes for *T. gondii* α- and β-tubulin, and the antigen P30, respectively, have been cloned and sequenced.[85,98] The genes for α- and β-tubulin were found to contain multiple introns, and were single copy. The presence of multiple introns is rare in the sequenced genes of parasitic protists, and the fact that the tubulin genes of *T. gondii* are single copy is unique among organisms studied to date. Tubulins are highly conserved proteins that are found in almost all eukaryotes. The two forms copolymerize to form the microtubules which are involved in many functional roles within the cell. It will therefore be interesting to see how the tubulin genes of *T. gondii* compare with those of closely related protists, and to see whether other housekeeping genes of *T. gondii* have unusual features. The P30 gene is also single copy, but does not contain introns.[85] The primary translation product of the P30 gene has a carboxy-terminal hydrophobic tail which is consistent with posttranslational cleavage and modification with a glycolipid anchor, consistent with it being a surface protein. Consistent with P30 being about 5% of the total cell proteins,[84] the P30 mRNA transcript is extremely abundant.

Johnson et al.[124] have recently cloned a gene fragment expressing the antigenic portion of the *T. gondii* NTPase (see Section III.D). The NTPase gene has a very low copy number, and its mRNA transcript is extremely abundant.

Two groups have reported *in vitro* translation of *T. gondii* mRNA in cell-free systems. Although Prince et al.[125] could not obtain mRNA that directed synthesis of polypeptides in a rabbit reticulocyte lysate system, they did get synthesis in an optimized wheat germ *in vitro* translation system. Johnson et al.[126] compared several RNA extraction methods and found one that gave good, intact RNA, the mRNA component of which worked very efficiently in a rabbit reticulocyte lysate assay. Both groups were able to immunoprecipitate the polypeptides produced with immune animal and human sera. Electrophoresis of the total cellular RNA extracted by Johnson et al.,[126] under nondenaturing conditions and after glyoxal denaturation, showed that the large and small subunits of *T. gondii* ribosomal RNA (rRNA) consisted of about 4250 and 1975 nucleotides, respectively. These sizes are similar to those of the large and small rRNA of several other parasitic protists such as *P. falciparum*, *P. berghei*, and *T. cruzi*. The value of the *T. gondii* large RNA fraction lies between that found for the large subunit rRNA of higher eukaryotes such as vertebrates and the values found for the same subunit in lower eukaryotes.

B. PHYLOGENY/TAXONOMY STUDIES

Recent findings on the molecular biology of *T. gondii* are also having an impact on the phylogeny and taxonomy of the organism. Speciation in the genus is still subject to debate, but most authors acknowledge that *T. gondii* is the only species that parasitizes most warm-blooded animals,[127] and it is probably the only species in the genus. However, classification of *Toxoplasma* at the genus and higher level is still problematic.

Based largely upon the possession of a variable number of structures constituting an apical complex, *Toxoplasma* and its related organisms have been placed in the phylum Apicomplexa Levine, 1970; class Sporozoasida Leuckart, 1879; subclass Coccidiasina Leuckart, 1879; order Eucoccidiorida Leger and Duboscq, 1910.[128] In this classification, the genus *Toxoplasma* is considered to be closely related to a range of other protistan parasites such as *Cryptosporidium, Plasmodium, Sarcocystis, Eimeria, Isospora, Hammondia,* and *Besnoitia*. However, in phylogenetic studies of such groups, the characters used must be truly homologous. Characters of uncertain homology cannot be used and, in groups such as parasitic protists, few truly homologous characters may be available for analysis. A good example here is the recent study of Barta.[129] Although rigorous in its parsimony analysis of characters, the study had to use characters present in most if not all of the organisms and a related outgroup, otherwise it could not have been performed. New techniques in molecular biology offer a solution to this philosophical problem.

The more precise molecular methods used to infer phylogeny include isoenzyme electrophoresis, immunological methods, DNA/DNA hybridization, protein and DNA sequencing. However, isoenzyme electrophoresis is restricted to the study of organisms that are closely related (less than 10 million years). Immunological methods and DNA/DNA hybridization that measure distances between taxa require the building of an n × n matrix (i.e., 10 taxa require 100 reactions) and all reference taxa must be included in every run. In addition, the results of these studies from different laboratories, are difficult to merge into a single coherent analysis. Protein sequencing and DNA sequencing of cloned genes do yield permanent data, but they are time consuming and technically difficult. These disadvantages have led to the use of small subunit rRNA (srRNA) to infer phylogenies. rRNA is ubiquitous, and although it has been highly conserved throughout evolution, the rate of nucleotide substitutions, deletions or insertions at various positions in the srRNA, in particular, is not constant. Some regions are totally conserved, unchanged over millions of years of evolution; other regions are highly variable, and the remainder is semi-conserved. Using a new technique (reviewed in Reference 130), Johnson et al. have determined the partial srRNA of *T. gondii* and several of the organisms considered to be closely related to it, in order to infer the phylogeny of *T. gondii*. These studies have found that although *T. gondii* and *Sarcocystis gigantea* are monophyletic,[131] there is no especially close relationship among *T. gondii, P. berghei,* and *Cryptosporidium* spp.[132,133]

Using cloned *T. gondii* rRNA gene probes and *Eimeria acervulina* total RNA, Johnson et al.[123] found identity in the restriction enzyme maps of the rRNA genes of *T. gondii* and *Hammondia hammondi*. The G + C content of *T. gondii* (50.4 ± 1.5%), was found to be similar to that of *H. hammondi* (49.9 ± 3.3%) and Johnson et al.[123] suggested that *H. hammondi* was a strain of *T. gondii*, or a separate *Toxoplasma* species. Levine[134] had previously suggested that *H. hammondi* should be a separate species of *Toxoplasma*, and some workers believe that the genus taxon *Toxoplasma* Nicolle and Manceaux, 1909, should become a junior synonym of *Isospora* Schneider, 1881.[8,121] However, *Toxoplasma* is so well known in the medical and veterinary literature that it is unlikely to be replaced.[135] Clearly, however, there is still doubt over the taxonomic classification of *Toxoplasma*. It is likely that more molecular phylogenetic studies will lead to a reconsideration of the genera now classified as apicomplexans.

Recently, Darde et al.[136] used isoelectrofocusing in polyacrylamide gels to investigate strain difference among seven *T. gondii* isolates. Fourteen enzymes gave identical electrophoretic patterns for all strains, but another four enzymes (aspartate amino transferase, glutathione reductase, amylase, and glucose phosphate isomerase) differentiated the isolates into three zymodemes. There were some indications of an approximate relationship between some biological characteristics (particularly pathogenicity and oocyst formation) and the enzyme patterns.

C. NUCLEIC ACID DETECTION

The problem with the traditional methods of diagnosing toxoplasmosis by antibody detection, such as length of time required to mount an antibody response and the interpretation of static high IgG titers in immunocompetent hosts, and the lack of an antibody response in immunosuppressed patients, has led to the development of tests to detect parasite antigen. However, it is becoming obvious that tests to detect antigen with polyclonal antibodies, and even tests that use MAb, have limitations.[137-140] Probes to detect the nucleic acid of infectious agents offer the potential to overcome many of these problems.[139,141-142] In particular, nucleic acid hybridization has been used to detect the DNA of parasitic protists such as *Leishmania*,[144]

	3323	3349
T. gondii	C-UUUAAC-CCGAAUGCCUAGUAGGCG	
Human	C-AUGAACGAGGAAUUCCCAGUAAGUG	
Murine	C-AUGAACGAGGAAUUCCCAGUAAGUG	

FIGURE 2. Alignment of the *T. gondii* srRNA primary sequences with the corresponding human and murine srRNA sequences. The numbering system is that of Dams et al.[152]

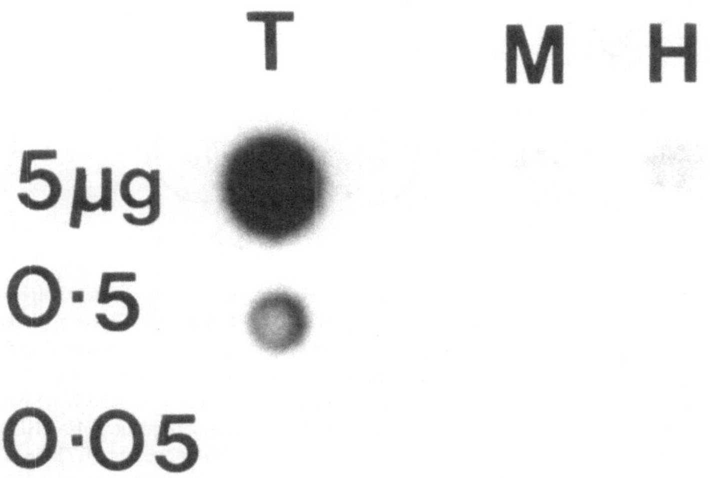

FIGURE 3. Autoradiograph of a ^{32}P-labeled synthetic oligonucleotide hybridized with decreasing amounts of *T. gondii* (T), human (H), and murine (M) total cellular RNA.

Trypanosoma,[145] *Babesia*,[146] and *Plasmodium*.[147] The use of synthetic oligonucleotides offers a theoretical increase in specificity over that obtained with genomic probes, and reduces hybridization time,[142,143] and a probe to detect rRNA offers an increase in sensitivity as rRNA is usually much more abundant than a gene.[143,149,150] A synthetic oligonucleotide complementary to a unique portion of the rRNA of *Plasmodium falciparum*,[148] *Mycoplasma pneumoniae*,[149] or *Proteus vulgarus*,[151] can be used to identify species in the genera *Plasmodium, Mycoplasma,* and *Proteus*, respectively.

From the nucleotide sequences of the srRNA of *T. gondii*,[132] humans,[152] and mice[152] (the natural and laboratory host of the parasite from which it most needs to be detected), Johnson and Fielke[153] identified a 25 base-pair long stretch of nucleotides that appeared ideal as a target in *T. gondii*, as it did not appear in human or murine srRNA. A 25 base-pair oligonucleotide with the sequence 5′-CGCCTACTAGGCATTCGGGTTAAAG-3′ was synthesized using phosphoramidite chemistry on an automated DNA synthesizer (Applied Biosystems, Foster City, CA). The corresponding srRNA sequences can be seen in Figure 2. When 5′ end labeled with ^{32}P, this oligonucleotide hybridized with *T. gondii* RNA, but not human nor murine RNA (Figure 3). Clearly, the sensitivity of the technique needs to be increased (by lowered hybridization stringency, different labeling techniques, etc.), before it could be considered for routine diagnosis, but the results to date suggest that this is a feasible diagnostic test, and likely to have more widespread use in the future.[154] The polymerase chain reaction (PCR)[155] is also likely to have great use in the detection of *T. gondii* nucleic acid, as it will in many fields of parasitology.[156] Burg et al.[157] have just established a PCR for the detection of *T. gondii*.

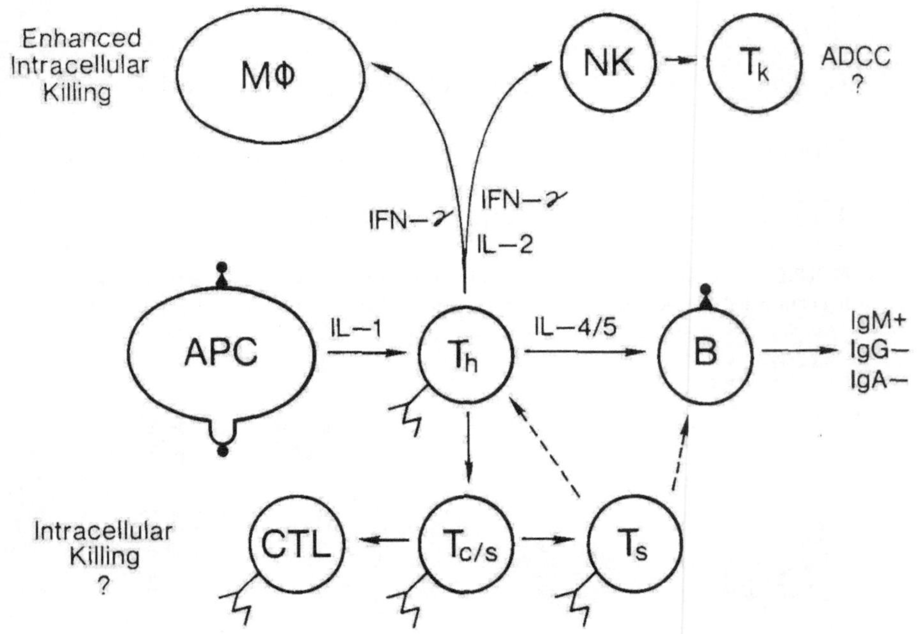

FIGURE 4. Summary of events associated with *Toxoplasma* infection. Antigens presented to helper T lymphocytes (T_h) with IL-1 will enable helper function resulting in IFN-γ, IL-4, and IL-5 release. These cytokines will help in activation of NK, macrophages (mø), and B lymphocytes. Suppressor cells are also initiated, and these will act on T_h and B cells (broken lines), inhibiting macrophage activation, NK, IgG, and possibly IgA synthesis. Suppressor macrophages releasing prostaglandin E_2 (PGE_2) may have direct effects on mø, NK, and B lymphocytes as well as down regulating antigen presentation, IL-1 release, and division of IL-2-dependent cells. PGE_2 will stimulate suppressor T cell activity. It is not known whether T_k or cytotoxic T lymphocytes (CTL) can kill parasites extracellularly, via antibody-dependent cell-mediated cytotoxicity (ADCC), or in infected cells (CTL). (Redrawn from Hughes, H. P. A., *ISI Atlas Sci. Immunol.*, 1, 185, 1988. With permission.)

V. IMMUNOLOGY

Knowledge on the immunology of *T. gondii* and toxoplasmosis up until 1982 has been extensively reviewed.[158-160] Britten and Hughes[161] compared the immunological aspects of toxoplasmosis with those of two other parasitic diseases, and Hughes has recently written an overview of toxoplasmosis in humans.[162] Therefore, this section will concentrate on areas of controversy within the immunology of toxoplasmosis, very recent developments in the field, and areas where research is likely to be directed in the future.

As our knowledge on the precise events in the immunological cascade (see Figure 4) occurring in toxoplasmosis increases, we are likely to need to look more at specific events rather than at the end result of the whole process. For example, in vaccine development we need to exploit anti-parasite responses and try to avoid irrelevant and anti-host responses. In addition, it is now quite clear that in considering the immune response to *T. gondii* we need to consider both the cell-mediated and humoral responses as they both play a role, although the former is likely to be the main effective response, in protection against toxoplasmosis. Therefore, we are likely to depend more on specific cell-mediated responses to *T. gondii* in considering who is immune to the disease, than we have in the past. Delayed hypersensitivity type skin tests have been unreliable until recently (see Section III.B), and it is likely therefore that although they are technically complex and time consuming, lymphocyte transformation type tests will find more of a use in the diagnostic immunology of toxoplasmosis.

A. LYMPHOCYTE TRANSFORMATION

Lymphocyte transformation was first used to assess the cell-mediated immune response to *T. gondii* in humans almost 20 years ago.[163,164] Although it has been used diagnostically in congenital toxoplasmosis,[165,166] in longitudinal studies of acute toxoplasmosis,[167,168] and in acute symptomatic and chronic asymptomatic toxoplasmosis,[169,171] there is often lowered or no responsiveness to *T. gondii* antigen in congenital toxoplasmosis[166] and in the acute phase of the infection compared with that found when the disease progresses into the chronic asymptomatic stage.[168] The kinetics of the depression early in the disease are consistent with the induction of a non-Leu 2 suppressor cell.[169] In a further report, Luft et al.[170] found that the induction of suppressor cell activity and the persistent increase in Leu 2-positive suppressor T cells correlated with the more severe and persistent clinical symptoms, whereas in asymptomatic patients, the number of Leu-2 positive suppressor T cells was normal and the induction of suppressor cell activity was minimal. They raised the possibility that these suppressor cells could have significance in prolonging the symptoms that may occur in acquired toxoplasmosis.

Lymphocyte transformation appears to be worthwhile as an indicator of cell-mediated immunity in healthy patients, for example, to test the effectiveness of future vaccines. The specificity of the test has been considered to be a problem, although this may be overcome by using appropriate antigen fractions of the parasite in the test.[171] In addition, as stated previously, the test requires technical expertise and is expensive. Nevertheless, I believe that tests of cell-mediated immunity to *T. gondii*, such as lymphocyte transformation, will become more common in future. The use of ES antigens increases the specificity of the test,[171] and at least one putative candidate antigen for a *T. gondii* vaccine, P30 (see Section III.A), has already been found to induce proliferation in the T cells of seropositive patients but not those of seronegative patients.[87]

B. RELEVANCE OF NATURAL KILLER CELLS

Early studies suggested a role for natural killer (NK) cells in the control of the early stages of toxoplasmosis, but this role is now being questioned.[162,172,173]

Most of the early studies on the role of NK in toxoplasmosis were performed by Kamiyama,[174-176] and Hauser and Sharma.[177-179] Increased NK cell activity in peritoneal exudate cells and spleen cells against YAC-1 tumor cells was found to peak 3 days after intraperitoneal injection of tachyzoites of virulent or relatively avirulent *T. gondii* strains in a range of mice.[174,177] Hauser et al.[177] found that mice chronically infected with *T. gondii* had significantly higher endogenous NK activity than did control mice in their peritoneal cells but not spleen cells, while Kamiyama and Hagiwara[174] found that NK cytotoxicity was markedly suppressed in both peritoneal and spleen cells, with maximum depression being 12 days postinfection. Further studies found that the subcellular parasite fraction of a sonicate prepared from RH strain tachyzoites enhanced the *in vitro* NK cell activity of human peripheral blood lymphocytes greater than that obtained with the parasite membrane fraction.[179] Both fractions were capable of enhancing murine peritoneal cell NK activity, whereas murine splenic NK cell activity was enhanced only by the membrane fraction.[178] Most of these observations were consistent with the hypothesis that NK cell activity was responsible for the control of parasite growth in the acute stage of infection. The finding that *T. gondii*-induced spleen cell NK anti-parasite activity was significant at all effector (nylon wool nonadherent spleen cells) to target cell (RH strain tachyzoites) ratios tested, and that the effect was mediated by direct contact between the host cell and parasite,[180] further strengthened this hypothesis.

However, more recent work has been consistent with the hypothesis that NK cells may play a small role, if any, in the control of toxoplasmosis. Hughes et al.[172] used cold target inhibition experiments to determine whether RH strain tachyzoites could inhibit NK cytolysis of YAC-1 tumor cells by binding to cell surface receptors on splenic NK cells. Cold YAC-1 cells inhibited the lysis, whereas RH strain tachyzoites did not, suggesting that the parasite did not bind to NK cells. In addition, using inoculation of mice as an indicator of parasite numbers, Hughes et al.[172] found no evidence of *in vitro* killing of tachyzoites by NK cells, in fact, more tachyzoites survived when cultured with NK cells than when cultured alone. Studies using beige mice, which are genetically deficient in NK cell activity, showed that C57BL/6*bg*/*bg*, and *bg*/+ mice, showed no difference in mean time to death caused by an oral ME49 strain oocyst challenge, compared with their +/+ littermates. Oral inoculation of mice with ME49 oocysts led to a pattern of NK enhancement and then depression,[172] similar to the results obtained previously using different routes of challenge.[174,177] However, the time scale associated with the oral route of infection was much reduced than that reported with other routes of challenge: NK levels had returned to normal levels by 15 days post-inoculation.

The results of Hughes and colleagues appear to question the relevance of NK cells in the control of toxoplasmosis. Some of the apparently controversial results may have been produced because of differences in the routes of challenge used to induce toxoplasmosis in the host animals. Pregnancy has been shown to lower augmentation of NK activity,[181] but even hormonal or physiological changes like those seen in pregnancy are extremely unlikely to have led to the apparently contradictory results in the studies discussed here. It is unlikely that obvious reasons will be found for the differences, but just as I have argued elsewhere for research on *T. gondii* vaccines,[81] it is important that laboratory animal studies on the immunology of toxoplasmosis are as realistic as possible. Consequently, the studies of Hughes using a natural route of infection with a natural challenge, oocysts, should probably be given considerable value when assessing the relevance of NK activity in the control of toxoplasmosis. However, clearly more work is needed in this area, particularly in the light of the recent finding of an antigen-specific subset of $CD4^-$, $CD8^+$, P30 responder T cells that have a direct anti-parasite effect on RH strain tachyzoites. This cytotoxicity is independent of antibody opsonization, lymphokine secretion, NK cell activity, and, apparently, MHC involvement as well.[182]

C. OXYGEN-DEPENDENT VS. OXYGEN-INDEPENDENT PARASITE KILLING

As mentioned in Section II.B, both oxygen-dependent and oxygen-independent processes can be involved in parasite killing. Hughes[42] has recently reviewed the oxidative killing of intracellular *T. gondii* mediated by macrophages. Reactive oxygen metabolites appear to be the first-line mechanism whereby mononuclear phagocytes kill *T. gondii*. Inhibition of tachyzoite growth by human monocytes and activated murine peritoneal macrophages is associated with the increased production of oxygen metabolites, and conversely, the parasites which escape killing inside normal macrophages avoid triggering the production of oxygen radicals and their intermediates.[43,183-185] The specific anti-parasite substances are probably the hydroxyl radical ($OH^\cdot$) and singlet oxygen ($^1O_2$), as RH strain tachyzoites in a cell-free model are resistant to superoxide anion (O_2^-) and hydrogen peroxide (H_2O_2).[186] The resistance may be due to the fact that virulent and low virulence *T. gondii* strains contain superoxide dismutase and catalase,[187] which reduce superoxide radicals to H_2O_2, and H_2O_2 to water and oxygen, respectively. However, the recent studies of Hughes et al.[188] imply a significant role for the OH radical and suggest that resistance to H_2O_2 is not totally dependent on the use of scavenging enzymes. A possible explanation for these findings is the induction of DNA-repair mechanisms in *T. gondii*, but this will need to be substantiated by further work.

In an extensive series of studies, Nathan et al. investigated cytokine activation of human monocyte-derived macrophage oxidative metabolism and anti-*T. gondii* activity. Nonspecific activation of macrophages can be influenced by interferon-α, interferon-β, interferon-γ, colony stimulating factor type 1, colony stimulating factor for granulocytes and macrophages, pluripotent colony stimulating factor, and migration inhibitory factor. Nathan et al.[189,190] investigated the ability of all these cytokines plus recombinant interleukin-2, and tumor necrosis factor. Only interferon-γ and partially purified migration inhibitory factor enhanced H_2O_2-releasing capacity, a measure of macrophage activation, but only interferon-γ also induced an anti-*T. gondii* activity in human macrophages. Two years later De Titto et al.[191] showed that recombinant tumor necrosis factor did not affect the growth of *T. gondii* in normal murine peritoneal macrophages or in human foreskin fibroblasts, and that it did not enhance the killing of tachyzoites by activated macrophages.

Hughes[42] has recently found a lymphokine distinct from interferon-γ and granulocyte macrophage colony stimulating factor, that is expressed by antigen or mitogen stimulated lymphocytes, which can activate bovine monocytes to kill intracellular *T. gondii*. This new lymphokine was more effective than interferon-γ in activating macrophages, and had an apparent molecular weight almost identical to that of tumor necrosis factor, 17.4 kDa.

The apparent failure of interleukin-2 to activate macrophages to kill *T. gondii in vitro* is interesting in the light of studies reported 2 years later,[192] showing that recombinant interleukin-2 administration enhanced the survival of C3H/He mice against an otherwise lethal *T. gondii* challenge. The apparent differences between the *in vitro* and *in vivo* effectiveness may have been the result of the amount of interleukin-2 used, or the fact that interleukin-2 has an anti-parasite effect other than that mediated by macrophage activation. Alternatively, the differences may have been due to the fact that splenic L3T4-T cells from *T. gondii*-infected C3H/He mice release interleukin-2 inhibiting factors.[193] During the early stages of *T. gondii* infection, the C3H/He macrophages lacked antigen presenting capacity, and a nonspecific B-cell mediated suppression of T cell proliferation was observed. These results suggested a possible role for B cells in the regulation of T cell immunity to *T. gondii*.

Very recent work has been aimed at elucidating the subset of T cells responsible for protection against toxoplasmosis. Using an *in vitro* model of human peripheral blood monocytes, Canessa et al.[194] found that RH strain tachyzoites attenuated by γ irradiation had to be processed by monocytes/macrophages in order to be presented to T cells. B cells failed to act as antigen presenting cells and activate T cells when cultured with tachyzoites. Under appropriate antigen stimulation, actively proliferating T cell clones from seropositive donors released interferon-γ which activated monocytes/macrophages to kill intracellular tachyzoites. The T cell clones had a $CD4^+$ surface phenotype and helper capacity on B cell differentiation and proliferation *in vitro*. Suzuki and Remington[195] also found that $L3T4^+$ (murine equivalent of $CD4^+$) T cells had a substantial role in resistance against toxoplasmosis. Using an *in vivo* murine model, they found that treatment with anti-Lyt-2.2 completely removed the protective ability, and that treatment with anti-Lyt-1.2 or anti-L3T4 partially diminished the protective ability, of immune T cells. These findings suggested dual regulation of resistance against *T. gondii* by the $Lyt\text{-}2^+$ and $Lyt\text{-}1^+$, $L3T4^+$ T cell subsets, but that $Lyt\text{-}2^+$ T cells are the principal mediators of the resistance. While it is still not clear how these cells mediate resistance, the release of interferon-γ apparently plays a key role. The recent work of Pfefferkorn[196-199] suggests that the anti-*T. gondii* activity of interferon-γ in human fibroblast tissue culture is the result of the induction of indoleamine 2,3-dioxygenase which depletes tryptophan and starves the tachyzoites of an essential amino acid. Interferon-γ also has an effect on tachyzoites *in vivo*. Intraperitoneal injection of *T. gondii* tachyzoites killed 50% of control mice so injected, whereas none of the mice challenged and given intravenous interferon-γ died.[200]

It is now clear that, as well as the oxygen-dependent killing mechanisms described above, intracellular tachyzoites can also be killed via oxygen-independent mechanisms. Monocytes from patients with hereditary myeloperoxidase deficiency have a significant anti-*T. gondii* defect that is abolished if the tachyzoites are first coated with eosinophil peroxidase.[201] The defect in the monocytes of patients with chronic granulomatous disease is more severe. The anti-*T. gondii* activity of these cells is not restored if the tachyzoites are first coated with eosinophil peroxidase,[201] but the addition of lymphokine or interferon-γ can stimulate cells from chronic granulomatous disease patients to exert near-normal levels of toxoplasmastatic activity and some toxoplasmacidal effects as well.[202,203] Cells, such as human,[197] hamster,[204] and murine[202] fibroblasts, which do not normally have an active phagocytic role, can kill intracellular tachyzoites. This oxygen-independent intracellular anti-*T. gondii* mechanism is stimulated by soluble lymphocyte products,[202] probably interferon-γ.[197,203] The respiratory burst of rat peritoneal macrophages is not triggered by the ingestion of tachyzoites, but other anti-*T. gondii* mechanisms kill the parasites within the cells.[198] Another rat cell line, alveolar macrophages, also possess a powerful anti-*T. gondii* mechanism that kills intracellular tachyzoites without significant involvement of toxic oxygen metabolites.[206] The work of Sibley et al.[207] suggested that the murine macrophage cell line J774G8, which normally produces only very low levels of oxygen intermediates, may kill intracellular tachyzoites by oxygen-independent mechanisms. Lymphokine treatment of these cells does lead to an elevated respiratory burst. It therefore appears that interferon-γ plays a very important role in both oxygen-dependent and oxygen-independent killing of *T. gondii*, although the oxygen-dependent mechanisms appear to be the major one leading to intracellular tachyzoite death.

VI. PATHOLOGY

A. HUMANS

1. Immunocompetent

The major symptom of acute toxoplasmosis in the adult is lymphadenopathy, while the pathology associated with congenital infection is noteworthy because of the extreme variability.

a. Lymphadenopathy

Although not definitively diagnostic, the lymph node appearance in toxoplasmosis is quite characteristic. The posterior cervical lymph nodes are most frequently involved.[208] The general structure is usually well preserved and germinal centers are frequent, with all components of the lymph node being recognized as reactive. So called "sinus histiocytosis" (the cells are in fact of the B cell lineage) is a striking feature and the large sinus cells frequently extend into follicles. In advanced cases the cells extend to the peripheral sinuses and may partly obscure the normal lymph node architecture. The lymphocyte proliferation often infiltrates and obliterates the node capsule, extending into the perinodal connective tissue.[208,209] The occurrence of epithelioid cells within germinal centers appears to be a striking feature of toxoplasmosis. However, many of the reactive features so typical of toxoplasmosis can occasionally also be seen in Hodgkin's disease.[210] Although tachyzoites and very rarely even tissue cysts are occasionally detected inside nodes,[211,212] the cellular reactions occurring in glandular toxoplasmosis are probably due to an immune driven hyperplasia. The lymphoid tissues of humans who die from toxoplasmosis are usually lymphocyte depleted.[213]

b. Congenital

The lesions associated with congenital toxoplasmosis are variable with respect to both organs involved and severity. This is probably due to a combination of a number of factors such as age of the fetus at infection, the actual number of organisms passing from the mother to the fetus and their virulence, the age at the time of autopsy, and the depth of the search for lesions in the autopsy. However, in almost all cases, the central nervous system is affected. Knowledge on the pathology of the lesions in human congenital infection up to 1982 has been extensively reviewed by Remington and Desmonts.[214] Since then, Dutton[215] has investigated the disease process occurring in the eye as the result of congenital infection. He suggests that inflammatory responses, to the release of parasites from intraretinal cysts, and against the retina enhanced by nonspecific inflammatory cell lysis, may be important pathogenetic mechanisms of the tissue damage in retinochorioditis caused by *T. gondii*. Frenkel has very recently updated knowledge on the pathophysiology of toxoplasmosis.[213]

2. Immunoincompetent

It has been known for many years that occasionally patients immunosuppressed as a result of organ transplantation or cancer therapy succumb to fatal toxoplasmosis.[216-219] However, interest in toxoplasmosis in immunocompromised patients has been renewed and heightened because of its potentially fatal effects in individuals immunosuppressed by the AIDS virus. Since the first reports in 1982,[220-222] numerous cases of toxoplasmosis in AIDS patients have been recorded. Levy et al.[223] list papers up to 1985 containing 103 cases, and Navia et al.[224] describe another 27 cases of their own. Toxoplasmosis in AIDS patients has also been the subject of several reviews,[225,226] so only a brief summary of the major points is called for here.

Cerebral toxoplasmosis in the compromised host is almost exclusively due to reactivation of a dormant infection.[227] The results of Vollmer et al.[228] are also consistent with this theory. They developed a murine model of superinfection in AIDS and found that selective depletion of $CD4^+$ T lymphocytes (which in AIDS is associated with central nervous system toxoplasmosis), had remarkably different effects in chronically or acutely infected mice. Chronically infected mice died after severe central nervous system toxoplasmosis and only minor systemic involvement, whereas the symptoms were reversed in acutely infected mice which developed severe systemic symptoms and only mild disease in the brain. Hofflin et al.[229] developed a murine model similar to that observed in the brains of humans with toxoplasmic encephalitis, by giving intracerebral injections of tachyzoites to immunosuppressed mice.

Symptomatic toxoplasmosis occurs in about 10% of all AIDS patients, but this figure is usually higher in Haitian AIDS patients probably due to the higher prevalence of *T. gondii* in warm, moist climates. Almost all AIDS patients with toxoplasmosis present with central nervous system symptoms, although retinochoroiditis,[230,231] pneumonia,[232] and orchitis[233] have been reported as occasionally briefly preceding the neurological symptoms. These neurological abnormalities may be focal (seizures, hemiparesis, hemiplegia, hemisensory loss, cerebellar tremor, homonymous, hemianopia, cranial nerve palsies, diplopia, blindness, personality change, and headaches which do not respond to analgesics), or generalized (weakness, myoclonus, confusion, lethargy, disorientation, and coma).[226]

Computed tomographic (CT) scans have been most beneficial in the diagnosis of cerebral toxoplasmosis complicating AIDS, although the technique does under represent the number of lesions actually present.[224] Double-dose CT scans and magnetic resonance imaging appear to be superior to the standard single-dose CT scan.[223,224,234,235] However, even these last two techniques are not pathognomonic for toxoplasmosis. The radiological patterns of change in toxoplasmic encephalitis can also be seen as a result of central nervous system lymphoma[236] or tuberculous abscess.[234] Because serological tests are rarely useful in AIDS, brain biopsy remains perhaps the only procedure to provide a definitive diagnosis of toxoplasmic encephalitis in AIDS. However, such an invasive procedure is not universally accepted in this situation,[237] and unfortunately a needle brain biopsy or aspiration does not appear to be as sensitive as an open excisional biopsy.[226,238]

Navia et al.[224] described the pathology in 16 patients, at postmortem, all of whom had multiple toxoplasmic abscesses, more frequently in the cerebral and cerebellar cortex. They grouped the lesions into three histological types (necrotizing, organizing, and chronic), based on the degree of tissue reaction. Necrotizing abscesses were the only finding in eight patients, five of whom had received no treatment, while the remainder had been treated for less than 3 days. Gross areas of softening and discoloration consisting of poorly demarcated areas of necrosis containing a variable number of petechiae, with surrounding edema, characterized the necrotizing abscesses. Vascular proliferation, mixed inflammatory cell infiltrates, and lipid-laden macrophages, were found in the necrotizing abscesses of six of the patients. Tissue cysts were present in all patients, but numerous in only three. Organizing abscesses were present in seven patients, all of whom had been treated for at least 2 weeks. These lesions consisted of a thin rim of tightly packed lipid-laden macrophages surrounding large, well demarcated areas of central coagulation necrosis. Three of these seven patients had tissue cysts and tachyzoites adjacent to their organizing abscesses. Chronic abscesses were found in six patients, five of whom also had organizing abscesses. These patients had been treated for at least 4 weeks. The chronic abscesses were small (usually less than 0.5 cm in diameter) cystic spaces that contained small numbers of lipid-laden macrophages with surrounding gliosis. In three of the patients, rare *T. gondii* cysts were noted in adjacent tissue.

Treatment of toxoplasmic encephalitis in AIDS (discussed in Section VIII) usually results in dramatic clinical and radiological improvement within 2 to 3 weeks[239] and the findings of Navia et al.[224] are consistent with the hypothesis that treatment of toxoplasmic encephalitis causes a change in the tissue pathology from the necrotizing, to the organizing, to the chronic abscesses, with the disappearance of the parasite. This is also consistent with the suggestion that brain necrosis depends less on host inflammatory responses than on a direct toxic effect of the proliferating organism related to impaired cellular immunity.[240]

B. ANIMALS

The pathology of naturally and experimentally acquired toxoplasmosis in animals has been extensively reviewed by Dubey.[127,241-244] The most significant fact about toxoplasmosis in animals is the similarity between the pathology of congenital toxoplasmosis in animals and that in humans.

The pathology of acute experimental toxoplasmosis caused by high-virulence strains in mice appears to consist of rapid lymphoid depletion of lymph nodes, spleen, and particularly the thymus,[245-249] usually leading to death. Low virulence strains also cause lymphoid depletion of the thymus reaching a peak at 10 to 15 days after infection, and by 30 days after infection the lymphocytes are replaced and the animal survives.[246] A lower blastogenic response to concanavalin A of lymphocytes from C57BL/6j mice, compared with that of BALB/c mice, 7, 14, or 30 days after infection with a low virulence *T. gondii* strain, appeared to correlate with an increased susceptibility of the C57BL/6j mice.[247] Jones et al.[248] found that a low virulence challenge of CBA/j mice induced a three- to fourfold increase in weight and cellularity of lymph nodes and spleen with maximum changes occurring 30 to 50 days after the infection. Both organ systems showed an increase in both mononuclear phagocytes and Lyt-2^+ T cells, significantly altering T cell/macrophage ratios. This was associated with decreases in *in vitro* cell proliferation to *T. gondii* antigen and concanavalin A. This latter type of infection is usually accompanied by development of tissue cysts in the brain and the establishment of an asymptomatic chronic infection. Major factors which may play a part in the eventual outcome of the infection are the virulence and numbers of the infecting strain,[250,251] the immunological and hormonal,[252-254] genetic,[255-257] and infection[258-260] status, of the animal. In addition, some species appear to be more susceptible than others to toxoplasmosis, probably because they have evolved in the absence of the parasite. Examples here are the Australian marsupials (discussed in References 261 and 262) and arboreal neotropical monkeys.[213] On the contrary, terrestrial neotropical animals tend to get infections that are asymptomatic.[213,263]

Dutton et al.[264,265] have recently investigated the ultrastructural pathology and clinicopathological features of a congenital murine model of ocular toxoplasmosis, in some detail.

VII. VACCINES

I have recently extensively reviewed the work towards the development of vaccines, particularly molecular vaccines, against toxoplasmosis in animals, and suggested it is likely that a commercial vaccine against toxoplasmosis in animals should be available within the next 5 years.[81] Hermentin and Aspock[266] highlighted the disadvantages of molecular vaccines and suggested that a successful vaccine against toxoplasmosis in humans was a long way off. These views are both likely to prove correct, given the very stringent conditions that apply to human vaccines.

Since the publication of the reviews mentioned above, three papers on vaccination against toxoplasmosis, worthy of comment, have been published. Darcy et al.[267,268] found that passive transfer of serum from +/+ Fischer rats vaccinated with ES antigens (see Section III.B) to their otherwise highly susceptible nu/nu littermates conferred a significant degree of protection against a highly virulent *T. gondii* challenge. Test rats had a mean survival time of 29.5 days after challenge with 10^5 RH strain tachyzoites, whereas control rats had a mean survival time of only 14.1 days. Parasite-specific IgE appeared to be involved in the protection since IgE-depleted sera were unable to afford protection. Only ES antigens or viable tachyzoites were able to induce a significant protective IgE response in +/+ rats. In addition, platelets bearing surface IgE, and to a lesser extent, eosinophil-rich populations from +/+ rats infected with *T. gondii* or immunized with ES antigens, were cytotoxic for *T. gondii in vitro*. Also, adoptive transfer of platelets from immune rats confers a significant degree of protection to nu/nu rats against an otherwise lethal challenge with irradiated RH strain tachyzoites. This work highlights the necessity for considering the humoral immune response, as well as the cell mediated immune response, in studies on vaccination against toxoplasmosis. This is further highlighted by the work of Eisenhauer et al.[269] They found that high titers of antibody alone did not protect mice against congenital infection. However treatment that produced high titers of *T. gondii* antibody acted synergistically with treatment to activate macrophages with *Proprionibacterium acnes*, allowing this latter treatment to confer significant protection against congenital transmission of the parasite.

An appeal for the vaccination of cats against toxoplasmosis has recently been made.[270] The rationale being that vaccination of the definitive host to prevent oocyst excretion would eventually control toxoplasmosis. However, unless every single cat in a large area was vaccinated within a short period of time, thereafter all kittens were treated shortly after birth, and stray or feral cats were kept out of the area, it is difficult to see this as being a viable control program. Even if all these conditions were achievable, and I think it is very unlikely that they would be, the soil in the area would still contain numerous infectious oocysts that could transmit the disease to humans for several years.

VIII. TREATMENT

Given that a vaccine against toxoplasmosis in humans is still some way off, and that it may not be effective in immunosuppressed patients such as those with AIDS, a resurgence of interest in anti-*T. gondii* drugs has occurred in the last few years. In particular, micromethods using parasite uptake of ^{3}H-uracil have been developed to look at the *in vitro* effects of drugs against the highly virulent RH strain of *T. gondii*.[271-273]

Most research to develop anti-*T. gondii* drugs has been aimed at treating humans, although some studies have been aimed at treatment of cats in an attempt to suppress oocyst excretion.[274-276] Monensin, a carboxylic ionophore, does totally suppress oocyst shedding in kittens when the drug is given from 2 days before, until 16 days after, oral challenge with tissue cysts.[276] However, as discussed in Section VII with regard to vaccination of cats, treatment of the definitive host is unlikely to prove a practical method to prevent toxoplasmosis in animals or humans.

The most common symptom of acute toxoplasmosis in humans, cervical lymphadenopathy, is usually self-resolving, and treatment should be instigated in only severe or persisting cases. In these situations, the most effective therapeutic regimen found so far is a combination of pyrimethamine and a sulfonamide, usually sulfadiazine, although sulfapyrazine, sulfamethazine, and sulfamerzine have activities similar to that of sulfadiazine.[277] Because pyrimethamine is a folic acid antagonist, folinic acid can be given to reduce the bone marrow depression. Prophylactic pyrimethamine and folinic acid on their own have recently been used in a study of toxoplasmosis in cardiac recipients.[278] Of 21 patients who were seronegative before the transplant and received an organ from a seropositive donor, four of the seven not given pyrimethamine acquired primary toxoplasmosis, whereas only 2 of the 14 treated patients developed an infection.

The toxicity of pyrimethamine means that other drugs have been sought as alternatives for treatment of congenital toxoplasmosis. Remington and Desmonts[215] have reviewed the recommendations for treatment of congenital toxoplasmosis up until 1982. Although spiramycin, a macrolide antibiotic, is less active against toxoplasmosis than is the combination of pyrimethamine and sulfadiazine, it has been used to treat pregnant women. Recently, Daffos et al.[279] gave spiramycin to pregnant women once maternal infection was diagnosed or suspected, and this was supplemented with pyrimethamine and sulfadoxine or sulfadiazine when toxoplasmosis was diagnosed in the fetus. Although this group contained only 15 patients, the neonatal outcomes suggested that prenatal therapy in women who wish to continue their pregnancies after the diagnosis of fetal toxoplasmosis reduces the severity of the disease manifestations.

As retinochoroiditis caused by *T. gondii* is usually self-limiting, treatment is generally only required for central, severe or persistent retinochoroiditis.[215] In these situations, the most rational treatment regimen appears to be an anti-*T. gondii* agent to reduce proliferation and dissemination of the parasite, and a steroid to diminish tissue damage caused by immune responses. The dangers associated with the administration of only corticosteroids are well known.[280,281] The anti-*T. gondii* drug of choice when central vision is threatened is clindamycin, although its real efficacy is still controversial. An in-depth review of the literature by Dutton[215] suggested that clindamycin did have some effect in shortening the healing time of ocular lesions. However, several of the studies reviewed were not controlled and some patients received therapy in addition to clindamycin. *In vitro* studies have suggested that clindamycin has little effect on RH strain tachyzoites,[272,273] and animal studies have produced conflicting results.[273] However, the drug is concentrated in the ocular choroid,[282] and until further studies are performed, clindamycin appears to be superior to pyrimethamine and sulfonamides, and spiramycin, for the treatment of ocular toxoplasmosis.[215]

Clindamycin has also been used for the treatment of *T. gondii* encephalitis in AIDS patients. It has been used alone to reduce the pathology and numbers of cysts in brains in a murine model of *T. gondii* encephalitis,[283] but has been given together with pyrimethamine in human AIDS patients.[284-286] This overcomes the adverse reactions that AIDS patients may get to sulfonamides.[287] The likelihood of such adverse reactions may be increased because treatment for *T. gondii* encephalitis in AIDS patients should be continued for the life of the patient. Sulfadiazine and pyrimethamine treatment of *T. gondii* encephalitis in AIDS patients results in a dramatic rapid improvement in the clinical and radiological picture.[224,228] However, once this regimen is stopped, recrudescence of the disease occurs within weeks in as many as 70% of patients.[238,288,289] This necessitates the life-long therapy.[238,288]

Two new macrolide antibiotics, azithromycin and roxithromycin, have been shown to have anti-*T. gondii* activity in murine models.[290,291] Although roxithromycin has been found to be less effective than sulfadiazine/pyrimethamine for treating cerebral toxoplasmosis in mice,[292] when given together with interferon, the combination is remarkably synergistic and reduced death of mice and the number of cysts in the brain.[293] Azithromycin has been shown to protect mice against death due to toxoplasmic encephalitis when given on its own.[290] Therefore, with further study, these agents may prove to be of benefit in treating AIDS patients with cerebral toxoplasmosis.

Trimetrexate, a lipid-soluble anti-folate, shows potent anti-*T. gondii* activity *in vitro* and *in vivo*, by inhibiting the parasite dihydrofolate reductase.[294] These studies suggest that trimetrexate given alone, or together with sulfonamide, may provide an effective alternative to pyrimethamine and sulfonamide for the treatment of toxoplasmosis. In particular, because trimetrexate penetrates the central nervous system, it may become important in the treatment of cerebral toxoplasmosis in AIDS patients.[295]

IX. CONCLUSION/FUTURE TRENDS

In the last 10 years we have learned a great deal about the protist, *T. gondii*. Much of this knowledge has been focused at the molecular level, and we can now dissect many of the processes which previously were only investigated on a macro scale. We are now investigating individual steps in the immune response to the parasite, specific receptors in the parasite-host cell interaction, and the nucleotide sequences of cloned genes from mutant parasite strains. The phylogeny of *T. gondii* is becoming clearer, and we have new and more effective drugs with which to treat toxoplasmosis. The antigenic structure of several life cycle stages of the parasite is known in some detail, and we are learning more about the genetics, of the parasite itself, and also of the immune response to the parasite.

Yet the answers to several basic questions still elude us. Why is the parasite obligately intracellular? No reasons have been found for its dependence on host cells. When incubated in cell-free medium the tachyzoite is capable of limited nucleic acid synthesis and respiration, yet normal multiplication of this stage of the parasite has been observed only within host cells. Active synthesis of macromolecules by the host cell is not necessary to support the growth of the tachyzoite, and even host cell protein-synthesis does not appear to be required. If we could determine why the parasite is obligately intracellular, we may find a mechanism to eradicate it.

In addition, another of the parasite's biological properties deserves comment. *T. gondii* has probably the biggest host range of any eukaryotic parasite. In most situations it is the perfect parasite, living in symbiotic harmony in the brain or skeletal muscles of its host. The advent of AIDS has upset this harmony, and has been a major factor in increasing the amount of research carried out on *T. gondii*, particularly at the molecular level.

In reviewing the recent literature on *T. gondii* I have attempted to place a personal emphasis on the relevance of some of the work, and not all workers in the area will agree with my suggestions on where the research should, or will, proceed. Clearly, many papers have been omitted because they add little to the area or are irrelevant, and unfortunately there are probably a few publications that should have been included but are not mentioned because I was unaware of them.

I believe that in the next few years we will make use of recombinant parasite antigens and synthetic oligonucleotides in the diagnosis of toxoplasmosis. Knowledge of the parasite's physiology and biochemistry should lead to the introduction of newer more effective drugs, and several groups are working on preventing the parasite from entering cells. Although I think phylogenetic studies will probably suggest a revision of the genus taxon *Toxoplasma*, this taxon is so well known that it is unlikely to disappear. I have already suggested that commercial vaccines against toxoplasmosis in animals will be available in the not too distant future, although vaccines for humans may be further off. *T. gondii* is an important parasite of humans and animals, and certainly deserves the scientific attention it has received in the recent past, and the increased attention I believe it will receive in the future.

REFERENCES

1. **Jira, J. and Kozojed, V.,** *Toxoplasmose 1968—1975,* Vols. 1 and 2, Gustav Fischer, Stuttgart, 1983.
2. **Long, P. L.,** *Biology of the Coccidia,* University Park Press, Baltimore, 1982.
3. **Nicolle, M. M. C. and Manceaux, L.,** Sur un Protozoaire nouveau du Gondi, *C.R. Acad. Sci.,* 148, 369, 1909.
4. **Hutchison, W. M., Dunachie, J. F., Siims, J. C., and Work, K.,** Coccidian-like nature of *Toxoplasma gondii, Br. Med. J.,* 1, 142, 1970.
5. **Frenkel, J. K., Dubey, J. P., and Miller, N. L.,** *Toxoplasma gondii* in cats: fecal stages identified as coccidian oocysts, *Science,* 167, 893, 1970.
6. **Sheffield, H. G. and Melton, M. L.,** *Toxoplasma gondii*: the oocyst, sporozoite, and infection of cultured cells, *Science,* 167, 892, 1970.
7. **Wieland, G. and Kuhn, D.,** Experimentelle *Toxoplasma* — infektionen bei der katze, *Berl. Munch. Tierarztl. Wochenschr.,* 83, 128, 1970.
8. **Overdulve, J. P.,** The identity of *Toxoplasma* Nicolle and Manceaux, 1909 with *Isosopora* Schneider, 1881, *Proc. K. Ned. Akad. Wet. Ser. C,* 73, 129, 1970.
9. **Jewell, M. L., Frenkel, J. K., Johnson, K. M., Reed, V., and Ruiz, A.,** Development of *Toxoplasma* oocysts in neotropical Felidae, *Am. J. Trop. Med. Hyg.,* 21, 512, 1972.
10. **Miller, N. L., Frenkel, J. K., and Dubey, J. P.,** Oral infections with *Toxoplasma* cysts and oocysts in felines, other mammals, and in birds, *J. Parasitol.,* 58, 928, 1972.
11. **Jacobs, L., Remington, J. S., and Melton, M. L.,** The resistance of the encysted form of *Toxoplasma gondii, J. Parasitol.,* 46, 11, 1960.
12. **Dubey, J. P. and Frenkel, J. K.,** Cyst-induced toxoplasmosis in cats, *J. Protozool.,* 19, 155, 1972.
13. **Dubey, J. P.,** Direct development of enteroepithelial stages of *Toxoplasma* in the intestines of cats fed cysts, *Am. J. Vet. Res.,* 40, 1634, 1979.
14. **Dubey, J. P., Miller, N. L., and Frenkel, J. K.,** Characterization of the new fecal form of *Toxoplasma gondii, J. Parasitol.,* 56, 447, 1970.
15. **Dubey, J. P., Miller, N. L., and Frenkel, J. K.,** The *Toxoplasma gondii* oocyst from cat feces, *J. Exp. Med.,* 132, 636, 1970.
16. **Yilmaz, S. M. and Hopkins, S. H.,** Effects of different conditions on duration of infectivity of *Toxoplasma gondii* oocysts, *J. Parasitol.,* 58, 938, 1972.
17. **Swartzberg, J. E. and Remington, J. S.,** Transmission of *Toxoplasma, Am. J. Dis. Child.,* 129, 777, 1975.
18. **Jackson, P. R., Honigberg, B. M., and Holt, S. C.,** Lectin analysis of *T. congolese* blood stream trypomastigote and culture procyclic surface saccharide by agglutination and electron microscope technic, *J. Protozool.,* 25, 471, 1978.
19. **Dwyer, D. M.,** Lectin binding saccharides on a parasitic protozoan, *Science,* 184, 471, 1974.
20. **Sethi, K. K., Rahmann, A., Pelster, B., and Brandis, H.,** Search for the presence of lectin-binding sites on *Toxoplasma gondii, J. Parasitol.,* 63, 1076, 1977.
21. **Johnson, A. M., McDonald, P. J., and Neoh, S. H.,** Molecular weight analysis of the major polypeptides and glycopeptides of *Toxoplasma gondii, Biochem. Biophys. Res. Commun.,* 100, 934, 1981.
22. **Mauras, G., Dodeur, M., Laget, P., Senet, J. M., and Bourrillon, R.,** Partial resolution of the sugar content of *Toxoplasma gondii* membrane, *J. Biochem. Biophys. Res. Commun.,* 97, 906, 1980.
23. **Crane, M. St. J. and Dvorak, J. A.,** Influence of monosaccharides on the infection of vertebrate cells by *Trypanosoma cruzi* and *Toxoplasma gondii, Mol. Biochem. Parasitol.,* 5, 333, 1982.
24. **Blackwell, J. M., Ezekowitz, R. A. B., Roberts, M. B., Channon, J. Y., Sim, R. B., and Gordon, S.,** Macrophage complement and lectin-like receptors bind *Leishmania* in the absence of serum, *J. Exp. Med.,* 162, 324, 1985.
25. **Mosser, D. M. and Edelson, P. J.,** The mouse macrophage receptor for C3bi (CR3) is a major mechanism in the phagocytosis of *Leishmania* promastigotes, *J. Immunol.,* 135, 2785, 1985.
26. **Finlay-Jones, J., Essery, M., Lowe, A., and Johnson, A. M.,** unpublished data, 1989.
27. **Pfefferkorn, E. R. and Pfefferkorn, L. C.,** Specific labeling of intracellular *Toxoplasma gondii* with uracil, *J. Protozool.,* 24, 449, 1977.
28. **Pulvertaft, R. J. V., Valentine, J. C., and Lane, W. F.,** The behaviour of *Toxoplasma gondii* on serum-agar culture, *Parasitology,* 44, 478, 1954.
29. **Bommer, W.,** The lifecycle of virulent *Toxoplasma* in cell cultures, *Aust. J. Exp. Biol. Med. Sci.,* 47, 505, 1969.
30. **Zaman, V. and Colley, F. C.,** Ultrastructural study of penetration of macrophages by *Toxoplasma gondii, Trans. R. Soc. Trop. Med. Hyg.,* 66, 781, 1972.
31. **Ryning, F. W. and Remington, J. S.,** Effect of cytochalasin D on *Toxoplasma gondii* cell entry, *Infect. Immun.,* 20, 739, 1978.

32. **Jones, T. C., Yeh, S., and Hirsch, T. G.,** The interaction between *Toxoplasma gondii* and mammalian cells I. Mechanism of entry and intracellular fate of the parasite, *J. Exp. Med.*, 136, 1157, 1972.
33. **Aikawa, M., Komata, Y., Asai, T., and Midorikawa, O.,** Transmission and scanning electron microscopy of host cell entry by *Toxoplasma gondii, Am. J. Pathol.*, 87, 285, 1977.
34. **Silva, S. R. L., Meirelles, S. S. L., and De Souza, W.,** Mechanism of entry of *Toxoplasma gondii* into vertebrate cells, *J. Submicrosc. Cytol.*, 14, 471, 1982.
35. **Nichols, B. A. and O'Connor, G. R.,** Penetration of mouse peritoneal macrophages by the protozoon *Toxoplasma gondii, Lab. Invest.*, 44, 324, 1981.
36. **Dvorak, J. A. and Howe, C. L.,** *Toxoplasma gondii* — Vertebrate cell interactions. I. The influence of bicarbonate ion, CO_2, pH and host cell culture age on the invasion of vertebrate cells *in vitro, J. Protozool.*, 24, 416, 1971.
37. **Kimata, I., Tanabe, K., Izumo, A., and Takada, S.,** Host cell ATP level and invasion of *Toxoplasma gondii, Trans. R. Soc. Trop. Med. Hyg.*, 81, 377, 1987.
38. **Lycke, E. and Norrby, R.,** Demonstration of a factor of *Toxoplasma gondii* enhancing the penetration of *Toxoplasma* parasites into cultured host cells, *Br. J. Exp. Pathol.*, 46, 189, 1965.
39. **Lycke, E., Carlberg, K., and Norrby, R.,** Interactions between *Toxoplasma gondii* and its host cell: function of the penetration enhancing factor of *Toxoplasma, Infect. Immun.*, 11, 853, 1975.
40. **Saffer, L. D., Kong Krug, S. A., and Schwartzman, J. D.,** The role of phospholipase in host cell penetration by *Toxoplasma gondii, Am. J. Trop. Med. Hyg.*, 40, 145, 1989.
41. **Locksley, R. M., Farkhauser, J., and Henderson, W. R.,** Alteration of leukotriene release by macrophages ingesting *Toxoplasma gondii, Proc. Natl. Acad. Sci. U.S.A.*, 82, 6922, 1985.
42. **Hughes, H. P. A.,** Oxidative killing of intracellular parasites mediated by macrophages, *Parasitol. Today*, 4, 340, 1988.
43. **Wilson, C. B., Tsai, Y., and Remington, J. S.,** Failure to trigger the oxidative metabolic burst by normal macrophages, *J. Exp. Med.*, 151, 328, 1980.
44. **Catterall, J. R., Sharma, S. D., and Remington, J. S.,** Oxygen-independent killing by alveolar macrophages, *J. Exp. Med.*, 163, 113, 1986.
45. **Gustafson, P. V., Agar, H. D., and Cramer, D. I.,** An electron microscope study of *Toxoplasma, Am. J. Trop. Med. Hyg.*, 3, 1008, 1954.
46. **Doran, D. J.,** Behaviour of coccidia in vitro, in *The Biology of the Coccidia*, Long, P. L., Ed., University Park Press, Baltimore, 1982, 253.
47. **Jones, T. C.,** Multiplication of *Toxoplasmas* in enucleate fibroblasts, *Proc. Soc. Exp. Biol. Med.*, 142, 1268, 1973.
48. **Sethi, K. K., Pelster, B., Pierkarski, G., and Brandis, H.,** Multiplication of *Toxoplasma gondii* in enucleated L cells, *Nature (New Biol.)*, 243, 255, 1973.
49. **Remington, J. S., Bloomfield, M. M., Russell, E., and Robinson, W. S.,** The RNA of *Toxoplasma gondii, Proc. Soc. Exp. Biol. Med.*, 133, 623, 1970.
50. **Goldman, M., Carver, R. K., and Sulzer, A. J.,** Reproduction of *Toxoplasma gondii* by internal budding, *J. Parasitol.*, 44, 161, 1958.
51. **Ogino, N. and Yoneda, C.,** The fine structure and mode of division of *Toxoplasma gondii, Arch. Ophthalmol.*, 75, 218, 1966.
52. **Ferguson, D. J. P. and Hutchison, W. M.,** Comparison of the development of a virulent and avirulent strains of *Toxoplasma gondii* in the peritoneal exudate of mice, *Ann. Trop. Med. Parasitol.*, 75, 539, 1981.
53. **Gallois, Y., Foussard, F., Girault, A., Hodbert, J., Tricaud, A., Mauras, G., and Motta, C.,** Membrane fluidity of *Toxoplasma gondii*: a fluorescence polarization study, *Biol. Cell*, 62, 11, 1988.
54. **Lainson, R.,** Observations on the development and nature of pseudocysts and cysts of *Toxoplasma gondii, Trans. R. Soc. Trop. Med. Hyg.*, 52, 396, 1958.
55. **Wanko, T., Jacobs, L., and Gavin, M. A.,** Electron microscope study of *Toxoplasma* cysts in mouse brain, *J. Protozool.*, 9, 235, 1962.
56. **Van Den Zypen, E. and Pierkarski, G.,** Ultrastructural differences between the so-called proliferative stage (RH strain, BK strain) and the so-called cystic stage (DX strain) of *Toxoplasma gondii, Zentralbl. Bakteriol. Orig. A*, 203, 495, 1967.
57. **Beyer, T. V., Siim, J. C., and Hutchison, W. M.,** Cytochemistry of *Toxoplasma gondii*. VI. Polysaccharides, lipids and phosphatases in the cyst forms, *Tsitologiia*, 19, 919, 1977.
58. **Jacobs, L.,** *Toxoplasma* and toxoplasmosis, *Adv. Parasitol.*, 5, 1, 1967.
59. **Jira, J., Zitora, D., and Princova, D.,** Kinetics of appearance of cysts and antibodies in experimental mouse toxoplasmosis, *Folia Parasitol. (Praha)*, 18, 295, 1971.
60. **Hogan, M. J., Yoneda, C., and Zwiegart, P.,** Growth of *Toxoplasma* strains in tissue culture, *Am. J. Ophthalmol.*, 51, 48, 1961.
61. **Stahl, W., Matsubayashi, H., and Akao, S.,** Effects of 6-mercaptopurine on cyst development in experimental toxoplasmosis, *Keio J. Med.*, 14, 1, 1965.
62. **Frenkel, J. K., Nelson, B. M., and Arias-Stella, J.,** Immunosuppression and toxoplasmic encephalitis, *Hum. Pathol.*, 6, 97, 1975.

63. **Brinkmann, V., Remington, J. S., and Sharma, S. D.,** Protective immunity in toxoplasmosis: correlation between antibody response, brain cyst formation, T-cell activation, and survival in normal B-cell-deficient mice bearing the H-2^k haplotype, *Infect. Immun.*, 55, 990, 1987.
64. **Frenkel, J. K., Dubey, J. P., and Hoff, R. L.,** Loss of stages after continuous passage of *Toxoplasma gondii* and *Besnoitia jellisoni*, *J. Protozool.*, 23, 421, 1976.
65. **Matsubayashi, H. and Akao, S.,** Morphological studies on the development of the *Toxoplasma* cyst, *Am. J. Trop. Med. Hyg.*, 12, 321, 1963.
66. **Sims, T. A., Hay, J., and Talbot, I. C.,** Host-parasite relationship in the brains of mice with congenital toxoplasmosis, *J. Pathol.*, 156, 255, 1988.
67. **Matsubayashi, H. and Akao, S.,** Immunoelectron microscopic studies on *Toxoplasma gondii*, *Am. J. Trop. Med. Hyg.*, 15, 486, 1966.
68. **Savin, A. B. and Feldman, H. A.,** Dyes as microchemical indicators of a new immunity phenomenon affecting a protozoon parasite *(Toxoplasma)*, *Science*, 108, 660, 1948.
69. **Johnson, A. M.,** The antigenic structure of *Toxoplasma gondii*: a review, *Pathology*, 17, 9, 1985.
70. **Hughes, H. P. A.,** Toxoplasmosis: the need for improved diagnostic techniques and accurate risk assessment, *Curr. Top. Microbiol. Immunol.*, 120, 105, 1985.
71. **Sabin, A. B.,** Toxoplasmic encephalitis in children, *JAMA*, 116, 801, 1941.
72. **Handman, E., Goding, J. W., and Remington, J. S.,** Detection and characterization of membrane antigens of *Toxoplasma gondii*, *J. Immunol.*, 124, 2578, 1980.
73. **Johnson, A. M., McDonald, P. J., and Neoh, S. H.,** Monoclonal antibodies to *Toxoplasma* cell membrane surface antigens protect mice from toxoplasmosis, *J. Protozool.*, 30, 351, 1983.
74. **Kasper, L. H., Crabb, J. H., and Pfefferkorn, E. R.,** Purification of a major membrane protein of *Toxoplasma gondii* by immunoabsorption with a monoclonal antibody, *J. Immunol.*, 130, 2407, 1983.
75. **Couvreur, G., Sadak, A., Fortier, B., and Dubremetz, J. F.,** Surface antigens of *Toxoplasma gondii*, *Parasitology*, 97, 1, 1988.
76. **Decoster, A., Darcy, F., and Capron, A.,** Recognition of *Toxoplasma gondii* excreted and secreted antigens by human sera from acquired and congenital toxoplasmosis: identification of markers of acute and chronic infection, *Clin. Exp. Immunol.*, 73, 376, 1988.
77. **Weiss, L. M., Udem, S., Tanowitz, A., and Wittner, M.,** Western blot analysis of the antibody response of patients with AIDS and *Toxoplasma* encephalitis: antigenic diversity among *Toxoplasma* strains, *J. Infect. Dis.*, 157, 7, 1988.
78. **Deletoille, P., Ovlaque, G., and Slizewicz, B.,** Advantage of using 35,000-molecular-weight protein for testing of *Toxoplasma gondii* immunoglobulin M, *J. Clin. Microbiol.*, 26, 796, 1988.
79. **Potasman, I. and Remington, J. S.,** Author's reply, *J. Clin. Microbiol.*, 26, 797, 1988.
80. **Dubremetz, J. F., Rodriguez, C., and Ferreira, E.,** *Toxoplasma gondii*: redistribution of monoclonal antibodies on tachyzoites during host cell invasion, *Exp. Parasitol.*, 59, 24, 1985.
81. **Johnson, A. M.,** *Toxoplasma* Vaccines, in *Veterinary Protozoan and Hemoparasite Vaccines*, Wright, I. G., Ed., CRC Press, Boca Raton, FL, 1989, chap. 9.
82. **Johnson, A. M., Haynes, W. D., Leppard, P. J., McDonald, P. J., and Neoh, S. H.,** Ultrastructural and biochemical studies on the immunohistochemistry of *Toxoplasma gondii* antigens using monoclonal antibodies, *Histochemistry*, 77, 209, 1983.
83. **Rodriguez, C., Afchain, D., Capron, A., Dissous, C., and Santoro, F.,** Major surface protein of *Toxoplasma gondii* (P30) contains an immunodominant region with repetitive epitopes, *Eur. J. Immunol.*, 15, 747, 1985.
84. **Santoro, F., Charif, H., and Capron, A.,** The immunodominant epitope of the major membrane tachyzoite protein (P30) of *Toxoplasma gondii*, *Parasite Immunol.*, 8, 631, 1986.
85. **Burg, J. L., Perelman, D., Kasper, L. H., Ware, P. L., and Boothroyd, J. C.,** Molecular analysis of the gene encoding the major surface antigen of *Toxoplasma gondii*, *J. Immunol.*, 141, 3584, 1988.
86. **Nagel, S. D. and Boothroyd, J. C.,** The major surface antigen, P30, or *Toxoplasma gondii* is anchored by a glycolipid, *J. Biol. Chem.*, 264, 5569, 1989.
87. **Khan, I. A., Eckel, M. E., Pfefferkorn, E. R., and Kasper, L. H.,** Production of γ interferon by cultured human lymphocytes stimulated with a purified membrane protein (P30) from *Toxoplasma gondii*, *J. Infect. Dis.*, 157, 979, 1988.
88. **Sibley, L. D., Krahenbuhl, J. L., and Adams, G. M. W., and Weidner, E.,** *Toxoplasma* modifies macrophage phagosomes by secretion of a vesicular network rich in surface proteins, *J. Cell Biol.*, 103, 867, 1986.
89. **Sibley, L. D. and Krahenbuhl, J. L.,** Modification of host cell phagosomes by *Toxoplasma gondii* involves redistribution of surface proteins and secretion of a 32 kDa protein, *Eur. J. Cell Biol.*, 47, 81, 1988.
90. **Hughes, H. P. A. and Van Knapen, F.,** Characterisation of a secretory antigen from *Toxoplasma gondii* and its role in circulating antigen production, *Int. J. Parasitol.*, 12, 433, 1982.
91. **Hughes, H. P. A.,** The antigenic structure of *Toxoplasma gondii*, *Lyon Med.*, 248, 135, 1982.

92. **Hughes, H. P. A.,** Characterization of the circulating antigen of *Toxoplasma gondii, Immunol. Lett.*, 3, 99, 1981.
93. **Roques, C., Bessieres, M. H., and Seguela, J. P.,** Caracterisation immunochimique des proteines des exo-antigenes provenant de differentes souches de *Toxoplasma gondii, Bull. Soc. Fr. Parasitol.*, 4, 79, 1986.
94. **Chumpitazi, B., Ambroise-Thomas, P., Cagnard, M., and Autheman, J. M.,** Isolation and characterization of *Toxoplasma* exo-antigens from *in vitro* culture in MRC5 and vero cells, *Int. J. Parasitol.*, 17, 829, 1987.
95. **Frenkel, J. K.,** Dermal hypersensitivity to *Toxoplasma* antigens (toxoplasmins), *Proc. Soc. Exp. Biol. Med.*, 68, 634, 1948.
96. **Rougier, D. and Ambroise-Thomas, P.,** Detection of toxoplasmic immunity by multipuncture skin test with excretory-secretory antigen, *Lancet*, 2, 121, 1985.
97. **Schwartzman, J. D., Krug, E. C., Binder, L. I., and Payne, M. R.,** Detection of the microtubule cytoskeleton of the coccidian *Toxoplasma gondii* and the hemoflagellate *Leishmania donovani* by monoclonal antibodies specific for β-tubulin, *J. Protozool.*, 32, 747, 1985.
98. **Nagel, S. D. and Boothroyd, J. C.,** The α- and β-tubulins of *Toxoplasma gondii* are encoded by single copy genes containing multiple introns, *Mol. Biochem. Parasitol.*, 29, 261, 1988.
99. **Schwartzman, J. D.,** Inhibition of a penetration-enhancing factor of *Toxoplasma gondii* by monoclonal antibodies specific for rhoptries, *Infect. Immun.*, 51, 760, 1986.
100. **Sadak, A., Taghy, Z., Fortier, B., and Dubremetz, J. F.,** Characterization of a family of rhoptry proteins of *Toxoplasma gondii, Mol. Biochem. Parasitol.*, 29, 203, 1988.
101. **Kimata, I. and Tanabe, K.,** Secretion by *Toxoplasma gondii* of an antigen that appears to become associated with the parasitophorous vacuole membrane upon invasion of the host cell, *J. Cell Sci.*, 88, 231, 1987.
102. **Sibley, L. D. and Sharma, S. D.,** Ultrastructural localization of an intracellular *Toxoplasma* protein that induces protection in mice, *Infect. Immun.*, 55, 2137, 1987.
103. **Sharma, S. D., Araujo, F. G., and Remington, J. S.,** *Toxoplasma* antigen isolated by affinity chromatography with monoclonal antibody protects mice against lethal infection with *Toxoplasma gondii, J. Immunol.*, 133, 2818, 1984.
104. **Asai, T., O'Sullivan, W. J., and Tatibana, M.,** A potent nucleoside triphosphate hydrolase from the parasitic protozoan *Toxoplasma gondii, J. Biol. Chem.*, 258, 6816, 1983.
105. **Asai, T., Kanzawa, T., Kobayashi, S., Takeuchi, T., and Kim, T.,** Do protozoa conceal a high potency of nucleoside triphosphate hydrolysis present in *Toxoplasma gondii, Comp. Biochem. Physiol.*, 85B, 365, 1986.
106. **Asai, T., Kim, T., Kobayashi, M., and Kojima, S.,** Detection of nucleoside triphosphate hydrolase as a circulating antigen in sera of mice infected with *Toxoplasma gondii, Infect. Immun.*, 55, 1332, 1987.
107. **Akao, S.,** Ultramicroscopic studies of the localization of adenosine triphosphate activity and H^3 glucose transport in *Toxoplasma gondii, Jpn. J. Parasitol.*, 18, 488, 1967.
108. **Handman, E. and Remington, J. S.,** Antibody responses to *Toxoplasma* antigens in mice infected with strains of different virulence, *Infect. Immun.*, 29, 215, 1980.
109. **Kasper, L. H. and Ware, P. L.,** Recognition and characterization of stage-specific oocyst/sporozoite antigens of *Toxoplasma gondii* by human antisera, *J. Clin. Invest.*, 75, 1570, 1985.
110. **Lunde, M. N. and Jacobs, L.,** Antigenic differences between endozoites and cystozoites of *Toxoplasma gondii, J. Parasitol.*, 69, 806, 1983.
111. **Kasper, L. H., Bradley, M. S., and Pfefferkorn, E. R.,** Identification of stage-specific sporozoite antigens of *Toxoplasma gondii* by monoclonal antibodies, *J. Immunol.*, 132, 443, 1984.
112. **Suzuki, Y., Thulliez, P., Desmonts, G., and Remington, J. S.,** Antigen (s) responsible for immunoglobulin G responses specific for the acute stage of *Toxoplasma* infection in humans, *J. Clin. Microbiol.*, 26, 901, 1988.
113. **Ware, P. L. and Kasper, L. H.,** Strain-specific antigens of *Toxoplasma gondii, Infect. Immun.*, 55, 778, 1987.
114. **Potasman, I., Araujo, F. G., and Remington, J. S.,** *Toxoplasma* antigens recognized by naturally occurring human antibodies, *J. Clin. Microbiol.*, 24, 1050, 1986.
115. **Potasman, I., Araujo, F. G., Desmonts, G., and Remington, J. S.,** Analysis of *Toxoplasma gondii* antigens recognized by human sera obtained before and after acute infection, *J. Infect. Dis.*, 154, 650, 1986.
116. **Potasman, I., Araujo, F. G., Thulliez, P., Desmonts, G., and Remington, J. S.,** *Toxoplasma gondii* antigens recognized by sequential samples of serum obtained from congenitally infected infants, *J. Clin. Microbiol.*, 25, 1926, 1987.
117. **Payne, R. A., Joynson, D. H. M., Balfour, A. H., Harford, J. P., Fleck, D. G., Mythen, M., and Saunders, R. J.,** Public health laboratory service enzyme-linked immunosorbent assay for detecting *Toxoplasma* specific IgM antibody, *J. Clin. Pathol.*, 40, 276, 1987.
118. **Verhofstede, C., Van Gelder, P., and Rabaey, M.,** The infection-stage-related IgG response to *Toxoplasma gondii* studied by immunoblotting, *Parasitol. Res.*, 74, 516, 1988.

119. **Neimark, H. and Blaker, R. G.**, DNA base composition of *Toxoplasma gondii* grown in vivo, *Nature*, 216, 600, 1967.
120. **Perrotto, J., Keister, D. B., and Gelderman, A. H.**, Incorporation of precursors into *Toxoplasma* DNA, *J. Protozool.*, 18, 470, 1971.
121. **Cornelissen, A. W. C. A., Overdulve, J. P., and Van Der Ploeg, M.**, Determination of nuclear DNA of five Eucoccidian parasites, *Isospora (Toxoplasma) gondii, Sarcocystis cruzi, Eimeria tenella, E. acervulina*, and *Plasmodium berghei*, with special reference to gamontogenesis and meiosis in *I.(T). gondii*, *Parasitology*, 88, 531, 1984.
122. **Borst, P., Overdulve, J. P., Weijers, P. J., Fase-Fowler, F., and Van Den Berg, M.**, DNA circles with cruciforms from *Isospora (Toxoplasma) gondii, Biochem. Biophys. Acta*, 781, 100, 1984.
123. **Johnson, A. M., Illana, S., Dubey, J. P., and Dame, J. B.**, *Toxoplasma gondii* and *Hammondia hammondi*: DNA comparison using cloned rRNA gene probes, *Exp. Parasitol.*, 63, 272, 1987.
124. **Johnson, A. M., Illana, S., McDonald, P. J., and Asai, T.**, The cloning, expression and nucleotide sequence of the gene fragment encoding the antigenic portion of the nucleoside triphosphate hydrolase of *Toxoplasma gondii*, *Gene*, 85, 215, 1989.
125. **Prince, J. B., Koven-Quinn, M. A., Remington, J. S., and Sharma, S. D.**, Cell free synthesis of *Toxoplasma gondii* antigens, *Mol. Biochem. Parasitol.*, 17, 163, 1985.
126. **Johnson, A. M., McDonald, P. J., and Illana, S.**, Characterization and in vitro translation of *Toxoplasma gondii* ribonucleic acid, *Mol. Biochem. Parasitol.*, 18, 313, 1986.
127. **Dubey, J. P. and Beattie, C. P.**, *Toxoplasmosis of Animals and Man*, CRC Press, Boca Raton, FL, 1989.
128. **Levine, N. D.**, Phylum II, Apicomplexa Levine 1970, in *Illustrated Guide to the Protozoa*, Lee, T. J., Hutner, S. H., and Bovee, E. C., Eds., Society for Protozoology, Lawrence, KS, 1985, 322.
129. **Barta, J. R.**, Phylogenetic analysis of the Class Sporozoea (Phylum Apicomplexa Levine 1970): evidence for the independent evolution of heteroxenous life cycles, *J. Parasitol.*, 75, 195, 1989.
130. **Johnson, A. M. and Baverstock, P. R.**, Rapid ribosomal RNA sequencing for phylogenetic analyses, *Parasitol. Today*, 5, 102, 1989.
131. **Johnson, A. M., Illana, S., Hakendorf, P., and Baverstock, P. R.**, Phylogenetic relationships of the apicomplexan protist *Sarcocystis* as determined by small subunit ribosomal RNA comparison, *J. Parasitol.*, 74, 847, 1988.
132. **Johnson, A. M., Murray, P. J., Illana, S., and Baverstock, P. J.**, Rapid nucleotide sequence analysis of the small subunit ribosomal RNA of *Toxoplasma gondii*: evolutionary implications for the Apicomplexa, *Mol. Biochem. Parasitol.*, 25, 239, 1987.
133. **Johnson, A. M., Fielke, R., Lumb, R., and Baverstock, P. R.**, Phylogenetic affinities of *Cryptosporidium* determined by ribosomal RNA comparison, *Int. J. Parasitol.*, in press.
134. **Levine, N. D.**, Taxonomy of *Toxoplasma*, *J. Protozool.*, 24, 36, 1977.
135. **Baker, J. R.**, The *Toxoplasma* tangle, *Parasitol. Today*, 3, 103, 1987.
136. **Darde, M. L., Bouteille, B., and Pestre-Alexandre, M.**, Isoenzymic characterization of seven strains of *Toxoplasma gondii* by isoelectrofocusing in polyacrylamide gels, *Am. J. Trop. Med. Hyg.*, 39, 551, 1988.
137. **Crowle, A. S.**, *Immunodiffusion*, Academic Press, New York, 1973, chap. 2.
138. **Zola, H.**, Speaking personally: monoclonal antibodies as diagnostic reagents, *Pathology*, 17, 53, 1985.
139. **Yolken, R. H.**, Nucleic acids or immunoglobulins: which are the molecular probes of the future?, *Mol. Cell. Probes.*, 2, 87, 1988.
140. **Payne, W. J., Marshall, D. L., Shockley, R. K., and Martin, W. J.**, Clinical laboratory applications of monoclonal antibodies, *Clin. Microbiol. Rev.*, 1, 313, 1988.
141. **Wetherall, B. L. and Johnson, A. M.**, Nucleic acid probes for *Campylobacter* species, in *Microbiol. Gene Probes*, Macario, A. J. L. and Macario, E. C., Eds., Academic Press, Orlando, FL, in press.
142. **Tenova, F. C.**, Diagnostic deoxyribonucleic acid probes for infectious diseases, *Clin. Microbiol. Rev.*, 1, 82, 1988.
143. **Eisenstein, B. I. and Engleberg, N. C.**, Applied molecular genetics: new tools for microbiologists and clinicians, *J. Infect. Dis.*, 153, 416, 1986.
144. **Wirth, D. F. and McMahon Pratt, D.**, Rapid identification of *Leishmania* species by specific hybridization of kinetoplast DNA in cutaneous lesions, *Proc. Natl. Acad. Sci. U.S.A.*, 79, 6999, 1982.
145. **Ashall, F., Yi-Chuck, D. A. M., Luquetti, A., and Miles, M. A.**, Radiolabeled total parasite DNA probe specifically detects *Trypanosoma cruzi* in mammalian blood, *J. Clin. Microbiol.*, 26, 576, 1988.
146. **McLaughlin, G. L., Edlind, T. D., and Ihler, G. M.**, Detection of *Babesia bovis* using DNA hybridization, *J. Protozool.*, 33, 125, 1986.
147. **Barker, R. J., Suebsaeng, L., Rooney, W., Alecrim, G. C., Dourado, H. V., and Wirth, D. F.**, Specific DNA probe for the diagnosis of *Plasmodium flaciparum* malaria, *Science*, 231, 1434, 1986.
148. **McCutchan, T. F., de la Cruz, V. F., Lal, A. A., Gunderson, T. H., Elwood, H. J., and Sogin, M. L.**, Primary sequences of two small subunit ribosomal RNA genes from *Plasmodium falciparum*, *Mol. Biochem. Parasitol.*, 28, 63, 1988.

149. **Gobel, V. B., Geiser, A., and Stanbridge, E. J.,** Oligonucleotide probes complementary to variable regions of ribosomal RNA discriminate between *Mycoplasma* species, *J. Gen. Microbiol.*, 133, 1969, 1987.
150. **Stahl, D. A.,** Evolution, ecology and diagnosis, *Biotechnology*, 4, 623, 1986.
151. **Haun, G. and Gobel, U.,** Oligonucleotide probes for genus-, species- and subspecies-specific identification of representatives of the genus *Proteus, FEMS Microbiol. Lett.*, 43, 187, 1987.
152. **Dams, E., Hendriks, L., Van de Peer, Y., Neefs, J.-M., Smits, G., Vandembempt, I., and De Wachter, R.,** Compilation of small ribosomal subunit RNA sequences, *Nucl. Acids Res.*, 16, r87, 1988.
153. **Johnson, A. M. and Fielke, R.,** unpublished data, 1988.
154. **Waters, A. P. and McCutchan, T. F.,** Diagnostic DNA amplification — not the complete answer, *Parasitol. Today*, 5, 9, 1989.
155. **Mullis, K. B. and Faloona, F. A.,** Specific synthesis of DNA *in vitro* via a polymerase-catalyzed chain reaction, in *Methods in Enzymology*, Wu, R., Ed., Acacemic Press, New York, 155, 335, 1987.
156. **de Bruijn, M. H. L.,** Diagnostic DNA amplification: no respite for the elusive parasite, *Parasitol. Today*, 4, 293, 1988.
157. **Burg, J. L., Grover, C. M., Pouletty, P., and Boothroyd, J. C.,** Direct and sensitive detection of a pathogenic protozoan, *Toxoplasma gondii*, by polymerase chain reaction, *J. Clin. Micro.*, 27, 1787, 1989.
158. **Sharma, S. D. and Remington, J. S.,** Macrophage activation and resistance to intracellular infection, in *Lymphokines and Lymphokines in Macrophage Activation*, Vol. 3, Pick, E., Ed., Academic Press, New York, 1981, 181.
159. **Frenkel, J. K.,** Experimental analysis of tissue immunity in toxoplasmosis, *Lyon Med.*, 248, 67, 1982.
160. **Krahenbuhl, J. L. and Remington, J. S.,** The immunology of *Toxoplasma* and toxoplasmosis, in *Immunology of Parasitic Infections*, Cohen, S. and Warren, K. S., Eds., Blackwell, London, 1982, 356.
161. **Britten, V. and Hughes, H. P. A.,** American trypanosomiasis, toxoplasmosis and leishmaniasis: intracellular infections with different immunological consequences, *Clin. Immunol. Allergy*, 6, 189, 1986.
162. **Hughes, H. P. A.,** Immunological aspects of toxoplasmosis in humans, *ISI Atlas Sci. Immunol.*, 1, 185, 1988.
163. **Tremonti, L. and Walton, B. C.,** Blast transformation and migration-inhibition in toxoplasmosis and leishmaniasis, *Am. J. Trop. Med. Hyg.*, 19, 49, 1970.
164. **Krahenbuhl, J. L., Gaines, J. D., and Remington, J. S.,** Lymphocyte transformation in human toxoplasmosis, *J. infect. Dis.*, 125, 283, 1972.
165. **Wilson, C. B., Desmonts, G., Couvreur, J., and Remington, J. S.,** Lymphocyte transformation in the diagnosis of congenital *Toxoplasma* infection, *N. Engl. J. Med.*, 302, 785, 1980.
166. **McLeod, R., Beem, O., and Estes, R. G.,** Lymphocyte anergy specific to *Toxoplasma gondii* antigens in a baby with congenital toxoplasmosis, *J. Clin. Lab. Immunol.*, 17, 149, 1985.
167. **Johnson, W. D.,** Chronological development of cellular immunity in human toxoplasmosis, *Infect. Immun.*, 33, 948, 1981.
168. **Anderson, S. E., Krahenbuhl, J. H., and Remington, J. S.,** Longitudinal studies of lymphocyte response to *Toxoplasma* antigens in humans infected with *T. gondii, J. Clin. Lab. Immunol.*, 2, 293, 1979.
169. **Luft, B. J., Kansas, G., Engleman, E. G., and Remington, J. S.,** Functional and quantitative alterations in T lymphocyte subpopulations in acute toxoplasmosis, *J. Infect. Dis.*, 150, 761, 1984.
170. **Luft, B. J., Pedrotti, P. W., Engleman, E. G., and Remington, J. S.,** Induction of antigen-specific suppressor T cells during acute infection with *Toxoplasma gondii, J. Infect. Dis.*, 155, 1033, 1987.
171. **Hughes, H. P. A., Connelly, C. A., Strangeways, J. M., and Hudson, L.,** Antigen specific lymphocyte transformation induced by secreted antigens from *Toxoplasma gondii, Clin. Exp. Immunol.*, 58, 539, 1984.
172. **Hughes, H. P. A., Kasper, L. H., Little, J., and Dubey, J. P.,** Absence of a role for natural killer cells in the control of acute infection by *Toxoplasma gondii* oocysts, *Clin. Exp. Immunol.*, 72, 394, 1988.
173. **Goyal, M., Ganguly, N. K., and Mahajan, R. C.,** Natural killer cell cytotoxicity against *Toxoplasma gondii* in acute and chronic murine toxoplasmosis, *Med. Sci. Res.*, 16, 375, 1988.
174. **Kamiyama, T. and Hagiwara, T.,** Augmented followed by suppressed levels of natural cell-mediated cytotoxicity in mice infected with *Toxoplasma gondii, Infect. Immun.*, 36, 628, 1982.
175. **Kamiyama, T. and Tatsumi, M.,** Effect of *Toxoplasma* infection on the sensitivity of mouse thymocytes to natural killer cells, *Infect. Immun.*, 42, 789, 1983.
176. **Kamiyama, T.,** *Toxoplasma*-induced activities of peritoneal and spleen natural killer cells from beige mice against thymocytes and YAC-1 lymphoma targets, *Infect. Immun.*, 43, 973, 1984.
177. **Hauser, W. E., Sharma, S. D., and Remington, J. S.,** Natural killer cells induced by acute and chronic *Toxoplasma* infection, *Cell. Immunol.*, 69, 330, 1982.
178. **Hauser, W. E., Sharma, S. D., and Remington, J. S.,** Augmentation of NK cell activity by soluble and particulate fractions of *Toxoplasma gondii, J. Immunol.*, 131, 458, 1983.
179. **Sharma, S. D., Verhoef, J., and Remington, J. S.,** Enhancement of human natural killer cell activity by subcellular components of *Toxoplasma gondii, Cell. Immunol.*, 86, 317, 1984.
180. **Hauser, W. E. and Tsai, V.,** Acute *Toxoplasma* infection of mice induces spleen NK cells that are cytotoxic for *T. gondii* in vitro, *J. Immunol.*, 136, 313, 1986.

181. **Luft, B. J. and Remington, J. S.,** Effect of pregnancy on augmentation of natural killer cell activity by *Corynebacterium parvum* and *Toxoplasma gondii, J. Immunol.*, 132, 2375, 1984.
182. **Khan, I. A., Smith, K. A., and Kasper, L. H.,** Induction of antigen-specific parasiticidal cytotoxic T cell splenocytes by a major membrane protein (P30) of *Toxoplasma gondii, J. Immunol.*, 141, 3600, 1988.
183. **Remington, J. S., Krahenbuhl, J. L., and Mendenhall, J. W.,** A role for activated macrophages in resistance to infection with *Toxoplasma, Infect. Immun.*, 6, 829, 1972.
184. **Jones, T. C., Len, L., and Hirsch, J. G.,** Assessment in vitro of immunity against *Toxoplasma gondii, J. Exp. Med.*, 141, 466, 1975.
185. **Murray, H. W. and Cohn, Z. A.,** Macrophage oxygen-dependent antimicrobial activity. III. Enhanced oxidative metabolism as an expression of macrophage activation, *J. Exp. Med.*, 152, 1596, 1980.
186. **Murray, H. W., Juangbhanich, C. W., Nathan, C. F., and Cohn, Z. A.,** Macrophage oxygen-dependent antimicrobial activity. II. The role of oxygen intermediates, *J. Exp. Med.*, 150, 950, 1979.
187. **Sibley, L. D., Lawson, R., and Weidner, E.,** Superoxide dismutase and catalase in *Toxoplasma gondii, Mol. Biochem. Parasitol.*, 19, 83, 1986.
188. **Hughes, H. P. A., Boik, R. J., Gerhardt, S. A., and Speer, C. A.,** Susceptibility of *Eimeria bovis* and *Toxoplasma gondii* to oxygen intermediates and a new mathematical model for parasite killing, *J. Parasitol.*, 75, 489, 1989.
189. **Nathan, G. F., Murray, H. W., Wiebe, M. E., and Rubin, B. Y.,** Identification of interferon-γ as the lymphokine that activates human macrophage oxidative metabolism and antimicrobial activity, *J. Exp. Med.*, 160, 600, 1984.
190. **Nathan, C. F., Prendergast, T. J., Wiebe, M. E., Stanley, E. R., Platzer, E., Remold, H. G., Welte, K., Rubin, B. Y., and Murray, H. W.,** Activation of human macrophages: comparison of other cytokines with interferon-γ, *J. Exp. Med.*, 160, 600, 1984.
191. **De Titto, E. H., Catterall, J. R., and Remington, J. S.,** Activity of recombinant tumor necrosis factor on *Toxoplasma gondii* and *Trypanosoma cruzi, J. Immunol.*, 137, 1342, 1986.
192. **Sharma, S. D., Hofflin, J. M., and Remington, J. S.,** In vivo recombinant interleukin 2 administration enhances survival against a lethal challenge with *Toxoplasma gondii, J. Immunol.*, 135, 4160, 1985.
193. **Brinkmann, V., Sharma, S. D., and Remington, J. S.,** Different regulation of the L3T4-T cell subset by B cells in different mouse strains bearing the H-2^k haplotype, *J. Immunol.*, 137, 2991, 1986.
194. **Canessa, A., Pistoia, V., Roncella, S., Merli, A., Melioli, G., Terragna, A., and Ferrarini, M.,** An in vitro model for *Toxoplasma* infection in man: interaction between $CD4^+$ monoclonal T cells and macrophages results in killing of trophozoites, *J. Immunol.*, 140, 3580, 1988.
195. **Suzuki, Y. and Remington, J. S.,** Dual regulation of resistance against *Toxoplasma gondii* infection by Lyt-2^+ and Lyt-1^+, L3T4^+ T cells in mice, *J. Immunol.*, 140, 3943, 1988.
196. **Pfefferkorn, E. R. and Guyre, P. M.,** Inhibition of growth of *Toxoplasma gondii* in cultured fibroblasts by human recombinant gamma interferon, *Infect. Immun.*, 44, 211, 1984.
197. **Pfefferkorn, E. R.,** Interferon-γ blocks the growth of *Toxoplasma gondii* in human fibroblasts by inducing the host cells to degrade tryptophan, *Proc. Natl. Acad. Sci. U.S.A.*, 81, 908, 1984.
198. **Pfefferkorn, E. R., Eckel, M., and Rebhun, S.,** Interferon-γ suppresses the growth of *Toxoplasma gondii* in human fibroblasts through starvation for tryptophan, *Mol. Biochem. Parasitol.*, 20, 215, 1986.
199. **Pfefferkorn, E. R.,** The mechanism by which interferon gamma blocks the growth of *Toxoplasma gondii* in cultured fibroblasts, in *Host-Parasite Cellular and Molecular Interactions in Protozoal Infections*, Vol. H 11, Chang, K. P. and Snary, D., Eds., Springer-Verlag, Berlin, 1987, 345.
200. **McCabe, R. E., Luft, B. J., and Remington, J. S.,** Effect of murine interferon gamma on murine toxoplasmosis, *J. Infect. Dis.*, 150, 961, 1984.
201. **Locksley, R. M., Wilson, C. B., and Klebanoff, S. J.,** Role of endogenous and acquired peroxidase in the toxoplasmacidal activity of murine and human mononuclear phagocytes, *J. Clin. Invest.*, 69, 1099, 1982.
202. **Murray, H., Byrne, G. I., Rothermel, C. D., and Cartelli, D. M.,** Lymphokine enhances oxygen-independent activity against intracellular pathogens, *J. Exp. Med.*, 158, 234, 1983.
203. **Murray, H. W., Rubin, B. Y., Carriero, S. M., Harris, A. M., and Jaffee, E. A.,** Human mononuclear phagocyte antiprotozoal mechanisms: oxygen-dependent vs. oxygen-independent activity against intracellular *Toxoplasma gondii, J. Immunol.*, 134, 1982, 1985.
204. **Chinchilla, M. and Frenkel, J. K.,** Mediation of immunity to intracellular infection (*Toxoplasma* and *Besnoitia*) within somatic cells, *Infect. Immun.*, 19, 999, 1978.
205. **McCabe, R. E. and Remington, J. S.,** Mechanisms of killing of *Toxoplasma gondii* by rat peritoneal macrophages, *Infect. Immun.*, 52, 151, 1986.
206. **Catterall, J. R., Sharma, S. D., and Remington, J. S.,** Oxygen-independent killing by alveolar macrophages, *J. Exp. Med.*, 163, 1113, 1986.
207. **Sibley, L. D., Krahenbuhl, J. L., and Weidner, E.,** Lymphokine activation of J774G8 cells and mouse peritoneal macrophages challenged with *Toxoplasma gondii, Infect. Immun.*, 49, 760, 1985.
208. **Gray, G. F., Kimball, A. C., and Kean, B. H.,** The posterior cervical lymph node in toxoplasmosis, *Am. J. Pathol.*, 69, 349, 1972.

209. **Stansfeld, A. G.**, The histological diagnosis of toxoplasmic lymphadenitis, *J. Clin. Pathol.*, 14, 565, 1961.
210. **Miettinen, M. and Franssila, K.**, Malignant lymphoma simulating lymph node toxoplasmosis, *Histopathology*, 6, 129, 1982.
211. **Argyle, J. C., Schumann, G. B., Kjeldsberg, C. R., and Athens, J. W.**, Identification of a *Toxoplasma* cyst by fine-needle aspiration, *Am. J. Clin. Pathol.*, 80, 256, 1983.
212. **Kean, B. H.**, Astronomic odds (1:22), *Am. J. Clin. Pathol.*, 81, 272, 1984.
213. **Frenkel, J. K.**, Pathophysiology of toxoplasmosis, *Parasitol. Today*, 4, 273, 1988.
214. **Remington, J. S. and Desmonts, G.**, Toxoplasmosis in *Infectious Diseases of the Fetus and Newborn Infant*, Remington, J. S. and Klein, J. O., Eds., W. B. Saunders, Philadelphia, 1983, chap. 5.
215. **Dutton, G. N.**, The causes of tissue damage in toxoplasmic retinochoroiditis, *Trans. Ophthalmol. Soc. U.K.*, 105, 404, 1986.
216. **Vietzke, W. M., Gelderman, A. H., Grimley, P. M., and Valsamis, M. P.**, Toxoplasmosis complicating malignancy. Experience at the National Cancer Institute, *Cancer*, 21, 816, 1968.
217. **Carey, R. M., Kimball, A. C., Armstrong, D., and Lieberman, P. H.**, Toxoplasmosis. Clinical experiences in a cancer hospital, *Am. J. Med.*, 54, 30, 1973.
218. **Roth, J. A., Siegel, S. E., Levine, A. S., and Berard, C. W.**, Fatal recurrent toxoplasmosis in a patient initially infected via a leukocyte transfusion, *Am. J. Clin. Pathol.*, 56, 601, 1971.
219. **Reynolds, E. S., Walls, K. W., and Pfeiffer, R. I.**, Generalized toxoplasmosis following renal transplantation, *Arch. Intern. Med.*, 118, 401, 1966.
220. **Vilaseca, J., Arnau, J. M., Bacardi, R., Mieras, C., Serrano, A., and Navarro, C.**, Kaposi's sarcoma and *Toxoplasma gondii* brain abscess in a Spanish homosexual, *Lancet*, i, 572, 1982.
221. Centers for Disease Control Task Force on Kaposi's Sarcoma and Opportunistic Infections, Epidemiologic aspects of the current outbreak of Kaposi's sarcoma and opportunistic infections, *N. Engl. J. Med.*, 306, 248, 1982.
222. **Luft, B. J., Conley, F., and Remington, J. S.**, Outbreak of central-nervous-system toxoplasmosis in Western Europe and North America, *Lancet*, i, 781, 1983.
223. **Levy, R. M., Bredesen, D. E., and Rosenblum, M. L.**, Neurological manifestations of the acquired immuno-deficiency syndrome (AIDS): experience at UCSF and review of the literature, *J. Neurosurg.*, 62, 475, 1985.
224. **Navia, B. A., Petito, C. K., Gold, J. W. M., Cho, E. S., Jordan, B. D., and Price, R. W.**, Cerebral toxoplasmosis complicating the acquired immune deficiency syndrome: clinical and neuropathological findings in 27 patients, *Ann. Neurol.*, 19, 224, 1986.
225. **Mills, J.**, *Pneumocystis carinii* and *Toxoplasma gondii* infections in patients with AIDS, *Rev. Infect. Dis.*, 8, 1001, 1986.
226. **Luft, B. J. and Remington, J. S.**, Toxoplasmic encephalitis, *J. Infect. Dis.*, 157, 1, 1988.
227. **Ruskin, J. and Remington, J. S.**, Toxoplasmosis in the compromised host, *Ann. Intern. Med.*, 84, 193, 1976.
228. **Vollmer, T. L., Waldor, M. K., Steinman, L., and Conley, F. K.**, Depletion of T-4$^+$ lymphocytes with monoclonal antibody reactivates toxoplasmosis in the central nervous system: a model of superinfection in AIDS, *J. Immunol.*, 138, 3737, 1987.
229. **Hofflin, J. M., Conley, F. K., and Remington, J. S.**, Murine model of intracerebral toxoplasmosis, *J. Infect. Dis.*, 155, 550, 1987.
230. **Parke, D. W. and Font, R. L.**, Diffuse toxoplasmic retinochoroiditis in a patient with AIDS, *Arch. Ophthalmol.*, 104, 571, 1986.
231. **Holland, G. N., Engstrom, R. E., Glasgow, B. J., Berger, B. B., Daniels, S. A., Sidikaro, Y., Harmon, J. A., Fischer, D. H., Boyer, D. S., Rao, N. A., Eagle, R. C., Kreiger, A. E., and Foos, R. Y.**, Ocular toxoplasmosis in patients with the acquired immunodeficiency syndrome, *Am. J. Ophthalmol.*, 106, 653, 1988.
232. **Catterall, J. R., Hofflin, J. M., and Remington, T. S.**, Pulmonary toxoplasmosis, *Am. Rev. Respir. Dis.*, 133, 704, 1986.
233. **Crider, S. R., Horstman, W. G., and Massey, G. S.**, *Toxoplasma* orchitis: report of a case and a review of the literature, *Am. J. Med.*, 85, 421, 1988.
234. **Donovan-Post, M. J., Chan, J. C., Hensley, G. T., Hoffman, T. A., Moskowitz, L. B., and Lippmann, S.**, *Toxoplasma* encephalitis in Haitian adults with acquired immunodeficiency syndrome: a clinical-pathologic-CT correlation, *Am. J. Neuroradiol.*, 4, 155, 1983.
235. **Britton, C. B. and Miller, J. R.**, Neurological complications in acquired immunodeficiency syndrome (AIDS), *Neurol. Clin.*, 2, 315, 1984.
236. **Edwards, K. R. and Pendlebury, W. W.**, Central nervous system lymphoma versus toxoplasmosis in a patient with AIDS, *N. Engl. J. Med.*, 317, 1540, 1987.
237. **Pitchenik, A. E., Fischl, M. A., and Walls, K. W.**, Evaluation of cerebral mass lesions in acquired immunodeficiency syndrome, *N. Engl. J. Med.*, 308, 1099, 1983.

238. **Wanke, C., Tuazon, C. U., Kovacs, A., Dina, T., Davis, D. O., Barton, N., Katz, D., Lunde, M., Levy, C., Conley, F. K., Lane, H. C., Fauci, A. S., and Masur, H.,** *Toxoplasma* encephalitis in patients with acquired immunodeficiency syndrome: diagnosis and response to therapy, *Am. J. Trop. Med. Hyg.*, 36, 509, 1987.
239. **Fauci, A. S., Masur, H., Gelmann, E. P., Markham, P. D., Hahn, B. H., and Lane, H. C.,** The acquired immunodeficiency syndrome: an update, *Ann. Intern. Med.*, 102, 800, 1985.
240. **Ghatak, N. R. and Sawyer, D. R.,** A morphologic study of opportunistic cerebral toxoplasmosis, *Acta Neuropathol. (Berl.)*, 42, 217, 1978.
241. **Dubey, J. P.,** A review of toxoplasmosis in pigs, *Vet. Parasitol.*, 19, 181, 1986.
242. **Dubey, J. P.,** Toxoplasmosis in dogs, *Canine Practice*, 12, 7, 1985.
243. **Dubey, J. P. and Towle, A.,** Toxoplasmosis in sheep: a review and annotated bibliography, miscellaneous publication number 10 of the Commonwealth Institute of Parasitology, 1986.
244. **Dubey, J. P.,** Toxoplasmosis in goats, *Agri-Practice*, March, 43, 1987, 348.
245. **Henry, L. and Beverley, J. K. A.,** The response of the mouse R-E system to infection with *Toxoplasma gondii, Virchows Arch. B*, 20, 55, 1976.
246. **Pelster, B., Piekarski, G., and Suzuki, N.,** Histopathological changes in the thymus of the white mouse after infection with *Toxoplasma gondii (in German), Z. Parasitenkd.*, 49, 113, 1976.
247. **McLeod, R., Van Le, L., and Remington, J. S.,** *Toxoplasma gondii*: lymphocyte function during acute infection in mice, *Exp. Parasitol.*, 54, 55, 1982.
248. **Jones, T. C., Alkan, S., and Erb, P.,** Spleen and lymph node cell populations, *in vitro* cell proliferation and interferon-γ production during the primary immune response to *Toxoplasma gondii, Parasite Immunol.*, 8, 619, 1986.
249. **Ito, S., Tsunoda, K., and Suzuki, K.,** Distribution of *Toxoplasma gondii*, RH strain, in infected mice as determined by the fluorescent antibody technique and the histopathology of toxoplasmosis, *Natl. Inst. Anim. Health Q.*, 7, 208, 1967.
250. **Olisa, E. G., Herson, J., Headings, V. E., and Poindexter, H. A.,** *Toxoplasma gondii*: survival time and variability in mouse host strains, *Exp. Parasitol.*, 41, 307, 1977.
251. **Araujo, F. G., Williams, D. M., Grumet, F. C., and Remington, J. S.,** Strain-dependent differences in murine susceptibility to *Toxoplasma, Infect. Immun.*, 13, 1528, 1976.
252. **Pung, O. J. and Luster, M. I.,** *Toxoplasma gondii*: decreased resistance to infection in mice due to estrogen, *Exp. Parasitol.*, 61, 48, 1986.
253. **Kittas, C. and Henry, L.,** Effect of sex hormones on the response of mice to infection with *Toxoplasma gondii, Br. J. Exp. Pathol.*, 61, 590, 1980.
254. **Luft, B. J. and Remington, J. S.,** Effect of pregnancy on resistance to *Listeria monocytogenes* and *Toxoplasma gondii* infections in mice, *Infect. Immun.*, 38, 1164, 1982.
255. **Williams, D. M., Grumet, F. C., and Remington, J. S.,** Genetic control of murine resistance to *Toxoplasma gondii, Infect. Immun.*, 19, 416, 1978.
256. **Jones, T. C. and Erb, P.,** H-2 complex-linked resistance in murine toxoplasmosis, *J. Infect. Dis.*, 151, 739, 1984.
257. **Johnson, A. M.,** Strain-dependent, route of challenge-dependent, murine susceptibility to toxoplasmosis, *Z. Parasitenkd.*, 70, 303, 1984.
258. **Copeland, D. and Grove, D. I.,** Effects of *Toxoplasma gondii* (Gleadle strain) on the host-parasite relationship in trichinosis, *Int. J. Parasitol.*, 9, 205, 1979.
259. **Reid, H. W., Buxton, D., Gardiner, A. C., Pow, I., Finlayson, J., and MacLean, M. J.,** Immunosuppression in toxoplasmosis: studies in lambs and sheep infected with louping-ill virus, *J. Comp. Pathol.*, 92, 181, 1982.
260. **Kloetzel, K., Chieffi, P. P., Faleiros, J. J., and Merluzzi Filho, T. J.,** Mortality and other parameters of concomitant infections in albino mice: the *Schistosoma-Toxoplasma* model, *Trop. Geogr. Med.*, 29, 407, 1977.
261. **Johnson, A. M., Roberts, H., and Munday, B. L.,** Prevalence of *Toxoplasma gondii* in wild macropods, *Aust. Vet. J.*, 65, 199, 1988.
262. **Johnson, A. M., Roberts, H., and Statham, P., and Munday, B. L.,** Serodiagnosis of acute toxoplasmosis in macropods, *Vet. Parasitol.*, 34, 25, 1989.
263. **Frenkel, J. K. and Sousa, O. E.,** Antibodies to *Toxoplasma* in panamanian mammals, *J. Parasitol.*, 69, 244, 1983.
264. **Dutton, G. N., McMenamin, P. G., Hay, J., and Cameron, S.,** The ultrastructural pathology of congenital murine toxoplasmic retinochoroiditis. II. The morphology of the inflammatory changes, *Exp. Eye Res.*, 43, 545, 1986.
265. **Dutton, G. N., Hay, J., Hair, D. M., and Ralston, J.,** Clinicopathological features of a congenital murine model of ocular toxoplasmosis, *Graefe's Arch. Clin. Exp. Ophthalmol.*, 224, 256, 1986.
266. **Hermentin, K. and Aspock, H.,** Efforts towards a vaccine against *Toxoplasma gondii*: a review, *Zentralbl. Bakteriol. Hyg. A*, 269, 423, 1988.

267. **Darcy, F., Deslee, D., Santoro, F., Charif, H., Auriault, C., Decoster, A., Duquesne, V., and Capron, A.**, Induction of a protective antibody-dependent response against toxoplasmosis by *in vitro* excreted/secreted antigens from tachyzoites of *Toxoplasma gondii, Parasite Immunol.*, 10, 553, 1988.
268. **Ridel, P. R., Auriault, C., Darcy, F., Pierce, R. J., Leite, P., Santoro, F., Neyrinck, J. L., Kusnierz, J. P., and Capron, A.**, Protective role of IgE in immunocompromised rat toxoplasmosis, *J. Immunol.*, 141, 978, 1988.
269. **Eisenhauer, P., Mack, D. G., and McLeod, R.**, Prevention of peroral and congenital acquisition of *Toxoplasma gondii* by antibody and activated macrophages, *Infect. Immun.*, 56, 83, 1988.
270. **Menning, E. L.**, Prenatal management and congenital toxoplasmosis, *New Engl. J. Med.*, 319, 373, 1988.
271. **Grossman, P. L. and Remington, J. S.**, The effect of trimethoprim and sulfamethoxazole on *Toxoplasma gondii* in vitro and in vivo, *Am. J. Trop. Med. Hyg.*, 28, 445, 1979.
272. **Mack, D. G. and McLeod, R.**, New micromethod to study the effect of antimicrobial agents on *Toxoplasma gondii*: comparison of sulfadoxine and sulfadiazine individually and in combination with pyrimethamine and study of clindamycin, metronidazole, and cyclosporin A, *Antimicrob. Agents Chemother.*, 26, 26, 1984.
273. **Harris, C., Salgo, M. P., Tanowitz, H. B., and Wittner, M.**, In vitro assessment of antimicrobial agents against *Toxoplasma gondii, J. Infect. Dis.*, 157, 14, 1988.
274. **Dubey, J. P. and Yeary, R. A.**, Anticoccidial activity of 2-sulfamoyl-4,4-diaminodiphenylsulfone, sulfadiazine, pyrimethamine and clindamycin in cats infected with *Toxoplasma gondii, Can. Vet. J.*, 18, 51, 1977.
275. **Sheffield, H. G. and Melton, M. L.**, Effects of pyrimethamine and sulfadiazine on the intestinal development of *Toxoplasma gondii* in cats, *Am. J. Trop. Med. Hyg.*, 25, 379, 1976.
276. **Frenkel, J. K. and Smith, D. D.**, Inhibitory effects of monensin on shedding of *Toxoplasma* oocysts by cats, *J. Parasitol.*, 68, 851, 1982.
277. **McCabe, R. E. and Remington, J. S.**, The diagnosis and treatment of toxoplasmosis, *Eur. J. Clin. Microbiol.*, 2, 95, 1983.
278. **Wreghitt, T. G., Hakim, M., Gray, J. J., Balfour, A. H., Stovin, P. G. I., Stewart, S., Scott, J., English, T. A. H., and Wallwork, J.**, Toxoplasmosis in heart and heart and lung transplant recipients, *J. Clin. Pathol.*, 42, 194, 1989.
279. **Daffos, F., Forestier, F., Capella-Pavlovsky, M., Thulliez, P., Aufrant, C., Valenti, D., and Cox, W. L.**, Prenatal management of 746 pregnancies at risk for congenital toxoplasmosis, *N. Engl. J. Med.*, 318, 271, 1988.
280. **O'Connor, G. R. and Frenkel, J. K.**, Dangers of steroid treatment in toxoplasmosis, *Arch. Ophthalmol.*, 94, 213, 1976.
281. **Nicholson, D. H. and Wolchok, E. B.**, Ocular toxoplasmosis in an adult receiving long-term corticosteroid therapy, *Arch. Ophthalmol.*, 94, 248, 1976.
282. **De Haan, R. M., Metzler, C. M., Schellenberg, D., and Vanderbosch, W. D.**, Pharmacokinetic studies of clindamycin phosphate, *J. Clin. Pharmacol.*, 13, 190, 1973.
283. **Hofflin, J. M. and Remington, J. S.**, Clindamycin in a murine model of *Toxoplasmic* encephalitis, *Antimicrob. Agents Chemother.*, 31, 492, 1987.
284. **Rolston, K. V. I. and Hoy, J.**, Role of clindamycin in the treatment of central nervous system toxoplasmosis, *Am. J. Med.*, 83, 551, 1987.
285. **Podzanczer, D. and Gudiol, F.**, Clindamycin in cerebral toxoplasmosis, *Am. J. Med.*, 84, 800, 1988.
286. **Westblom, T. U. and Belshe, R. B.**, Clindamycin therapy of cerebral toxoplasmosis in an AIDS patient, *Scand. J. Infect. Dis.*, 20, 561, 1988.
287. **Gordin, F. M., Simon, G. L., Wofsy, C. B., and Mills, J.**, Adverse reactions to trimethoprim-sulfamethoxazole in patients with the acquired immunodeficiency syndrome, *Ann. Intern. Med.*, 100, 495, 1984.
288. **Leport, C., Raffi, F., Matherson, S., Katlama, C., Regnier, B., Saimot, A. G., Marche, C., Vedrenne, C., and Vilde, J. L.**, Treatment of central nervous system toxoplasmosis with pyrimethamine/sulfadiazine combination in 35 patients with the acquired immunodeficiency syndrome, *Am. J. Med.*, 84, 94, 1988.
289. **Haverkos, H. W.**, Assessment of therapy for *toxoplasma* encephalitis, *Am. J. Med.*, 82, 107, 1987.
290. **Araujo, F. G., Guptill, D. R., and Remington, J. S.**, Azithromycin, a macrolide antibiotic with potent activity against *Toxoplasma gondii, Antimicrob. Agents Chemother.*, 32, 755, 1988.
291. **Chan, J. and Luft, B. J.**, Activity of roxithromycin (RV28965), a macrolide, against *Toxoplasma gondii* infection in mice, *Antimicrob. Agents Chemother.*, 31, 1147, 1987.
292. **Chang, H. R. and Pechere, J. C. F.**, Effects of roxithromycin on acute toxoplasmosis in mice, *Antimicrob. Agents Chemother.*, 31, 1147, 1987.
293. **Hofflin, J. M. and Remington, J. S.**, In vivo synergism of roxithromycin (RU965) and interferon against *Toxoplasma gondii, Antimicrob. Agents Chemother.*, 31, 346, 1987.
294. **Kovacs, J. A., Allegra, C. J., Chabner, B. A., Swan, J. C., Drake, J., Lunde, M. N., Parrillo, J. E., and Masur, H.**, Potent effect of trimetrexate, a lipid-soluble antifolate, on *Toxoplasma gondii, J. Infect. Dis.*, 155, 1027, 1987.
295. **Tuazon, C. V.**, Toxoplasmosis in AIDS patients, *J. Antimicrob. Chemother.*, 23 (Suppl. A), 77, 1989.

Chapter 8

CRYPTOSPORIDIUM: INFECTIONS IN MAN AND DOMESTICATED ANIMALS

William L. Current and Byron L. Blagburn

TABLE OF CONTENTS

I. INTRODUCTION

Organisms of the genus *Cryptosporidium* are small (2 to 6 μm, depending on stage of life cycle) coccidian parasites that invade and then replicate within the microvillous region of epithelial cells lining the digestive and respiratory organs of vertebrates.[1-5] These obligate intracellular protozoans were recognized and named over 80 years ago[6-8] and have remained until recently nothing more than a biomedical curiosity. Prior to 1980, infections with species of *Cryptosporidium* were considered rare in animals and in man they were thought to be the result of a little-known opportunistic pathogen of immune deficient individuals outside its normal host range. Beginning in 1982, our concept of these protozoan parasites changed to that of important, widespread causes of diarrheal illness in several mammalian hosts, especially calves, lambs, and humans. Several species of *Cryptosporidium* have also been recognized as primary pathogens causing intestinal and respiratory disease in chickens, turkeys, and quails.

At the time of this writing, no effective therapy for cryptosporidiosis has been identified; thus, the finding of this parasite in the immunocompromised host, especially patients with AIDS, usually carries an ominous prognosis. Reports of infections of the respiratory tract[9] and biliary tree[10] demonstrate that the developmental stages of this protozoan are not confined to the gastrointestinal tract and suggest that *C. parvum* may be an under-reported cause of respiratory and biliary tract disease, especially in the immune deficient patients.

Recent recognition of the importance of *Cryptosporidium* spp. as pathogens of man and his domesticated animals can be confirmed easily by the number of relevant publications that have appeared in the biomedical literature. Less than 30 papers addressing these parasites were published prior to 1980; however, at the time of this writing, more than 950 papers on *Cryptosporidium* spp. and cryptosporidiosis exist. Among the many recent papers are several reviews of the biology of *Cryptosporidium* spp. infecting man and his domesticated animals.[2,4,11] Following discussions of the history, taxonomy, and life cycles of *Cryptosporidium* spp., this necessarily brief chapter will address the most important aspects of mammalian and avian cryptosporidiosis. The portion on mammalian cryptosporidiosis will concentrate on *C. parvum* and diseases it produces in human beings whereas the section on avian cryptosporidiosis will emphasize the parasites of commercially grown poultry and the diseases they produce.

II. HISTORY AND TAXONOMY

It is possible that in 1895 Clarke[12] was first to observe a species of *Cryptosporidium* which he described as "swarm spores lying upon the gastric epithelium of mice". In

TABLE 1
The Taxonomic Classification of *Cryptosporidium*

Classification	Name	Biological characteristics
Phylum	Apicomplexa	Invasive forms have apical complex with polar rings, rhoptries, micronemes, conoid, and subpellicular microtubules
Class	Sporozoasida	Locomotion of invasive forms by body flexion, gliding, or undulation
Subclass	Coccidiasina	Life cycle with merogony, gametogony, and sporogony
Order	Eucoccidiorida	Merogony present; in vertebrate hosts
Suborder	Eimeriorina	Male and female gametes develop independently
Family	Cryptosporidiidae	Homoxenous (one host life cycle), with developmental stages just under the membrane of the host cell; oocyst without sporocysts and with four sporozoites; microgametes without flagella

retrospect, the microbes he observed were probably the motile merozoites of *C. muris*, the type species named and described approximately 12 years later by the well-known American parasitologist, E. E. Tyzzer.[6] This protozoan infecting the gastric epithelium of laboratory mice was placed in a new genus (*Cryptosporidium* = hidden sporocysts) because, unlike the previously known coccidia, the oocyst of this parasite did not have sporocysts surrounding the sporozoites. Three years later, Tyzzer[7] described some of the life cycle stages of *C. muris* and in 1912 he described much of the life cycle of a second species, *C. parvum*, found in the small intestine of laboratory mice.[8] During the ensuing 70 years, approximately 19 additional species of *Cryptosporidium* were named from a variety of vertebrate hosts;[13] however, only a few of these named species, including the two originally described by Tyzzer, are now considered valid.[2,4]

Interest in *Cryptosporidium (C. parvum)* by the veterinary medical profession has increased significantly since this protozoan was first reported in 1971 to be associated with bovine diarrhea.[14] Numerous case reports from many different animals now occupy the biomedical literature and one species, *C. parvum*, is recognized as an important cause of neonatal diarrhea in calves and lambs.[1,2,5] Another species, *C. baileyi*, is now recognized as an important cause of respiratory disease in poultry.[15-17]

The first cases of human cryptosporidiosis were reported in 1976,[18,19] and subsequent reports were rare until it was recognized that *Cryptosporidium* (now believed to be *C. parvum*) can produce a short-term diarrheal illness in immunocompetent persons and a prolonged, life-threatening, cholera-like illness in immune deficient patients, especially those with the acquired immune deficiency syndrome (AIDS).[2,4,20] Additional details of many of the historical events outlined above can be found in review papers published between 1983 and 1989.[1-5,11]

The taxonomic classification of small intracellular protozoans assigned to the genus *Cryptosporidium* is presented in Table 1. Species of *Plasmodium*, causing malaria in man, are in the same order (Euccodiorida) but in a different suborder (Haemospororina) than are species of *Cryptosporidium*. More closely related to *Cryptosporidium* spp. are the other true coccidia (suborder Eimeriorina), *Isospora belli, Sarcocystis* spp., and *Toxoplasma gondii*, that infect human beings and *Eimeria* spp. that infect other mammals and birds. Most species of *Cryptosporidium* named in the biomedical literature following Tyzzer's description of the genus were done so with the assumption that these coccidia were as host specific as the closely related (taxonomically) species of *Eimeria* infecting mammals and birds. However, studies conducted in the early 1980s demonstrated little or no host specificity for "species" of *Cryptosporidium* isolated from mammals. This lack of host specificity exhibited by mammalian isolates prompted Tzipori et al.[21] to consider *Cryptosporidium* as a single species genus. A more realistic approach was presented later by Levine[13] who consolidated the 20 named parasites into four species; one each for those infecting fishes (*C. nasorum*), reptiles

(*C. crotali*), birds (*C. meleagridis*), and mammals (*C. muris*). Information available at the time of this writing suggests that this consolidation is not entirely correct. *Cryptosporidium crotali* is now recognized as a species of *Sarcocystis*, a genus of coccidian parasites found commonly in snakes. At least two valid species, *C. baileyi* and *C. meleagridis*, can infect chickens and turkeys.[16] There are also at least two valid species infecting mammals (*C. parvum* infecting the small intestine and *C. muris* infecting the stomach) and on the basis of oocyst structure it is *C. parvum*, not *C. muris*, that is associated with all of the well-documented cases of cryptosporidiosis in mammals.[22] Therefore, at the time of this writing, the species with oocysts measuring 4 to 5 μm that produces clinical illness in man and other mammals should be referred to as *C. parvum*, or *Cryptosporidium* sp. if there is not enough morphologic, life-cycle, and/or host specificity data to relate it to Tyzzer's original description. This conservative approach has been adopted realizing that careful studies of proposed differences in host specificity, sites of infection, and pathogenicity among mammalian isolates[2,4,5] may result in the validation of additional species. In light of the present uncertainties in the taxonomy of *Cryptosporidium* spp., the designation of a particular parasite obtained from a mammalian host as an isolate rather than a strain is preferable.

III. LIFE CYCLES

Studies of different isolates (human and calf) of *C. parvum* in the suckling mouse model,[23] in chicken embryos,[24] and in cell cultures[25] revealed that the life cycle of this parasite (Figure 1) is similar to that of other true coccidia (e.g., *Eimeria* and *Isospora* spp.) infecting mammals in that it can be divided into six major developmental events: excystation (release of infective sporozoites), merogony (asexual multiplication within host cells), gametogony (formation of micro- and macrogametes), fertilization, oocysts wall formation, and sporogony (sporozoite formation). The life cycle of human and calf isolates of *C. parvum* differs somewhat from that of other monoxenous (one host in life cycle) coccidia (*Eimeria* and *Isospora* spp.). Each intracellular stage of *C. parvum* resides within a parasitophorous vacuole confined to the microvillous region of the host cell, whereas comparable stages of *Eimeria* or *Isospora* spp. occupy parasitophorous vacuoles usually deep (perinuclear) within the host cells. Oocysts of *C. parvum* undergo sporogony while they are within the host cells and are infective when released in the feces, whereas oocysts of *Eimeria* or *Isospora* spp. do not sporulate until they are passed from the host and exposed to oxygen and temperatures below 37°C. Studies using experimentally infected mice have also shown that approximately 20% of the oocysts of *C. parvum* within host enterocytes do not form a thick, two-layered, environmentally resistant oocyst wall; the four sporozoites of this autoinfective stage are surrounded only by a single unit membrane. Soon after being released from the host cell, the membrane surrounding the four sporozoites ruptures and these invasive forms penetrate into the microvillous region of other enterocytes and reinitiate the life cycle.[23] The majority (approximately 80%) of the oocysts of *C. parvum* found in enterocytes of suckling mice were similar to those of *Eimeria* and *Isospora* spp. in that they develop thick, environmentally resistant oocyst walls and are passed in the feces. Thick-walled oocysts are the life cycle forms that transmit the infection from one host to another. The presence of autoinfective, thin-walled oocysts and type I meronts that can recycle are believed to be the life cycle features of *C. parvum* responsible for the development of severe infections in hosts exposed to only a small number of thick-walled oocysts, and for persistent, life-threatening disease in immune deficient persons who are not exposed repeatedly to these environmentally resistant forms. Ultrastructural features of some of the developmental stages of *Cryptosporidium* in enterocytes of the experimentally infected host are shown in Figures 2 and 3.

Subsequent studies[16] of *C. baileyi* in experimentally infected chickens revealed that this species has a life cycle similar to that described above for *C. parvum* in the suckling mouse

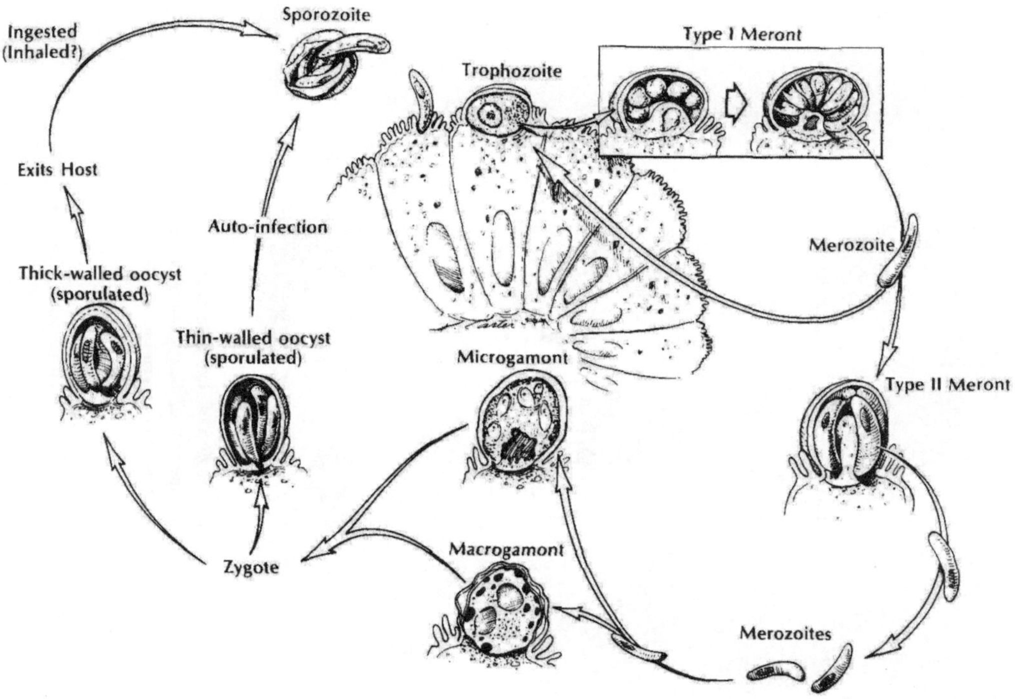

FIGURE 1. Diagrammatic representation of the proposed life cycle of *C. parvum* as it occurs in the mucosal epithelium of an infected mammalian host. Most (approximately 80%) of the zygotes formed after fertilization of the microgamete by the microgametes (released from microgamont) develop into environmentally resistant thick-walled oocysts containing four sporozoites at the time they are released from the host cell. These environmentally resistant oocysts are the life cycle forms that transmit the infection from one host to another. A smaller percentage of zygotes (approximately 20%) do not form a thick, two-layered oocyst wall, they only have a unit membrane surrounding the four sporozoites. These thin-walled oocysts represent autoinfective life cycle forms that can maintain the parasite in the host without repeated oral exposure to the thick-walled oocysts present in the environment. The life cycle of *C. baileyi*, infecting chickens, differs from the one shown in that this parasite has an additional type (type III) meront that is derived from type II merozoites. (Drawing by Kip Carter, University of Georgia.)

model. The major difference in the life cycle of these two species is that *C. baileyi* has three distinct types of meronts rather than the two types found in *C. parvum*.

IV. EPIDEMIOLOGY/EPIZOOITIOLOGY

Studies of experimental infections in farm and laboratory animals clearly demonstrate that *Cryptosporidium* spp. are transmitted by environmentally resistant oocysts that are fully sporulated and infective at the time they are passed from the host.[2,20] As long as the thick, two-layered wall remains intact, *Cryptosporidium* spp. oocysts are remarkably resistant to most common disinfectants and they can survive for months if kept cold and moist. One study designed to evaluate the efficacy of commercial disinfectants demonstrated that ammonia (50% or higher), and formalin (10% or higher) for 30 min can kill *C. baileyi* oocysts.[26] When these disinfectants and others used commonly in hospitals, laboratories, and farms were evaluated at the lower concentrations recommended by the manufacturers, none were effective against *Cryptosporidium* spp. oocysts. Freeze-drying and exposure (30 min) to temperatures above +60°C and below −20°C have also been reported to kill *Cryptosporidium* sp. oocysts (Anderson 1985, Tzipori 1983). Most *C. parvum* oocysts stored at 4°C in 2.5% (weight/volume) aqueous potassium dichromate remain viable for 3 to 4 months, and some may remain infective for cell cultures and suckling mice for more than 1 year.[2] A procedure used routinely in several laboratories to sterilize *Cryptosporidium* spp. oocysts

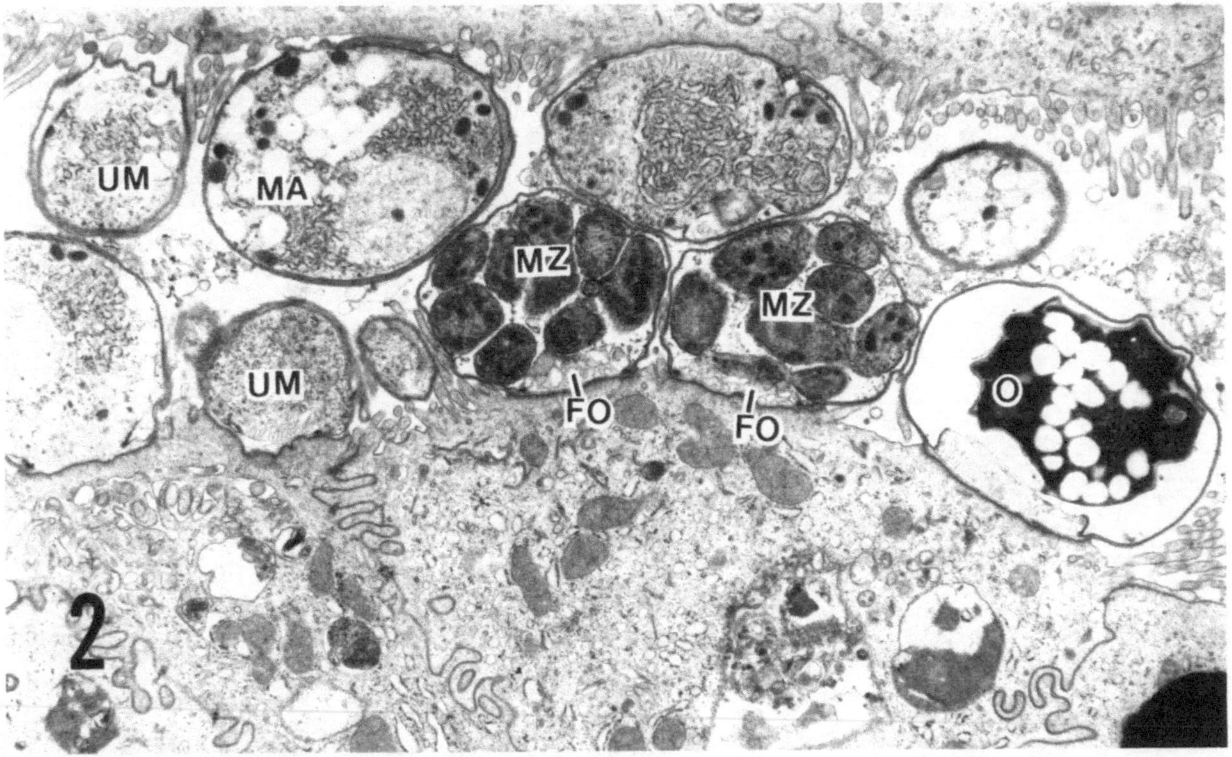

FIGURE 2. Transmission electron micrograph of developmental stages of *C. parvum* in the microvillous region of ileal enterocytes of an experimentally infected mouse. Two of the meronts contain merozoites (MZ) and a feeder organelle (FO) at the base of the parasitophorous vacuole; one macrogamete (MA) contains the characteristic amylopectin granules near the center and wall-forming bodies near the periphery. One thick-walled oocyst (O) and several uninucleate meronts (UM) are also shown.

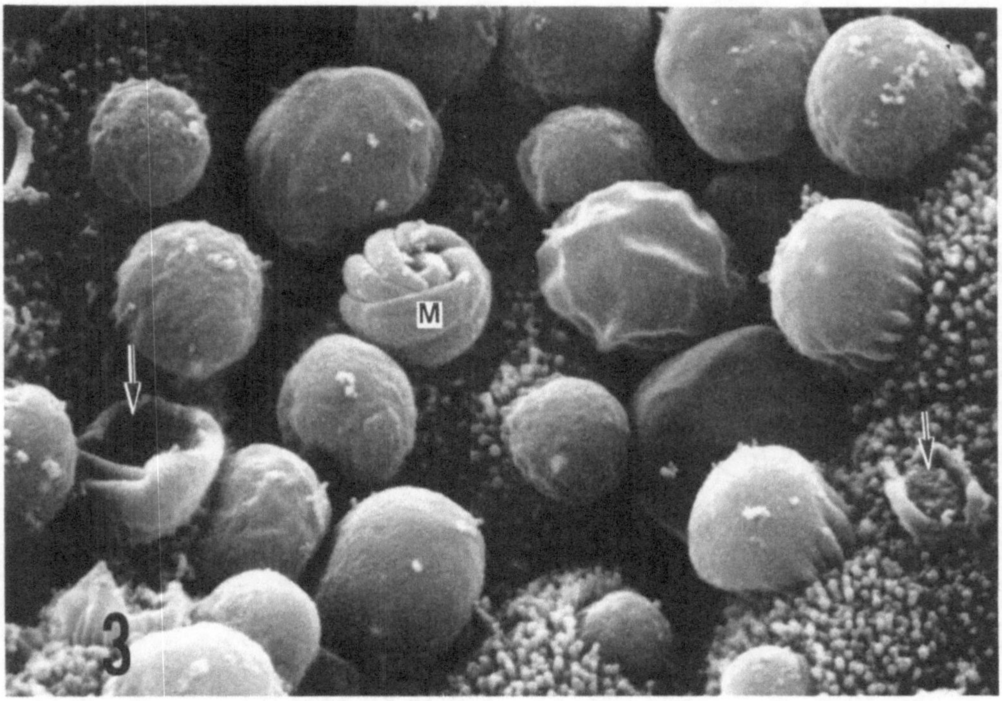

FIGURE 3. Scanning electron micrograph of intestinal mucosa containing many developmental stages of *Cryptosporidium*. The host cell and parasitophorous vacuole membrane of one of the meronts (M) was removed during processing allowing visualization of the merozoites. Arrows point to craters left in the microvillous border after parasites are released from the host cells.

prior to obtaining viable sporozoites by *in vitro* excystation involves incubation in 10 to 25% commercial bleach (full strength bleach is 5.25% sodium hypochlorite) for 10 to 15 min in an ice bath. Since oocysts of *Cryptosporidium* spp. are resistant to this sterilization procedure, routine chlorination of drinking water should have little or no effect on their viability.

V. MAMMALIAN CRYPTOSPORIDIOSIS

At the time of this writing there appears to be two valid species of *Cryptosporidium* infecting mammals, *C. parvum* which is responsible for all well documented cases of intestinal and respiratory cryptosporidiosis in numerous mammalian hosts and *C. muris* which infects the gastric epithelium of rodents and cattle. Oocysts of these two species are shown diagrammatically in Figure 4. It is not the intent of this necessarily brief chapter to discuss all of the mammalian species that have been reported to harbor these two parasites; however, some of the hosts in which *C. parvum* and *C. muris* produce clinical disease of importance to human and veterinary medicine will be addressed. More complete discussions of the more than 40 different mammals reported as hosts for *C. parvum* and *C. muris* can be found in published reviews.[1-5,13]

A. HUMAN CRYPTOSPORIDIOSIS CAUSED BY *C. PARVUM*

1. Sources of Human Infection

Data published from several laboratories during the early 1980s demonstrated that calves are a source of human infection.[20,28,29] Companion animals such as rodents, puppies, and kittens may also serve as reservoir hosts.[20] These findings in conjunction with reports of more than 40 mammals may harbor the parasite[3] and the realization that *C. parvum* readily

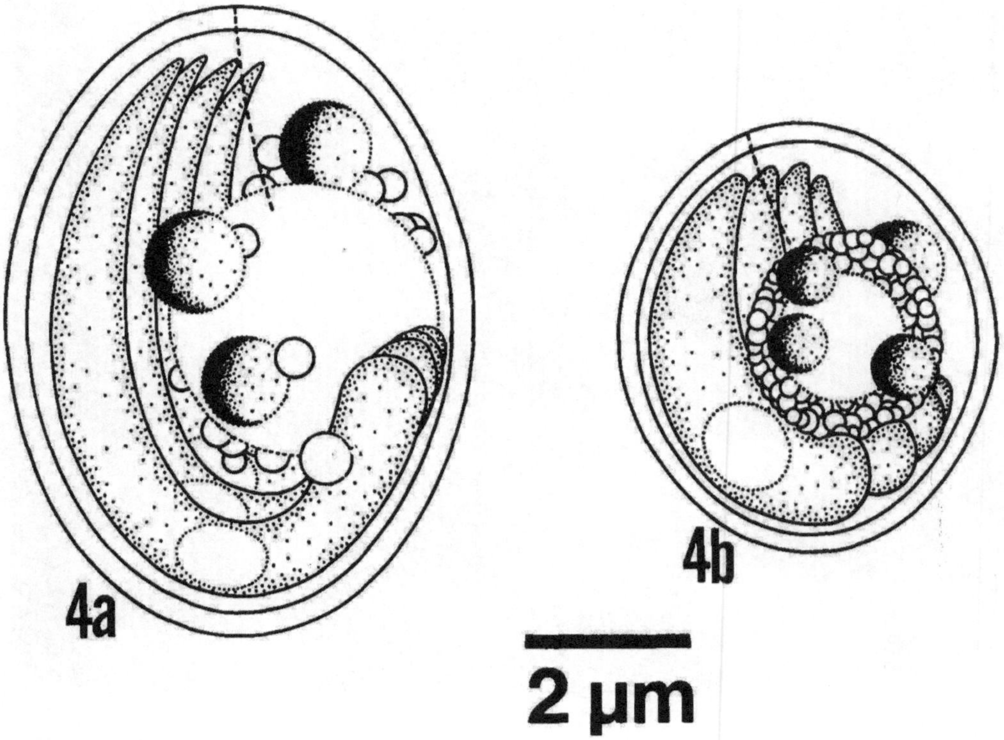

FIGURE 4. Composition line drawings of oocysts of two species of *Cryptosporidium* from mammalian hosts. (a) *C. muris* which infects the gastric glands in the stomach of cattle, mice, and perhaps other rodents. (b) *C. parvum* which infects the mucosal epithelium of the small intestine and respiratory tract of numerous mammalian hosts.

crosses host species barriers, led to the concept that most human infections are a result of zoonotic transmission. This view is probably correct for persons living and working in environments where exposure to fecal contamination from potential reservoir hosts is likely. However, zoonotic transmission does not explain the large number of infections reported from persons living and working in urban areas where exposure to animal feces is minimal. Evidence available at the time of this writing indicates that person-to-person transmission of cryptosporidiosis is common. In 1983, an accidental laboratory infection demonstrated that a human isolate of *C. parvum* could be transmitted from one person to another.[30] Since that time, outbreaks of cryptosporidiosis have been reported among children in day-care centers,[31-33] hospital-acquired infections have been investigated,[34] at least two large water-borne outbreaks have been well documented,[35,36] and this protozoan is now recognized as a cause of travelers' diarrhea.[37,38]

After the first large water-borne outbreak[33] was investigated, we (Current and Schaefer, unpublished data) demonstrated that oocysts of *C. parvum* can be recovered from water samples by high volume filters that were designed to trap cysts of the enteric protozoan, *Giardia lamblia*. Application of similar filtration techniques in conjunction with immunofluorescent detection methods has since been used to demonstrate *Cryptosporidium* sp. oocysts in surface and drinking waters, and sewage effluent samples obtained from different geographic regions of the U.S. and from several other countries.[39,40]

The epidemiologic features of cryptosporidiosis emphasized above — transmission by environmentally resistant cysts (oocysts), existence of numerous potential reservoir hosts for zoonotic transmission, documentation of person-to-person transmission in settings such as day-care centers, and documentation of waterborne transmission — are similar to those of human giardiasis that were revealed during the past decade. *Cryptosporidium parvum* is now gaining the recognition it deserves as an important, widespread cause of diarrheal illness in man.

2. Prevalence

Prevalence data contained in published surveys relying on standard stool examination techniques to detect *C. parvum* oocysts are quite variable from one geographic location to another. Direct comparison of these data is often difficult because study populations may not be comparable and because different stool sampling and oocyst detection procedures were used. In spite of these difficulties, a data base is being compiled from which a limited understanding of the geographic distribution and prevalence of human cryptosporidiosis is beginning to emerge.

A review of 36 large-scale surveys of selected populations, such as children and adults seeking medical attention for diarrhea and other gastrointestinal symptoms, demonstrates that *Cryptosporidium* sp. is associated with diarrheal illness in most areas of the world and that the prevalence of cryptosporidiosis is highest in poorly developed regions.[4] For example, prevalence rates reported in surveys from Europe (1 to 2%) and North America (0.6 to 4.3%) are lower than those reported in surveys from Asia, Australia, Africa, and Central and South America (3 to 20%). In most of the surveys reviewed by Fayer and Ungar,[4] *Cryptosporidium* sp. was the most common parasite found and, in several, this protozoan was considered to be the most significant of all known enteropathogens causing diarrheal illness. Other findings common to many of the surveys were that children usually had a significantly higher prevalence than did adults and that infections were often seasonal, with a higher prevalence during warmer, wetter months. Another interesting finding from the standpoint of infection control was that a small number of oocysts may be present in feces for up to 2 weeks following resolution of diarrhea.

Several, more recent reviews[11,41] of the published reports of cryptosporidiosis in persons residing in industrialized and developing countries support the overall conclusions presented above, and may also provide a more global view of the prevalence of human infection. Crawford and Vermund[11] compared the worldwide occurrence of *Cryptosporidium* infection compiled by Navin[42] from studies prior to 1985 with that obtained from studies published after 1985. Data compiled from the pre- and post-1985 studies were similar. Navin reported that studies prior to 1985 suggested that the overall occurrence of *Cryptosporidium* infection in individuals with diarrhea was 2.5% (19 of 7779) for persons living in industrialized countries and 7.2% (82 of 1135) for persons residing in developing countries. The more recent studies summarized by Crawford and Vermund suggested an infection rate for individuals with diarrheal illness was 2.2% (285 of 11,716) for individuals in industrialized countries and 8.5% (532 of 6295) for individuals in developing countries. Estimates provided by Walsh and Warren[38] suggest that in Asia, Africa, and Latin America alone there are as many as five billion eposides of diarrhea and five to ten million diarrhea-associated deaths annually. If the estimates of Walsh and Warren[38] are accurate and if the *Cryptosporidium* prevalence data summarized by Navin[42] and Crawford and Vermund[11] are correct, then one may predict 360 million to 425 million *Cryptosporidium* infections annually in persons living in Asia, Africa, and Latin America.

Limited serologic surveys also support the concept that *Cryptosporidium* infection is common, especially in developing countries. For example, approximately 64% of 389 children and adults in Lima, Peru and 84 children in Maracaibo and Caracas, Venezuela had serologic evidence for previous infection, i.e., their sera contained antibodies (IgG and/or IgM) specific for *Cryptosporidium*.[44]

In light of the epidemiologic information just reviewed, it is important that health care professions emphasize the importance of *Cryptosporidium* in training programs so that cryptosporidiosis is considered in the differential diagnosis of diarrheal illness. This educational role should be approached aggressively because of the common occurrence of the disease, because of the large number of potential reservoir hosts, and because persons with impaired immune function may develop life-threatening cryptosporidiosis.

3. Clinical Features

The most common clinical feature of cryptosporidiosis in immunocompetent and immunocompromised persons is diarrhea; it is this symptom that most often leads to diagnosis. Characteristically, the diarrhea is profuse and watery, it may contain mucus but rarely blood and leukocytes, and it is often associated with weight loss. Other less common clinical features include abdominal pain, nausea, and vomiting, and low-grade fever (<39°C). Occasionally, nonspecific symptoms such as myalgia, weakness, malaise, headache, and anorexia occur. Severity of these symptoms may wax and wane in individuals and usually parallel the intensity of oocyst shedding. Both the duration of symptoms and the outcome typically vary according to the immune status of the host. AIDS patients usually experience a prolonged, life-threatening illness, whereas most immunocompetent persons experience a short-term illness with complete, spontaneous recovery. However, the clinical presentation of gastrointestinal cryptosporidiosis does not always fit within one of these two divergent categories. Persons with the clinical and laboratory features of AIDS have been reported to clear infections after several months of diarrhea, and individuals reported to be immunocompetent have had infections lasting more than a month.[2] Mild and asymptomatic infections have been reported in immunocompetent persons and in several patients with AIDS.[2]

Immunocompetent persons — Most of the 18 cases of cryptosporidiosis in immunocompetent humans reported prior to 1983 and the numerous cases reported since then[2-4,11,41] describe a self-limited, cholera-like or flu-like gastrointestinal illness. The most common symptoms reported are profuse, watery diarrhea (cholera-like), and abdominal cramping, nausea and vomiting, low-grade fever, and headache (flu-like). In their review of the symptoms reported for 586 persons in 36 large-scale surveys, Fayer and Ungar[4] reported that diarrhea was the most commonly listed clinical feature (92%), followed by nausea and vomiting (51%), abdominal pain (45%), and low-grade fever (63%).

In most well-nourished persons, diarrheal illness due to *C. parvum* infections lasts from 3 to 12 days. Occasionally, these patients may require fluid replacement therapy, and occasionally the diarrheal illness may last for more than 2 weeks. In poorly nourished children with cryptosporidiosis, oral and parenteral rehydration therapy are often required because of excessive fluid loss that may last more than 3 weeks. One study[45] from a hospital in Jerusalem revealed that children with diarrhea and *Cryptosporidium*-positive stools were significantly more malnourished than children with diarrhea and no *Cryptosporidium* sp. oocysts in their stools. Also, children with severe malnutrition and with *Cryptosporidium* sp. oocysts in their stools had a significantly longer duration of diarrhea than similarly malnourished children without cryptosporidiosis. Diarrheal illness is a major cause of morbidity and mortality, especially in young children living in developing countries. Based on the limited prevalence data from stool and serologic surveys (reviewed above), it is likely that *Cryptosporidium* plays an important role in the overall health status of these children.

Immunodeficient persons — In the most severely immunocompromised host, such as patients with AIDS, diarrheal illness due to *Cryptosporidium* infection of the gastrointestinal tract becomes progressively worse with time and may be a major factor leading to death. It is believed that the infection usually begins with organisms colonizing the ileum or jejunum and develops into a life-threatening condition when a large portion of the gastrointestinal mucosa is covered with what has been described as a monolayer of parasites. Fluid loss in patients with AIDS and cryptosporidiosis is often excessive; 3 to 6 l of diarrheic stool per day is common, and as much as 17 l of watery stool per day has been reported.

In the immune deficient patient, *Cryptosporidium* sp. infections are not always confined to the gastrointestinal tract. Acute and gangrenous cholecystitis in AIDS patients has been attributed to *Cryptosporidium* sp. infections of biliary tree and gall bladder epithelium.[10] Respiratory tract infections with this parasite have been associated with chronic coughing, dyspnea, bronchitis, and pneumonitis in persons with AIDS and in an infant with severe combined immune deficiency.[9,46] In children, symptomatic intestinal and respiratory infections with *Cryptosporidium* sp. have also been associated with the acute phase of measles, a cause of transient immunosuppression.[47] The role of *C. parvum* as a cause of respiratory illness in persons whose immune function is compromised because of congenital or acquired immune deficiencies or because of malnutrition and/or other infectious diseases remains to be determined.

4. Pathogenicity

At the time of this writing, the pathophysiologic mechanisms of *Cryptosporidium*-induced diarrhea are poorly defined. Studies in germ-free calves monoinfected with *C. parvum* suggest that malabsorption and impaired digestion in the small bowel coupled with malabsorption in the large intestine are major factors responsible for diarrhea in calves with cryptosporidiosis.[48] This malabsorption and impaired digestion may result in an overgrowth of intestinal microflora, a change in osmotic pressure across the gut wall, and an influx of fluid into the lumen of the intestine. Malabsorption and impaired digestion has also been reported in humans infected with *C. parvum*. However, the secretory (often described as cholera-like) diarrhea common to most immune deficient patients with cryptosporidiosis suggests a toxin-mediated hypersecretion into the gut. Definitive, systematic studies are needed to determine the mechanisms by which *C. parvum* and its metabolites or toxins may alter normal intestinal function of a susceptible animal model.

5. Treatment

The lack of an effective treatment for cryptosporidiosis in previously healthy, immunocompetent persons has not resulted in a great deal of concern within the biomedical community since, in such patients, the duration of diarrhea is almost always less than 20 days and since clinical symptoms and oocyst shedding generally resolve spontaneously. However, reports[45,47] demonstrating an association between cryptosporidiosis and severe malnutrition in children may warrant a change of perception. Regardless of one's present perception, if a safe and effective therapy were available, most clinicians would probably treat the severe, short-term diarrheal illness that often develops in immunocompetent persons following oral exposure to oocysts.

Because immunocompromised persons often develop a prolonged life-threatening infection following exposure to the parasite, an effective therapy is desperately needed for the treatment of cryptosporidiosis in this patient population. To date, treatment of cryptosporidiosis in immune deficient persons has been frustrating and unsuccessful in most cases. No controlled studies have been published and all therapeutic information is based on isolated reports. The list of unsuccessful attempts to treat cryptosporidiosis in immunocompromised persons is growing rapidly and these attempts include the use of more than 80 different drugs.[2-4,49,50] Of the antimicrobial agents used to treat AIDS patients with cryptosporidiosis, spiramycin has been the only one suspected of having efficacy; it was reported to control diarrhea in some individuals but not in others. This macrolide antibiotic may help control the diarrhea in some patients treated for cryptosporidiosis while in an early stage of AIDS but does not appear to have any effect on the course of clinical cryptosporidiosis in patients who have progressed into the later stages of AIDS.[50]

In the absence of an effective treatment for cryptosporidiosis, supportive therapy appears to be only intervention available to most clinicians. Oral and parenteral rehydration therapy is often required by both immunodeficient and immunocompetent persons, especially young children with severe cryptosporidial diarrhea. Parenteral nutrition may also help sustain the nutritional status of some patients with persistent cryptosporidiosis.

Since immune status of the host appears to be the major factor determining the severity and duration of infection following oral exposure to *C. parvum* oocysts[51] and since an effective therapy is not available,[3] immunologic intervention may be a reasonable approach to the control of cryptosporidiosis. Discontinuation of immunosuppressive chemotherapy, allowing restoration of immune function, has resulted in complete resolution of intestinal cryptosporidiosis in several patients.[52] One other approach directed toward restoration of immune function has also been reported to be of some value in treating cryptosporidiosis in AIDS patients. In one small study, five of eight patients with AIDS and prolonged cryptosporidiosis exhibited a marked decrease in the number of daily bowel movements and developed formed stools following oral administration of an uncharacterized dialyzable extract prepared from lymph node cells obtained from calves immune to *C. parvum* infection.[53] Oocyst shedding was eradicated from the stools of four patients; two of the four remained parasite free following therapy and the other two had subsequent relapses after the termination of therapy. Additional studies are needed to confirm the findings reported for a small number of AIDS patients and to determine the feasibility and utility of transfer factor therapy in the treatment of immunocompromised hosts with persistent cryptosporidiosis.

Although there appears to be *Cryptosporidium*-specific IgA, IgM, and IgG responses detectable in the sera by ELISA and IFA procedures,[44,54-56] it is likely that these antibodies play only a minor role in protective immunity. Because the parasite appears to be confined to the mucosal surface and because numerous studies have failed to demonstrate a protective role for serum antibodies against closely related species of coccidia,[57] it is more probable that secretory antibodies coupled with cell-mediated immune mechanisms are responsible for the clearance of parasites from the infected mucosa and for rendering the immunocompetent host resistant to reinfection. We are not aware of studies addressing the presence or function of secretory antibodies in hosts during or following infection. Mata et al.[58] reported that breast-fed infants in Costa Rica had a significantly lower incidence of cryptosporidiosis than did age-matched babies in the same study populations that were fed artificial diets, and they proposed that lactogenic immunity may play an important role in controlling *C. parvum* infections. This concept was subsequently tested in several studies to determine if antibodies in milk or colostrum can prevent or abrogate intestinal infections with *C. parvum*.

Colostrum or milk from dairy cows that are exposed naturally to the parasite does not appear to protect calves or humans from *C. parvum* infection; however, colostrum from hyperimmunized cows may provide some protection. We have routinely administered oocysts of *C. parvum* to 1-day-old calves along with the first of 3 l of colostrum. The subsequent course of infection was not affected by the presence of high titers of *Cryptosporidium*-specific antibody in the colostrum; the parasites excyst, invade, and then replicate within the intestinal mucosa when high levels of colostrum antibody are present in the gut lumen. The fact that most calves will experience cryptosporidiosis while they are nursing from cows, most of which have colostrum antibodies to *C. parvum*, also supports the concept that natural exposure to the parasite does not result in significant lactogenic immunity. Oral administration of colostrum from a nonimmunized, naturally exposed dairy cow that contained antibodies to *C. parvum* did not alter the course of infection in an AIDS patients with cryptosporidiosis.[59] A similar lack of lactogenic immunity was also reported in infant mice whose dams were immunized by oral inoculation of *C. parvum* oocysts.[60]

In contrast to the above reports, several studies indicate that colostrum obtained from cows hyperimmunized with oocyst + sporozoite antigens of *C. parvum* may protect humans and mice from cryptosporidiosis. Tzipori et al.[61] reported that three immune deficient patients recovered from cryptosporidiosis within 3 to 5 days after initiation of oral administration of hyperimmune cow colostrum produced by immunizing pregnant cows with concentrated *C. parvum* oocyst/sporozoite antigens. Two of the patients had subclinical infections following treatment and the other remained free of infection for several months after the treatment was stopped. Fayer et al.[62] demonstrated that hyperimmune bovine colostrum, obtained from cows immunized with purified *C. parvum* oocysts, neutralized sporozoites and protected mice from oocyst challenge. Significantly fewer stages of *C. parvum* were found in suckling mice that were given whey from hyperimmune colostrum (undiluted or diluted 1:20 or 1:50) before and after oocyst inoculation than in mice given whey from control colostrum. Significantly fewer stages were also found in mice following intrarectal inoculation of sporozoites incubated in hyperimmune whey (diluted 1:20 or 1:50) than in mice receiving sporozoites incubated in similar dilutions of control whey. Fayer et al.[63] also reported that this hyperimmune colostrum provided prophylaxis against cryptosporidiosis in calves. Calves given hyperimmune colostrum 2 days after oral inoculation of *C. parvum* oocysts had less diarrhea and shed oocysts for a shorter period of time than did calves given non-hyperimmune colostrum 2 days after oocyst inoculation. Although these studies indicate that some component of hyperimmune bovine colostrum may exhibit anti-*Cryptosporidium* activity, additional studies are needed to further define the role of lactogenic immunity in preventing and treating cryptosporidiosis and to isolate and characterize the components of hyperimmune colostrum or milk that are responsible for the reported protective effects.

B. BOVINE CRYPTOSPORIDIOSIS

Two sizes of *Cryptosporidium* sp. oocysts, presumably representing two distinct species, have been isolated from the feces of cattle (Figure 4). Since these oocysts are morphologically indistinguishable from those of the two species that Tyzzer originally described from mice,[7,8] and since they infect different regions of the gut (*C. parvum*, small intestine; *C. muris*, stomach) the names *C. parvum* and *C. muris* have been applied to these coccidian parasites in cattle.[22] It is generally accepted that *C. parvum* is a common cause of diarrheal illness in young calves and that this parasite can be readily transmitted to other mammalian hosts, including humans.[1-5] A second species, *C. muris*, is found less commonly in the abomasum (stomach) of calves and adult cattle, and has been associated with reduced weight gains.[64,65]

1. Neonatal Cryptosporidiosis Caused by *C. parvum*

The role of *Cryptosporidium* sp. (*C. parvum*) as a cause of gastrointestinal illness in calves was first reported in 1971 by Panciera et al.[14] From 1976 to 1979 papers published by researchers in Iowa,[66] Canada,[67] and South Dakota[68] presented data to support the concept that *C. parvum* is a primary pathogen that can cause diarrheal illness in young calves. All of these early reports relied on the recognition of endogenous stages of the parasite in the microvillous border of tissues obtained at the time of necropsy. With the development of techniques to demonstrate oocysts in the feces, numerous reports[3,4,69-73] of individual infections, herd outbreaks, and regional surveys indicate that infections with *C. parvum* are common among calves through the world. Surveys of the agents causing calf diarrhea in the U.K. have shown that *C. parvum* is second only to rotavirus as a major enteropathogen.[74,75]

Most cases of cryptosporidiosis in calves have been reported as a diarrheal illness in animals 5 to 15 days of age that results in low to moderate mortality. Infections of *C. parvum* may occur in conjunction with other enteropathogens of the neonatal calf diarrhea complex, such as rotavirus, coronavirus, and enterotoxogenic *E. coli*. In such multiple infections, the enteropathogens may act in combination to increase morbidity and mortality. High mortality may also occur when calves with cryptosporidiosis are exposed to extreme cold temperatures.[68,76] Apparently, malabsorption and fluid loss due to cryptosporidiosis may prevent calves fed diets that are formulated for warm weather from meeting the additional energy demands of extreme cold. Thus, postmortem examination of these animals reveals that death is often due to starvation. When cryptosporidiosis is associated with high calf mortality, concurrent infections with other enteropathogens, cold weather, and perhaps poor management practices should also be suspected. Additional aspects of the pathogenesis of cryptosporidiosis are discussed above as it relates to human cryptosporidiosis also applies to the infection in the small intestine of calves. The role of *C. parvum* as a cause of respiratory disease in calves has not been established as it has for immunocompromised humans.

At the time of this writing, an effective treatment for neonatal bovine cryptosporidiosis has not been reported; thus antibiotics that are used commonly to treat bacterial pathogens causing calf scours are of no value in controlling *C. parvum* infections. Because much of the morbidity and mortality is due to fluid loss and a reduction of nutrient uptake by the damaged intestinal mucosa, animals with cryptosporidial diarrhea should be given all of the milk, milk replacer, or other fluids that they will drink. Since infection is spread by the fecal-oral route from one calf to another, management practices which minimize fecal contamination may be beneficial. Resistance of *C. parvum* oocysts to chemical treatment (discussed above) makes the use of disinfectants impractical, especially on the farm.

2. Cryptosporidiosis Caused by *C. muris*

A species of *Cryptosporidium* whose oocysts are morphologically identical to those of *C. muris*, described by Tyzzer in 1912,[8] is the agent of a disease known as abomasal (gastric, peptic gland) cryptosporidiosis.[64,65] This organism produces lesions of dilation of peptic glands in the abomasum of calves and adult cattle. Infected cattle may shed oocysts in their feces for many (4 to 6 or more) months. A severely affected abomasum may weigh two to three times more than an abomasum from a noninfected bovine of the same age and size. Histologic examination of such heavy infections often reveals that most peptic glands contain numerous developmental stages of the parasite; one report described all glands as being full of developing parasites.[65] In such severely parasitized cattle, lower weight gains (a decrease of up to 33%) have been noted when compared to noninfected penmates.[65]

The prevalence and distribution of *C. muris* in cattle and its overall role as an agent causing economic loss to the bovine industry remain undefined. The parasite has been reported from calves and/or adult cattle in Washington state, Idaho, Nevada, California, and Alabama.[22,64,65] Using an acid-fast staining technique with a sensitivity of approximately 86% of that provided by Sheather's sugar flotation, Anderson[65] reported that 54 of 1216 (4.4%) feeder cattle in eight lots in Idaho had *C. muris* oocysts in their feces. Using the same diagnostic procedure, approximately 1.6% of the 12,069 samples obtained from calves and cows in 21 dairy herds in Idaho, Nevada, and California contained *C. muris* oocysts. Additional studies are needed to more clearly define the impact of *C. muris* on beef and dairy production.

C. CRYPTOSPORIDIOSIS IN OTHER MAMMALS

1. Sheep and Goats

As in cattle, cryptosporidiosis in sheep and goats is primarily a disease of the neonate and this pathogen appears to be present wherever these animals are raised commercially. In lambs and to some extent in goats, cryptosporidiosis is usually sporadic but may be the cause of high mortality.[77,78] Because of its apparent sporadic nature, the economic impact of cryptosporidiosis on individual flocks may be considerable; however, its overall impact on sheep and goat production is undefined.

With respect to morbidity and mortality, neonatal cryptosporidiosis appears to be more severe in lambs that it is in cattle. Natural infections in neonates are characterized by diarrhea lasting 2 to 12 days, sometimes accompanied by anorexia, poor growth, stiffness, hyperpnea, slow gait, limb muscle fasciculations, and depression.[4] In very young lambs, concurrent infections of *C. parvum* and enterotoxogenic *E. coli* or rotavirus may be particularly severe.[78] Other factors thought to contribute to the morbidity and mortality of cryptosporidiosis in lambs are cold weather and/or inadequate collostral and/or milk production by the dam. These factors when combined with *C. parvum* infection may result in death due to inadequate nutrient uptake by the affected intestine.

Although there are occasional reports of deaths in young kids due to cryptosporidiosis, almost no information is available on the overall prevalence of *C. parvum* in goats. In France, diarrhea is the most common reported cause of death in young goats and *C. parvum* is the most frequently identified causative agent.[79] It is clear that additional studies are needed to determine the importance of *C. parvum* as a cause of diarrheal disease in sheep and goats.

2. Pigs

Although the literature contains reports of naturally occurring *Cryoptosporidium* sp. infections in pigs,[80,81] clinical illness due to cryptosporidiosis in commercially grown swine appears to be uncommon.[82] This is somewhat puzzling since experimental infections of *C. parvum* show a range of clinical responses from subclinical to severe[82-84] and since other coccidian parasites (*Eimeria* and *Isospora* spp.) are common in commercially grown pigs. Severe infections in pigs are associated with clinical signs of vomiting, diarrhea, and anorexia, clinical signs attributed to extensive mucosal damage caused by the parasite. In heavily infected regions of the lower small intestine villi are shortened, fused, and cross-bridged, and numerous inflammatory cells can be found in the lamina propria.[85] Experimental infections have also been induced in the trachea and conjunctival epithelium following inoculation of oocysts of human and bovine origin.

3. Horses

Cryptosporidiosis was first reported in horses as a cause of severe diarrhea leading to the death of Arabian foals with severe combined immunodeficiency syndrome.[86] Since the first report several papers been published suggesting cryptosporidiosis may be rare or very sporadic in immunocompetent horses. Based on fecal or histologic examination, 0 of 23 foals without clinical signs in Colorado,[87] 0 of 14 diarrheic mares and foals in Ohio,[88] and 2 of 2 diarrheic foals in Canada[89] were infected with *Cryptosporidium* sp. More recently, Angus[78] presented some preliminary findings of an ongoing survey suggesting that *Cryptosporidium* sp. may be the most common nonbacterial pathogen involved in diarrhea outbreaks among top bloodstock in Britain and Ireland. In the first year of the survey only 6 of 246 (2.4%) of the fecal samples from healthy foals and yearlings contained *C. parvum* oocysts. However, 20 of 92 (22%) of diarrheic as well as healthy foals and yearlings sampled the following year were infected. These preliminary findings demonstrate that immunocompetent foals can develop patent infections and suggest that *C. parvum* may contribute to morbidity and mortality in young foals. These findings are particularly notable when one considers the immense potential value of some thoroughbred foals.

4. Domesticated Deer and Nondomesticated Ruminants

Within 4 weeks of the capture of 82 red deer calves at a research station in Scotland, 56 had diarrhea that lasted 2 to 14 days and 20 died. Oocysts (presumably *C. parvum*) from 27 of 34 calves with diarrhea and from 11 of 22 healthy calves.[90] Since this initial report it has become clear that the practice of artificially rearing deer calves to accustom them to domestic life tends to result in digestive complications and neonatal infections, including cryptosporidiosis, resulting in high mortality.[78,90] A more recent prospective study conducted on a deer farm in the east of Scotland confirmed that rising levels of environmental contamination with *C. parvum* oocysts cause primary infections and deaths late in the calving period.[91] From these investigations it also became clear that sporadic deaths associated with *Cryptosporidium* infections occurred in naturally reared deer calves. Thus, as new deer farms are set up, outbreaks are being diagnosed. Details of two such outbreaks involving high calf mortality have been reviewed by Angus.[78] The importance of cryptosporidiosis in red deer will continue to increase because the industry is expanding in Britain; within the past decade there has been the addition of least 250 herds with a total of up to 10,000 breeding hinds. Additional studies are needed to determine the overall impact of cryptosporidiosis on the red deer industry and its importance as a pathogen of other deer species.

Cryptosporidiosis may also be a cause of sporadic outbreaks of diarrhea in young exotic ruminants housed in wild animal parks and zoos. One such documented outbreak occurred at the San Diego Wild Animal Park and involved diarrheal illness, dehydration, and emaciation in 44 ruminants less than 21 days of age.[92] Among the infected animals were species of *Antilope, Oryx, Addax,* and *Hippotragus.*

5. Mammals Serving as Potential Reservoir Hosts

Mice — As stated in the introduction, the two valid species of *Cryptosporidium* infecting mammals (*C. parvum* in the small intestine, cecum, and colon; *C. muris*, in the stomach) were first found in mice.[6-8] Since these early reports, suckling mice have proven to be a good laboratory animal model for *C. parvum* infections. However, studies reviewed elsewhere,[51] indicate that all strains (inbred, outbred, immune deficient, and germ-free) of laboratory mice tested to date are difficult to infect once they are more than 3 weeks of age. In contrast, adult wild mice have been reported to be easy to infect with *C. parvum*.[93] These observations suggest that genetically based as well as age related factors may be responsible for determining susceptibility or resistance to *C. parvum* in mice. It is possible that, with respect to susceptibility to *C. parvum*, the laboratory mouse of today is quite different genetically from those used by Tyzzer during the early 1900s in his detailed study[8] of the life cycle of this parasite and that wild mice may be an important potential reservoir of infection for humans and domesticated mammals. Others have proposed that a calf-mouse-calf may exist in nature.[93] If this is true, one could make the same argument for a calf-mouse-human cycle.

Companion and wild animals — In addition to mice, numerous other species of animals have been reported to be infected with *Cryptosporidium* spp. Of particular interest to physicians, veterinarians, and others interested in the epidemiology of cryptosporidiosis are clinical (usually diarrhea) and nonclinical infections in companion animals such as rodents, puppies, and kittens; and in wild animals such as rabbits, rats, squirrels, beavers, raccoons, foxes, and coyotes.[1-5]

VI. AVIAN CRYPTOSPORIDIOSIS

A. HISTORICAL ASPECTS OF AVIAN CRYPTOSPORIDIOSIS

Cryptosporidium was first seen in avians in 1929 when Tyzzer described it from the cecal epithelium of asymptomatic chickens.[94] He felt that the observed parasites were conspecific with *C. parvum*, which he had described in 1912.[8] Tyzzer's publication[94] lacked detailed descriptions of either endogenous or exogenous cryptosporidial developmental stages. This isolate was later named *C. tyzzeri*.[95]

Slavin[96] was the first to associate the disease with the presence of *Cryptosporidium* in avian hosts. He described asexual and sexual stages of *Cryptosporidium* from the terminal one third of the small intestine of 10- to 14-day-old turkey poults. Slavin concluded that *C. meleagridis* was responsible for diarrhea and unthriftiness, but rarely resulted in severe economic loss or death. Assuming host specificity within the genus, he named it *C. meleagridis*. This species remains valid in the current literature.[13]

Current et al.[16] regarded biological characteristics of *Cryptosporidium* recovered from broiler chickens as unique and named it *C. baileyi*. They were of the opinion that oocyst morphometrics, sites of endogenous development, and susceptibility of other hosts to experimental *C. baileyi* infections were sufficiently different from *C. meleagridis* to justify its species status.

Reports of both respiratory and enteric cryptosporidial infections in quail are appearing with increased frequency in the scientific literature. Disease, observed in both common quail and bobwhite quail, may be accompanied by extreme morbidity and substantial mortality. Studies pursuant to characterization of quail cryptosporidiosis are ongoing in several laboratories. Results of these studies may establish an additional *Cryptosporidium* species infecting quail hosts.

At this time, the two additional named species of avian *Cryptosporidium, C. anserinum*[97] and *C. tyzzeri*[95] are considered invalid because of lack of detailed information on oocyst morphology, host specificity, or endogenous development.[13]

B. BIOLOGICAL CHARACTERISTICS OF AVIAN *CRYPTOSPORIDUM* SPP.

1. Hosts and Host Specificity

Among avian hosts, cryptosporidial developmental stages have been observed in the tissues or feces of chickens (*Gallus gallus*),[94,98,99] turkeys (*Meleagris gallopavo*),[96,100-104] bobwhite quail (*Colinus virginianus*),[105-107] common quail (*Coturnix coturnis*),[108,109] peafowl (*Pavo cristatus*),[110] guinea fowl (*Numida meleagris*) and chukar partridges (*Alectoris graeca*), (Blagburn and Lindsay, unpublished data) pheasants (*Phasianus colchicus*),[108,111-113] junglefowl (*Gallus sonnaratii*),[114] domestic ducks (*Anas platyrhynchos*),[16,115-117] muscovy ducks (*Cairina moschata*),[118] domestic geese (*Anser anser*),[16,97] red-lored parrots (*Amazona autumnalis*),[119] budgerigars (*Melopsittacus undulatus*),[108,120] macaws (*Ara* sp.).[120] canaries (*Serinus canarius*),[117] Tundra swans (*Cygnus* sp.),[120] black-throated finches (*Peophila cincta*),[121] and lovebirds (*Agapornis* sp.).[122] Identification of cryptosporidial species infecting each of the above hosts was not attempted in most instances. However, results of attempted cross-transmission studies allow some preliminary conclusions regarding avian cryptosporidial host affinities. *Cryptosporidium baileyi* appears to be loosely host specific and easily transmitted among several galliform and anseriform hosts. Oocysts of *C. baileyi* isolated from broiler chickens are infective for turkeys, pheasants, guinea fowl, chukar partridges, domestic ducks, domestic geese, and muscovy ducks.[16,116,118,123,124] Few studies report attempted transmission of *C. meleagridis* to other susceptible avian hosts. Oocysts isolated from the feces of naturally infected domestic turkeys were infectious for broiler chickens and domestic turkeys.[101,125] Parasites were observed in the small intestine, cecum, colon, and cloaca of recipient chickens and in the small intestine, cecum, colon, or bursa of Fabricius of recipient turkeys.

Little is known regarding the species status or host affinities of quail isolates of *Cryptosporidium*. The authors are aware of a single report of successful experimental transmission of *Cryptosporidium* of quail origin (*C. coturnix*) to chickens.[108] Recipient chickens harbored bursal infections only, suggesting that the infecting species may have been *C. baileyi* (see below, site specificity).

Results of attempts to infect mammals with oocysts of avian origin indicate that avian cryptosporidial species do not develop in experimentally inoculated mammalian hosts.[16,108,118] Conversely, however, when oocysts of *C. parvum* were given intratracheally to chickens, developmental stages were observed in the tracheas at necropsy.[126]

2. Sites of Development and Site Specificity

Avian cryptosporidia have been observed in the enteric, respiratory, and renal epithelium of infected birds.[108,114,121,123,127,128] Within the enteric tract, cryptosporidia are seen in the salivary glands, proventriculus, small intestine, cecum, colon, cloaca, and bursa of Fabri-

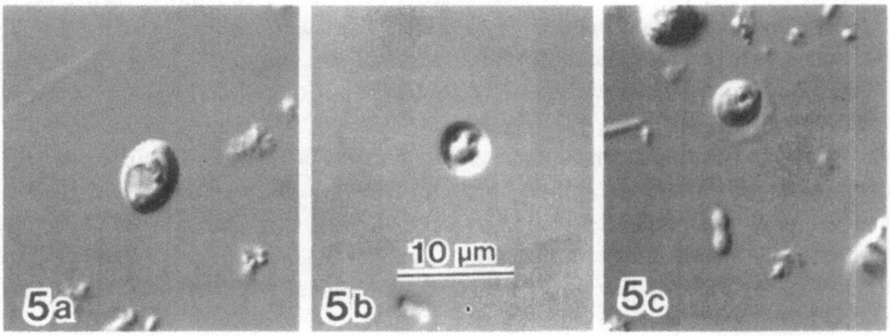

FIGURE 5. Nomarski interference contrast photomicrographs of oocysts of *Cryptosporidium* spp. from fresh fecal flotations from avian hosts. (a) Oocysts of *C. baileyi* in chicken feces. (b) Oocyst of *C. meleagridis* in turkey feces. (c) Oocyst of *Cryptosporidium* sp. in quail feces.

cius.[94,96,104,117,123,127] In the respiratory tract and oculonasal tissues, parasites are observed in the nasal chamber, palatine cleft, turbinates, infraorbital sinuses, conjunctiva, larynx, trachea, all levels of the bronchi, and in the air sacs.[116,123,127,128] Within the excretory tract, cryptosporidial stages were present in the ureters, collecting ducts, collecting tubules, and distal convoluted tubules.[121,128]

Recognized cryptosporidial species appear to have preferred sites of development. *Cryptosporidium baileyi* is apparently restricted to development in the respiratory tract and the distal gastrointestinal tract including the cecum, distal colon, cloaca, and bursa of Fabricius.[16,123,127] *Cryptosporidium meleagridis* develops predominantly in the small intestine and is seen only rarely, and in small numbers, in the distal gastrointestinal tract or respiratory tract.[96,101,129] It is not known which species of avian *Cryptosporidium* is responsible for naturally occurring respiratory cryptosporidiosis in turkeys.[102-104]

Little is known regarding the tissue specificities of quail *Cryptosporidium*, although during natural disease outbreaks, parasites may be found in both the respiratory and enteric tracts.[105-109]

3. Oocyst Morphology and Morphometrics

Differentiation of cryptosporidial species in avian hosts is often determined based on morphometrics of sporulated oocysts in saline mounts, stained smears, or flotation preparations. Following are morphometric descriptions of avian species for which there is detailed information.

Cryptosporidium baileyi — Oocysts of *C. baileyi* are fully sporulated when passaged in feces (Figure 5a). Four crescent-shaped sporozoites surround a large compact oocyst residuum. Sporozoites are without refractile bodies. The oocyst wall is about 0.5 µm thick, colorless and has no micropyle. Sporocysts are not present. Oocysts are approximately 6.8 × 5.0 µm. The oocyst length/width ratio is approximately 1.33.[127]

Cryptosporidium meleagridis — Oocysts of *C. meleagridis* are also fully sporulated when passaged in feces (Figure 5b). Sporozoites are without refractile bodies and are arranged as described above for *C. baileyi*. Oocysts of *C. meleagridis* are approximately 5.2 × 4.6 µm. The oocyst length/width ratio is approximately 1.13.[125] Results of statistical analyses indicate that means lengths, widths, and length/width ratios for *C. baileyi* and *C. meleagridis* oocysts are significantly different ($p \leq 0.01$).[125] These results substantiate our observations that oocysts of the two species may be differentiated morphometrically.

Quail *Cryptosporidium* — Oocysts are fully sporulated when passaged in feces (Figure 5c). Sporozoites are without refractile bodies and are arranged as described for *C. baileyi*. Oocysts are approximately 5.0 × 4.3 µm. The oocyst length/width ratio is approximately

1.19 (Blagburn, unpublished data). Oocysts of quail *Cryptosporidium* are morphologically similar to those of *C. meleagridis*.

4. Endogenous Stages and Cycles of Development

Few comprehensive studies investigating endogenous development of avian *Cryptosporidium* species have been published. Most published studies detail morphological aspects of endogenous development retrospectively from tissues collected during outbreaks of clinical disease. The endogenous developmental cycle of *C. baileyi* in broiler chickens has been studied and is as follows. The prepatent period varies from 3 to 5 days and the patent period from 4 to 24 days.[16,127] Endogenous development proceeds in the ileum, large intestine, cloaca, and bursa of Fabricius. Few developmental stages are seen in the ileum during early phases of development, and only for a short time. Stages are most numerous in the cloaca and bursa of Fabricius. Type I meronts contain eight merozoites and appear at about 12 h postinoculation (PI). Type II meronts contain four merozoites and appear at about 48 h PI. Type III meronts contain eight merozoites and appear at about 72 h PI. Gamonts are first seen at 96 hours PI. Oocysts are present in feces at 72 h PI. Both thick- and thin-walled oocysts are produced, suggesting that an autoinfective cycle exists for *C. baileyi* as was described for *C. parvum*.[3] We assume that development of *C. baileyi* in the respiratory tract is similar to that reported following oral inoculation, but detailed studies are lacking.

Itakura[130] described the ultrastructure of some cryptosporidial developmental stages in the bursa of Fabricius, cecum, and trachea of naturally infected broiler chickens. His descriptions were similar ultrastructurally to those previously described for *C. parvum*.[23]

The life cycle stages of *C. meleagridis* were described by Slavin[96] from naturally infected turkey poults. Development occurred in the terminal one third of the small intestine. He described trophozoites, a single type of meront containing eight merozoites (there are no doubt additional generations or types), gamonts, and oocysts. Although Slavin was incorrect in his assumptions regarding anterior ends and attachment sites of merozoites, and presumed absence of endogenous sporulation of oocysts, his description was remarkably precise, and correlated well with Tyzzer's original description of the endogenous development of *C. parvum*.[8]

Morphological descriptions of the endogenous stages of *Cryptosporidium* spp. recovered from the enteric tract of naturally infected quail[105] were similar to those described for *C. meleagridis*,[96] but details of the life cycle await further study.

5. Prevalence of *Cryptosporidium* in Avian Hosts

Few studies have been conducted to establish prevalences of *Cryptosporidium* sp. in avian hosts. Published reports represent retrospective[131,132] or direct[98,131] examinations of chicken sera or formalin-fixed tissues, direct examination of Auromine-O stained chicken feces,[120] and examination of turkey feces by sucrose flotation.[101] A survey of banked serum samples from broilers in Delaware, Maryland, and Virginia indicated that *Cryptosporidium* was present in 38% of 42-day-old birds, and 50% of 63-day-old birds on an extended grow-out program.[131] Direct serological examination of birds from that same region yielded a 22% prevalence among 454 broiler flocks.[131] Histological examination of the bursas of 139 broilers from Scotland resulted in a 18.7% prevalence.[98] A similar retrospective histologic examination conducted in Georgia resulted in a prevalence of 6.4% among 1065 broilers.[132] Interestingly, *Eimeria* spp. were the only protozoan parasites diagnosed more frequently that *Cryptosporidium*. A survey of chicken feces from broilers in North Carolina, using the Auromine-O procedure, resulted in positive rates of 27.3% (9/33) in broilers, 10% (3/30) in broiler-breeders, and 5.9% in layers.[120] Examination of the feces of 17- and 24-day-old turkeys in Iowa, using a sucrose flotation procedure, resulted in prevalences of 80% (18/20) and 38% (38/100), respectively.[101] We are not aware of published prevalence studies involving *Cryptosporidium* in quail hosts.

6. Immunological and Biochemical Characterizations of Avian *Cryptosporidium* spp.

Little information is available regarding immunological reactions and antigens associated with avian cryptosporidia. Current[17] demonstrated that 2-week-old broiler chickens are capable of clearing *C. baileyi* endogenous stages from the cloaca and bursa of Fabricius 14 to 16 days after inoculation with 1×10^6 oocysts. Birds were refractory to challenge with 2×10^6 oocysts, 5 days after initial cessation of shedding. Resistance to challenge was accompanied by the appearance of *C. baileyi*-specific antibodies, demonstrable by both an indirect fluorescent antibody test (IFAT) and an enzyme-linked immunosorbent assay (ELISA). Current[17] also demonstrated that immune sera from challenged birds recognized more than ten oocyst-sporozoite antigens after separation using gel electrophoresis.

Broiler chickens also display age resistance to *C. baileyi*. Lindsay et al.[133] demonstrated that as age at inoculation increases, susceptibility to respiratory disease decreases. This is accompanied by a concurrent increase in the prepatent period and a decrease in the patent period.

Cryptosporidial infections may alter host responses to heterologous antigens. Broiler chickens harboring 7-day-old *C. baileyi* infections had significantly increased antibody titers to Newcastle disease virus (NDV). Delayed hypersensitivity reactions to NDV were decreased. No significant differences in infectious bursal disease virus titers were observed between *C. baileyi* infected and control birds.[15]

Avian and mammalian cryptosporidial oocysts appear to share common antigens. A monoclonal antibody generated against a *C. parvum* oocyst wall antigen, cross-reacted with antigens on oocysts of both *C. meleagridis* and quail *Cryptosporidium*.[134] Additionally, human sera containing antibodies specific for *C. parvum* antigens, react with oocysts of *C. baileyi* using an ELISA (Ungar; personal communication). Similar cross-reactions were reported by Tzipori and Campbell[135] when they observed that 22 of 25 chicken sera reacted with *C. parvum* using an IFAT.

7. Miscellaneous Experimental Studies

a. Site Specificity

Site specificity of *C. baileyi* in broiler chickens was investigated by inoculating oocysts intracloacally into 2-day-old broiler chicks.[136] Subsequent infections developed in the cloaca and bursa of Fabricius, without development in the respiratory tract. This suggests that avian respiratory infections are likely acquired following inhalation or aspiration of oocysts, and not by circulatory migration of intestinal stages. Similar studies were conducted to determine sites of infection and clinical sequelae following intravenous, intraabdominal, and ocular inoculation of *C. baileyi* oocysts.[137,138] Intravenous inoculation failed to induce infections in either the enteric or respiratory tracts. Intraabdominal inoculation resulted in both respiratory and enteric infections indistinguishable from those induced by oral or intratracheal inoculation of oocysts. Following ocular inoculation, developmental stages of *C. baileyi* were observed on the conjunctival epithelium of 3 or 14 chickens and 0 of 7 turkeys. Neither clinical signs nor ocular lesions were observed. Cryptosporidial developmental stages were also observed in the clocacas of 14 of 14 of the chicks and 4 of 7 of the turkeys following ocular inoculation.[138]

b. Coinfections with Other Avian Pathogens

Interactions of *C. baileyi*, infectious bronchitis virus (IBV), *Escherichia coli* and infectious bursal disease virus (IBDV) were investigated in 7-day-old broiler chickens.[139] Results indicated that both IBV and *E. coli* enhanced the severity of *C. baileyi*-induced respiratory disease. Both agents also decreased the density of *C. baileyi* in the respiratory tract. These results suggest that *C. baileyi* might be overlooked in virus-infected birds because of its lower density. Observable interactions between *C. baileyi* and IBDV were not noted.

In a similar co-infection study, inoculation of 4-day-old specific-pathogen-free (SPF) chickens with a moderately pathogenic strain of IBDV and *C. baileyi* concurrently, resulted in more severe bursal lesions, more infected birds, and greater numbers of cryptosporidia in tissues than in birds given *C. baileyi* alone.[140] Experimental infections in 5-day-old SPF chickens with *C. baileyi* and reovirus (strain 2035) concurrently, resulted in significantly decreased weight gains over the 21-day study period, compared to noninoculated control birds or birds receiving reovirus alone.[141]

c. *In Vitro and In Ovo Cultivation and Excystation*

Lindsay et al.[142] attempted to cultivate *C. baileyi* in cell cultures following sporozoite inoculations. Sporozoites did not undergo development in six primary avian cell cultures, six primary mammalian cell cultures, or in four established mammalian cell lines. *Cryptosporidium baileyi* was successfully cultivated in the chorioallantoic membranes of chicken embryos and eight other avian embryos that were inoculated.[142] Characteristics of endogenous development appeared similar to that described for *in vivo* development in broiler chickens.[16,127] Maintenance of *C. baileyi* for 20 passages in chicken embryos did not affect its pathogenicity in 2-day-old broiler chicks. Likewise, maintenance of *C. baileyi* for 10 passages in chicken embryos did not affect its development or pathogenicity in 4-day-old turkey poults.[142]

Excystation studies have shown that sporozoites of *C. baileyi* excyst in greatest numbers in 0.75% sodium taurocholate at 40°C for 120 min. Trypsin, necessary for excystation of closely related *Eimeria* spp., was not necessary for excystation of *C. baileyi* sporozoites.[143]

C. PATHOGENETIC ASPECTS OF NATURAL AND EXPERIMENTAL CRYPTOSPORIDIOSIS IN AVIAN HOSTS

1. Chickens

Cryptosporidiosis in chickens occurs as a respiratory, intestinal, or renal disease. Respiratory disease is observed most commonly and is the most severe of the three forms.[98,99,144,145] Clinical signs in affected birds consist of respiratory distress characterized by depression, rales, coughing, sneezing, or dyspnea. Lesions commonly observed include excessive accumulations of mucus in the oculonasal cavities, trachea, lungs, and air sacs. Severely affected birds may be sternally recumbent and unable to move. The lumens of organs of the respiratory tract contain a mucocellular exudate consisting of exfoliated epithelial cells, macrophages, heterophils, lymphocytes, plasma cells, and developing parasites. Infected epithelial cells may be hyperplastic, hypertrophied, or metaplastic. Intraepithelial heterophils, exostosing heterophils, and hyperemia of vessels are often present. Focal or diffuse denudement of tracheal cilia results in decreased or complete loss of respiratory mucociliary function.

Experimentally induced respiratory cryptosporidiosis (*C. baileyi*) results in clinical signs, and gross and microscopic lesions identical to those described for naturally occurring disease (Figures 6a—d).[15,127,142,146] Transient weight depressions are observed during the acute (2 to 3 weeks postinfection) disease, although birds that recover do not differ significantly in weight from noninoculated control birds.[15]

In most instances, the presence of cryptosporidia in the intestine and associated organs of chickens is not accompanied by clinically apparent disease.[15,16,118,127,133,141] In some reports,[147,148] diarrhea, weakness, cachexia, or failure to thrive accompanies infections. In such cases, lesions include atrophic or misshapen villi, with accompanying infiltration of inflammatory cells into the lamina propria,[147] or enlarged, thickened villi, covered by a thick tenacious mucoid material.[148] Bursal cryptosporidiosis can result in epithelial cell hypertrophy or hyperplasia,[98,149,150] and may be associated with increased broiler mortality.[151]

We are aware of a single report of renal cryptosporidiosis in chickens.[128] Parasites, found in the ureters, and in the collecting ducts and distal convoluted tubules of the kidneys, were associated with epithelial cell hyperplasia and inflammatory cell infiltration.

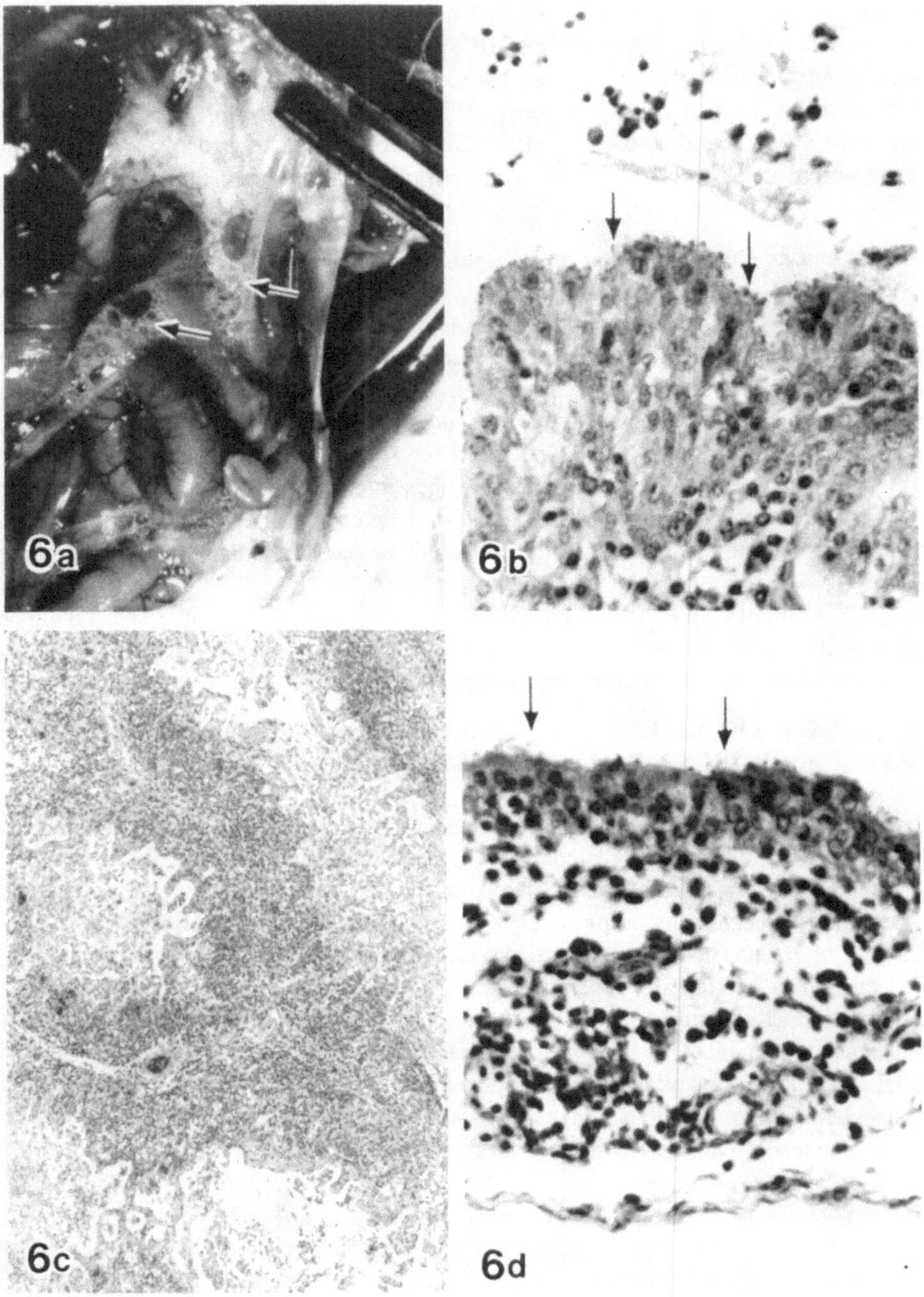

FIGURE 6. One photograph and three photomicrographs illustrating important aspects of the pathology of respiratory cryptosporidiosis in broiler chickens following intratracheal inoculation. (a) Abdominal airsacculitis 14 days after inoculation. The mucosal surface of the air sacs are covered with a foamy, gray, fluid exudate (arrows). (b) Secondary bronchitis in a broiler 14 days after inoculation. Note the presence of exudate in the lumen, developmental stages of the parasite in the epithelial surface (arrows), and mucosal thickening due to epithelial hyperplasia and inflammatory cell infiltration. Magnification × 333. (c) Lung from a broiler 14 days after inoculation. Bronchopneumonia is evident in the tertiary bronchi and air capillaries. Magnification × 67. (d) Air sac from a broiler with *C. baileyi*-induced airsacculitis. The air sac membrane is thickened by lymphocytes, plasma cells, and histiocytes. Developmental stages of the parasite are seen on the surface of the respiratory epithelium (arrows). Magnification × 333.

2. Turkeys

Naturally occurring respiratory[100,102-104] and intestinal[101,129,152] cryptosporidial infections have been reported from turkeys. Clinical signs of respiratory disease, and gross and microscopic lesions are similar to those observed in chickens. In one report, however, infections in the infraorbital sinuses resulted in bilateral facial swelling and hyperplasia of the sinus epithelium.[103] Experimentally induced respiratory cryptosporidiosis in turkeys was achieved by inoculation of *C. baileyi* oocysts intratracheally into 6-day-old poults,[123] and by introduction of sinus exudate from naturally infected birds into 1-day-old poults (donor parasite unknown).[103] Recipient birds developed respiratory disease[123] and sinusitis,[103] indistinguishable from naturally occurring disease.

Intestinal cryptosporidial infections in turkeys may result in moderate to severe diarrhea,[129,152] mortality,[152] or lack of observable clinical signs.[101,124] The presence of a companion reovirus may have been responsible for mortalities observed in one of the above reports.[152] Postmortem examination of *Cryptosporidium*-infected turkeys with diarrhea revealed a small intestine that was pale and contained cloudy, mucoid contents. Villi were moderately atrophic, crypts were hyperplastic, and infected segments of small intestine contained a mixed mucosal inflammatory infiltrate.[129] Experimentally induced intestinal cryptosporidial infections in turkeys using *C. baileyi* from broiler chickens or uncharacterized turkey isolates failed to induce clinical disease.[101,123,124]

3. Quail

Natural or experimental cryptosporidiosis has been reported from bobwhite quail[105-107] and common quail.[108,109] Infections are seen in both the respiratory[108,109] and intestinal[105-107] tracts. Clinical signs and lesions associated with *Cryptosporidium* induced respiratory disease include depression, sneezing, congested nasal mucosa, excess mucus in the trachea, epithelial hyperplasia and deciliation, dilatation of mucus glands, and infiltration of the lamina propria with inflammatory cells. Outbreaks of respiratory disease may be accompanied by high morbidity and moderate mortality.[108]

Intestinal disease usually results in high mortality in birds from hatch to 12 days.[105,106] Diarrhea appears as a clear to brown fluid, or may be white and watery. Villus atrophy, sloughed villi, or detached enterocytes are prominent histopathological features. *Cryptosporidium* may induce disease alone[105] or in concert with reovirus.[106,107] An apparent synergism occurs between *Cryptosporidium* and reovirus in co-infected birds, because greater numbers of *Cryptosporidium* and enhanced shedding, as well as greater desemination of virus, were noted.[107] Further studies are needed to fully characterize morphological, epizootiological, developmental, and pathogenetic aspects of quail cryptosporidiosis.

4. Other Galliformes

Cryptosporidial infections have been reported from the respiratory and oculonasal tissues of pheasants[108,111,112] and peafowl,[110] and from the renal and oculonasal tissues of jungle fowl.[114] Respiratory disease in pheasants consisted of rhinitis, laryngitis, and tracheitis, characterized by depression, quiet gurgling respiration, and a flocculant nasal discharge.[111] Cryptosporidial infection in the peafowl was accompanied by gurgling respiration, coughing, sneezing and a serous oculonasal discharge.[110] The jungle fowl died suddenly from renal failure. At necropsy, cryptosporidial organisms were present in pale and swollen kidneys, and in thick mucus in the nasal cavity.[114] Because other contaminating microorganisms were cultured from several anatomic sites, renal failure could not be attributed to *Cryptosporidium*.

5. Miscellaneous Avians

Developmental stages of *Cryptosporidium* have been reported from a variety of nongalliform avians. In most cases, neither clinical signs nor demonstrable lesions were associated with cryptosporidial infections. A summary of infections in nongalliform avian hosts is given in Table 2.

TABLE 2
***Cryptosporidium* Infections in Non-Galliform Avian Hosts**

Host	Site of infection	Natural/experimental infection	Clinical signs	Ref.
		Anseriformes		
Mallard duck	Respiratory tract	Natural	Yes	108
Domestic duck	Bursa, cloaca	Experimental	No	16, 116
	Respiratory tract	Experimental	Yes (mild)	116
	Bursa	Natural	?	117
	Conjunctiva	Natural	No	115
Muscovy duck	Bursa, cloaca	Experimental	No	118
Domestic goose	Large intestine	Natural	No	97
	Bursa	Experimental	No	16
Tundra swan	Feces	Natural	No	120
		Passeriformes		
Black-throated finch	Kidney	Natural	?[a]	121
Canary	Proventriculus	Natural	?	117
		Psittaciformes		
Lovebird	Small intestines	Natural	Yes	122
Red-lored parrot	Cloaca	Natural	No	119
Macaw	Feces	Natural	No	120
Budgerigar	Respiratory tract	Natural	Yes	108
	Feces	Natural	No	120

[a] ? = Other agents present or relationship of *Cryptosporidium* to clinical signs could not be determined.

D. TREATMENT AND CONTROL OF AVIAN CRYPTOSPORIDIOSIS

An effective cryptosporidicidal drug has not been identified, despite several attempts to treat or prevent avian cryptosporidiosis in a variety of hosts. Neither chlortetracycline nor amprolium had a demonstrable effect on *Cryptosporidium*-induced sinusitis in turkeys.[103] Oxytetracycline, neomycin, and furazolidone were equally ineffective in controlling an outbreak of intestinal cryptosporidiosis in bobwhite quail.[105] Oxytetracycline failed to control an outbreak of respiratory cryptosporidiosis in peafowl chicks, despite continuous treatment in the drinking water.[110] In an experimental study designed to evaluate halifuginone, salinomycin, lasalocid, and monensin against experimentally induced respiratory and bursal/cloacal cryptosporidiosis (*C. baileyi*) in broiler chicks, Lindsay et al.[146] found that none prevented infections. Birds treated with each of the drugs developed signs of respiratory disease following intratracheal inoculation with oocysts, despite continuous administration of drugs.

Failure to control avian cryptosporidiosis with anticoccidials has led investigators to evaluate the ability of premise disinfectants to render oocysts nonviable. As mentioned previously, Sundermann et al.[26] evaluated several commercial disinfectants for their ability to inactivate *C. baileyi* oocysts. Exposure of oocysts to 50% (v/v) commercial ammonia or bleach in water for at least 30 min prevented excystation of a majority of *C. baileyi* sporozoites. Likewise, recurrence of *Cryptosporodiosis*-induced diarrhea and mortality in bobwhite quail was prevented by thoroughly cleaning cages and by exposing them to commercial bleach.[105]

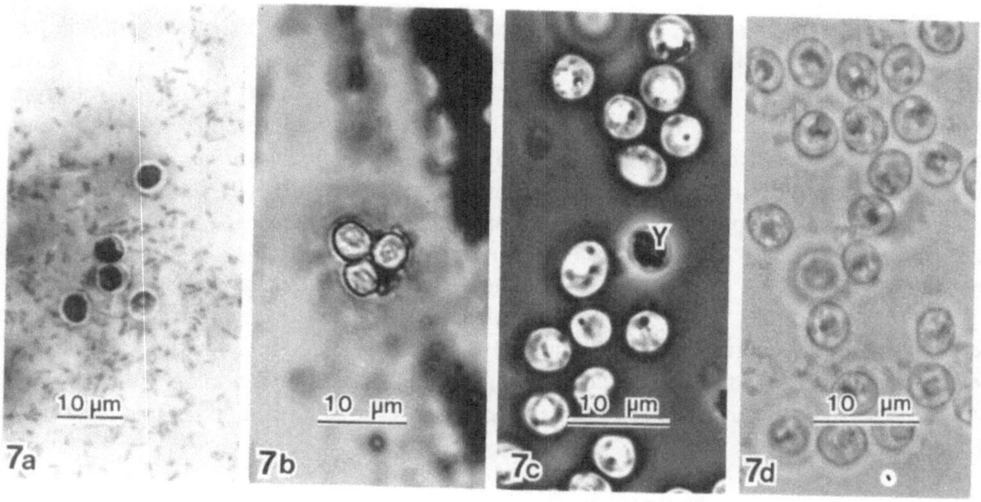

FIGURE 7. Light micrographs of oocysts of *Cryptosporidium parvum* demonstrated by three common stool diagnostic techniques. (a) Acid fast staining[154,155] results in oocysts appearing bright red against the counter stained background. Yeast stain darkly with the background stain. (b) Carbol fuschin negative staining[156] resulting in bright, refractive oocysts which do not take up the stain. Yeast stain darkly. (c) Sheathers sugar flotation and phase-contrast microscopy[20] results in bright, refractive oocysts containing dark granules. Yeast cells (Y) are not bright and refractile. (d) Sheather's sugar flotation and brightfield microscopy results in translucent oocysts that often contain one or more dark granule. With brightfield microscopy, *Cryptosporidium* spp. oocysts are often difficult to distinguish from yeast cells.

VII. DIAGNOSIS

A. MAMMALIAN CRYPTOSPORIDIOSIS

Prior to 1980, diagnosis of mammalian cryptosporidiosis required identification of the small spherical life cycle stages of *C. parvum* in the microvillous region of the intestinal mucosa obtained by biopsy or at the time of necropsy and subsequently processed for examination by light or electron microscopy. Such invasive and time-consuming procedures are no longer required since a variety of techniques have been developed for identifying *Cryptosporidium* spp. oocysts in fecal specimens.

For the diagnosis of cryptosporidiosis, stool specimens should be submitted as fresh material or in 10% formalin or sodium acetate-acetic acid-formalin (SAF) preservatives. Most recommended stains for *Cryptosporidium* spp. oocysts cannot be performed on stools preserved in polyvinyl alcohol (PVA) fixative. The routine stains (trichrome, iron hematoxylin) used for stool diagnosis of other parasites are not acceptable for the identification of *Cryptosporidium* spp. oocysts. Several widely used techniques (Figure 7) for demonstrating *Cryptosporidium* spp. oocysts in fecal specimens from humans and animals are modified acid-fast staining,[154,155] negative staining,[156] and Sheather's sugar flotation.[20,29,69] Although the last two procedures are useful in the research laboratory, acid-fast staining is usually the method of choice for the clinical microbiology laboratory. Considerable experience is often required with the concentration and staining methods to obtain an accurate diagnosis. For this reason, immunofluorescent antibody (IFA) procedures employing *Cryptosporidium*-specific polyclonal or monoclonal antibodies have been developed to aid in the identification of oocysts in stool specimens.[40,157]

In the research laboratory and in the veterinary diagnostic laboratory, *Cryptosporidium* spp. infections can also be monitored by examining mucosal scrapings as fresh preparations in a balanced salt solution[16] or as Giemsa-stained smears.[153] Examination of infected cells with Nomarski interference contrast microscopy is particularly useful for identifying various developmental stages of the parasite in fresh mucosal scrapings,[16,23] cultured cells,[25] and the chorioallantoic membrane of chicken embryos.[24]

The use of serodiagnostic techniques for monitoring exposure to *Cryptosporidium* has thus far been limited to a few laboratories. Antibodies specific to *Cryptosporidium* have been detected, using an IFA procedure, in sera obtained from persons who recovered from confirmed infections[54,55] and an IFA assay has been used for the presumptive diagnosis of cryptosporidiosis in two clusters of cases.[34,35] Specific anti-*Cryptosporidium* IgG and/or IgM were also detected, by an enzyme linked immunosorbent assay (ELISA), in the sera of 95% of patients with cryptosporidiosis at the time of medical presentation and in 100% within 2 weeks of presentation.[56] Several serologic surveys have reported that more than 50% of persons with no known infection may have anti-*Cryptosporidium* IgG, suggesting recent exposure to the parasite.[34,44,158] Additional evaluations are needed to confirm the utility of these serologic procedures for diagnosing and monitoring infections, for determining the prevalence of prior exposure in selected study populations, and to determine if there is any correlation between the presence of *Cryptosporidium*-specific serum antibodies and resistance to reinfection.

B. AVIAN CRYPTOSPORIDIOSIS

Active infections of the respiratory, intestinal, or urinary tracts of avian hosts can be diagnosed by identification of developmental stages of the parasite in fresh or stained mucosal scrapings, in histologic sections or by demonstrating oocysts in exudative fluids or feces. Histologic, cytologic, and oocyst detection procedures outlined above for mammalian cryptosporidiosis are also useful for diagnosing and monitoring avian cryptosporidiosis. Cytologic examination of tracheal imprints after Diff Quick or acid fast staining provides a rapid, reliable, and economically feasible means of diagnosing respiratory cryptosporidiosis.[153] This procedure can also be used to detect *Cryptosporidium* in other tissues. In addition to fecal samples, diagnostic specimens can also be obtained by swabbing the cloacal or tracheal epithelium with a moist cotton-tipped applicator.[165] Oocysts are removed from the microvillous region of the infected mucosa and are trapped in the cotton fibers. The swabs can then be placed in tubes containing 1 ml of water of fixative and transported to the laboratory. In the laboratory, oocysts can be removed from the cotton by the use of a vortex mixer and then subjected to the concentration or staining procedures mentioned above.

Serological tests are useful for determination of serological prevalence, but have not been widely used for confirmation of clinically apparent disease. Developed assays include IFAT and ELISA.[17,131]

REFERENCES

1. **Angus, K. W.,** Cryptosporidiosis in man, domestic animals, and birds: a review, *J. R. Soc. Med.*, 76, 62, 1983.
2. **Current, W. L.,** *Cryptosporidium*: its biology and potential for environmental transmission, *CRC Crit. Rev. Environ. Control.*, 17, 21, 1986.
3. **Current, W. L.,** *Cryptosporidium* spp., in *Parasitic Infections in the Compromised Host*, Walzer, P. D. and Genta, R. M., Eds., Marcel Dekker, New York, 1989, 281.
4. **Fayer, R. and Ungar, B. L. P.,** *Cryptosporidium* spp. and cryptosporidiosis, *Microbiol. Rev.*, 50, 458, 1986.
5. **Tzipori, S.,** Cryptosporidiosis in animals and humans, *Microbiol. Rev.*, 47, 84, 1983.
6. **Tyzzer, E.,** A sporozoan found in the peptic glands of the common mouse, *Proc. Soc. Exp. Biol.*, 5, 12, 1907.
7. **Tyzzer, E. E.,** An extracellular coccidium, *Cryptosporidium muris* (gen et sp nov.,) of the gastric glands of the common mouse, *J. Med. Res.*, 23, 487, 1910.
8. **Tyzzer, E. E.,** *Cryptosporidium parvum* (sp. nov.), a coccidium found in the small intestine of the common mouse, *Arch. Protistenkd.*, 26, 394, 1912.

9. **Forgacs, P., Tarshis, A., Ma, P., Federman, M., Mele, L., Silverman, M. L., and Shea, J. A.,** Intestinal and bronchial cryptosporidiosis in an immunodeficient homosexual man, *Ann. Intern. Med.*, 99, 793, 1983.
10. **Pitlik, S., Fainstein, V., Rios, A., Guarda, L., Mansell, P. W. A., and Hersh, E. M.,** Cryptosporidial cholecystitis, *N. Engl. J. Med.*, 308, 976, 1983.
11. **Crawford, F. G. and Vermund, S. H.,** Human cryptosporidiosis, *CRC Crit. Rev. Microbiol.*, 16, 113, 1988.
12. **Clarke, J. J.,** A study of coccidia met with in mice, *J. Microsc. Sci.*, 37, 277, 1895.
13. **Levine, N. D.,** Taxonomy and review of the coccidian genus *Cryptosporidium* (Protozoa, Apicomplexa), *J. Protozool.*, 31, 94, 1984.
14. **Panciera, R. J., Thomassen, R. W., and Garner, F. M.,** Cryptosporidial infection in a calf, *Vet. Pathol.*, 8, 479, 1971.
15. **Blagburn, B. L., Lindsay, D. S., Giambrone, J. J., Sundermann, C. A., and Hoerr, F. J.,** Experimental cryptosporidiosis in broiler chickens, *Poult. Sci.*, 66, 442, 1987.
16. **Current, W. L., Upton, S. J., and Haynes, T. B.,** The life cycle of *Cryptosporidium baileyi* n. sp. (Apicomplexa, Cryptosporidiidae) infecting chickens, *J. Protozool.*, 33, 289, 1986.
17. **Current, W. L. and Snyder, D. B.,** Development of and serologic evaluation of acquired immunity to *Cryptosporidium baileyi* by broiler chickens, *Poult. Sci.*, 67, 720, 1988.
18. **Miesel, J. L., Perera, D. R., Meligro, C., and Rubin, C. E.,** Overwhelming watery diarrhea associated with *Cryptosporidium* in an immunosuppressed patient, *Gastroenterology*, 70, 1156, 1976.
19. **Nime, F. A., Burek, J. D., Page, D. L., Holscher, M. A., and Yardley, J. H.,** Acute enterocolitis in a human being infected with the protozoan *Cryptosporidium*, *Gastroenterology*, 70, 592, 1976.
20. **Current, W. L., Reese, N. C., Ernst, J. V., Bailey, W. S., Heyman, M. B., and Weinstein, W. M.,** Human cryptosporidiosis in immunocompetent and immunodeficient persons: studies of an outbreak and experimental transmission, *N. Engl. J. Med.*, 308, 1252, 1983.
21. **Tzipori, S., Angus, K. W., Campbell, I., and Gray, E. W.,** *Cryptosporidium*: evidence for a single species genus, *Infect. Immun.*, 30, 884, 1980.
22. **Upton, S. J. and Current, W. L.,** The species of *Cryptosporidium* (Apicomplexa: Cryptosporidiidae) infecting mammals, *J. Parasitol.*, 71, 625, 1985.
23. **Current, W. L. and Reese, N. C.,** A comparison of endogenous development of three isolates of *Cryptosporidium* in suckling mice, *J. Protozool.*, 33, 98, 1986.
24. **Current, W. L. and Long, P. L.,** Development of human and calf *Cryptosporidium* in chicken embryos, *J. Infect. Dis.*, 148, 1108, 1983.
25. **Current, W. L. and Haynes, T. B.,** Complete development of *Cryptosporidium* in cell culture, *Science*, 224, 603, 1984.
26. **Sundermann, C. A., Lindsay, D. S., and Blagburn, B. L.,** Evaluation of disinfectants for ability to kill avian *Cryptosporidium* oocysts, *Comp. Anim. Prac.*, November, 36, 1987.
27. **Anderson, B. C.,** Moist heat inactivation of *Cryptosporidium* sp., *Am. J. Publ. Health*, 75, 1433, 1985.
28. **Anderson, B. C., Donndelinger, T., Wilkins, R. M., and Smith, J.,** Cryptosporidiosis in a veterinary student, *J. Am. Vet. Med. Assoc.*, 180, 408, 1982.
29. **Reese, N. C., Current, W. L., Ernst, J. V., and Bailey, W. S.,** Cryptosporidiosis of man and calf: a case report and results of experimental infections in mice and rats, *Am. J. Trop. Med. Hyg.*, 31, 226, 1982.
30. **Blagburn, B. L. and Current, W. L.,** Accidental infection of a researcher with human *Cryptosporidium*, *J. Infect. Dis.*, 148, 772, 1983.
31. **Alpert, G., Bell, L. M., Kirkpatrick, C. E., Budnick, L. D., Campos, J. M., Friedman, H. M., and Plotkin, S. A.,** Cryptosporidiosis in a day care center, *N. Engl. J. Med.*, 311, 860, 1984.
32. **Anon.,** Cryptosporidiosis among children attending day care centers — Georgia, Pennsylvania, Michigan, California, New Mexico, *Morbid. Mortal. Week. Rep.*, 33, 599, 1984.
33. **Driscoll, M. S., Thomas, V. L., and Sanford, B. A.,** *Cryptosporidium* infection in day care centers, *Drug Intell. Clin. Pharm.*, 22, 636, 1988.
34. **Koch, K. L., Phillips, D. L., and Current, W. L.,** Cryptosporidiosis in hospital personnel: evidence for person-to-person transmission, *Ann. Intern. Med.*, 102, 593, 1984.
35. **D'Antonio, R. G., Winn, R. E., Taylor, J. P., Gustafson, T. L., Current, W. L., Rhodes, M. M., Gary, G. W., and Zajac, R. A.,** A waterborne outbreak of cryptosporidiosis in normal hosts, *Ann. Intern. Med.*, 103, 886, 1985.
36. **Hayes, E. B., Matte, T. D., O'Brien, T. R., McKinley, T. W., Logsdon, G. S., Rose, J. B., Ungar, B. L. P., Word, D. M., Pinsky, P. F., Cummings, M. L., Wilson, M. A., Long, E. G., and Hurwitz, E. S.,** Large community outbreak of cryptosporidiosis due to contamination of a filtered public water supply, *N. Engl. J. Med.*, 320, 1372, 1989.
37. **Hokipii, L., Pohjola, S., and Jokipii, A. M. M.,** *Cryptosporidium*: a frequent finding in patients with gastrointestinal symptoms, *Lancet*, 1, 358, 1983.
38. **Soave, R. and Map, P.,** Cryptosporidiosis: travelers' diarrhea in 2 families, *Arch. Intern. Med.*, 145, 70, 1985.

39. **Madore, M. S., Rose, J. B., Gerba, C. P., Arrowood, M. J., and Sterling, C. R.,** Occurrence of *Cryptosporidium* oocysts in sewage effluents and selected surface waters, *J. Parasitol.*, 73, 702, 1987.
40. **Stibbs, H. H. and Ongereth, J.,** Detection of *Cryptosporidium* oocysts in fecal smears and river water, *Proc. 60th Annu. Meet. Am. Soc. Parasitol.*, Athens, GA, abstr. no. 100, 1985.
41. **Garcia, L. S. and Current, W. L.,** Cryptosporidiosis: clinical features and diagnosis, *CRC Crit. Rev. Lab. Sci.*, in press.
42. **Navin, T. R.,** Cryptosporidiosis in humans: a review of recent epidemiologic studies, *Eur. J. Epidemiol.*, 1, 77, 1985.
43. **Walsh, J. A. and Warren, K. S.,** Selective primary care. An interim strategy for disease control in developing countries, *N. Engl. J. Med.*, 301, 967, 1979.
44. **Ungar, B. L. P., Gilman, R. H., Lanata, C. F., and Perez-Schael, I.,** Seroepidemiology of *Cryptosporidium* infection in two Latin American populations, *J. Infect. Dis.*, 157, 551, 1988.
45. **Sallon, S., Deckelbaum, R. J., Schmid, I. I., Harlap, S., Baras, M., and Spira, D. T.,** *Cryptosporidium*, malnutrition, and chronic diarrhea in children, *Am. J. Dis. Child.*, 142, 312, 1988.
46. **Lewis, I. J., Hart, C. A., and Baxby, D.,** Diarrhoea due to *Cryptosporidium* in acute lymphoblastic leukemia, *Arch. Dis. Child.*, 60, 60, 1985.
47. **Bogaerts, J., Lepage, P., Rouvonoy, D., and Vandepitte, J.,** *Cryptosporidium* spp., a frequent cause of diarrhea in Central Africa, *J. Clin. Microbiol.*, 20, 874, 1984.
48. **Heine, J., Pholenz, J. F. L., Moon, H. W., and Woode, G. N.,** Enteric lesions and diarrhea in gnotobiotic calves monoinfected with *Cryptosporidium* species, *J. Infect. Dis.*, 150, 768, 1984.
49. **Hart, A. and Baxby, D.,** Management of cryptosporidiosis, *J. Antimicrob. Chemother.*, 15, 3, 1985.
50. **Soave, R.,** Diagnosis, management, and prognosis of human cryptosporidiosis, *Proc. 34th Annu. Meet. Am. Soc. Trop. Med. Hyg.*, Miami, FL, abstr. no. 135, 1985.
51. **Current, W. L. and Bick, P. W.,** The immunobiology of *Cryptosporidium* spp., *Pathol. Immunopathol. Res.*, 8, 141, 1989.
52. **Miller, R. A., Holmberg, R. E., and Clausen, C. R.,** Life-threatening diarrhea caused by *Cryptosporidium* in a child undergoing therapy for acute lymphocytic leukemia, *J. Pediatr.*, 103, 256, 1983.
53. **Louie, E., Borkowsky, W., Klesius, P. H., Haynes, T. B., Gordon, S., Bonk, S., and Lawrence, H. S.,** Treatment of cryptosporidiosis with oral bovine transfer factor, *Clin. Immunol. Immunopathol.*, 44, 329, 1987.
54. **Campbell, P. N. and Current, W. L.,** Demonstration of serum antibodies to *Cryptosporidium* sp. in normal and immunodeficient humans with confirmed infections, *J. Clin. Microbiol.*, 18, 165, 1983.
55. **Casemore, D. P.,** The antibody response to *Cryptosporidium*: development of a serological test and its use in a study of immunologically normal persons, *J. Infect.*, 14, 125, 1987.
56. **Ungar, B. L. P., Soave, R., Fayer, R., and Nash, T. E.,** Enzyme immunoassay detection of immunoglobulin M and G antibodies to *Cryptosporidium* in immunocompetent and immunocompromised persons, *J. Infect. Dis.*, 153, 570, 1986.
57. **Rose, M. E.,** Immunity to *Eimeria* infections, *Vet. Immunol. Immunopathol.*, 17, 33, 1987.
58. **Mata, L., Bolanos, H., Pizarro, D., and Vives, M.,** Cryptosporidiosis en ninos de Costa Rica: estudio transversal y longitudinal, *Rev. Biol. Trop.*, 32, 129, 1984.
59. **Saxon, A. and Weinstein, W.,** Oral administration of bovine colostrum anti-cryptosporidia antibody fails to alter the course of human cryptosporidiosis, *J. Parasitol.*, 73, 413, 1987.
60. **Moon, H. W., Woodmansee, D. B., Harp, J. A., Abel, S., and Ungar, B. L. P.,** Lacteal immunity to enteric cryptosporidiosis in mice: immune dams do not protect their suckling pups, *Infect. Immun.*, 56, 649, 1988.
61. **Tzipori, S., Robertson, D., and Chapman, C.,** Remission of diarrhea due to cryptosporidiosis in an immunodeficient child treated with hyperimmune bovine collostrum, *Br. Med. J.*, 293, 1276, 1986.
62. **Fayer, R., Perryman, L. E., and Riggs, M. W.,** Hyperimmune bovine colostrum neutralizes *Cryptosporidium* sporozoites and protects mice against oocyst challenge, *J. Parasitol.*, 75, 151, 1989.
63. **Fayer, R., Anderson, C., Ungar, B. L. P., and Blagburn, B. L.,** Efficacy of hyperimmune bovine colostrum for prophylaxis of cryptosporidiosis in calves, *J. Parasitol.*, 75, in press.
64. **Anderson, B. C. and Bulgin, M. S.,** Enteritis caused by *Cryptosporidium* in calves, *Vet. Med.*, 76, 865, 1981.
65. **Anderson, B. C.,** *Cryptosporidium* spp. in cattle, in *Cryptosporidiosis: Proc. 1st Inter. Workshop*, Angus, K. W. and Blewett, D. A., Eds., Animal Disease Research Association, Edinburgh, Scotland, 1988, 55.
66. **Pholenz, J., Bemrick, W. J., Moon, H. W., and Chevelle, N. F.,** Bovine cryptosporidiosis: a transmission electron microscopic study of some stages in the life cycle and of the host-parasite relationship, *Vet. Pathol.*, 15, 417, 1978.
67. **Morin, M., Lariviere, S., and Lallier, R.,** Pathological and microbiological observations made on spontaneous cases of acute neonatal calf diarrhea, *Can. J. Comp. Med.*, 40, 228, 1976.
68. **Bergeland, M. E., Johnson, D. D., and Shave, H.,** Bovine cryptosporidiosis in the north central United States, *Am. Assoc. Vet. Lab. Diag., 22nd Annu. Proc.*, 131, 1979.

69. **Anderson, B. C.,** Patterns of shedding of cryptosporidial oocysts in Idaho calves, *J. Am. Vet. Med. Assoc.*, 178, 982, 1981.
70. **Meuten, D. J., Van Kruiningen, H. J. H., and Lein, D. H.,** Cryptosporidiosis in a calf, *J. Am. Vet. Med. Assoc.*, 165, 1974.
71. **Powell, H. S., Holscher, M. A., and Heath, J. E.,** Bovine cryptosporidiosis, *Vet. Med. Small Anim. Clin.*, 71, 205, 1976.
72. **Pearson, G. R. and Logan, E. F.,** Demonstration of cryptosporidia in the small intestine of a calf by light, transmission electron, and scanning electron microscopy, *Vet. Rec.*, 103, 212, 1978.
73. **Snodgrass, D. R., Angus, K. W., and Gray, D. W.,** Cryptosporidia associated with rotavirus and *Escherichia coli* in an outbreak of calf scour, *Vet. Rec.*, 106, 458, 1980.
74. **Angus, K. W.,** Cryptosporidiosis in domestic animals and humans, *In Practice*, 9, 47, 1987.
75. **Reynolds, D. J., Morgan, J. H., Chanter, N., Jones, P. W., Bridger, J. C., Debney, J. C., and Bunch, K. J.,** Microbiology of calf diarrhoea in southern Britain, *Vet. Rec.*, 119, 34, 1986.
76. **Haynes, T. B., Lauerman, L. H., Klesius, P. H., Mitchell, F. M., Long, I. R., and Ellis, A. C.,** Cryptosporidiosis in newborn calves, *Am. Assoc. Vet. Parasitol. 29th Ann. Proc.*, (Abstr.), 66, 1984.
77. **Angus, K. W., Appleyard, W. T., Menzies, J. D., Campbell, I., and Sherwod, D.,** An outbreak of diarrhoea associated with cryptosporidiosis in naturally reared lambs, *Vet. Rec.*, 110, 129, 1982.
78. **Angus, K. W.,** Mammalian Cryptosporidiosis: a veterinary perspective, in *Cryptosporidiosis: Proc. 1st Int. Workshop*, Angus, K. W. and Blewett, D. A., Eds., Animal Disease Research Association, Edinburgh, Scotland, 1988, 43.
79. **Yvore, P., Esnault, A., Naciri, M., Leclerc, C., Bind, J. L., Contrepois, M., Levieux, D., and Laporte, J.,** Enquete epidemiologique sue les diarrheas neonatales des chevreaux dans les elevages de Touraine. Colloque International sur les maladies de la chevre, Institute National de la Researche Agronomique, Niort, France, 1984, 437.
80. **Links, I. J.,** Cryptosporidial infection of piglets, *Aust. Vet. J.*, 58, 60, 1982.
81. **Bergeland, M. E.,** Necrotic enteritis in nursing piglets, *Am. Assoc. Vet. Lab. Diag.*, 20, 151, 1977.
82. **Tzipori, S., Smith, M., Maken, T., and Halpin, C.,** Enterocolitis in piglets caused by *Cryptosporidium* sp. purified from calf feces, *Vet. Parasitol.*, 11, 1121, 1982.
83. **Moon, H. W. and Bemrick, W. J.,** Fecal transmission of calf cryptosporidia between calves and pigs, *Vet. Pathol.*, 18, 248, 1981.
84. **Tzipori, S., McCartney, E., Lawson, G. H. K., and Rowland, A. C.,** Experimental infection of piglets with *Cryptosporidium*, *Res. Vet. Sci.*, 31, 358, 1981.
85. **Kennedy, G. A., Kreitner, G. L., and Strafus, A. C.,** Cryptosporidiosis in three pigs, *J. Am. Vet. Med. Assoc.*, 170, 348, 1977.
86. **Gibson, J. A., Gill, M. W. M., and Huber, M. J.,** Cryptosporidiosis in Arabian foals with severe combined immunodeficiency, *Aust. Vet. J.*, 60, 378, 1983.
87. **Snyder, S. P., England, J. J., and McChesney, A. E.,** Cryptosporidiosis in immunodeficient Arabian foals, *Vet. Pathol.*, 15, 12, 1978.
88. **Reinemeyer, C. R., Kline, R. C., and Stauffer, G. D.,** Absence of *Cryptosporidium* oocysts in faeces of neonatal foals, *Equine Vet. J.*, 16, 217, 1984.
89. **Gajadhar, A. A., Caron, J. P., and Allen, J. R.,** Cryptosporidiosis in two foals, *Can. Vet. J.*, 26, 132, 1985.
90. **Tzipori, S., Angus, K. W., Campbell, I., and Sherwood, D.,** Diarrhea in young red deer associated with infection with *Cryptosporidium*, *J. Infect. Dis.*, 144, 170, 1981.
91. **Angus, K. W.,** Cryptosporidiosis in red deer, *Publ. Vet. Deer Soc.*, 3, 3, 1988.
92. **VanWinkle, T. J.,** Cryptosporidiosis in young artiodactls, *J. Am. Vet. Med. Assoc.*, 187, 1170, 1985.
93. **Klesius, P. H., Hanes, T. B., and Malo, L. K.,** Infectivity of *Cryptosporidium* sp. isolated from wild mice for calves and mice, *J. Vet. Med. Assoc.*, 189, 192, 1986.
94. **Tyzzer, E. E.,** Coccidiosis in gallinaceous birds, *Am. J. Hyg.*, 10, 269, 1929.
95. **Levine, N. D.,** *Protozoan Parasites of Domestic Animals and of Man*, Burgess, Minneapolis, MN, 1961.
96. **Slavin, D.,** *Cryptosporidium meleagridis* (sp. nov.), *J. Comp. Pathol.*, 65, 262, 1955.
97. **Proctor, S. J. and Kemp, R. L.,** *Cryptosporidium anserinum* sp. n., (sporozoa) in a domestic goose *Anser anser* L. from Iowa, *J. Protozool.*, 21, 664, 1974.
98. **Randall, C. J.,** Cryptosporidiosis of the bursa of Fabricius and trachea in broilers, *Avian Pathol.*, 11, 95, 1982.
99. **Itakura, C., Goryo, M., and Umemura, T.,** Cryptosporidial infections in chickens, *Avian Pathol.*, 13, 487, 1984.
100. **Ranck, F. M., Jr. and Hoerr, F. J.,** Cryptosporidia in the respiratory tract of turkeys, *Avian Dis.*, 31, 389, 1987.
101. **Woodmansee, D. B., Pavlasek, I., Pohlenz, J. F. L., and Moon, H. W.,** Subclinical cryptosporidiosis in turkeys in Iowa, *J. Parasitol.*, 74, 898, 1988.
102. **Tarwid, N. J., Cawthorn, R. J., and Riddell, C.,** Cryptosporidiosis in the respiratory tract of turkeys in Saskatchewan, *Avian Dis.*, 29, 528, 1985.

103. **Glisson, J. R., Brown, T. P., Brugh, M., Page, R. K., Kleven, S. H., and Davis, R. B.,** Sinusitis in turkeys associated with respiratory cryptosporidiosis, *Avian Dis.*, 28, 783, 1984.
104. **Hoerr, F. J., Ranck, F. M., and Hastings, T. F.,** Respiratory cryptosporidiosis in turkeys, *J. Am. Vet. Med. Assoc.*, 173, 1591, 1978.
105. **Hoerr, F. J., Current, W. L., and Haynes, T. B.,** Fatal cryptosporidiosis in quail, *Avian Dis.*, 30, 421, 1986.
106. **Ritter, G. D., Ley, D. H., Levy, M., Guy, J., and Barnes, H. J.,** Intestinal cryptosporidiosis and reovirus isolation from bobwhite quail (*Colinus virginianus*) with enteritis, *Avian Dis.*, 30, 603, 1986.
107. **Guy, J. S., Levy, M. G., Ley, D. H., Barnes, H. J., and Gerig, T. M.,** Experimental reproduction of enteritis in bobwhite quail (*Colinus virginianus*) with *Cryptosporidium* and reovirus, *Avian Dis.*, 31, 713, 1987.
108. **O'Donoghue, P. J., Tham, V. L., de Saram, W. G., Paul, K. L., and McDermott, S.,** *Cryptosporidium* infections in birds and mammals and attempted cross-transmission studies, *Vet. Parasitol.*, 26, 1, 1987.
109. **Tham, V. L., Kniesberg, S., and Dixon, B. R.,** Cryptosporidiosis in quails, *Avian Pathol.*, 11, 619, 1982.
110. **Mason, R. W. and Hartley, W. J.,** Respiratory cryptosporidiosis in a peacock chick, *Avian Dis.*, 24, 771, 1980.
111. **Whittington, R. J. and Wilson, J. M.,** Cryptosporidiosis of the respiratory tract in a pheasant, *Aust. Vet. J.*, 62, 284, 1985.
112. **Randall, C. J.,** Conjunctivitis in pheasants associated with cryptosporidial infection, *Vet. Rec.*, 118, 211, 1985.
113. **O'Donoghue, P. J.,** *Cryptosporidium* infections in man, animals, birds, and fish, *Aust. Vet. J.*, 62, 253, 1985.
114. **Randall, C. J.,** Renal and nasal cryptosporidiosis in a junglefowl *(Gallus sonnaratii), Vet. Rec.*, 119, 130, 1986.
115. **Mason, R. W.,** Conjunctival cryptosporidiosis in a duck, *Avian Dis.*, 30, 598, 1986.
116. **Lindsay, D. S., Blagburn, B. L., Sundermann, C. A., and Hoerr, F. J.,** Experimental infections in domestic ducks with *Cryptosporidium baileyi* isolated from chickens, *Avian Dis.*, 33, 69, 1989.
117. **Tsai, S. S., Ho, L. F., Chang, C. F., and Chu, R. M.,** Cryptosporidiosis in domestic birds, *Chin. J. Microbiol. Immunol.*, 16, 307, 1983.
118. **Lindsay, D. S., Blagburn, B. L., and Sundermann, C. A.,** Host specificity of *Cryptosporidium* sp. isolated from chickens, *J. Parasitol.*, 72, 565, 1986.
119. **Doster, A. R., Mahaffey, E. A., and McClearen, J. R.,** Cryptosporidia in the cloacal coprodeum of Red-Lored parrots *(Amazona autumnalis), Avian Dis.*, 23, 654, 1979.
120. **Ley, D. H., Levy, M. G., Hunter, L., Corbett, W., and Barnes, H. J.,** Cryptosporidia-positive rates of avian necropsy accessions determined by examination of auramine O-stained fecal smears, *Avian Dis.*, 32, 108, 1988.
121. **Gardiner, C. H. and Imes, G. D., Jr.,** *Cryptosporidium* sp. in the kidneys of a black-throated finch, *J. Am. Vet. Med. Assoc.*, 185, 1401, 1984.
122. **Belton, D. J. and Powell, I. B.,** Cryptosporidiosis in lovebirds (*Agapornis* sp.), *N.Z. Vet. J.*, 35, 15, 1987.
123. **Lindsay, D. S., Blagburn, B. L., and Hoerr, F. J.,** Experimentally induced infections of turkeys with *Cryptosporidium baileyi* isolated from chickens, *Am. J. Vet. Res.*, 48, 104, 1987.
124. **Bermudez, A. J., Ley, D. H., Levy, M. G., Ficken, M. D., Guy, J. S., and Gerig, T. M.,** Intestinal and bursal cryptosporidiosis in turkeys following inoculation with *Cryptosporidium* isolated from commercial poults, *Avian Dis.*, 32, 445, 1988.
125. **Lindsay, D. S., Blagburn, B. L., and Sundermann, C. A.,** Morphometric comparison of the oocysts of *Cryptosporidium meleagridis* and *C. baileyi*, *Proc. Helminthol. Soc. Wash.*, 56, 91, 1989.
126. **Lindsay, D. S., Blagburn, B. L., and Ernest, J. A.,** Experimental *Cryptosporidium parvum* infections in chickens, *J. Parasitol.*, 73, 242, 1987.
127. **Lindsay, D. S., Blagburn, B. L., Sundermann, C. A., Hoerr, F. J., and Ernest, J. A.,** Experimental *Cryptosporidium* infections in chickens: oocyst structure and site specificity, *Am. J. Vet. Res.*, 47, 876, 1986.
128. **Nakamura, K. and Abe, F.,** Respiratory (especially pulmonary) and urinary infections of *Cryptosporidium* in layer chickens, *Avian Pathol.*, 17, 703, 1988.
129. **Goodwin, M. A., Steffens, W. L., Russell, I. D., and Brown, J.,** Diarrhea associated with intestinal cryptosporidiosis in turkeys, *Avian Dis.*, 32, 63, 1988.
130. **Itakura, C., Nakamura, T., Umemura, T., and Goryo, M.,** Ultrastructure of cryptosporidial life cycle in chicken host cells, *Avian Pathol.*, 14, 237, 1985.
131. **Snyder, D. B., Current, W. L., Russek-Cohen, E., Gorham, S. L., Mallinson, E. T., Marquardt, W. W., and Savage, P. K.,** Serologic prevalence of *Cryptosporidium* in Delmarva broiler flocks, *Poult. Sci.*, 67, 730, 1988.

132. **Goodwin, M. A. and Brown, J.**, Histologic incidence and distribution of *Cryptosporidium* sp. in chickens: 68 cases in 1986, *Avian Dis.*, 32, 365, 1988.
133. **Lindsay, D. S., Blagburn, B. L., Sundermann, C. A., and Giambrone, J. J.**, Effect of broiler chicken age on susceptibility to experimentally induced *Cryptosporidium baileyi* infections, *Am. J. Vet. Res.*, 49, 1412, 1988.
134. **Darlington, M. V. and Blagburn, B. L.**, The Merifluor™ immunofluorescent detection procedure is nonspecific for *Cryptosporidium* species, in *Proc. Annu. Meet.* Southeastern Soc. Parasitologists, abstr. no. 34, 1988.
135. **Tzipori, S. and Campbell, I.**, Prevalence of *Cryptosporidium* antibodies in 10 animal species, *J. Clin. Microbiol.*, 14, 455, 1981.
136. **Lindsay, D. S. and Blagburn, B. L.**, *Cryptosporidium* sp. infections in chickens produced by intracloacal inoculation of oocysts, *J. Parasitol.*, 72, 615, 1987.
137. **Lindsay, D. S., Blagburn, B. L., Sundermann, C. A., Hoerr, F. J., and Giambrone, J. J.**, *Cryptosporidium baileyi*: effects of intra-abdominal and intravenous inoculation of oocysts on infectivity and site of development in broiler chickens, *Avian Dis.*, 31, 841, 1987.
138. **Lindsay, D. S., Blagburn, B. L., Hoerr, F. J., and Giambrone, J. J.**, Experimental *Cryptosporidium baileyi* infections in chickens and turkeys produced by ocular inoculation of oocysts, *Avian Dis.*, 31, 355, 1987.
139. **Blagburn, B. L., Hoerr, F. J., Lindsay, D. S., Giambrone, J. J., Lockaby, S. B., and Sundermann, C. A.**, Interactions of *Cryptosporidium baileyi* with viral and bacterial co-pathogens in broiler chickens, in *Proc. 63rd Annu. Meet. Am. Soc. Parasitologists*, 1988, 60.
140. **Levy, M. G., Ley, D. H., Barnes, H. J., Gerig, T. M., and Corbett, W. T.**, Experimental cryptosporidiosis and infectious bursal disease virus infection of specific-pathogen-free chickens, *Avian Dis.*, 32, 803, 1988.
141. **Guy, J. S., Levy, M. G., Ley, D. H., Barnes, H. J., and Gerig, T. M.**, Interaction of reovirus and *Cryptosporidium baileyi* in experimentally infected chickens, *Avian Dis.*, 32, 381, 1988.
142. **Lindsay, D. S., Sundermann, C. A., and Blagburn, B. L.**, Cultivation of *Cryptosporidium baileyi*: studies with cell cultures, avian embryos, and pathogenicity of chicken embryo-passaged oocysts, *J. Parasitol.*, 74, 288, 1988.
143. **Sundermann, C. A., Lindsay, D. S., and Blagburn, B. L.**, In vitro excystation of *Cryptosporidium baileyi* from chickens, *J. Protozool.*, 34, 28, 1987.
144. **Goodwin, M. A., Latimer, K. S., Brown, J., Stette, W. L., Martin, P. W., Resurrecion, R. S., Smeltzer, M. D., and Dickson, T. G.**, Respiratory cryptosporidiosis in chickens, *Poult. Sci.*, 67, 1684, 1988.
145. **Dhillon, A. S., Thacker, H. L., Dietzel, A. V., and Winterfield, R. W.**, Respiratory cryptosporidiosis in broiler chickens, *Avian Dis.*, 25, 744, 1981.
146. **Lindsay, D. S., Blagburn, B. L., Sundermann, C. A., and Ernest, J. A.**, Attempted chemoprophylaxis of cryptosporidiosis in chickens using halofuginone, salinomycin, lasalocid, and monensin, *Am. J. Vet. Res.*, 48, 354, 1987.
147. **Goodwin, M. A.**, Small intestinal cryptosporidiosis in a chicken, *Avian Dis.*, 32, 844, 1988.
148. **Gharagozlou, M. J. and Khodashenas, M.**, Cryptosporidiosis in a native rooster with proliferative enteritis, *Arch. Vet.*, 17, 129, 1989.
149. **Fletcher, O. J., Munnell, J. F., and Page, R. K.**, Cryptosporidiosis in the bursa of Fabricius of chickens, *Avian Dis.*, 19, 630, 1975.
150. **Ogimoto, K., Inamoto, T., Soga, T., and Itakura, C.**, Experimental infection of chickens with *Cryptosporidium, Zentralbl. Bakteriol. Hyg. A*, 264, 343, 1987.
151. **Gorham, S. L., Mallinson, E. T., Snyder, D. B., and Odor, E. M.**, Cryptosporidia in the bursa of Fabricius — A correlation with mortality rates in broiler chickens, *Avian Pathol.*, 16, 205, 1987.
152. **Wages, D. P. and Fricken, M. D.**, Cryptosporidiosis and viral hepatitis in turkeys, *Avian Dis.*, 33, 191, 1989.
153. **Latimer, K. S., Goodwin, M. A., and Davis, M. K.**, Rapid cytologic diagnosis of respiratory cryptosporidiosis in chickens, *Avian Dis.*, 32, 826, 1988.
154. **Garcia, L. S., Bruckner, D. A., Brewer, T. C., and Shimzu, R. Y.**, Techniques for the recovery and identification of *Cryptosporidium* oocysts from stool specimens, *J. Clin. Microbiol.*, 18, 185, 1983.
155. **Ma, P. and Soave, R.**, Three step stool examination for cryptosporidiosis in ten homosexual men with protracted watery diarrhea, *J. Infect. Dis.*, 147, 824, 1983.
156. **Current, W. L.**, Human cryptosporidiosis, *N. Engl. J. Med.*, 309, 614, 1983.
157. **Garcia, L. S., Brewer, T. C., and Bruckner, D. A.**, Fluorescent detection of *Cryptosporidium* oocysts in human fecal specimens by using monoclonal antibodies, *J. Clin. Microbiol.*, 25, 119, 1987.
158. **Tzipori, S. and Campbell, I.**, Prevalence of *Cryptosporidium* antibodies in 10 animal species, *J. Clin. Microbiol.*, 14, 455, 1981.

Chapter 9

CARYOSPORA: BIOLOGY

Steve J. Upton and Christine A. Sundermann

TABLE OF CONTENTS

I. INTRODUCTION

Except for those organisms of medical, veterinary, or economic importance, there is a substantial lack of information of even the most basic biology of most species of coccidia. This is especially true for coccidian genera containing few species, since the majority infect vertebrates of little interest to most biologists. Life cycles of these "lesser coccidia" have traditionally been thought to be similar to species of *Eimeria* and *Isospora*. Recent studies, however, suggest a number of exceptions to our traditional concepts of apicomplexan biology. These studies have revealed some genera have exceptional development, e.g., *Atoxoplasma, Calyptospora, Caryospora, Elleipsisoma, Frenkelia, Hammondia,* and *Neospora*. The interest that these unusual apicomplexans have generated recently is typified by the fact that in the present book there are chapters on common coccidia, as well as chapters on more exotic coccidia, such as *Caryospora* spp. Although a number of fundamental questions still need to be answered concerning the basic biology of the genus *Caryospora*, enough information is now available to provide preliminary insights into this unusual, and perhaps heterologous taxon.

II. HISTORY AND CLASSIFICATION

The genus *Caryospora* Léger, 1904 is currently the third most speciose of the Eimeriidae. About 40 members are assigned to the genus at present, most of which occur in snakes and raptors (Table 1). The type species is *Caryospora simplex* Léger, 1904, originally described from *Vipera aspis* (Serpentes: Viperidae) in France.[1] As noted in an earlier review,[2] the original spelling of the genus was *Karyospora*, which was apparently a printing error since the current spelling of the genus was utilized by Léger during a paper where endogenous development in snakes was reported.[3] This coccidian was later redescribed from *V. x. xanthina*,[4] and has also been found in *V. a. aspis* and *V. r. russelli* in zoos in North America.[2] It should be noted, however, that a slightly different morphologic type was found recently in *V. kaznakovi* from West Caucasia, which may be a separate species.[5] If so, it may be as true a form of *C. simplex* as the parasite we have been working with.

III. LIFE CYCLES

Until the early 1980s, all species of *Caryospora* were considered monoxenous, with merogony, gamogony, and formation of oocysts occurring in the intestine. Early researchers studied the life cycle of *C. simplex* in the intestinal tract of *Vipera aspis*,[3,6] which supported the above hypothesis. However, studies on *C. bubonis* in raptors first suggested that an indirect, predator-prey life cycle may be the predominant way that some *Caryospora* spp. complete development.[7,8] In these studies, the great horned owl, *Bubo virginianus*, not only developed patent infections after ingesting oocysts, but also after ingesting mice inoculated orally 4 weeks previously with oocysts. Because *C. bubonis* had a facultative heteroxenous life cycle, it was suggested that *Caryospora* spp. be placed within the Sarcocystidae.[7]

Independently, the crotalid coccidian, *C. bigenetica*, was also found to possess a facultatively indirect life cycle.[9] Mice inoculated orally with oocysts not only transmitted the parasite to a coccidia-free Massasauga, but were also found to support parasite stages within the dermis of the cheek and tongue, which included gamogony and sporogony. Following sporogony *in situ*, sporozoites penetrated new cells and formed dormant hypnozoites, the host cells of which were termed caryocysts. Caryocysts were presumed to represent the stage infective for snakes. These researchers elected to retain the *Caryospora* within the Eimeriidae.[9]

Although some *Caryospora* spp. have been shown to be capable of utilizing rodents as reservoirs of infection, it should be noted that natural infections in rodents have never been

TABLE 1
Named and Unnamed Species of *Caryospora*

Species	Host(s)	Ref.
	Aves	
Charadriiformes		
argentati	*Larus argentatus*	45-47
undata [syn. *C. undulata* Poelma and Strik, 1966, *lapsus*]	*Larus argentatus* *Lunda cirrhata*,[2] *Uria aalage*	48,49
Falconiformes		
arcayae	*Buteo magnirostris*	49
falconis	*Falco peregrinus*, *Falco subbuteo*, *Falco tinnunculus*	28,50-53
kutzeri [syn. *C. henryae* of Yakimoff & Matschulsky, 1936, in part] [syn. *C.* sp. of Kutzer et al., 1980] [syn. *C. falconis* Schellner & Rodler, 1971]	*Falco tinnunculus*, *Falco mexicanus*, *Falco biarmicus*, *Falco jugger*, *Falco cherrug*, *Falco rusticolus*, *Falco peregrinus*, *Falco subbuteo*[3]	24,26,51,54, 55
neofalconis	*Falco mexicanus*, *Falco subbuteo*, *Falco biarmicus*, *Falco peregrinus*, *Falco tinnunculus*[3]	26,55
sp. [syn. *C. Henryae*, in part]	*Milvus migrans*	55
tremula [syn. *Eumonospora tremula* Allen, 1933]	*Catharates aura*	27,56
uptoni	*Buteo jamaicensis*	25,57
Passeriformes		
gloriae	*Dives atroviolaceus*	58
jiroveci [syn. *C.* sp. of Černa, 1973]	*Erithacus rubecula*	59,60
sp.	*Diphyllodes magnificus*	61
Strigiformes		
bubonis	*Bubo virginianus*	7,8,62
henryae [syn. *Isospora henryi* Yakimoff & Matikachwili, 1932]	*Bubo bubo*	53,55
sp. [non. *C. falconis* of Wetzel & Enigk, 1939]	*Athene noctua*	52
strigis	*Tyto alba*	63
	Mammalia	
microti[4] [syn. *C.* sp. of Saxe, 1952]	*Microtus pennsylvanicus*	33,34

TABLE 1 (continued)
Named and Unnamed Species of *Caryospora*

Species	Host(s)	Ref.
Sauria		
ernsti	*Anolis carolinensis*	30
gekkonis	*Gekko gecko*	29
Serpentes		
bengalensis	*Enhydris enhydris*	64
bigenetica	*Crotalus horridus*, *Crotalus adamanteus*, *Agkistrodon contortrix*, *Sistrurus catenatus*	21,40,65, 9,17,19, 23,39,43
brasiliensis	*Philodryas aestivus*, *Philodryas olfersi*, *Philodryas nattareri*, *Leimadophis poecilogyrus*	38,66
cobrae	*Naja naja*	67
colubris	*Coluber viridiflavus*	68
corallae	*Corallus canius*	69
demansiae	*Demansia psammophis*	36
dendrelaphis	*Dendrelaphis punctulatus*	37
duszynskii	*Elaphe guttata*, *Elaphe obsoleta*	44,70
hermae[1]	*Psammphis sibilans*	35
japonicum	*Natrix trigrina*	71,72
jararacae [syn. *C. jararacae* Carini, 1939, *lapsus*]	*Bothrops jararaca*	73
lampropeltis	*Lampropeltis calligaster*	74
legeri [syn. *C. legeti* Matubayasi, 1937, *lapsus*]	*Psammophis sibilans*	75
najadae	*Coluber najadum*	76
najae	*Naja nigricollis*	77
peruensis	*Oxybellis argenteus*	78
psammophi	*Psammophis sibilans*	35
simplex [syn. *Karyospora simplex* Leǵer, 1904, *lapsus*]	*Vipera aspis*, *Viper xanthina*, *Vipera palestinae*, *Vipera russelli*,	1,3,4,6,16, 20,23,41,42, 79
sp.	*Vipera kaznakovi*	5
telescopis	*Telescopus fallax*	5
weyerae[1]	*Psammophis sibilans*	35
zuckermanae	*Coluber ravergieri*	35
Testudina		
cheloniae	*Chelonia mydas*	31,32

[1] Probably synonyms of *C. psammophi*; all three may be synonyms of *C. legeri*.
[2] New host record.
[3] Experimental host.
[4] Probably abnormal oocysts of *Isospora mcdowelii*.

documented. Therefore, the assumption that rodents serve as natural reservoirs of *Caryospora* spp. is questioned. Several papers have, however, reported incidental findings of anomalous, non-Sarcocystidae-like coccidia developing in unusual sites in mammals. These include reports of an unusual coccidian developing in the epididymis of wapiti, *Cervus canadensis nelsoni*,[10] a coccidium forming meronts, gametes, and unsporulated oocysts in the genitalia of the golden hampster, *Cricetus cricetus*,[11] a report of an unknown species of coccidium in the mammary tissue of water shrews, *Sorex palustris navigator*,[12] and several reports of subcutaneous dermal coccidiosis in canines.[13-15] Although some of these reports may represent species of *Caryospora*, other genera including *Neospora* and *Toxoplasma* are also potential candidates.

A. *CARYOSPORA* spp. IN SNAKES

Caryospora spp. from snakes generally represent either one of two basic morphologic types, based predominantly on features of the oocyst wall (Figure 1). All, or nearly all, lack a micropyle and oocyst residuum and have sporocysts with a Stieda body. However, in the first morphologic type, the oocyst wall is relatively smooth and appears either as a single layer or faintly bilayered. The layers of the oocyst wall cannot be separated easily. This form is best typified by the type species, *C. simplex*, as well as *C. bigenetica* and *C. jararacae* (Figure 1). In the second type, oocysts possess a distinctly bilayered wall where the layers are separated easily by coverslip pressure. This type is typified by *C. duszynskii* and *C. peruensis* (Figure 1) and, perhaps, *C. zuckermanae* as well. The outer surface of the oocyst wall is rough and heavily pitted and oocysts tend to be slightly larger than the first type. Thus far, this type has only been found in terrestrial colubrids.

The first to describe endogenous development of a caryosporan was Léger,[3] when he reported developmental stages of *C. simplex* from the intestinal epithelium of naturally infected *Vipera aspis*. In histologic sections, at least two generations of meronts, as well as gametes and unsporulated oocysts were found. Developmental stages were located predominantly in the posterior one half of the middle portion of the small intestine. Lavier also found the parasite in *V. aspis* and attempted to transmit the infection to three 18-day-old vipers by oral inoculation of oocysts.[6] Although snakes were examined at approximately weekly intervals, no developmental stages were found. However, oocysts from the original inoculum were reported to be low in number and had sporulated slowly,[6] which implies that the parasites may have been nonviable. Recent studies have shown that oocysts of *C. simplex* obtained from the feces of *Vipera xanthina* are perfectly capable of infecting *V. palestinae*.[16] Oocysts of *C. simplex* are not transmissible to *Crotalus adamanteus*, which suggests that *C. simplex* may be genus specific.[17]

Developmental stages of *C. bigenetica* occur in the epithelium of the duodenum and jejunum of both naturally and experimentally rattlesnakes.[9] Like *C. simplex*, at least two generations of meronts, as well as gametes and unsporulated oocysts can be found. Oocysts, as well as experimentally infected mice, have been shown to be capable of transmitting infections to uninfected rattlesnakes.[9]

Both *C. simplex* and *C. bigenetica* have been shown to be capable of establishing snake-mouse-snake, mouse-to-mouse, and mouse-cotton rat infections, the latter two of which occur by cannibalism.[16,18-20] As with oocyst-induced infections, merogony, gamogony, and sporogony all occur during rodent-rodent transmission.[19] In cases where parasite stages have been transmitted back to snakes, the prepatent period was >30 days postinoculation (DPI), also tending to explain the failure by Lavier[6] to transmit infections to uninfected snakes since his animals were examined at intervals of 1 to 3 weeks PI. Because patency generally lasts for many months or even years in snakes, it is likely that at least one asexual generation is capable of recycling, although intermittent ingestion of oocysts cannot be excluded.

Recently, the basic biology of both *C. bigenetica* and *C. simplex* have been examined

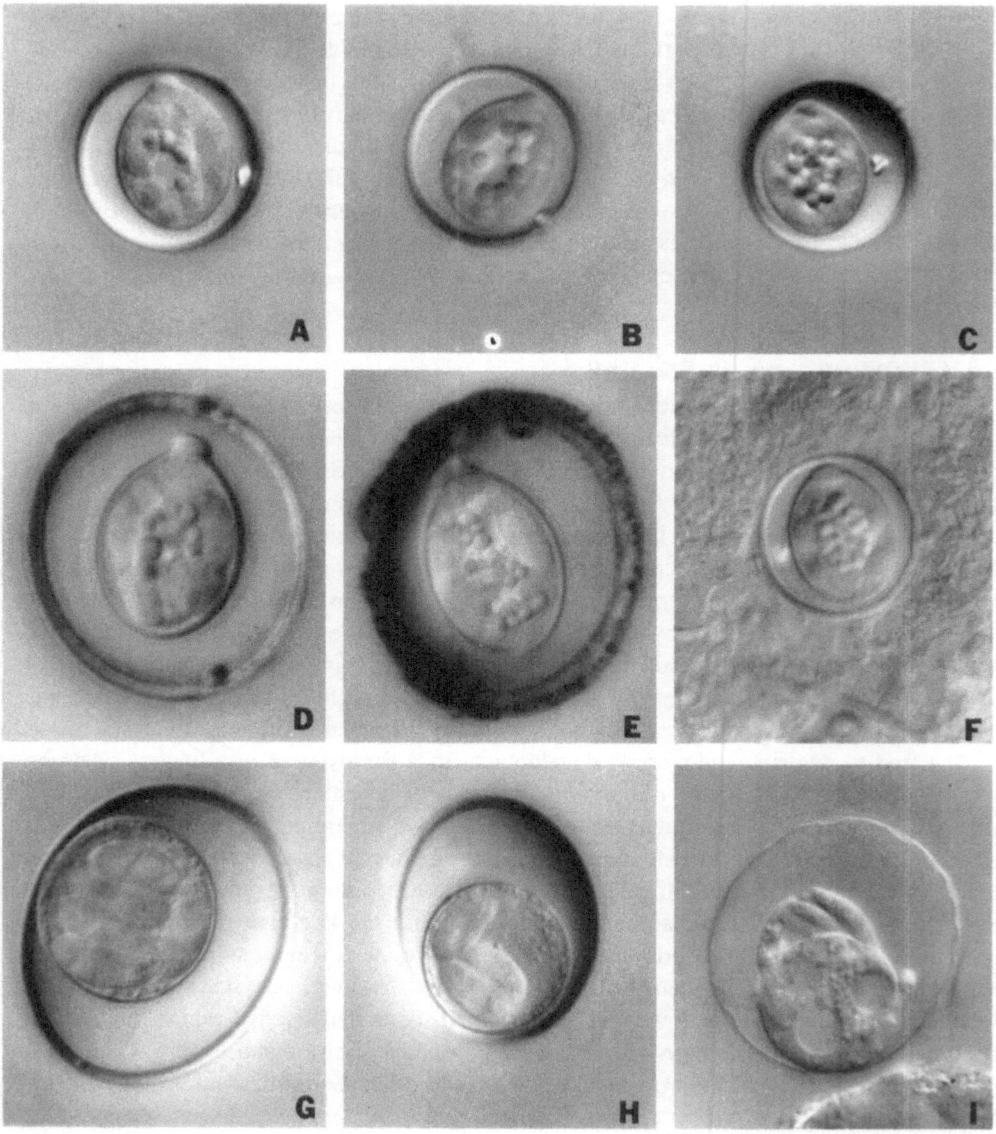

FIGURE 1. Nomarski interference contrast photomicrographs of sporulated oocysts of various *Caryospora* spp. A—F, magnification × 1800. G—I, magnification × 1260. A. *C. simplex* from *Vipera palestinae*. B. *C. bigenetica* from *Crotalus adamanteus*. C. *C.* sp., perhaps *C. jararacae*, from *Bothrops lateralis*. D. *C. duszynskii* from *Elaphe obsoleta*. E. *C. peruensis* from *Oxybelis argenteus*. F. *C. ernsti* from *Anolis carolinensis*. G. *C.* sp. from *Buteo swainsonii*. H. *C.* sp. from *Buteo platypterus*. I. *C. undata* from *Linda cirrhata*.

in more detail, especially those stages occurring in rodents.[17,19-23] Following ingestion of sporulated oocysts, excystation occurs within the intestinal lumen. Infections can be greatly enhanced by inoculating sporocysts rather than oocysts orally since the oocyst wall on only a small portion of parasites passing through the gut of a rodent appears to be removed.[23] Although the early fate and route of migration of sporozoites is not known, parasite stages are routinely found in fibroblasts and perhaps other cells of the dermal tissues of the face, tongue, scrotum, and base of the ears as early as 8 DPI. Heavy infections result in parasite stages being found in a variety of other tissues also in cotton rats, including lung, testicles, epididymis, rectum, footpads, and bone marrow.[21] Depending upon the species, two to three generations of merogony occur followed by gamogony about 10 DPI. Zygotes sporulate into

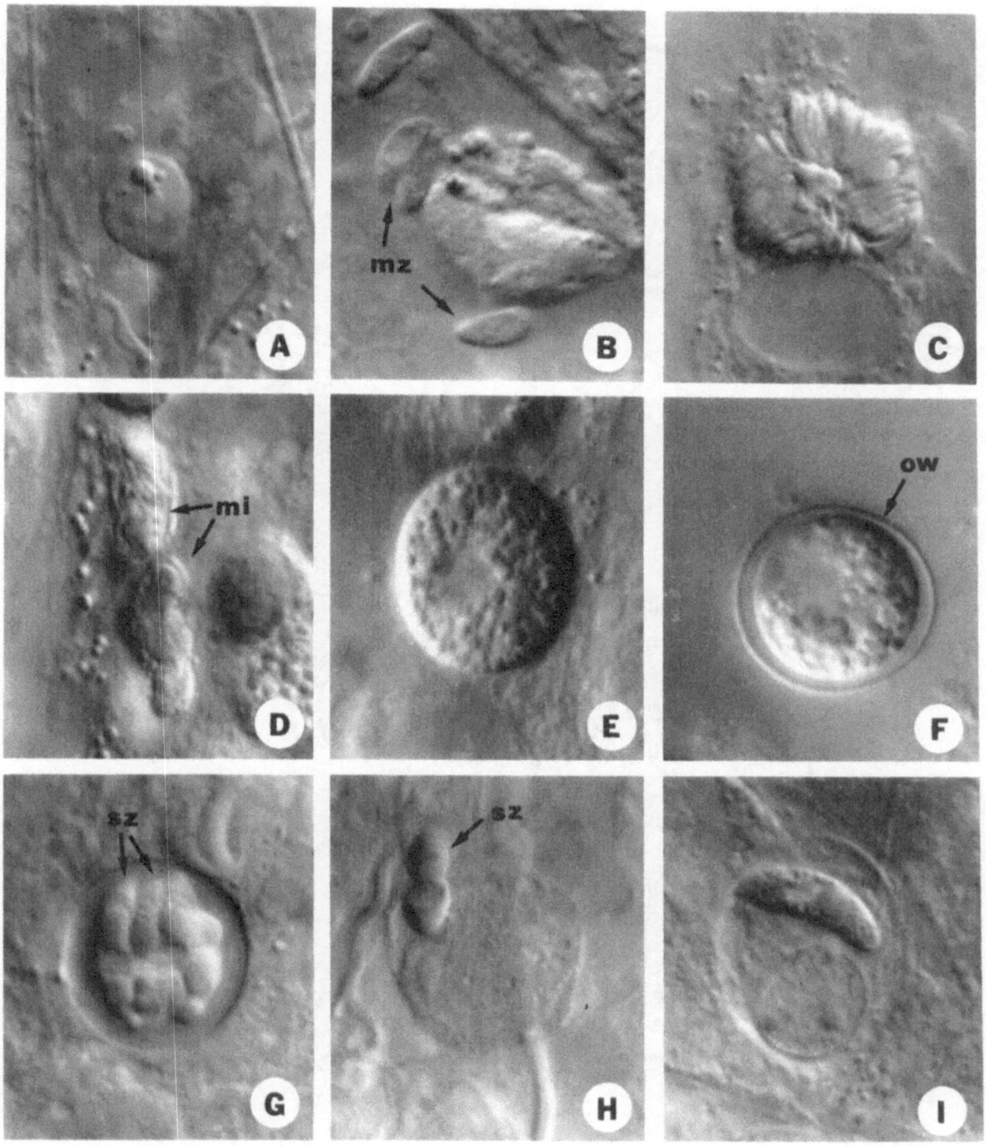

FIGURE 2. Nomarski interference contrast photomicrographs of endogenous stages of *Caryospora bigenetica* in cheek tissue of experimentally infected mice, *Mus musculus* (Magnification ×1825.) A. Undifferentiated meront, 8 DPI. B. Merozoites, 8 DPI. C. Type II meront, 10 DPI. D. Microgametocytes, 12 DPI. E. Macrogamete, 12 DPI. F. Unsporulated oocyst removed from surrounding tissues, 12 DPI. G. Sporulated oocyst, 13 DPI. H. Free sporozoite penetrating host cell, 14 DPI. I. Caryocyst, 20 DPI. Abbreviations: mi, microgametes; mz, merozoites, ow, oocyst wall; sz, sporozoite.

thin-walled oocysts containing eight sporozoites surrounding a sporocyst residuum, which are all enclosed within a thin, membrane-like sporocyst. Sporozoites exit from these thin walled oocysts, penetrate new cells, and form hypnozoites within macrophages and fibroblasts. Dormant sporozoites are fully viable even after 15 months and will leave the caryocyst by incubation in trypsin-bile salt solutions. The entire life cycle of these heteroxenous *Caryospora* spp. can best be illustrated in Figures 2 and 3.

Not all species of *Caryospora* appear capable of infecting rodents and transmitting the infection back to the primary host, however. Developmental stages of *Caryospora duszynskii* have not been found in infected mice experimentally,[4] and experimental transmission attempts

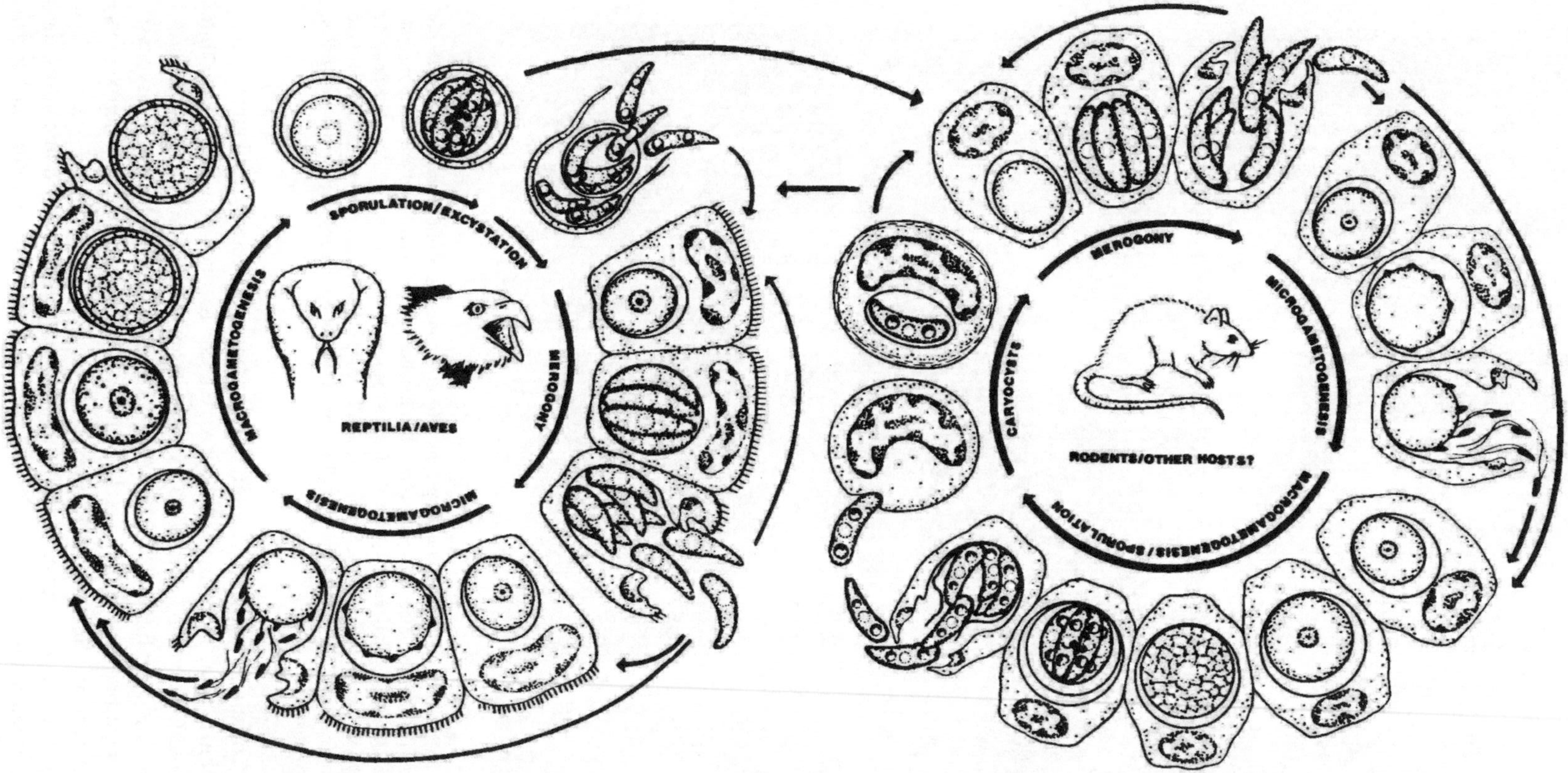

FIGURE 3. Schematic of proposed life cycle of the genus *Caryospora*. Note that only for *C. bigenetica* and *C. simplex* has the life cycle been characterized well enough to fit into the proposed scheme.

using *C. najae, C. colubris,* and *C. corallae* have also failed.[5] Therefore, either other types of secondary hosts are involved or some species of *Caryospora* are monoxenous and will probably warrant eventual placement in a separate genus.

B. *CARYOSPORA* spp. IN AVIAN HOSTS

Most species of *Caryospora* infecting avian hosts occur in raptors (Table 1). With a few exceptions, oocysts are generally large, spherical to ovoid, lack a micropyle and oocyst residuum, and have a simple spherical or subspherical sporocyst that lacks a Stieda body and encloses eight stubby sporozoites and a residuum (Figure 1). Because oocysts and sporocysts lack many useful morphologic features, differentiation of species is often difficult.

After ingesting oocysts of *C. bubonis*, the great horned owl, *Bubo virginianus*, was shown to become patent 8 to 13 DPI.[7,8] Uninfected owls fed mice inoculated *per os* 4 weeks previously with oocysts also passed the parasite in the feces 8 to 10 days post-feeding. These data were the first to show the heteroxenous nature of the genus *Caryospora*, although the exact location of dormant stages in mice is still not known for any species of avian *Caryospora*. One plausible explanation for failure to find developmental stages in histological sections from rodents is that sporozoites of avian *Caryospora* spp. may penetrate murine tissues and become dormant without development, similar to the isosporans of canids and felids. The low numbers of oocysts used in the inoculum would have made identification of parasite stages difficult.

Caryospora bubonis localizes in the posterior one third of the small intestine, where two types of meronts, gamonts, and unsporulated oocysts can be found.[7,8] The preferred site for *C. kutzeri* in common kestrels, *Falco tinnunculus*, is also the posterior small intestine[24] whereas gametes of *C. uptoni* were found throughout the small intestine of *Buteo jamaicensis*.[25]

Cross-transmission experiments have shown that although various avian *Caryospora* may not always be species specific, those infecting raptors tend to be genus specific. Cawthorn and Stockdale[7] failed to transmit oocysts of *C. bubonis* to long eared owls (*Asio otus*), a short eared owl (*Asio flammeus*), and domestic chicks (*Gallus domesticus*). Likewise, *C. kutzeri* and *C. neofalconis* were found in, and could be transmitted to, a variety of species of *Falco* spp., whereas oocysts were not transmissible to *Accipiter gentilis, Asio otus, Bubo bubo, Buteo buteo,* or *Milvus milvus*.[26] *Caryospora tremula* is not infective for domestic chickens.[27] The only study reporting a raptor caryosporan to cross generic boundaries is where *C. falconis* was experimentally transmitted to little owls, *Athene noctua*.[28] Details of the experimental protocol are lacking and we suspect that the owl may have been naturally infected prior to the experiment.

In Passeriform birds, *Caryospora* spp. tend to be similar in structure to avian isosporans and may represent aberrations from avian *Isospora*/*Atoxoplasma* ancestors. Basically, oocysts are subspherical and possess a single sporocyst with a Stieda body. No life cycle data have been reported for any of these species, although host feeding habits would lead one to suspect that they represent monoxenous forms.

In Charadriform birds, two species of *Caryospora, C. argentati* and *C. undata*, have been reported. Although little data are available for the former species, *C. undata* is unusual because the oocyst wall is thin and membrane-like, with small indentations similar to the surface of a golf ball (Figure 1). The sporocyst wall is very thin and membrane-like, and may represent a transitional form between the *Caryospora* and *Tyzzeria*. The species is also interesting because it apparently has little host specificity and has been reported from several genera (Table 1). Based on host feeding preferences, one would suspect that these are monoxenous parasites as well, although fish as intermediate hosts is an alternative.

C. *CARYOSPORA* spp. IN MISCELLANEOUS HOSTS

Two species of *Caryospora* have been described from Sauria (Table 1). *Caryospora gekkonis* was described from *Gekko gecko* in India[29] and *C. ernsti* from *Anolis carolinensis* in North America (Figure 1).[30] Little is known of either species, although meronts, gamonts, and unsporulated oocysts were found in the anterior portion of the small intestine of the latter species.[2]

Caryospora cheloniae occurs in green sea turtles, *Chelonia mydas mydas*, and has only been reported from mariculture-reared turtles in the West Indies. Based on structure and endogenous development, its only link with the other members of the genus appears to be that it possesses eight sporozoites enclosed within a single sporocyst. It was first reported in 1974 where it caused an epidemic of disease and mortality in newly hatched turtles.[31] Clinical signs included emaciation and lethargy. Later, endogenous stages of the parasite were described from the intestine of the *C. mydas*, which included asexual reproduction by transverse binary fission, gamonts, and unsporulated oocysts.[32] No other information is available on this pathogen, although it should be noted that it may be indigenous elsewhere and is thought to have been introduced inadvertently into the mariculture facility.

Caryospora microti was reported from the eastern meadow vole, *Microtus pennsylvanicus*[33,34] and is the only species reported thus far from a mammalian host. Based on the description, and because the parasite has not been reported since, there is every reason to believe that this species is invalid and was the result of abnormal sporulation of the oocysts of *Isospora mcdowelli*.

IV. PATHOGENESIS

A. INFECTIONS IN SNAKES

Developmental stages of *Caryospora simplex* occur above the nucleus of intestinal epithelial cells,[3,6] which appears also to be true for several other *Caryospora* spp. of snakes.[35-38] Léger[36] failed to observe any pathology associated with infections whereas Lavier[6] observed edema, dilation of blood vessels, hemorrhage, exfoliative desquamation, ulceration, and inflammation associated with heavy infections. Although the lesions were extensive, the small size and supranuclear position of the parasite led to the conclusion that the lesions were probably not caused by *C. simplex*.[6] No other *Caryospora* spp. have been reported to cause pathogenic lesions in snakes.

B. INFECTIONS IN AVIAN HOSTS

No intestinal lesions were observed in *Bubo virginianus* inoculated *per os* with 1.08×10^5 oocysts of *C. bubonis*.[7] However, both *C. neofalconis* and *C. kutzeri* may be pathogenic for falcons, with experimental inoculations of 1 to 3×10^3 oocysts resulting in slight to heavy diarrhea, listlessness, and anorexia.[26] Symptoms appeared more pronounced in peregrine falcons that in common kestrels. Successful treatment was observed using amprolium daily for 6 days with a dosage of 30 mg/kg body weight.[26] In contrast, 1 to 2.5×10^5 oocysts of *C. kutzeri* did not result in clinical signs or pathologic lesions in European kestrels, *Falco tinnunculus*.[24]

C. INFECTIONS IN TURTLES

Caryospora cheloniae appears to be a significant pathogen of green sea turtles under conditions of mariculture. The disease is associated with mortality, resulting in serious losses of young turtles about 30 days after hatching.[31] The greatest histopathology is found associated with the hindgut, and includes dilation of the lumen, denuding of the tips of the villi, thinning of the intestinal wall, epithelial hyperplasia at the margins of denuded mucosa, hemorrhage, and inflammation.[32] Tube-like casts of cellular debris and caseous masses of oocysts can impact the gut late in the disease.[31]

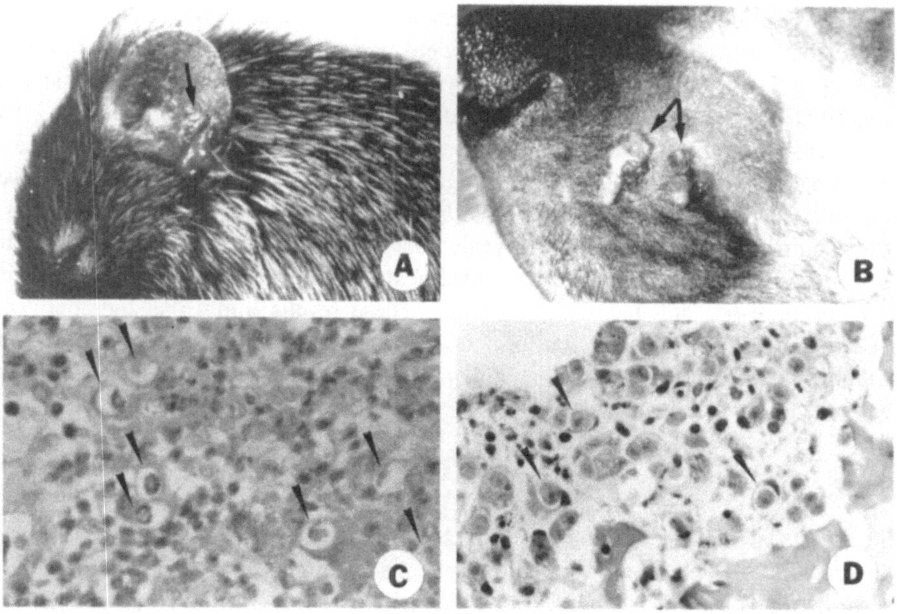

FIGURE 4. Pathology associated with *Caryospora bigenetica* infections. A. Cotton rat with hemorrhagic ear lesion (arrow), 12 DPI. (Magnification ×0.8.) B. Dog with weeping dermal lesions on muzzle (arrows), 10 DPI. (Magnification ×0.5.) C. H & E histological section of oocysts (arrowheads) within cheek tissue of cotton rat, 12 DPI. (Magnification ×190.) D. Gamonts (arrowheads) within bone marrow of cotton rat, 13 DPI. (Magnification ×300.)

D. EXPERIMENTAL INFECTIONS IN RODENTS

The most obvious result of experimentally infecting rodents with *Caryospora simplex* and *C. bigenetica* is swollen muzzles 7 to 14 DPI, which is most pronounced 8 to 12 DPI.[9,17,19-21,23] Additional gross clinical response to infection includes swelling of the tongue, ears, scrota, and footpads. Animals become less active during this time, often becoming anorexic. In cotton rats, infections are apparently more pathogenic than in mice, especially in males. Two-more × 10^5 parasites results in hemorrhagic skin lesions, cauliflower ear, and death (Figure 4).[21] Parasites also become disseminated into a variety of additional tissues in cotton rats, including lungs, testicles, epididymis, rectum, and bone marrow.

Histologically, parasite stages are generally seen in fibroblast-like cells, and probably other cells, beginning 7 to 8 DPI (Figure 2). Meronts are found 8 to 11 DPI, followed 10 to 12 DPI by gametes.[17,23] Sporulated oocysts are seen 12 to 14 DPI and caryocysts, which consist of sporozoites within macrophages, first appear 12 to 13 DPI (Figure 2). Infected tissues appear to be infiltrated with a variety of inflammatory cells, including neutrophils, eosinophils, and mononuclear leukocytes[17] All indications suggest *C. bigenetica* to be more pathogenic to mice than *C. simplex*.

E. EXPERIMENTAL INFECTIONS IN OTHER HOSTS

Although *Neospora caninum* is known to cause dermal coccidiosis, it has been suggested that *Caryospora* spp. may also be the causative agent of some cases of extraintestinal coccidiosis.[17] To our knowledge, the first report of this phenomenon was from a dog in Italy and the parasite was named "*Coccidium nudum*".[13] More recently, subcutaneous nodular coccidiosis has been reported from dogs in Missouri and Georgia, where dermal meronts and gametes were noted.[14,15] The animal studied by Shelton et al.[15] died later of acute canine distemper, a virus known to be immunosuppressive while the other wandered off from the owner's residence and never returned so follow-up examinations could not be performed.[14]

In support of the above hypothesis, experimentally induced subcutaneous coccidiosis has been induced in canids,[39] where animals were first pretreated with methylprednisolone followed by oral inoculation of 1×10^7 oocysts of *Caryospora bigenetica*. Clinical signs appeared 10 DPI, when gross lesions consisted of periocular swelling, swollen muzzle with visible external lesions, ocular mucoid discharge, nasal edema, inflammation of the ears and abdominal skin, and spongy footpads (Figure 4). Smears of scrotum, footpads, eyelid, and muzzle revealed numerous gamonts of *C. bigenetica* at this time. Examination of tissues from these animals at 20 and 67 DPI revealed caryocysts in the eyelid, muzzle, footpad, and abdominal skin. Control animals remained healthy and uninfected, as apparently did a nonimmunosuppressed dog fed oocysts *per os*. These results imply that members of the genus *Caryospora* have the potential to cause a disseminated cutaneous coccidiosis given an immunosuppressed host. *Caryospora bigenetica* appears not to be the causative agent of equine protozoal myeloencephalitis.[40]

V. ULTRASTRUCTURE

Ultrastructurally, most developmental stages of *Caryospora* spp. appear similar to that of other coccidians. Meronts are formed in host cells resembling fibroblasts or macrophages and merozoites possess a typical apical complex (Figure 5). During microgametogenesis, a well-developed residium is present and contains a large mass of amylopectin-like granules in addition to electron-dense bodies. Microgametes have two flagella each, which consist of a typical 9 + 2 microtubule arrangement, an osmiophilic and elongate nucleus, and a mitochondrion with tubular cristae. Near the end of the microgamete is a curled, hooked anlagen.

Macrogametes also appear similar to other coccidia. The large, central nucleus is surrounded by numerous amylopectin granules and electron-dense bodies of varying sizes that are apparently wall-forming bodies (Figure 5). As development proceeds, the cytoplasm separates from the parasitophorous vacuole by a thickened membrane that overlies a thicker layer of material of moderate electron density. After fertilization has presumably occurred, the outer layers of the zygote increase in thickness and the unsporulated oocyst readily collapses during processing by routine electron microscopy methods. Oocysts may be found either free in the extracellular matrix or still within a parasitophorous vacuole.

The final stage, the caryocyst, is easily recognizable at the ultrastructural level. The host cell is spherical and surrounded by a thick (0.35 to 1.0 μm), fibrous, PAS-positive covering, which is closely appressed to the host cell plasma membrane (Figure 5).[22] The sporozoite within has typical coccidian structure, with such features as an apical complex, anterior and posterior refractile bodies, micropore, posterior pore, amylopectin granules, lipid inclusions, and 24 subpellicular microtubules. A plasma membrane and two pellicular membranes envelope the sporozoite which may or may not be located within a parasitophorous vacuole. Even at the ultrastructural level it is unclear whether the outer fibrous layer over the caryocyst is produced by the host cell or the extracellular matrix.

VI. *IN VITRO* CULTIVATION

Some *Caryospora* spp. are able to develop well in cell cultures and have been found to be suitable models to study coccidial development *in vitro*, albeit working with venomous snakes tends to be unpopular with colleagues in adjacent laboratories. Sporozoites of *C. simplex* excyst over a wide range of temperatures[41] and, when inoculated onto human fetal lung cells, merogony, gamogony, and formation of unsporulated oocysts occur.[42] Oocysts could not be induced to sporulate in cell culture media or in 2.5% (w/v) $K_2Cr_2O_7$ solution, however. When murine 3T3 cells are used, we also find additional stages of merogony that

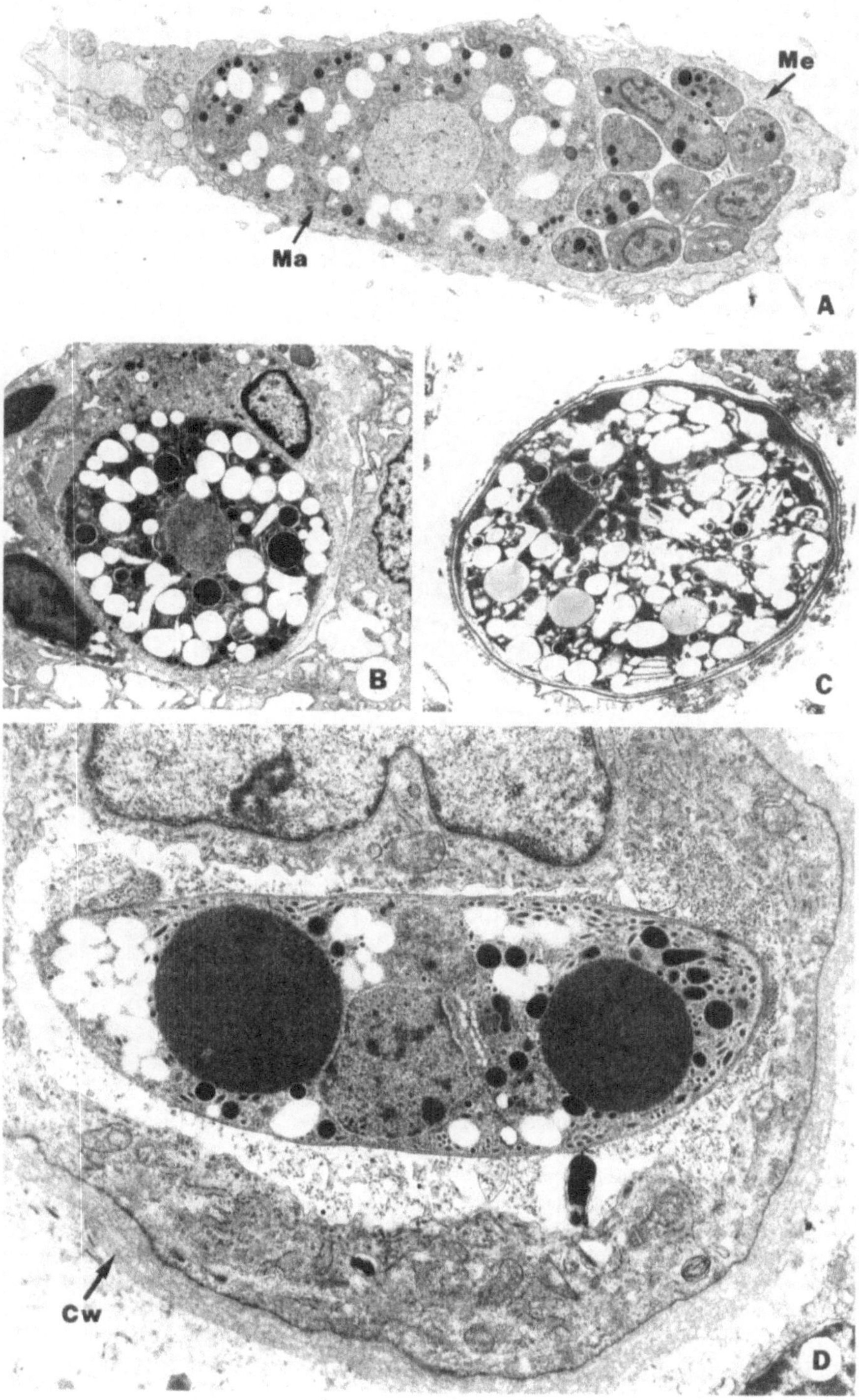

FIGURE 5. Transmission electron micrographs of developmental stages of *Caryospora bigenetica* in scrotal tissue of experimentally infected cotton rats. A. Macrogamont (Ma) and meront (Me) in host cell, 10 DPI. (Magnification ×5040.) B. Macrogamete, 11 DPI. (Magnification ×3600.) C. Unsporulated oocyst, 12 DPI. (Magnification ×3300.) D. Caryocyst containing dormant sporozoite, surrounded by fibrous caryocyst wall (Cw), 16 DPI. (Magnification ×10,800.) (From Sundermann, C. A. and Lindsay, D. L., *J. Protozool.*, 36, 81, 1989. With permission.)

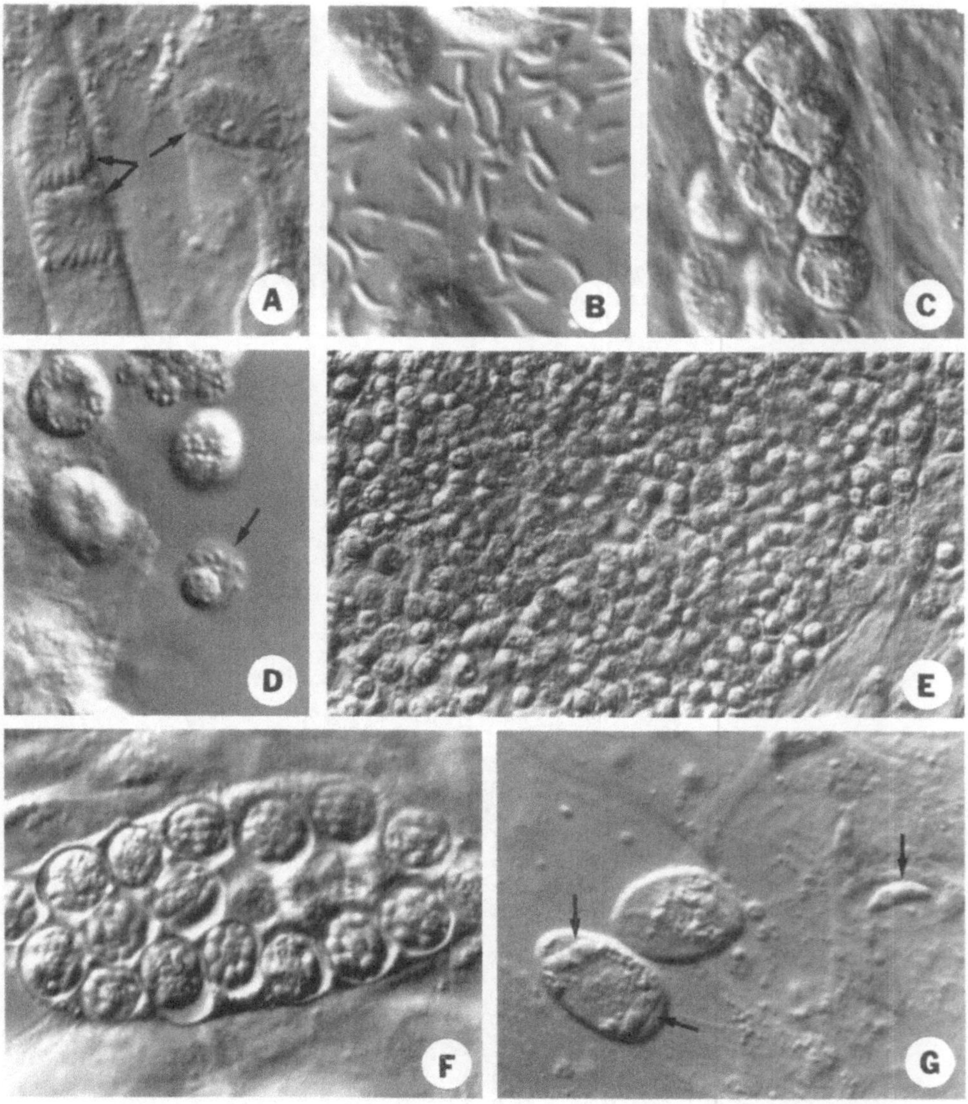

FIGURE 6. *In vitro* developmental stages of *Caryospora bigenetica* in human fetal lung cell culture. A. Meronts (arrows), 2 DPI. (Magnification ×668.) B. Free merozoites among cells, 3 DPI. (Magnification ×712.) C. Macrogametes, 9 DPI. (Magnification ×668.) D. Microgametocyte (arrow), 9 DPI. (Magnification ×668.) E. Monolayer of cells heavily infected with unsporulated and sporulated oocysts, 11 DPI. (Magnification ×170.) F. Sporulated oocysts within host cells, 11 DPI. (Magnification ×668.) G. Sporozoites (arrows) liberated from *in vitro* derived oocysts, 16 DPI. (Magnification ×668.) The two sporozoites on the left are within a caryocyst. (From Sundermann, C. A., Lindsay, D. S., Tibbs, R. E., Jr., and Bailey, M. A., *J. Protozool.*, 35, 465, 1988. With permission.)

appear to involve endodyogeny of Type I meronts at 12 to 14 DPI (unpublished). Although a mixture of Type I and Type II merozoites collected from supernatants of infected cell cultures was not infective to mice following peritoneal injections,[42] results need to be reconfirmed.

Caryospora bigenetica is capable of developing completely in human fetal lung cells and in primary cotton rat testicle and kidney cells (Figure 6).[43] Sporozoites from snake-derived oocysts penetrated cells and underwent two generations of merogony, followed by gamogony. By 10 DPI, fertilized macrogametes developed into oocysts that sporulated intracellularly. Sporozoites excysted *in situ*, exited host cells, and entered new cells to form caryocysts by 16 DPI (Figure 6). Infectivity of *in vitro*-derived merozoites and oocysts was demonstrated by rodent inoculations. These results suggest that the genus *Caryospora* may be a good candidate for a continual cultivation system, where gamogony and sporogony might be studied in detail.

The only other species of *Caryospora* to be studied *in vitro* is *C. duszynskii*. Sporozoites penetrated human fetal lung cells but did not undergo further development.[44] Infected cells changed from spreading fibroblasts to ovoid cells, but sporozoites within remained viable throughout the duration of the study, 29 days. The behavior of this parasite *in vitro* suggests that sporozoites from oocysts may simply penetrate cells in a paratenic host and become dormant as hypnozoites.

REFERENCES

1. **Léger, L.**, Protozoaires, parasites des Vipères *(Vipera aspis), Bull. Men. Ass. Fr. Avanc. Sci.*, 9, 268, 1904.
2. **Upton, S. J., Current, W. L., and Barnard, S. M.**, A review of the genus *Caryospora* Léger, 1904 (Apicomplexa: Eimeriidae), *Syst. Parasitol.*, 8, 3, 1986.
3. **Léger, L.**, *Caryospora simplex*, coccidie monosporée et la classification des coccidies, *Arch. Protistenkd.*, 22, 71, 1911.
4. **Upton, S. J., Ernst, J. V., Barnard, S. M., and Current, W. L.**, Redescription of the oocysts of *Caryospora simplex* (Apicomplexa: Eimeriidae) from an Ottoman viper, *Vipera xanthina xanthina, Trans. Am. Microsc. Soc.*, 102, 258, 1983.
5. **Matuschka, F.-R.**, *Caryospora telescopis* sp. n. (Apicomplexa: Eimeriidae) from the cat snake, *Telescopus fallax* (Serpentes: Colubridae), and *Caryospora* spec. from the West Caucasian viper, *Vipera kaznakovi* (Serpentes: Viperidae), *Arch. Protistenkd.*, 132, 93, 1986.
6. **Lavier, S.**, Sur une coccidie atypique de *Vipera aspis, Ann. Parasitol.*, 17, 37, 1939.
7. **Cawthorn, R. J. and Stockdale, P. H. G.**, The developmental cycle of *Caryospora bubonis* Cawthorn and Stockdale, 1981 (Protozoa: Eimeriidae) in the great horned owl, *Bubo virginianus* (Gmelin), *Can. J. Zool.*, 60, 152, 1982.
8. **Stockdale, P. H. G. and Cawthorn, R. J.**, The coccidian *Caryospora bubonis* in the Great Horned Owl *(Bubo virginianus), J. Protozool.*, 28, 255, 1981.
9. **Wacha, R. S. and Christiansen, J. L.**, Development of *Caryospora bigenetica* n. sp. (Apicomplexa, Eimeriidae) in rattlesnakes and laboratory mice, *J. Protozool.*, 29, 272, 1982.
10. **Hrudka, F., Cawthorn, R. J., and Haigh, J. C.**, The occurrence of coccidia (Eimeriidae) in epididymal semen of a wapiti *(Cervus canadensis nelsoni), Can. J. Zool.*, 61, 1693, 1983.
11. **Arcay, L.**, Genital coccidiosis in the golden hampster *(Cricetus cricetus), in Molecular and Biochemical Parasitology. Parasites — Their World and Ours*, Müller, M., Gutteridge, W., and Kohler, P., Eds., Proc. 5th Int. Congr. Parasitology (Toronto), August 7—14, Elsevier, Toronto, 1982, 404.
12. **Cable, R. M. and Conaway, C. H.**, Coccidiosis of mammary tissue in the water shrew, *Sorex palustris navigator, J. Parasitol.*, 39, (Suppl.) 30, 1953.
13. **Marcone, G.**, Sporozoen-Dermatosen des hundes, *Z. Infekt. Parasitenkd. Hyg. Haust.*, 4, 5, 1908.
14. **Sangster, L. T., Styler, E. L., and Hall, G. A.**, Coccidia associated with cutaneous nodules in a dog, *Vet. Pathol.*, 22, 186, 1985.
15. **Shelton, G. C., Kintner, L. D., and MacKintosh, D. O.**, A coccidia-like organism associated with subcutaneous granulomata in a dog, *J. Am. Vet. Med. Assoc.*, 152, 263, 1968.
16. **Upton, S. J. and Barnard, S. M.**, Experimental transmission of *Caryospora simplex* (Apicomplexa: Eimeriidae) to Palestine vipers, *Vipera xanthina palestinae* (Serpentes: Viperidae), *J. Protozool.*, 33, 129, 1986.
17. **Upton, S. J. and Barnard, S. M.**, Development of *Caryospora bigenetica* (Apicomplexa: Eimeriorina) in experimentally infected mice, *Int. J. parasitol.*, 17, 15, 1988.
18. **Lindsay, D. S. and Sundermann, C. A.**, Recent advances in the biology of the coccidian genus *Caryospora, J. Vet. Parasitol.*, 3, 1, 1989.
19. **Sundermann, C. A., Lindsay, D. S., Blagburn, B. L., Bailey, M. A., and Wester, E. E.**, Serial transmission of *Caryospora bigenetica* Wacha & Christiansen, 1982 (Apicomplexa: Eimeriidae) between different species of rodents, *J. Parasitol.*, 75, 327, 1989.

20. **Upton, S. J., Lindsay, D. S., Current, W. L., and Barnard, S. M.,** Mouse-to-mouse transmission of *Caryospora simplex* (Apicomplexa: Eimeriidae), *J. Parasitol.*, 71, 395, 1985.
21. **Lindsay, D. S., Sundermann, C. A., Blagburn, B. L., Upton, S. J., and Barnard, S. M.,** Fatal *Caryospora bigenetica* (Apicomplexa: Eimeriidae) infections in cotton rats, *Sigmodon hispidus*, *J. Parasitol.*, 74, 838, 1988.
22. **Sundermann, C. A. and Lindsay, D. S.,** Ultrastructure of in vivo-produced caryocysts containing the coccidian *Caryospora bigenetica* (Apicomplexa: Eimeriidae), *J. Protozool.*, 36, 81, 1989.
23. **Upton, S. J., Current, W. L., Ernst, J. V., and Barnard, S. M.,** Extraintestinal development of *Caryospora simplex* (Apicomplexa: Eimeriidae) in experimentally infected mice, *Mus musculus*, *J. Protozool.*, 31, 392, 1984.
24. **Schaffrath-Böer, A.,** Untersuchungen über den endogenen Entwicklungszyklus von *Caryospora kutzeri* (Böer, 1982). Dissertation from the Institute für Parasitologie der Tierärztlichen Hochschule Hannover, 1983.
25. **Lindsay, D. S. and Blagburn, B. L.,** *Caryospora uptoni* and *Frenkelia* sp.-like coccidial infections in red-tailed hawks, *Buteo borealis*, *J. Wildl. Dis.*, 25, 407, 1989.
26. **Böer, B.,** Untersuchungen über das Vorkommen von Kokzidien bei Greifvögeln und über die Entwicklung von zwei *Caryospora*-Arten der Falken (*Caryospora neofalconis* n. sp. und *Caryospora kutzeri* n. sp.), Dissertation from the Institute für Parasitologie der Tierärzlichen Hochschule Hannover, 1982.
27. **Allen, E. A.,** *Eumonospora tremula* gen. et sp. nov., a coccidium from the intestine of the turkey buzzard, *Cathartes aura septentrionalis* Wied, *Trans. Am. Microsc. Soc.*, 52, 192, 1933.
28. **Wetzel, R. and Enigk, K.,** Beobachtungen uber parasitare Erkrankungen der Falken und Vorschlage zu ihrer Bekampfung, *Jahrb. Dtsch. Falkenorden*, 1, 24, 1939.
29. **Chakravarty, M. and Kar, A. B.,** Observations on two reptilian coccidia, *J. R. Asiat. Soc. Bengal Sci.*, 12, 3, 1947.
30. **Upton, S. J., Current, W. L., and Barnard, S. M.,** A new species of *Caryospora* (Apicomplexa: Eimeriorina) from the green lizard, *Anolis carolinensis*, *Trans. Am. Microsc. Soc.*, 103, 245, 1984.
31. **Rebell, G., Rywlin, A., and Ulrich, G.,** Coccidiosis in the green turtle (*Chelonia mydas*) in mariculture, *Proc. 5th Meet. World Maricul. Soc.*, 1974, 197.
32. **Leibovitz, L., Rebell, G., and Boucher, G. C.,** *Caryospora cheloniae* sp. n.: a coccidial pathogen of mariculture-reared green sea turtles (*Chelonia mydas mydas*), *J. Wildl. Dis.*, 14, 269, 1978.
33. **Saxe, L. H.,** Observations on coccidia from *Microtus pennsylvanicus*, *Proc. Soc. Protozool.*, 3, 13, 1952.
34. **Saxe, L. H., Levine, N. D., and Ivens, V.,** New species of coccidia from the meadow mouse, *Microtus pennsylvanicus*, *J. Protozool.*, 7, 61, 1960.
35. **Bray, R. S.,** On the parasitic protozoa of Liberia. III. Three new species of *Caryospora* Léger, 1904 from the hissing sand snake *Psammophis sibilans phillipsi* and a note on a *Caryospora* in *Coluber ravergieri nummifer* from Israel, *J. Protozool.*, 7, 314, 1960.
36. **Cannon, L. R. G.,** *Caryospora demansiae* sp. nov. (Sporozoa: Eimeriidae) from an Australian snake *Demansia psammophis* (Elapidae), *Parasitology*, 57, 221, 1967.
37. **Cannon, L. R. G. and Rzepczyk, C. M.,** A new species of *Caryospora* Léger, 1904 (Sporozoa: Eimeriidae) from an Australian snake, *Parasitology*, 69, 197, 1974.
38. **Lainson, R. and Shaw, J. J.,** Coccidia of Brazilian snakes: *Isospora decipiens*, *Eimeria micruri*, *E. liophi* and *E. leimadophi* spp. n., with redescriptions of *Caryospora brasiliensis* Carini, 1932 and *Eimeria poecilogyri* Carini, 1933, *J. Protozool.*, 20, 358, 1973.
39. **Sundermann, C. A., Blagburn, B. L., Swango, L. J., and Boosinger, T.,** Dermal coccidiosis in dogs, *J. Ala. Acad. Sci.*, 59, 3, 1988.
40. **Fayer, R. and Dubey, J. P.,** Comparative epidemiology of coccidia: clues to the etiology of equine protozoal myeloencephalitis, *Int. J. Parasitol.*, 17, 615, 1987.
41. **Upton, S. J., Current, W. L., Barnard, S. M., and Ernst, J. V.,** *In vitro* excystation of *Caryospora simplex* (Apicomplexa: Eimeriorina), *J. Protozool.*, 31, 293, 1984.
42. **Upton, S. J., Haynes, T. B., Current, W. L., and Barnard, S. M.,** Development of *Caryospora simplex* (Apicomplexa: Eimeriidae) from sporozoites to oocysts in human embryonic lung cell culture, *J. Protozool.*, 31, 398, 1984.
43. **Sundermann, C. A., Lindsay, D. S., Tibbs, R. E., and Bailey, M. A.,** Complete development of *Caryospora bigenetica* (Apicomplexa: Eimeriidae) in vitro, *J. Protozool.*, 35, 465, 1988.
44. **Lindsay, D. S., Sundermann, C. A., and Blagburn, B. L.,** Caryocyst-like host cell formation by *Caryospora duszynskii* (Apicomplexa: Eimeriidae) in human fetal lung cell cultures, *J. Protozool.*, 35, 32, 1988.
45. **von Frank, C.,** *Plasmodium (Haemamoeba)* und Coccidia: ein Beitrag zur Protozoenfauna verschiedener Vögel aus Ostosterreich, *Z. Angew. Zool.*, 65, 195, 1978.
46. **Rachmatullyna-Batyrshyna, N. K. and Svanbaev, S. K.,** Infections of wild birds with coccidia, *Trans. Inst. Zool. Alma-Ata Acad. Sci. Kaz. USSR*, 33, 5, 1972.

47. **Schwalbach, G.**, Untersuchungen und Beobachtungen an Coccidien der Gattungen *Eimeria, Isospora* und *Caryospora* bei Vögeln mit einer Beschreibung von sechzehen neuen Arten, *Arch. Protistenkd.*, 104, 431, 1959.
48. **Poelma, F. G. and Strik, W. J.**, *Caryospora undata*, Schwalbach 1959, a coccidium from the intestine of a Guillemot, *Uria allge aalge*, Potoppidan, 1763, *Tijdschr. Diergeneeskd.*, 91, 917, 1966.
49. **Volcán, G. S. and Medrano, C. E.**, Esporozoario aves Falconiformes en el estado Bolivar, Venezuela *Caryospora arcayae* n. sp. (Apicomplexa: Eimeriidae), in *Cuadernos de Geografia Medica de Guayana*, Vol. 1, Volcan, G. S., Ed., Universidad de Oriente and Sociedad Parasitologica Venezolana, Venezuela, 1984, 1.
50. **Von Frey, H. and Von Kutzer, E.**, Zur disgnostik heimischer Greïfvogelund Eulenparasiten, *Prakt. Tierarzt.*, 10, 894, 1982.
51. **Von Kutzer, E., Von Frey, H., and Kotremba, J.**, Zur Parasitenfauna österreichischer Greifvögel (Falconiformes), *Angew. Parasitol.*, 21, 183, 1980.
52. **Wetzel, R. and Enigk, K.**, *Caryosporà falconis* n. sp. (Eimeridea) aus dem Wanderfalken, *Sitzungsber. Ges. Naturforsch. Freunde. Berlin*, p. 6, 1937.
53. **Yakimoff, W. L. and Matikaschwili, I. L.**, La coccidie du hibou, *Bull. Soc. Pathol. Exot.*, 25, 1045, 1932.
54. **Schellner, H.-P. and Rodler, S.**, Ein trematoden- und Kokzidienbefall bei einem Luggerfalken *(Falco jugger)*, *Jahrb. Dtsch. Falkenorden*, 1970/71, 90, 1971.
55. **Yakimoff, W. L. and Matschulsky, S. N.**, XVI — On the coccidia of birds of prey, *J. R. Microsc. Soc.*, 56, 372, 1936.
56. **Hoare, C. A.**, On the systematic status of *Eumonospora tremula* Allen, 1933 (Coccidiida Eimeriidae), *Trans. Am. Microsc. Soc.*, 53, 6, 1934.
57. **Lindsay, D. S. and Blagburn, B. L.**, *Caryospora uptoni* n. sp. (Apicomplexa: Eimeriidae) from red-tailed hawks *(Buteo jamaicensis borealis)*, *J. Parasitol.*, 72, 762, 1986.
58. **Pellérdy, L.**, Three new coccidia parasitic in Cuban birds (Protozoa: Sporozoa), *Acta Zool. Acad. Sci. Hung.*, 13, 227, 1967.
59. **Černa, Ž.**, Interesting coccidians from wild birds, *Proc. Soc. Protozool.*, 20, 535, 1973.
60. **Černa, Ž.**, Two new coccidians from passeriform birds, *Folia Parasitol.*, 23, 277, 1976.
61. **Varghese, T. and Yayabu, R.**, A survey of coccidian and helminth parasites of birds in Papua New Guinea with special reference to the Birds of Paradise, in *Wildlife Diseases of the Pacific Basin and other Countries*, Proc. 4th Int. Conf. Wildl. Dis. Assoc., Aug. 25—28, 1981, Sydney, Australia, Fowler, M. E., Ed., Wildlife Disease Association, Sydney, 1981, 102.
62. **Cawthorn, R. J. and Stockdale, P. H. G.**, Description of *Eimeria bubonis* sp. n. (Protozoa: Eimeriidae) and *Caryospora bubonis* sp. n. (Protozoa: Eimeriidae) in the great horned owl, *Bubo virginianus* (Gmelin), of Saskatchewan, *Can. J. Zool.*, 59, 170, 1981.
63. **Von Gottschalk, C.**, Beitrag zür Faunistik der Vogelkokzidien Thüringens und Sachsens, *Beitr. Vogelkd. Leipzig*, 18, 61, 1972.
64. **Mandal, R.**, *Caryospora bengalensis* sp. n. and *Eimeria fibrilosa* sp. n., new coccidia (Protozoa: Eucoccida) from a fresh-water snake, *Enhydris enhydris* (Schneider), *Acta Protozool.*, 15, 405, 1976.
65. **Stibbs, H. H. and Ongerth, J. E.**, Immunofluorescence detection of *Cryptosporidium* oocysts in fecal smears, *J. Clin. Microbiol.*, 24, 517, 1986.
66. **Carini, A.**, *Caryospora brasiliensis* n. sp., Parasita do intestino da cobra verde, *Chlorosoma aestivum*, *Rev. Biol. Hyg.*, 3, 85, 1932.
67. **Nandi, R.**, *Caryospora cobrae* n. sp. from the Indian cobra *Naja naja* Linnaeus, *J. Protozool.*, 32, 204, 1985.
68. **Matuschka, F.-R.**, Description of *Caryospora colubris* sp. n. (Apicomplexa: Eimeriidae) from the western whip snake, *Coluber viridiflavus* (Serpentes: Colubridae), *Can. J. Zool.*, 62, 1525, 1984.
69. **Matuschka, F.-R.**, *Caryospora corallae* n. sp. (Apicomplexa: Sporozoea, Eimeriidae) from the emerald tree boa *Corallus caninus* (Serpentes: Boidae), *J. Protozool.*, 31, 358, 1984.
70. **Upton, S. J., Current, W. L., and Barnard, S. M.**, A new species of *Caryospora* (Apicomplexa: Eimeriorina) from *Elaphe* spp. (Serpentes: Colubridae) of the southeastern and central United States, *Trans. Am. Microsc. Soc.*, 103, 240, 1984.
71. **Matubayasi, H.**, On a new coccidia, *Koidzumiella natrix* n. g., n. sp., parasitic in Japanese snakes; with some remarks on two other ophidian coccidia of the genus *Caryospora* and *Isospora*, *Keio-Igaku*, 16, 1719, 1936.
72. **Matubayasi, H.**, Studies on parasitic protozoa in Japan II. Coccidia parasitic in snakes; with special remarks on *Tyzzeria (Koidzumiella) natrix*, parasitic in *Natrix tigrina*, *Annot. Zool. Jpn.*, 16, 255, 1937.
73. **Carini, A.**, Sobre uma cariospora da Jararaca, *Arch. Biol.*, 23, 41, 1939.
74. **Anderson, D. R., Duszynski, D. W., and Marquardt, W. C.**, Three new coccidia (Protozoa: Telosporea) from kingsnakes, *Lampropeltis* spp., in Illinois, with a redescription of *Eimeria zamenis* Phisalix, 1921, *J. Parasitol.*, 54, 577, 1968.

75. **Hoare, C. A.,** Studies on some new ophidian and avian coccidia from Uganda, with a revision of the classification of the Eimeriidoa, *Parasitology,* 25, 359, 1933.
76. **Matuschka, F.-R.,** *Caryospora najadae* sp. n. (Apicomplexa: Eimeriidae) from Dahl's whip snake, *Coluber najadum* (Serpentes: Colubridae), *Z. Parasitenkd.*, 72, 181, 1986.
77. **Matuschka, F.-R.,** *Caryospora najae* sp. n. (Apicomplexa: Sporozoea, Eimeriidae) from the spitting cobra, *Naja nigricollis pallida* (Serpentes: Elapidae), *J. Parasitol.*, 68, 1149, 1982.
78. **Upton, S. J., Freed, P. S., and Freed, M. G.,** Two new species of coccidia (Apicomplexa: Eimeriorina) from Peruvian reptiles, *Trans. Am. Microsc. Soc.*, 108, 96, 1989.
79. **Augustine, P. C. and Danforth, H. D.,** Hybridoma antibodies reveal differences in antigenic composition of coccidia refractile bodies, Abstr. 60th Annu. Meet. Am. Soc. Parasitol., August 4—8, Athens, GA, 1985, 56.

Chapter 10

GENETICS OF COCCIDIA AND GENETIC MANIPULATIONS

Martin W. Shirley and Thomas K. Jeffers

TABLE OF CONTENTS

I. INTRODUCTION

In comparison with some of the other parasitic protozoa, in particular the malarial parasites, relatively little new knowledge on the genetics of the coccidial parasites has been obtained during the past 5 years. For example, a search of appropriate computer data bases revealed the publication of only a few papers during the past 10 years and virtually all of these papers relate to techniques and/or experimental approaches that have been described previously.

Genetic studies on the coccidia are inherently difficult for several reasons. These include the complex nature of the life cycles, the absence of straightforward methods by which, with the exception of *Toxoplasma*, cloned populations can be obtained, and the absence of techniques to isolate purified gametes. Consequently, only relatively simple experimental protocols have been undertaken, such as cross-fertilization between strains of the same species (with complementary markers) in the same host. The limitations of these experimental strategies mean, for example, that as yet there are few data on the inheritance patterns of genetic markers in cloned progeny of the crosses. The supposition that coccidia may inherit genes in a manner consistent with those expected for a eukaryotic organism with haploid and diploid stages can arguably at best be inferred from the results of cross-fertilization experiments with *Toxoplasma* and of similar, but more comprehensive studies on the malarial parasites.[1]

Another fundamental aspect of the genetics of the coccidia to be resolved, but little investigated during the past 20 years, is an accurate determination of the number of chromosomes that the coccidia possess. However, the application of some of the new molecular biological techniques for the characterization of large DNA molecules either alone or more especially in combination with DNA hybridization procedures, is likely to prove rewarding for researchers interested in this area of coccidial genetics.

In view of our previous reviews of this subject[2,3] we will herein focus principally on recent publications of relevant research and also attempt to indicate the new techniques and experimental approaches which are likely to be used with success.

II. DEFINITIONS

Definitions of the major terms given below are based in part on the views of Joyner et al.[4] who reviewed some of the problems of the nomenclature of the *Eimeria*.

- Species — A group of actually or potentially interbreeding populations which is reproductively isolated from other such groups.
- Sample — Part of a coccidial population collected on a unique occasion at a specified location.
- Isolate — Viable organisms present in experimental animal hosts, avian embryos or cultured cells after the introduction of a sample. An isolate is not necessarily monospecific and not all organisms in the sample will necessarily appear.
- Strain — A population of a single species obtained from an isolate and maintained by serial passage, having regularly reproducible characteristics which are stable within the limits of the tests used. Although the term is extensively used to describe populations of coccidial parasites and, arguably, conveniently differentiates between "isolates" and "lines", Walliker[1] did not use the term to define similar populations of malaria. Walliker,[5] moreover, in a consideration of "strains" argued that malarial parasites be regarded as a "population of organisms in which interbreeding by random mating of gametes probably occurs at high frequency, resulting in the production of novel gene combinations" and that drug resistance "be perceived as the spread of resistance genes

rather than of resistant strains". However, the general acceptance of the word, together with the relative ease with which populations can be established from infections with single oocysts, will probably ensure that the term "strain" is enduring for the coccidia.

- Line — A population derived from a strain as a consequence of passage in the laboratory under different conditions (e.g., passage in eggs) or by selection for, say, drug-resistance or early completion of the life cycle.
- Clone — A population of parasites which is the progeny of a single asexual life cycle stage, e.g., a sporozoite of *Eimeria* or tachyzoite of *Toxoplasma*. Some care is required for the use of this term as it is sometimes inferred, incorrectly, to describe populations derived from single oocysts.[6]

III. EVOLUTION

Coccidial parasites have for some years been considered to be eukaryotes because they possess a cytoplasm delineated into membrane bound organelles such as nuclei, mitochondria, and Golgi.[7]

The characteristics of ribosomes from *E. tenella* were determined by Wang,[8] who considered that they possessed some prokaryotic features including sensitivity to antibiotics that usually had no inhibitory effects on the cytoplasmic ribosomes of eukaryotic organisms. Pasternak et al.,[9] however, concluded that the features of both ribosomal and translatable messenger RNA from *E. tenella* were consistent with those from a lower eukaryote. The molecular weights of the RNAs of the large and small ribosomal subunits (1.27 and 0.65 $\times 10^6$, respectively) were similar to those of *Toxoplasma* (1.3 and 0.77 $\times 10^6$, respectively) and, moreover, showed greater similarity to those from lower eukaryotes (e.g., the yeast *Saccharomyces carlsbergensis* 1.21 and 0.72 $\times 10^6$, respectively) than to those from prokaryotes. In addition, almost all the translatable mRNA was characterized by the addition of more than 15 adenylate residues to the 3′ end.

The presence of multiple introns in the α and β tubulin genes of *Toxoplasma*[10] and in a gene for a major surface antigen of *E. tenella*[11] is also consistent with the features of eukaryotic organisms.

The development of techniques to determine the sequence of nucleotides of DNA and RNA has provided a further means to assess the phylogeny of organisms. Hasegawa et al.,[12] for example, considered the phylogenic relationships of several eukaryotes by investigation of the sequence of ribosomal RNA (rRNA) genes. As far as is known the procedure has only been applied to the coccidia by Johnson and co-workers[13-15] (see also Chapter 7). Johnson et al.[13,15] characterized the sequence of the small subunit ribosomal RNA (srRNA) of *T. gondii* and *S. gigantea*, respectively, and phylogenetic analyses of the sequences indicated that *Toxoplasma* and *Sarcocystis* are monophyletic and are not closely related to any other organism (including the malarial parasite *P. berghei*), for which srRNA sequences are available.

The taxonomic relationship between *T. gondii* and *Hammondia hammondi* was investigated by Johnson et al.[14] DNA extracted from each of three strains of *T. gondii* and one of *H. hammondi* was cut with a series of ten restriction endonucleases and then probed with radiolabeled cloned rRNA probes. The autoradiograph patterns of the four parasites were identical for each of the enzymes tested and it was suggested that these results, together with almost identical guanine + cytosine contents of DNA from the two genera (see Table 1), were consistent with the notion that *H. hammondi* be classified as a separate species within the genus *Toxoplasma*.[16] The taxonomy of these parasites clearly needs further attention.

Very little is known about the molecular features of DNA from the mitochondria of coccidia. Borst et al.[17] however, working with tachyzoites of *T. gondii* isolated a closed circular duplex DNA fraction which they characterized by electron microscopy. Much of

TABLE 1
G + C Content of DNA from Different Species and Strains of Coccidia

Parasite	G + C Content (%)	Ref.
H. hammondi	49.0	Johnson et al.[23]
T. gondii	53	Neimark and Blaker[24]
T. gondii	57	Perrotto et al.[25]
T. gondii (RH strain)	49.9	Johnson et al.[23]
T. gondii (ME49 strain)	50.4	Johnson et al.[23]
E. maxima	48	Lee and Fernando[26]
E. tenella	52	Lee and Fernando[26]

the DNA consisted of circles with a contour length of 12 μm and contained a small cruciform structure with 0.5 μm tails. Borst et al.[17] considered that the cruciform was not present *in vivo* and that the 12 μm circles possibly represented the mitochondrial DNA. Moreover, if all the mitochondrial DNA were composed of 12 μm circles, Borst et al.[17] calculated from the data of Cornelissen et al.[18] that each cell of *T. gondii* should contain about 76 such circles of DNA. This number is similar to that found in other eukaryotes.[19]

IV. CYTOCHEMISTRY

A. DNA CONTENT

Cornelissen et al.,[20] in a prelude to quantitative nuclear DNA studies, examined the fluorescence emission of nuclear DNA in different life cycle stages of *T. gondii, E. tenella* and *S. cruzi* after the parasites had been stained with Feulgen-Pararosaniline (SO2). The nuclei of all stages examined, including cystozoites, proliferative parasites, young and multinucleated schizonts, merozoites, microgamonts, microgametes, macrogamonts, and macrogametes of *T. gondii*; sporozoites, merozoites, schizonts, microgamonts, and macrogametes of *E. tenella*; and metrocytes and cystozoites of *S. cruzi*, showed a typically bright red fluorescence.

Cornelissen et al.[18] subsequently determined the DNA content of some of these stages, either by direct microfluorometry (measurement of the total amount of fluorescence emitted by individual Fuelgen stained nuclei using excitation with UV light), or by scanning of photomicrograph negatives produced either by fluorescence emission of Fuelgen stained nuclei or by absorption photography of the same Fuelgen stained nuclei.

The most comprehensive study was on *T. gondii* and the findings indicated that all stages contained a haploid amount of DNA with the exception of fully mature macrogametes that were presumed to be diploid (Table 2). Cornelissen et al.,[18] however, in their study on *E. tenella* did not examine macrogametes of *E. tenella*, but the DNA contents of sporozoites, second generation merozoites and microgametes were similar to each other (see Table 3). The data obtained from these studies were consistent with the notion proposed earlier that meiosis in the coccidia occurred during sporogony with the diploid state represented by the zygotes (e.g., Canning and Morgan[21] and Cheissen et al.[22] who studied *E. tenella* and *Hepatozoan domerguei* and *E. intestinalis*, respectively).

The absolute amounts of DNA per cell in different life cycle stages of *T. gondii, E. tenella, E. acervulina, P. berghei*, and *S. cruzi* have been determined by Cornelissen et al.[20] with human metaphase chromosome number 21 serving as an internal control (Table 3). Values obtained for sporozoites of the two species of *Eimeria* (approximately 76×10^{-15} g DNA/nucleus) were ten times lower than those deducible from the data of Wang and Stotish[27] who calculated a total of 5.8×10^{-12} g of DNA per sporulated oocyst of *E. tenella*; equivalent to 725×10^{-15} g per sporozoite. In view of the greater sizes of many

TABLE 2
Relative Amounts of DNA Per Cell in Successive Stages of *T. gondii* and *E. tenella*

Parasite and life cycle stage	Number of cells examined	Relative amount of DNA
T. gondii		
M3 merozoite	20 (4)	112.8
Microgametes	84 (68)	100.0
M2 merozoite	138 (126)	96.9
Young macrogametes	26 (0)	107.8
Mature macrogametes	8 (0)	217.1
Sporozoites	38 (11)	96.7
E. tenella		
Second generation merozoites	40	107.4
Microgametes	20	100.0
Sporozoites	9	96.5

Note: Cells were fixed and stained in Geimsa prior to hydrolysis (1h) and Feulgen staining. Photomicrograph-negatives were produced by fluorescence emission photography or by absorption photography (number of cells examined by the latter procedure are given in parentheses).

From Cornelissen, A. W. C. A. et al., *Parasitology*, 88, 531, 1984. With permission.

TABLE 3
Absolute Amounts of DNA in Different Life-Cycle Stages of Coccidia

Parasite	Stage	DNA Content ($\times 10^{-15}$g)
T. gondii	Microgamete	96.0
	Sporozoite	98.7
E. acervulina	Sporozoite	75.5
E. tenella	Sporozoite	75.5
	Microgamete	77.6
S. cruzi	Cresent cystozoite	216.0

Note: Human chromosome 21 was used as a control for each analysis.

From Cornelissen, A. W. C. A. et al., *Parasitology*, 88, 13, 1984. With permission.

of the chromosomes in *Eimeria* compared with those of *Plasmodium*[28] it is of interest to note that Cornelissen et al.[20] observed three times more DNA in *E. acervulina* and *E. tenella* than in *P. berghei* with extrapolated genome sizes of 7.0×10^7 and 2.53×10^7 bp, respectively.[29]

Cells were fixed and stained in Geimsa prior to hydrolysis (1 h) and Feulgen staining. Photomicrographic negatives were produced by fluorescence emission photography or by absorption photography (numbers of cells examined by the latter procedure are given in parentheses).

B. SEX DIFFERENTIATION IS PHENOTYPICALLY DETERMINED

Shirley and Millard[30] and Lee et al.,[31] who obtained viable oocysts from infections with single sporozoites of *E. tenella*, and Pfefferkorn et al.[32] who recovered viable oocysts from cats given the progeny of clones derived from single proliferative parasites (tachyzoites) grown in microwells of human diploid cells, provided definitive evidence that differentiation of the sexes in these parasites was phenotypically determined.

Further studies have now been done with *Toxoplasma* which both confirm the earlier results and establish the point(s) in the life cycle at which the parasites become sexually differentiated.

Cornelissen and Overdulve[33] inoculated mice with single oocysts, single sporocysts, single sporozoites, single proliferative parasites, single cysts recovered from the brain or single cystozoites of *T. gondii* (all isolated by the use of a de Fonbrune micromanipulator)[30] and obtained patent infections in 28, 32, 21, 8, 54, and 7% of the animals, respectively. Cats given the progeny of any of the single stages produced oocysts which sporulated normally and were infective for mice. These results demonstrate that asexual life cycle stages of *T. gondii* can give rise to both male and female gametes, thus providing further evidence to support the notion that sex differentiation in *T. gondii* is phenotypically controlled. This suggests that sex determination is not the result of segregation of sex chromosomes or sex genes, but is triggered by environmental factors stimulating differential gene expression in genetically identical cells.

Cornelissen and Overdulve[33] suggested that the environmental "trigger(s)" in *T. gondii* were likely to be in the final host, acting on a stage between the cystozoite and M2 and M3 merozoites which, from the results of DNA studies,[20] were presumed to be sexually differentiated.

The nature of the factors responsible for the differentiation of the different coccidian life cycle stages is totally unknown but it is clear that sex differentiation in *Eimeria* can occur early in the life cycle. A precocious line of *E. tenella* is characterized by only a single generation of schizogony before the appearance of gametes.[34]

It is also of interest to note that Jones et al.[35] observed that the addition of interferon to *in vitro* cultures of *T. gondii* apparently stimulated the differentiation of tachyzoites into bradyzoites.

The nature and mechanisms of the regulators of sex differentiation are obviously topics worthy of investigation and *Toxoplasma* would seem to be the coccidian of choice given that it can be grown *in vitro* with relative ease.

V. GENETIC STUDIES

A. CLONING OF *EIMERIA* AND *TOXOPLASMA*

Further clones of *Eimeria* have been produced *in vivo* from infections with single parasites that were visually isolated by the use of a de Fonbrune micromanipulator.[30] The use of a de Fonbrune micromanipulator, or a similar instrument, for obtaining single cells clearly provides more accuracy than the use of dilution methods, especially if the cells are transferred between two micropipettes as a further step to ensure their purity.

Chapman and Rose[36] recovered viable oocysts of *E. tenella* from 4/18 chickens (22%) each given a single sporozoite inoculated directly into the cecum. Shirley and Millard,[37] using another method, obtained oocysts from 1/8 chickens (13%) that had been given, by mouth, a single sporocyst which contained only one sporozoite after a brief incubation in an excystation medium. The attraction of the latter method is that it precludes the need for surgery. Cornelissen and Overdulve[33] in studies on *T. gondii*, obtained clones in 3/14 (21%) of mice each given a single sporozoite by intraperitoneal inoculation, i.e., the success rate was similar to that with the *Eimeria*. Asexual stages (tachyzoites) of *T. gondii*, unlike those of eimerian parasites, can be serially passaged *in vitro* and, by inoculating individual microwells of human diploid fibroblasts with dilutions of tachyzoites, calculated to contain

Population A Population B

Concurrent infection of host

↓

Recovery of progeny

↓

Selection of recombinant population

↓

Characterization of recombinant population

typically by enzyme electrophoresis and/or

chemotherapy studies

FIGURE 1. Scheme used for the cross-fertilization of coccidial parasites.

between 0.3 and 3 PFU, Pfefferkorn and Pfefferkorn[38] and Pfefferkorn et al.[32] were able to recover cloned populations.

Populations of *Eimeria* have also been obtained from infections of chickens with single sporocysts. Shirley,[39] for example, maintained a line of *E. maxima* by serial passage of single sporocysts and, during 13 generations, obtained infections in 21/133 (16%) of the birds inoculated. However, such populations cannot be regarded as clones because the timing of the final nuclear division in the oocyst has not been determined definitively. From the results of a series of electron microscopical studies on the ultrastructure of *E. brunetti* during zygote formation and sporulation, Ferguson et al.[40-42] suggested that the final nuclear division occurred within the four developing sporoblasts which were formed simultaneously by invagination of the limiting membrane. Apparently, there was no evidence to suggest the presence of more than one nucleus in the early sporoblast, despite the scrutiny of several hundred thin sections of sporulating oocysts (Ferguson, personal communication). In contrast, two nuclei at opposite ends of the late sporoblast stages were observed in the terminal process of sporulation.[40] Thus the nuclei of the two eimerian sporozoites within a sporocyst might be the products of a final nuclear division which occurs after the separation of the sporoblasts by a unit membrane, in which case they would be genetically identical. However, if as is suggested by canning and Anwar,[43] the final nuclear division were to occur before the formation of sporoblasts, each of the two sporozoites within a sporocyst would be derived from genetically dissimilar nuclei.

B. CROSS FERTILIZATION EXPERIMENTS

A prerequisite for studies on cross fertilization is the availability of parasites which exhibit intraspecific variation. Commonly used genetic markers are therefore drug resistance, isoenzymes and the trait of precocious development (i.e., early completion of the life cycle). Most experiments have been undertaken with *Eimeria* because populations with complementary genetic markers are relatively easy to obtain and concurrent infections of chickens with two populations are simple to establish.

All experiments described to date share a common protocol (see Figure 1). Thus, infection of the host, to facilitate cross-fertilization of gametes, is followed by collection of the progeny. A recombinant population can then be selected during further passage, usually by the imposition of drug pressure (resistance to an anticoccidial drug is the genetic marker most amenable for the selection of recombinant progeny), or the progeny of the cross may be analyzed directly (which is possible for *T. gondii* which can be grown and cloned *in vitro*). A major limitation to studies of this type is that, compared with similar work on malaria, the derivation of sufficient clones from progeny of the cross for comprehensive genetic analyses is difficult for coccidia other than *Toxoplasma* (see above). Thus, with the exception of the studies by Pfefferkorn and Pfefferkorn[44] and Pfefferkorn and Kasper,[45] analyses of the inheritance patterns of the different markers in cloned progeny of the genetic cross have not been undertaken.

C. GENETIC DIVERSITY AMONG THE COCCIDIAL PARASITES

1. Enzyme Variation

Enzymes are direct products of genes and electrophoretic variation of enzymes is therefore a direct reflection of genetic variation. Comparatively few coccidial enzymes have been extensively characterized although O'Donoghue et al.,[46] working with *Sarcocystis* spp., obtained excellent resolution with 15 to 39 enzymes analyzed on cellulose acetate gels. Glucose phosphate isomerase (GPI), lactate dehydrogenase (LDH) and malate dehydrogenase (MDH) are the enzymes that have been studied most comprehensively.

The numbering system commonly used for reference to the enzyme variants of *Eimeria* (e.g., GP1-1) from *E. tenella*) is for convenience only.[47] New variants are now given the next number in sequence and the numerical code thus increasingly reflects the order in which the variants were found.

Intraspecific variation in the electrophoretic mobility patterns of GP1, LDH, or MDH has been clearly demonstrated in four of the seven species of *Eimeria* from the chicken.

Two forms of GP1 have been characterized in *E. acervulina*[48] and Kucera (personal communication), furthermore, obtained evidence for a second allele of LDH. Shirley[49] described two variants of GPI in *E. mitis* and Chapman,[50] who used the technique of enzyme electrophoresis for the first time to identify isolates of *Eimeria* recovered from commercial poultry houses, isolated a population characterized by a presumptive new variant of the same enzyme. This finding was confirmed by Shirley et al.[51] who assigned to it the reference number GPI-11. The first evidence of intraspecific variation in *E. necatrix* was provided by Shirley[52] who characterized five strains and found that each possessed identical forms of GPI and MDH but one of two variant forms of LDH. Two variants of GPI in *E. tenella* (coded GPI-1 and GPI-9) have been described[48] and more recent studies have confirmed these findings. Shirley et al.[53] undertook a study of electrophoretic variation of GPI, LDH, and MDH in 70 isolates of *E. tenella* recovered from commercial poultry houses around the world. All possessed the same form of LDH and MDH and, with one exception, only one of the two known forms of GPI. The most common form of the enzyme was GPI-9 which was found in 64 isolates, including all of those from the U.K., Europe, and New Zealand. Only one isolate, from the Philippines, possessed both GPI-1 and GPI-9. The comparative rarity of isolates with GPI-1, either alone or in combination with GPI-9, contrasts with the results of similar, but less extensive, studies on *E. acervulina* and *E. mitis*. These species are also characterized by intraspecific variation of GPI but isolates with a combination of different types of GPI appear to be relatively common.[54]

In contrast to the extent of variation found with the four species above, virtually no intraspecific variation of enzymes characterizes the other species from the chicken. Variation of phosphoglucomutase was seen in only one field isolate of *E. maxima*[55] but no variation in any enzyme studied so far has been observed in any laboratory strains or field isolates of *E. brunetti* and *E. praecox*.

Nakamura et al.[56] characterized LDH and GPI from one strain each of *E. acervulina, E. maxima, E. mitis, E. necatrix, E. praecox* and six strains of *E. tenella* and obtained results similar to those summarized by Shirley and Rollinson.[48] Two variants of GPI were observed in the strains of *E. tenella* and were subsequently referred to as GPI-1 and GPI-9 by Nakamura et al.[57]

Only a few studies on enzyme variation have been undertaken with other coccidia and most work has been done with *Sarcocystis*. Atkinson and Collins[58] characterized three enzymes and O'Donoghue et al.,[46] in a more comprehensive study, of seven species, characterized 15 enzymes from cystozoites recovered from macroscopic or microscopic cysts taken from hosts which included sheep, cattle and goats.

2. Drug Resistance

Resistance of populations of *Eimeria* and *Toxoplasma* to different drugs has provided some of the most useful genetic markers and similar conclusions have been drawn from the results of cross-fertilization experiments with both parasites.

Drug resistance is likely to be a much more complicated genetic trait than enzyme variation, in that the expression of resistance may depend upon multiple genes and essential gene linkages within resistant organisms. In order to be a useful genetic marker, drug resistance must be homogeneously expressed within the respective strain of coccidia, and this "marker strain" must be resistant to sufficiently high concentrations of the respective drug such that the drug may be used as a developmental barrier to differentiate recombinant progeny from the respective drug-sensitive parental strains of coccidia (see Figure 1). Through the use of drug concentrations that optimize selection pressure, it has been possible to develop from wild-type populations of coccidia, strains that homogeneously express a high degree of resistance to certain anticoccidial drugs.[2] These strains have been useful in studies on the inheritance of anticoccidial drug resistance (see Section 4 below).

3. Early Development of Oocysts

The selection of a line of *Eimeria (E. tenella)* with a shortened endogenous life cycle was first achieved by Jeffers[59] by repeated selection for the first oocysts to be produced during an infection. Since then, other similar lines of all species from the chicken and *E. intestinalis* from the rabbit have been developed in several laboratories.[60] The selected populations are referred to as "precocious lines".[61,62]

As far as is known, all precocious lines described hitherto were selected from parent strains which were the progeny of one or more oocysts. Thus it is not known whether a precocious line can be obtained from a cloned population. To investigate this possibility, selection of a precocious line was undertaken using a population of *E. praecox* that was derived from a single sporozoite by Shirley and Millard.[37]

A reduction in the prepatent time of the parasite from 84 to 64 to 67 h was achieved within 14 generations of selection in an apparently stepwise manner similar to that observed for noncloned parental populations (see Table 4).

One generation of schizont (usually the final generation) is deleted or substantially depleted in the precocious lines in comparison with their parents,[63] but the molecular biological events that inevitably occurred as a result of repeated selection for early development of oocysts have not, as yet, been investigated. Jeffers[64] provided the first demonstration that the trait of precociousness could be used as a genetic marker from the results of cross-fertilization experiments which utilized a precocious, drug-sensitive line and a drug-resistant strain of normal development.

4. Patterns of Inheritance

The first evidence of genetic recombination in coccidia was obtained by Jeffers[65] by crossing strains of *E. tenella* differing in resistance to amprolium and decoquinate, respectively. Cross-fertilization between two populations with three complementary genetic markers was first reported by Chapman[66] who used lines of *E. tenella* characterized by resistance (or sensitivity) to the anticoccidial drugs amprolium, decoquinate, and robenidine.

Recombination between three distinct genetic markers, viz, precociousness or normal development, resistance or sensitivity to the anticoccidial drug arprinocid, and isoenzymes of GP1, was demonstrated for the first time by Sutton et al.[67] who worked with *E. acervulina*.

TABLE 4
Selection of a Precocious Line from a Cloned Parent Strain of *E. praecox*

Passages of cloned parent	Time when oocysts collected (h)
1	96—144
2	82—85
3	80—84
4	80—84
5	79—81
6	75—80
7	75—80
8	69—72
9	64—73
10	120
11	67—70
12	64—70
13	67—70
14	64—67

The results of this experiment provided evidence for the stability of the trait for precocious development in the precocious line of *E. acervulina* and these findings were similar to those of Jeffers[64] for the precocious line of *E. tenella*. Genetic analyses of enzyme variation within *E. tenella* were first reported by Nakamura et al.[57] in a study in which both isoenzymes (GPI-1 and GPI-9) and resistance or sensitivity to decoquinate were used as genetic markers. The F_1 progeny of the cross were passaged in the presence of decoquinate and genetic recombination was confirmed by the presence in the F_2 (drug resistant) progeny of both GPI-1 and GPI-9. Six single oocysts taken from the F_2 generation were serially passaged three times and then characterized by enzyme electrophoresis. One population showed the phenotype of GPI-1, four had GPI-9 and one had the phenotype of both GPI-1 and GPI-9. The latter population was further passaged and subjected to electrophoresis with subsequent densitometric analyses of GPI within the gels. The results showed that both isoenzymes were maintained in almost equal ratios after 3, 6, and 10 generations.

The further significance of this study was that it provided, for the first time, a clear demonstration that the progeny of a single oocyst could express the phenotype of both parents.

Recombination between mutants of *T. gondii* resistant to fluorodeoxyuridine (FUDR) or adenine arabinoside (araA) was described by Pfefferkorn and Pfefferkorn.[44] Kittens were fed concurrently the brains from mice chronically infected with either mutant (see general scheme above), and oocysts deriving from the sexual phase of the infection were collected from the feces. The results of plaque assays, in which tachyzoites were cultured in cells grown in media with different combinations of the drugs, revealed the presence of recombinant parasites that were resistant to both drugs. Moreover, a further assay showed that the progeny of the cross also contained doubly drug sensitive parasites and that both populations (doubly resistant or doubly sensitive recombinants) comprised about 12% of the total population.

As Pfefferkorn and Pfefferkorn[44] pointed out, the use of *T. gondii* for genetic recombination experiments has distinct advantages over some of the other coccidia. One major advantage not offered by studies on *Eimeria*, is that the progeny of the cross can be analyzed by plaque assays in cultured cells with the attendant generation of reliable quantitative data. Pfefferkorn and Pfefferkorn[44] were thus able to consider their results against two models for the ploidy of the different parasite life cycle stages, and concluded that their data were consistent with the model that all stages of *T. gondii* are haploid except the zygote and unsporulated oocyst.

The availability of an *in vitro* assay system coupled with a mathematical derivation of the data, permitted Pfefferkorn and Kasper[45] to conduct quantitative recombination experiments in which cats were simultaneously infected with drug-resistant strains of *T. gondii* resistant to FUDR, araA, sinefungin (SF), or hydroxyurea (HU). The four-way crosses yielded recombinants with all possible pairs of resistance genotypes (with the single exception of resistance to HU and FUDR) at the frequencies predicted for unlinked genes. Failure to observe recombination between HU and FUDR was shown to be caused by the phenotypic suppression of HU resistance by FUDR resistance. No recombinants arising from the four-way cross were characterized by resistance to more than two drugs, a result which suggests that only one microgamete can successfully fertilize a macrogamete.

The results of the recent cross-fertilization experiments with *Eimeria* and *Toxoplasma* have clearly confirmed earlier findings that recombination between strains with complementary genetic markers is readily demonstrable. The mechanisms involved are not known but, given the failure of Joyner and Norton[68,69] to demonstrate recombination between strains of *E. maxima* resistant to clopidol or methyl benzoquate, it is probable that recombination arises through the reassortment of unlinked genes on heterologous chromosomes. However, it remains possible that the recombination could result from the crossing over of linked genes on homologous chromosomes.

6. Molecular Biology

The next major advances in our understanding of the genetics of the coccidial parasites will undoubtedly accrue as a consequence of the application of some of the new molecular biological techniques. For example, the separation and characterization of large DNA molecules up to 5 megabases (mb) in size by pulsed field gel electrophoresis (PFGE) and its modifications has proved to be particularly rewarding to studies on the genetics of many microorganisms.[70,71] Application of this technique to studies on the malarial parasites, for example, has revealed not only the number of chromosomes in asexual stages (14 in *P. falciparum*) and the existence of size polymorphisms in some chromosomes, but it has also facilitated the physical mapping of chromosomes with the assignment of some genes to individual chromosomes.[72]

There are only two reports of karyotype analyses of the coccidia. Mead et al.[73] reported the use of field inversion gel electrophoresis (FIGE) for the separation of chromosome-sized DNA from *Cryptosporidium parvum* and *C. baileyi*. Five isolates of *C. parvum* of bovine, equine, or human origin were each characterized by a similar migration pattern of five bands of DNA which differed from that of a *C. baileyi* isolate obtained from chickens and passaged in embryonating eggs. The bands of DNA ranged in molecular weight between 1.4 and 3.3 mb and, while it is likely that they are chromosomes, the authors did not obtain definitive evidence for their identity by hybridization (after Southern transfer) to chromosome-specific probes, such as cloned telomeric genes.

A superior technique for determining the molecular karyotype of an organism is the contour-clamped homogeneous electric field (CHEF) gel system.[74] Conditions of gel preparation and electrophoresis can be adjusted to facilitate the resolution of DNA molecules with a size in excess of 7 mb.[75] The CHEF gel system was used by Shirley et al.,[28] in a study on *E. tenella* of Australian origin. From direct staining with ethidium bromide of gels run under different conditions, followed by hybridization with gene probes, at least 12 chromosomes were identified of which the two smallest had sizes (from comparison with those of *P. falciparum*) of about 1.1 and 1.4 mb. All other chromosomes were in excess of about 2.5 mb but, because of their large size in comparison to the karyotype of *P. falciparum*, it was not possible to assess accurately the sizes of the biggest chromosomes which might be above 5 mb pairs.

The results of these first studies clearly demonstrate that the coccidia possess more chromosomes than were observed from the previous cytochemical studies on the nuclei of microgametes or oocysts. Estimates for the number of chromosomes in haploid stages of *Eimeria* have included five for both *E. maxima*[43,76] and *E. tenella*.[43]

The application of molecular biological techniques to studies on both the genetics and mechanisms of drug-resistance in the coccidia is also likely to be rewarding as detailed information on the underlying molecular events associated with drug-resistance is not available to date. It remains possible that an understanding of the basis of resistance to a given anticoccidial drug may permit more rational decisions to be made to overcome the problem and/or to prolong the useful life of the product. It will be of special interest to pursue an experimental approach similar to that described recently for the human malarial parasite *P. falciparum*.

A multidrug resistance phenotype has been described in tumor cells of human origin which are characterized by cross-resistance to multiple compounds that are unrelated to the selecting agent in structure, cellular target and mode of action.[77] The over expression of a high molecular weight plasma membrane glycoprotein (P-glycoprotein) has been shown to correlate with multiple drug resistance in some tumor model systems. P-glycoprotein appears to form a channel in the plasma membrane and it transports drugs out of cells using energy derived from the hydrolysis of ATP. The process of drug removal can be reversed by a diverse group of compounds which appear to be a competitive inhibitor of drug transport. Interestingly, similar results were obtained by Martin et al.[78] who showed that strains of *P. falciparum* resistant to chloroquine also became sensitive to the action of the drug when grown in the presence of verapamil. Using a probe derived from a human tumor cell line, Foote et al.[79] were able to isolate a gene for multiple drug resistance (MDR) in *P. falciparum* which they mapped to chromosome 5. In lines of *P. falciparum* that were resistant to the antimalarial drug chloroquine, amplification of the MDR gene had occurred. In view of these findings, Foote et al.[79] speculated that the gene for MDR could mediate resistance of *P. falciparum* to chloroquine. Given the economic importance of drug-resistance of *Eimeria* from the chicken, especially the more recent emergence of isolates that are resistant (and cross-resistant) to the ionophores, the search for an MDR gene and its expression in these parasites would seem to be particularly worthwhile.

Many genes from the malarial parasite *P. falciparum* have been cloned and sequenced and most have been found to contain tandem oligonucleotide repeats in single or multiple blocks.[80] Genes from *Eimeria* have similarly been characterized and at least one has been found to contain a repeat sequence. A cDNA clone from *E. acervulina* encoded an antigen which possessed a single block of 15 repeated amino acid sequences, of which 14 were identical.[81]

The differential identification of parasites through the use of gene probes, that bind to DNA with homologous sequences, has been amply demonstrated from the results of studies on several genera of protozoa including *Leishmania*,[82] *Giardia*,[83] and *Plasmodium*.[84] However, as far as is known, no DNA probes that could prove useful for the rapid identification of species, and the possible discrimination between strains, have yet been described for the coccidia. Such probes would make extremely valuable laboratory tools and, moreover, could be suitable for the direct identification of parasites present in host tissues.

VII. CONCLUDING REMARKS

Studies on the genetics of the coccidia continue to be few in comparison to the amount of work done on many other aspects of the biology of these parasites. However, the broad features of the genetics of the life-cycle and the manner in which traits are inherited have been examined. Much is known about genetic variation within some coccidia, especially species of *Eimeria* from the chicken, and use of some of the new molecular biological techniques is beginning to make an impact on our understanding of parasite genetics.

It is now to be expected that studies on molecular genetics will become the dominant area of research into coccidial genetics and the results are likely to amply demonstrate the "plasticity" of the organisms and reveal some of the mechanisms that the parasites invoke to survive in potentially hostile environments. For example, resistance of the avian coccidia to many different classes of anticoccidial drugs is well documented yet the underlying molecular mechanisms have not been investigated. That the coccidia may amplify certain genes to develop resistance, or that they may alter their molecular karyotype in the same way as the malaria have been shown to increase their chromosome number from 14 to 15 in response to attack by pyrimethamine,[85] are fascinating possibilities which deserve investigation.

The characterization of drug-resistant parasites by molecular karyotype analyses is a line of investigation that impinges on a major practical problem, but other more fundamental aspects of the biology of the coccidia need to be similarly studied. For example, is the absolute number of chromosomes conserved between representatives of the different genera and does inter- and intraspecific variation occur in the sizes of the chromosomes?

The derivation of physical maps of chromosomes could prove a useful adjunct for studies on the development of genetically engineered vaccines, a major step being the identification, selection, and characterization of antigens that can induce protective immune responses.

REFERENCES

1. **Walliker, D.,** The genetic basis of diversity in malarial parasites, *Adv. Parasitol.*, 217, 22, 1983.
2. **Jeffers, T. K.,** Genetics of coccidia and the host response, in Long, P. L., Boorman K. N., and Freeman, B. M. Eds., *Avian Coccidiosis*, British Poultry Science, Ltd., Edinburgh, 1978, 51.
3. **Jeffers, T. K. and Shirley, M. W.,** Genetics, specific and infraspecific variation, in *The Biology of the Coccidia*, Long, P. L., Ed., University Park Press, Baltimore, 1982, chap. 3.
4. **Joyner, L. P., Canning, E. U., Long, P. L., Rollinson, D., and Williams, R. B.,** A suggested terminology for populations of coccidia *(Eimeriorina)*, particularly of the genus *Eimeria* (Protozoa: Apicomplexa), *Parasitology*, 77, 27, 1978.
5. **Walliker, D.,** Genetics of malaria parasites, in *The Biology of Parasitism*, Englund, P. T. and Sher, A., Eds., Alan R. Liss, New York, 1988, 457.
6. **Hosek, K. E., Todd, K. S., and Kuhlenschmidt, M. S.,** Improved method for high-yield excystation and purification of infective sporozoites of *Eimeria* spp., *J. Protozool.*, 35, 583, 1988.
7. **Chobotar, B. and Scholtyseck, E.,** Ultrastructure, in *The Biology of the Coccidia*, Long, P. L. Ed., University Park Press, Baltimore, 1982, chap. 4.
8. **Wang, C. C.,** The prokaryotic characteristics of *Eimeria tenella* ribosomes, *Comp. Biochem. Physiol.*, 61B, 571, 1978.
9. **Pasternak, J., Winkfein, R. J., and Fernando, M. A.,** Ribosomal and translatable messenger RNA of *Eimeria tenella*, *Mol. Biochem. Parasitol.*, 3, 133, 1981.
10. **Nagel, S. D. and Boothroyd, J. C.,** The α and β tubulins of *Toxoplasma gondii* are encoded by single copy genes containing multiple introns, *Mol. Biochem. Parasitol.*, 29, 261, 1988.
11. **Files, J. G., Paul, L. S., and Gabe, J. D.,** Identification and characterization of the gene for a major surface antigen of *Eimeria tenella*, in *Molecular Strategies of Parasitic Invasion*, Agabian, N., Goodman, M., and Nogueira, N., Eds., Alan R. Liss, New York, 1987, 713.
12. **Hasegawa, M., Iida, Y., Yano, T., Takaiwa, F., and Iwabuchi, M.,** Phylogenetic relationships among eukaryotic kingdoms inferred from ribosomal RNA sequences, *J. Mol. Evol.*, 22, 32, 1985.
13. **Johnson, A. M., Murray, P. J., Illana, S., and Baverstock, P. J.,** Rapid nucleotide sequence of the small subunit ribosomal RNA of *Toxoplasma gondii:* evolutionary implications for the Apicomplexa, *Mol. Biochem. Parasitol.*, 25, 239, 1987a.
14. **Johnson, A. M., Illana, S., Dubey, J. P., and Dame, J. B.,** *Toxoplasma gondii* and *Hammondia hammondi:* DNA comparison using cloned rRNA gene probes, *Exp. Parasitol.*, 63, 272, 1987b.
15. **Johnson, A. M., Illana, S., Hakendorf, P., and Baverstock, P. J.,** Phylogenetic relationships of the apicocomplexan protist *Sarcocystis* as determined by small subunit ribosomal RNA comparison, *J. Parasitol.*, 72, 847, 1988.

16. **Levine, N. D.,** Taxonomy of *Toxoplasma*, *J Parasitol.*, 24, 36, 1977.
17. **Borst, P., Overdulve, J. P., Whijers, P. J., Fase-Fowler, P., and van den Berg, M.,** DNA circles with cruciforms from *Isopora (Toxoplasma) gondii*, *Biochim. Biophys. Acta*, 78, 100, 1984.
18. **Cornelissen, A. W. C. A., Overdulve, J. P., and Van der Ploeg, M.,** Determination of nuclear DNA of five Eucoccidian parasites, *Isospora (Toxoplasma) gondii, Sarcocystis cruzi, Eimeria tenella, E. Acervulina* and *Plasmodium berghei*, with special reference to gamontogenesis and meiosis in *I. (T.) gondii*, *Parasitology*, 88, 531, 1984.
19. **Grimes, G. W., Mahler, H. R., and Perlman, P. S.,** Nuclear gene dosage effects on mitochondrial mass and DNA, *J. Cell Biol.*, 61, 565, 1974.
20. **Cornelissen, A. W. C. A., Overdulve, J. P., and Van der Ploeg, M.,** Cytochemical studies on nuclear DNA of four eucoccidian parasites, *Isospora (Toxoplasma) gondii, Eimeria tenella, Sarcocystis cruzi* and *Plasmodium berghei*, *Parasitology*, 88, 13, 1984.
21. **Canning, E. U. and Morgan, K.,** DNA synthesis, reduction, and elimination during life cycles of the eimeriine coccidian, *Eimeria tenella* and the haemogregarine, *Heptatozoon domerguei*, *Exp. Parasitol.*, 38, 217, 1975.
22. **Cheissen, E. M., Rozonav, J. M., and Kudriavtsev, B. N.,** Fluorescence microscopic study on the content of nucleic acids in merozoites and microgametes of *Eimeria intestinalis* from the intestine of rabbit, *Acta Protozool.*, 5, 389, 1968.
23. **Johnson, A. M., Dubey, J. P., and Dame, J. B.,** Purification and characterization of *Toxoplasma gondii* tachyzoite DNA, *Aust. J. Exp. Biol. Med. Sci.*, 64, 351, 1986.
24. **Neimark, H. and Blaker, R. G.,** DNA base composition of *Toxoplasma gondii* grown *in vivo*, *Nature*, 216, 600, 1967.
25. **Perrotto, J., Keister, D. B., and Gelderman, A. H.,** Incorporation of precursors into *Toxoplasma* DNA, *J. Protozool.*, 18, 470, 1971.
26. **Lee, E.-H. and Fernando, M. A.,** Characterization of coccidial DNA: An aid in species identification, *Z. Parasitenkd.*, 53, 129, 1977.
27. **Wang, C. C. and Stotish, R. L.,** Changes of nucleic acids in the oocysts of *Eimeria tenella* during sporulation, *J. Protozool.*, 22, 438, 1975.
28. **Shirley, M. W., Kemp, D. J., Prowse, S. J., and Pallister, J.,** A molecular karyotype of *Eimeria tenella* as revealed by contour-clamped homogeneous electric field gel electrophoresis, *Mol. Biochem. Parasitol.*, 38, 169, 1990.
29. **Clarke, L. E., Messer, L. I., Greenwood, N. M., and Wisher, M. H.,** Isolation of amp3 genomic recombinants coding for antigens of *Eimeria tenella*, *Mol. Biochem. Parasitol.*, 22, 79, 1987.
30. **Shirley, M. W. and Millard, B. J.,** Some observations on the sexual differentiation of *Eimeria tenella* using single sporozoite infections in chicken embryos, *Parasitology*, 73, 337, 1976.
31. **Lee, E.-H., Remmler, O., and Fernando, M. A.,** Sexual differentiation in *Eimeria tenella* (Sporozoa: Coccidia), *J. Parasitol.*, 66, 172, 1977.
32. **Pfefferkorn, E. R., Pfefferkorn, L. C., and Colby, E. D.,** Development of gametes and oocysts in cats fed cysts derived from cloned trophozoites of *Toxoplasma gondii*, *J. Parasitol.*, 63, 158, 1977.
33. **Cornelissen, A. W. C. A. and Overdulve, J. P.,** Sex determination and sex differentiation in coccidia: gametogony and oocyst production after monoclonal infection of cats with free-living stages and intermediate host stages of *Isospora (Toxoplasma) gondii*, *Parasitology*, 90, 35, 1985.
34. **McDougald, L. R. and Jeffers, T. K.,** *Eimeria tenella* (Sporozoa: Coccidia): gametogony following a single asexual generation, *Science*, 192, 282, 1976.
35. **Jones, T. C., Bienz, K. A., and Erb, P.,** *In vitro* cultivation of *Toxoplasma gondii* cysts in astrocytes in the presence of gamma interferon, *Infect. Immun.*, 51, 147, 1986.
36. **Chapman, H. D. and Rose, M. E.,** Cloning of *Eimeria tenella* in the chicken, *J. Parasitol.*, 72, 605, 1986.
37. **Shirley, M. W. and Millard, B. J.,** A novel method for cloning *Eimeria praecox* in the chicken, *J. Parasitol.* in press.
38. **Pfefferkorn, E. R. and Pfefferkorn, L. C.,** *Toxoplasma gondii*: isolation and preliminary characterization of temperature-sensitive mutants, *Exp. Parasitol.*, 39, 365, 1976.
39. **Shirley, M. W.,** Maintenance of *Eimeria maxima* by serial passage of single sporocysts, *J. Parasitol.*, 56, 172, 1980.
40. **Ferguson, D. J. P., Birch-Andersen, A., Hutchison, W. M., and Siim, J. Chr.,** Observation on the ultrastructure of the late sporoblast and the initiation of sporozoite formation in *Eimeria brunetti*, *Acta Pathol. Microbiol. Scand.*, 85, 110, 1977.
41. **Ferguson, D. J. P., Birch-Andersen, A., Hutchison, W. M., and Siimm J. Chr.,** Light and electron microscopy on the sporulation of the oocysts of *Eimeria brunetti*. I. Development of the zygote and formation of the sporoblasts, *Pathol. Microbiol. Scand.*, 86, 1, 1978.
42. **Ferguson, D. J. P., Birch-Andersen, A., Hutchison, W. M., and Siim, J. Chr.,** Light and electron microscopy on the sporulation of the oocysts of *Eimeria brunetti*. II. Development into the sporocyst and formation of the sporozoite, *Pathol. Microbiol. Scand.*, 86, 13, 1978.

43. **Canning, E. U. and Anwar, M.,** Studies on meiotic division in cocidial and malarial parasites, *J. Protozool.*, 15, 290, 1968.
44. **Pfefferkorn, E. R. and Pfefferkorn, L. C.,** *Toxoplasma gondii:* genetic recombination between drug resistant mutants, *Exp. Parasitol.*, 50, 305, 1980.
45. **Pfefferkorn, E. R. and Kasper, L. H.,** *Toxoplasma gondii:* genetic crosses reveal phenotypic suppression of hydroxyurea resistance by fluorodeoxyuridine resistance, *Exp. Parasitol.*, 55, 207, 1983.
46. **O'Donoghue, P. J., Adams, M., Dixon, B. R., Ford, G. E., and Baverstock, P. R.,** Morphological and biochemical correlates in the characterization of *Sarcocystis* spp., *J. Protozool.*, 33, 114, 1986.
47. **Shirley, M. W.,** Enzyme variation in *Eimeria* species of the chicken, *Parasitology*, 71, 369, 1975.
48. **Shirley, M. W. and Rollinson, D.,** Coccidia: the recognition and characterization of populations of *Eimeria*, in *Problems in the Identification of Parasites and Their Vectors*, Vol. 17, British Society of Parasitology Symposium, Blackwell Scientific, Oxford, 1979, 7.
49. **Shirley, M. W.,** A reappraisal of the taxonomic status of *Eimeria mivati*, Edgar and Seibold, 1964, by enzyme electrophoresis and cross immunity tests, *Parasitology*, 78, 221, 1979.
50. **Chapman, H. D.,** The use of enzyme electrophoresis for the identification of the species of *Eimeria* present in field isolates of coccidia, *Parasitology*, 85, 437, 1982.
51. **Shirley, M. W., Jeffers, T. K., and Long, P. L.,** Studies to determine the taxonomic status of *Eimeria mitis*, Tyzzer 1929 and *E. mivati*, Edgar and Seibold 1964, *Parasitology*, 87, 185, 1983.
52. **Shirley, M. W.,** Some observations on the biology of five strains of *Eimeria necatrix*, *Z. Parasitenkd.*, 71, 287, 1985.
53. **Shirley, M. W., Chapman, H. D., Kucera, J., Jeffers, T. K., and Bedrnik, P.,** Studies on the enzyme variation and pathogenicity of recent field isolates of *Eimeria tenella*, *Res. Vet. Sci.*, 46, 79, 1989.
54. **Shirley, M. W.,** New methods for the identification of species and strains of *Eimeria*, in *Research in Avian Coccidiosis, Proc. Georgia Coccidiosis Conf.*, McDougald, L. R., Joyner, L. P., and Long, P. L., Eds., University of Georgia Press, Athens, 1986, 13.
55. **Shirley, M. W.,** Electrophoretic variants of phosphoglucomutase in different species and strains of *Eimeria* from the chicken, in *Avian Coccidiosis*, Long, P. L., Boorman, K. N., and Freeman, B. M., Eds., British Poultry Science, Ltd., 1978, 127.
56. **Nakamura, T., Konishi, T., and Kawaguchi, H.,** Isozymes of chicken coccidia in Japan, *Jpn. Vet. Sci.*, 48, 587, 1986.
57. **Nakamura, T., Konishi, T., Kawaguchi, H., and Hayashi, Y.,** Glucose phosphate isomerase isoenzyme as genetic markers for lines of *Eimeria tenella*, *Parasitology*, 96, 281, 1988.
58. **Atkinson, E. M. and Collins, G. H.,** Electrophoretic studies on three enzymes from *Sarcocystis* species in sheep, *Syst. Parasitol.*, 2, 213, 1981.
59. **Jeffers, T. K.,** Immunization against *Eimeria tenella* using an attenuated strain, *Proc. 15th World's Poultry Congr.*, New Orleans, 105, 1974.
60. **Jeffers, T. K.,** Attenuation of coccidia — a review, in *Research in Avian Coccidiosis, Proc. Georgia Coccidiosis of Conf.*, McDougald, L. R., Joyner, L. P., and Long, P. L., Eds., University of Georgia Press, Athens, 1986, 482.
61. **Jeffers, T. K.,** Attenuation of *Eimeria tenella* through selection for precociousness, *J. Parasitol.*, 61, 1083, 1975.
61a. **Licios, D., Coudert, P., and Behagia, S.,** Some biological characteristics of a precocious line of *Eimeria intestinalis*, *Parasitol. Res.*, in press.
62. **McDonald, V., Ballingall, S., and Shirley, M. W.,** A preliminary study of the nature of infection and immunity in chickens given an attenuated line of *Eimeria acervulina*, *Parasitology*, 84, 21, 1982.
63. **McDonald, V. and Shirley, M. W.,** The asexual development of precocious lines of Eimeria spp in the chicken, in *Research in Avian Coccidiosis, Proc. Georgia Coccidiosis of Conf.*, McDougald, L. R., Joyner, L. P., and Long, P. L. Eds., University of Georgia Press, Athens, 1986, 502.
64. **Jeffers, T. K.,** Genetic recombination of precociousness and anticoccidial drug resistance in *Eimeria tenella*, *Z. Parasitenkd.*, 50, 251, 1976.
65. **Jeffers, T. K.,** Genetic transfer of anticoccidial drug resistance in *Eimeria tenella*, *J. Parasitol.*, 60, 900, 1974.
66. **Chapman, H. D.,** Development by genetic recombination of a line of *Eimeria tenella* resistant to robenidine, decoquinate and amprolium, *Z. Parasitenkd.*, 70, 437, 1984.
67. **Sutton, C. A., Shirley, M. W., and McDonald, V.,** Genetic recombination of markers for precocious development, arprinocid-resistance and isoenzymes of glucose phosphate isomerase in *Eimeria acervulina*, *J. Parasitol.*, 72, 965, 1986.
68. **Joyner, L. P. and Norton, C. C.,** Transferred drug resistance in *Eimeria maxima*, *Parasitology*, 71, 385, 1975.
69. **Joyner, L. P. and Norton, C. C.,** Further observations on the genetic transfer of drug resistance in *Eimeria maxima*, *Parasitology*, 74, 205, 1977.

70. **Schwartz, D. C. and Cantor, C. R.,** Separation of yeast chromosome-sized DNAs by pulsed field gel gradient electrophoresis, *Cell,* 37, 67, 1984.
71. **Vollrath, D. and Davis, R. W.,** Resolution of DNA molecules greater than 5 megabases by contour-clamped homogeneous electric fields, *Nucl. Acids Res.,* 15, 7865, 1987.
72. **Kemp, D. J., Thompson, J. K., Walliker, D., and Corcoran, L.,** Molecular karyotype of *Plasmodium falciparum:* conserved linkage groups and expendable histidine-rich protein genes, *Proc. Natl. Acad. Sci. U.S.A.,* 84, 7672, 1987.
73. **Mead, J. R., Arrowood, M. J., Current, W. L., and Sterling, C. R.,** Field inversion gel electrophoretic separation of *Cryptosporidium* spp. chromosome-sized DNA, *J. Parasitol.,* 74, 366, 1988.
74. **Fulton, T. R., Bowcock, A. M., Smith, D. R., Daneshvar, L., Cavalli-Sforza, L. L., and Donis-Keller, H.,** A 12 megabase restriction map at the cystic fibrosis locus, *Nucl. Acids Res.,* 17, 271, 1989.
75. **Orbach, M. J., Vollrath, D., Davis, R. W., and Yanofsky, C.,** An electrophoretic karytoype of *Neurospora crassa, Mol. Cell. Biol.,* 8, 1469, 1988.
76. **Scholtyseck, E.,** Untersuchungen ubee de Kernverhaltnisse und das Wachstum bei Coccidiomorphen unter besonderer Berucksichtigung von *Eimeria maxima, Z. Parasitenkd.,* 22, 428, 1963.
77. **Bradley, G., Juranka, P. F., and Ling, V.,** Mechanism of multidrug resistance, *Biochim. Biophys. Acta,* 948, 87, 1988.
78. **Martin, S. K., Odoula, A. M., and Milhous, W. K.,** Reversal of chloroquine resistance in *Plasmodium falciparum* by verapamil, *Science,* 235, 899, 1987.
79. **Foote, S. J., Thompson, J. K., Cowman, A. F., and Kemp, D. J.,** Amplification of the multidrug-resistance gene in some chloroquine resistant isolates of *Plasmodium falciparum, Cell,* 57, 921, 1989.
80. **Kemp, D. J., Coppel, R. L., and Anders, R. F.,** Repetitive proteins and genes of malaria, *Annu. Rev. Microbiol.,* 41, 181, 1987.
81. **Jenkins, M. C.,** A cDNA encoding a merozoite surface protein of the protozoan *Eimeria acervulina* contains tandem-repeated sequences, *Nucl. Acids Res.,* 16, 9863, 1988.
82. **Wirth, D. F. and McMahon-Pratt, D.,** Rapid identification of *Leishmania* species by specific hybridization of kinetoplast DNA in cutaneous lesions, *Proc. Natl. Acad. Sci. U.S.A.,* 79, 6999, 1982.
83. **Nash, T. E., McCutchan, T. F., Keister, D., Dame, J. B., Conrad, J. D., and Gillin, F. D.,** Restriction-endonuclease analysis of DNA from 15 *Giardia* isolates obtained from humans and animals, *J. Infect. Dis.,* 152, 64, 1985.
84. **Coppel, R. L., Saint, R. B., Stahl, H. D., Langford, C. J., Brown, G. V., Anders, R. F., and Kemp, D. J.,** *Plasmodium falciparum:* differentiation of isolates with DNA hybridization using antigen gene probes, *Exp. Parasitol.,* 60, 82, 1985.
85. **Cowman, A. F. and Lew, A. M.,** Antifolate drug selection results in duplication and rearrangement of chromosome 7 in *Plasmodium chaubaudi, EMBO J.,* submitted.

Chapter 11

RECENT ADVANCES IN THE *IN VITRO* CULTIVATION OF THE COCCIDIA

Richard G. Strout and Dennis M. Schmatz

TABLE OF CONTENTS

I. INTRODUCTION

The *in vitro* cultivation of the coccidia is becoming increasingly important as immunologists, biologists, biochemists, and molecular biologists begin to ask new questions in regard to the growth, development, and control of these organisms. Although coccidian parasites are present in almost every member of the animal kingdom, they are for the most part only a major problem in domesticated animals where high density housing has enhanced the infective process. Coccidial infections in chickens have been one of the most important areas due to the type of dense housing used for chickens and the new demands on the poultry industry fueled by the increased consumption of chicken meat throughout the world due to price, new products and as an alternative to red meat.

In addition to domesticated animals, several of the coccidia have been a problem in humans including *Isospora belli*, *Isospora natalensis*, *Cryptosporidium parvum*, and *Toxoplasma gondii*. Although serious disease rarely occurs via infections with these organisms in immunocompetent individuals (with the exception of the effect of *T. gondii* on the newborn fetus), the AIDS epidemic has generated renewed interest in these organisms. The immunocompromised status of the AIDS patient allows these organisms to persist and in many cases causes the eventual death of the patient. Therefore, new methods for the prevention and treatment of these infections have become extremely important.

Since comprehensive reviews of the *in vitro* cultivation of the coccidia have already appeared in two previous works,[1,2] the objective of this chapter is to update developments in this area since Doran's last review in 1982.[2] Most of the work which has appeared in the literature since this time has been on the *Eimeria* and *T. gondii.* and therefore much of this chapter will focus on these organisms. Recent developments in the *in vitro* cultivation of *Cryptosporidium* and *Caryospora* are discussed elsewhere in this volume. Unfortunately few other coccidia have been propagated *in vitro.*

One of the greatest limitations of the *in vitro* culture system is the inability to propagate several of the species of chicken coccidia in cell culture. Of the nine species of *Eimeria* that parasitize the chicken intestine, six are considered to be economically important. Of these, only *E. tenella* is propagated routinely, and usually only through the asexual stages. Asexual development has been accomplished with *E. mivati*,[3] *E. brunetti*,[4] *E. necatrix*,[5] and *E. acervulina*.[6] There have been no reports of successful cultivation of *E. maxima*, *E. praecox*, *E. hagani*, and *E. brunetti*. Considerable success has been achieved with turkey species *E. meleagrimitis*, *E. adenoides*, *E. dispersa*, and *E. gallopayonis*.[7-9] *T. gondii* tachyzoites can be propagated and maintained in tissue culture quite effectively, although the sexual stages and oocysts can only be obtained from infected cats.

It is still necessary to design a culture system which can efficiently reproduce the entire life cycle of these organisms *in vitro*. Although *E. tenella* has been propagated in embryonic chicken kidney cell cultures from sporozoite to oocyst, the yield of oocysts is poor and has never been efficient enough to be a tool for cloning or selecting mutant parasites. This problem is also true for all of the other *Eimeria* since the yield of most of the stages is quite limited and has required investigators to depend on infected animals as the source of parasite material.

It would be even more desirable to have an axenic culture system to produce coccidia; however, the parasitic nature of these organisms makes this goal more of a dream than an objective. Of course, the key to such an effort would be an understanding of why these organisms are parasites and the host's role in their survival.

There is a great need to know more about the basic biology of these parasites — their life cycles, their effects on the host and what they require from the host for survival. Such basic information is difficult to extract from *in vivo* studies due to the complex array of events taking place in the host animal. In contrast, propagation in cell culture simplifies the study of intracellular parasites since the infection of a specific host cell type can be monitored in a more defined environment.

II. CONTRIBUTIONS OF *IN VITRO* CULTURE

A better understanding of the coccidia has been developed through the use of tissue culture since Patton's original report of the successful cultivation of *E. tenella in vitro*.[10] Fayer[11] used *in vitro* cultures to determine that *Sarcocystis* was actually a member of the coccidia by inoculating cultures with organisms isolated from the muscle of an infected grackle and observing the stages of the life cycle. McDougald and Jeffers[12] used the culture system to monitor the development of a precocious strain of *E. tenella* obtained through the selection of prematurely produced oocysts from the chicken. By observing the development of these parasites in chick kidney cell cultures, they discovered that the first generation merozoites went directly to the sexual stage and that the elimination of second generation merogony accounted for the shortened life cycle.

Once development of even a portion of the life cycle of the *Eimeria in vitro* was achieved, it was obvious that the technique could be potentially useful for the identification of new anticoccidial agents. To confirm this, Fayer and McLaughlin[13] added amprolium to cultures of *E. adenoides* in bovine embryonic kidney cells and found a significant reduction in the number of developing schizonts. Since then considerable attention has been given to developing new methods for identifying new anticoccidial drugs *in vitro*. There are obvious advantages to using *in vitro* screening of candidate compounds, including the small amount of compound required in comparison to what is needed for testing *in vivo*. The anticoccidial activity of many known drugs can be detected *in vitro* at levels between 1 and 50 μg/ml.[14] Ryley and Wilson[15] demonstrated the efficacy of robenidine and methyl benzoquate at levels as low as 0.003 and 0.0003 ppm, respectively. *In vitro* activity of the ionophores such as monensin, lasalocid, and salinomycin can also be detected at similar levels.[16]

In vitro studies can also provide useful information on the stage specificity and the mode of action of anticoccidials and can be used to determine if a drug is coccidiocidal or coccidiostatic.[17] Using tissue culture, it has been confirmed that nicarbazin specifically inhibited the development of second generation schizogony,[14] methyl benzoquate inhibits the development of sporozoites once they are intracellular and robenidine inhibits further development of first generation schizonts.[15,18] *In vitro* cultures have also been used to understand the mode of action of some anticoccidial drugs. Ryley and Wilson showed that the addition of thiamine to the culture medium resulted in the reversal of the anticoccidial activity of amprolium.[19] *In vitro* evaluation of drug effects on *T. gondii* has also been done on a limited basis using a plaque assay facilitated by the rapid growth and host cell destruction which occurs with the tachyzoite stage.[20] Ultrastructural studies of drug treated cells in cell culture have been used to determine which parasite organelles might be specifically affected by a compound. Morphological alteration of sporozoites exposed to ionophores was first described by Pfeiffer[21] who noted the extensive ballooning effect and disruption of the parasite membrane induced by low levels of monensin. Similar effects on intracellular *E. tenella* sporozoites were observed when cultures were treated with lasalocid[22] and narasin,[23] suggesting that the anticoccidial ionophores were all working in a similar manner — disrupting the integrity of the parasite membrane. It is unclear whether the ionophores work by directly killing the extracellular sporozoites, the intracellular stages or both. Recent evidence suggests that this may vary with the particular species of *Eimeria* in question and its location in the gut.[24]

In the early days of culturing the coccidia *in vitro* there were two major objectives; understanding the biology of these organisms and the screening of drugs to identify potential anticoccidial agents. These goals have expanded and changed with many of the recent studies focusing on the immune response to coccidial infections and increasing interest in the parasite-host cell interaction. Experiments demonstrating the effect of cytokines on the intracellular development of *T. gondii*,[25,26] *E. bovis*,[26] and *E. tenella*[27] have shed new light on possible mechanisms by which the host may be controlling coccidial infections. Further investigations into the mechanism of gamma interferon (γ-IFN) killing of intracellular *T. gondii* demonstrated a rapid depletion of tryptophan from intracellular pools in host cells pretreated with γ-IFN, starving the parasite of this essential amino acid.[25] The antiparasitic effect of γ-IFN could be reversed by adding increased quantities of tryptophan to the treated cultures. However, this mechanism may be unique to the cell line used for these studies. Monoclonal antibodies raised against the coccidia have also been evaluated in cell culture to determine stage specificity and possible mechanisms for humoral immunity.[28]

Another area of great interest is the parasite-host cell interaction. Knowledge concerning the means by which the parasite finds, attaches to, and penetrates host cells may lead to a better understanding of the site and host specificity commonly observed with these organisms. In culture, sporozoites will penetrate a variety of cell types, indicating that the mechanism of attaching to and penetrating host cells must involve a common characteristic present in most animal cells. This information may be beneficial to those interested in the parasite-host cell interaction as a target for developing new chemotherapeutics or a vaccine since such a mechanism may have broad spectrum applications. In contrast, once a sporozoite has penetrated a cell its subsequent development appears to be critically dependent on host cell type. The consensus is that the host cell must be an epithelial cell and with a few exceptions must be from the host species. Even under these circumstances many intracellular sporozoites fail to develop. The signal or nutrient required to trigger the transition of sporozoite to trophozoite has never been identified.

Many of the studies which focus on the parasite-host interaction involve the use of lectins or their respective ligands (usually monosaccharides) to identify receptors involved in the attachment process.[29,30] However, it is still unclear what the results of such studies reveal since charge interactions rather than ligand-receptor may also be playing a role in some of these studies.[31,32] Further work is required in this area to confirm the exact nature of the inhibition of penetration observed with selected sugars and lectins. It has also been recently suggested that phospholipase A_2 (PLA_2) plays a role in the penetration process of *T. gondii* into human fibroblasts.[33] It was reported that treatment of tachyzoites with PLA_2 enhances the infection process and treatment with PLA_2 inhibitors reduces infection. Similar experiments pretreating the host cells with these reagents confirmed the parasite-specific effects.

III. FACTORS WHICH INFLUENCE THE SUCCESS OF *IN VITRO* CULTURE

A. AGE AND STRAIN OF PARASITES

Based on studies with *E. tenella* it appears that the strain of parasite can greatly influence the success of *in vitro* propagation. Considerable differences have been found between strains with regard to penetration, development, and oocyst production.[34] Generally, the more virulent strains of *E. tenella* are also more proliferative in tissue culture, although one should keep in mind the difficulties which are encountered when attempting to compare two strains of *Eimeria* in a parallel study. One critical factor is the age of the parasites. Unless both organisms were collected from birds and sporulated at the same time, and stored under the same conditions, a direct comparison cannot be made. It is well known that the age of the oocysts greatly influences the viability of these organisms. Although oocysts can be stored at 4°C for extended periods of time, it is not recommended that they be used for tissue culture beyond 1 month. The only alternative is continuous passage of organisms in animals to maintain fresh oocyst stocks. In order to preserve the original isolate for more reproducible results, freshly excysted sporozoites should be cryopreserved in liquid nitrogen and used, rather than oocysts, to inoculate animals on a monthly basis. This practice will help maintain the characteristics of the initial isolate since it is not subject to alterations which can occur from to serial passage *in vivo*. Unfortunately, an efficient method for cryopreserving oocysts has never been reported. Strains of *E. tenella* that have provided consistently good results include Lilly 65, Beltsville, and Merck LS18. Continuous passage of *T. gondii* tachyzoites in mice or in cell culture does not appear to dramatically alter their infectivity or morphological appearance.

B. STAGE AND SOURCE OF PARASITES

Tissue culture infections may be initiated with either sporozoites or merozoites and in the case of *T. gondii,* tachyzoites may also be used. Sporulated oocysts will not excyst in tissue culture and therefore do not initiate an infection. Sporozoites must be excysted *in vitro* from sterile sporulated oocysts while merozoites may be collected from the lumen of infected animals during an infection. *E. tenella* second generation merozoites may be collected from the lumen of the cecae on days 5 to 7. The obvious problem with a merozoite inoculum for cell culture is contamination with intestinal contents, blood, and the lack of sterility for tissue culture. Merozoites can be partially purified from the large particular debris by passage over a nylon wool column. To control bacterial growth, antibiotics such as penicillin/streptomycin or gentamicin are added. Fungal contamination can be controlled with fungizone (containing amphotericin B), although it has been observed that fungizone kills sporozoites of *E. tenella* at 10.0 μg/ml and reduces the number of sporozoites in culture after 24 h at 1.0 μg/ml.[35] Because of this antiparasitic or toxic effect, the amount of fungizone that may be used is limited and fungal contaminations become a major problem when attempting to use merozoites as inoculum for cell culture; therefore sporozoites are the preferred inoculum. However, in some cases the merozoites of various species of coccidia can be propagated successfully *in vitro* where the sporozoites cannot.

Sporozoites for the poultry coccidia may be isolated by initially breaking the oocyst wall to release the sporocysts. This is achieved by grinding the oocysts in a tissue homogenizer or by disrupting them with glass beads. This is not required for the mammalian coccidia where the oocyst can be opened by enzymatic treatment. In both cases the released sporocysts must be treated with bile components and trypsin to facilitate the release of the sporozoites. Once excysted the sporozoites are capable of infecting the tissue culture cells, although sporozoite preparations are always contaminated with debris from the excystation process. Since this debris is known to have a toxic effect on host cells, it is desirable to remove this material. In addition immunological and biochemical studies require pure sporozoites.

C. PURIFICATION OF SPOROZOITES

Many methods have been devised for cleaning up sporozoite preparations, although most have limitations in the purity and/or yield. One of the first methods described was the use of a glass bead column.[36] The recovery rate was generally poor and sporocysts and sporozoites could not be efficiently separated. Leukopac fiber columns have also been tried with some success yielding 80 to 95% of the sporozoites although 25% of the sporocysts were still present in the final sporozoite preparation.[37] DEAE-52 anion exchange columns have been used and have good yields and purity,[38] although the columns begin to clog with debris when large quantities of parasites are being purified. Another method which was described is centrifugal elutriation which gave relatively pure sporozoites but the yields were only around 50 to 60%.[39] In addition, this method was not suitable for routine use.

Other methods employing gradient centrifugation to purify sporozoites have also been described. Wisher and Rose[40] purified *E. tenella* sporozoites using a metrizamide gradient and while the purity of these preparations was good the recoveries were only around 60%. More recently, Dulski and Turner[41] described a method using a Percoll gradient. Good yields (90 to 95%) and purity were obtained by this method and the technique is rather simple for Percoll gradients do not have to be constructed since they automatically form during centrifugation. The method involves two steps: first, the separation of sporocysts from oocysts and oocyst hulls, and second, the separation of sporozoites from intact sporocysts and sporocyst shells. The method is simple, rapid, and the organisms remain fully viable. It is also interesting to note that the authors used a vital dye staining technique originally reported for other protozoan parasites to determine viability. The method reported by Jackson[42] utilizes the selective staining of dead parasites with ethidium bromide and live organisms with fluorescein diacetate. Under fluorescent microscopy, dead sporozoites appear red from absorbing ethidium bromide and live forms appear green from the metabolic processing of the fluorescein diacetate.

D. SIZE OF INOCULUM

The number of sporozoites used in the inoculum has a direct bearing on the percentage of intracellular parasites which subsequently develop. Many studies of intracellular parasites use the parasite to host cell ratio as a measure of the level of infection. Although this has some bearing on the development of the coccidia as well, it has been found that there is a limited number of sporozoites which can be present per milliliter of culture medium above which both the parasites and host cells are adversely affected. In studies using *E. tenella*, approximately 2×10^5 sporozoites/ml of culture medium is the limit. This number of sporozoites, regardless of the number of host cells, results in maximum development.[43] Levels below this will result in reduced levels of development, although in some cases this is preferred since it leaves more viable host cells available for later stages of the life cycle. The number of sporozoites in the inoculum should be chosen to give the desired results in terms of the length of culture and the stage(s) of the life cycle under study.

E. HOST CELL TYPE

Another important parameter is the choice of host cell. *T. gondii* tachyzoites can be propagated in almost any cell type; this is not the case for the *Eimeria* which usually require primary host cells with the exception of a few epithelial-like cell lines such as Madin-Darby Bovine Kidney (MDBK)[10] or embryonic bovine tracheal (EBTr) cells,[44] which will only support first generation schizogony. For the avian coccidia, primary chicken kidney cells or liver cells from 18- to 21-day embryos are commonly used, although kidney cells from 1- to 3-week-old chicks have also been used with similar success.[45] Attempts to propagate several species of coccidia in kidney cultures from quail, pheasant, partridge, bovine, ovine, porcine, and human sources have had some success.[46-48] It is interesting that many of the *Eimeria* will develop to first generation merozoites in MDBK or EBTr cells but not in any other cell lines. *E. tenella* merozoites from MDBK cultures can be used to infect chickens per cloaca confirming that they are viable. In contrast, second generation merozoites from chickens will not infect MDBK cells, suggesting that these cells are an inappropriate host for merozoite infections (Schmatz and Gnozzio, personal observation).

It would be preferable to use a cell line for culturing the *Eimeria*, although at present true avian epithelial cell lines are not available. Quist and Strout (personal observation) did achieve first generation schizogony of *E. tenella* in a transformed cell line of chicken macrophages, HD_{11}; but the development was limited and the cells detached from the substrate easily.

F. HOST CELL OVERGROWTH

The health of the host cells is also important for successful cultivation of intracellular parasites. Unfortunately, healthy monolayers contain dividing cells whether the culture is primary or transformed. This, combined with the relatively long life cycle of the coccidia, means that host cell overgrowth and the subsequent peeling of the monolayer will always be a problem. In the past subconfluent monolayers were used for the initial infection with sporozoites and became confluent during the culture period. This method has not always been successful since subconfluent monolayers have less target cells than a confluent monolayer and, in addition, it is well known that the percentage of intracellular sporozoites which develop in subconfluent monolayers is much lower than in confluent monolayers. This implies that confluent monolayers create a better microenvironment for the parasite. Another problem with subconfluent monolayers is the continuous diluting out of the percentage of infected cells.

One method which has been used to prevent host cell overgrowth is the reduction of the amount of serum in the culture medium. However, this has a negative effect on the host cells and therefore has had very limited success. Another approach has been to irradiate the host cell monolayer with sublethal levels of gamma irradiation prior to infecting,[49] a method originally developed for the intracellular cultivation of *Trypanosoma cruzi* and the exoerythrocytic stages of malaria.[50] An optimal dose of radiation must be determined for the host cell type used since different cell types are more sensitive to the treatment. A sublethal dose of 1000 rads inhibits replication of MDBK cells for at least 7 days while the cells remain fully capable of supporting the intracellular development of *E. tenella* to first generation merozoites. This method also works with primary chicken kidney cell cultures where *E. tenella* sporozoites will go the complete life cycle to oocysts (Pittarelli and Schmatz, unpublished).

Irradiated monolayers have consistently been shown to have much better levels of infection. There are several possible explanations for this phenomenon. It is known that radiation treatment causes host cell hypertrophy. The increased size of the cell makes them larger targets for parasite invasion and thus there are more intracellular sporozoites per host cell. In addition, the increased size of the cytoplasm may eliminate premature rupturing of the host cell due to the physical size of the developing meront. Also, it is known that irradiation blocks mitosis, causing the eventual accumulation of cells in the G2 phase of the cell cycle. G2 cells have been shown to be more susceptible to infection by *Toxoplasma gondii* and *T. cruzi*.[51]

Host cell division can also be inhibited by several drugs such as colcemid, vinblastine or ^{3}H-thymidine; however, these methods require the complete removal of the drug to prevent effects on the parasite, while pretreatment of the monolayer with gamma radiation is much cleaner and is specifically targeted at the host cells.

G. FOCAL DEVELOPMENT

A common observation when culturing most coccidia in cell culture is the small number of intracellular sporozoites which go on to schizogony. Often those parasites which do develop commonly appear in adjacent cells or the same cell. This focal development makes quantitation of parasitism difficult since one must examine a large number of microscopic fields to evaluate growth. One possible explanation for focal development is that the adjacent cells are a clonal population of more susceptible cells, i.e., fibroblast type cells in primary cultures will not support the growth of the *Eimeria* while epithelial cells will. If a more susceptible epithelial cell type exists, colonies of these cells may occur in focal clumps, and thus explain the patchy type of development seen with the *Eimeria*. Unfortunately the cloning of a more susceptible host cell from primary culture is difficult and impractical due to the limited life span of these cells. This is not the case with cell lines such as MDBK cells. To test this hypothesis, clones of MDBK cells were isolated and tested for their ability to support the intracellular development of *E. tenella*.[52] The results of this study suggested that certain clones were more susceptible to infection with sporozoites and were better hosts for the development of the intracellular parasites. One clonal line, MDBK/4, consistently had 2.5 times as many developing parasites per host cell as the parent MDBK cell line.[52] The ability to select more susceptible cells suggests that cultures may contain cells with varying degrees of susceptibility. Another possible explanation for focal development is the microenvironment created in the more confluent areas of the culture.

H. CHOICE OF CULTURE MEDIUM

The choice of culture medium is also important for successful cultivation of the coccidia. Early studies on the development of *E. tenella* were limited to medium that would support good growth of the host cells. The first nutrients used included Eagle's minimal essential medium (MEM) supplemented with 10% fetal or newborn bovine serum for the embryonic chicken kidney cell cultures and Eagle's Basal Medium (BME) supplemented with up to 10% bovine serum for the MDBK cells.[10] Using various combinations of culture media, Doran and Augustine showed the influence of medium on the extent of development of *E. tenella* in cell culture.[46] Certain combinations resulted in much better oocyst yields. This experiment led to further investigations into the effects of various nutrients on the development of *E. tenella* in culture.

Because of the obligate nature of the coccidia, it is difficult to study the specific nutrient requirements of the parasites themselves since any changes in the culture medium may have an effect on the parasite and/or the host cell. Doran and Augustine[53] conducted studies to determine the specific nutrients in medium 199 required for the development of *E. tenella* in chicken kidney cell cultures. By eliminating individual components from the culture medium, it was found that all of the vitamins in medium 199 were required for schizogony while gametogony did not require calciferol or folic acid. Latter and Holmes[54] examined the development of *E. tenella* in chick liver cells using six defined media without serum or supplements. They found that parasite development through schizogony and gametogony occurred with media NCTC 135, CMRL 1066, Waymouth's MAB 752/1, Waymouth's MAB 87/3, Ham's F12, and Newman and Tytell's serumless medium. Only schizogony was achieved in RPMI 1640 and MCDB 104 media. Latter and Holmes also conducted studies to determine the role of individual medium components and found that D-biotin, folic acid, *i*-inositol, thiamine HCl, vitamin E, menodione, paraaminobenzoic acid (pAbA), and vitamin B_{12} were important for first generation schizogony and oocyst formation. Second generation schizogony was reduced when D-biotin, *i*-inositol, thiamine HCl, menodione, vitamin B_{12}, and calciferol were eliminated from the medium.[54] Using a radiolabeled uracil-uridine assay to monitor the growth of the parasites and host cells, Schmatz[52] found that none of the above vitamins were required for the development of *E. tenella* through first generation schizogony in MDBK cells grown in Eagle's MEM. In fact, the elimination of the entire vitamin mixture had no negative effects on parasite development. It is important to keep in mind that this model is quite different from the ones used in previous studies since different host cells were used and the method of determining development was not visual.

Sodium pyruvate enhanced the development of *E. tenella* in MDBK cells but increasing the glucose content had no effect.[52] In another study, individual amino acids were eliminated from MDBK cultures to determine if the parasite growth could be affected. In many cases the amino acid depletions had adverse effects on both the parasite and host cell. However, the removal of arginine, L-glutamine, isoleucine, leucine, and tryptophan all had much more dramatic effects on the development of the intracellular parasites.[52] Sofield and Strout[55] also studied amino acid requirements in chick kidney cell cultures and reported that glutamine, isoleucine, methionine, tryptophan, and tyrosine were important for asexual development of *E. tenella*. It is interesting to note that both studies had similar results despite the very different host cell types which were used.

The finding that glutamine was important for parasite development is not unexpected since it plays a key role as a nitrogen donor in the biosynthesis of many metabolic intermediates. It is also important to note that L-glutamine is unstable in culture medium and has a half-life of 3 weeks at 4°C and 1 week or less at 36.5°C[56] and therefore should be added to culture medium immediately prior to use and added during long term culture.

It is difficult to determine exactly which components in the culture medium are critical when serum is used as one of the components. The use of dialyzed serum can eliminate low molecular weight components but the culture medium is still not fully defined. Sanford[57] described the development of *E. tenella* from sporozoites to oocysts in Waymouth's 87/3 medium without any serum or undefined supplements. This medium was originally developed to culture primary mouse bone cells and later liver cells.[58] It proved to be equally useful for culturing chicken kidney and liver cells. It is one of the least complex defined media in terms of its components and therefore appeared to be a good starting point for developing a better understanding of the nutrient requirements of the parasite. Sanford[57] evaluated the influence of eleven formulated medium components as well as two other supplements not in the formulation. Cell growth curves were compared to parasite development via microscopic evaluation over a 7-day period.

The study showed that increasing the standard levels of insulin, ascorbic acid, choline, thiamine, riboflavin, nicotinimide, and glutathione in the medium by twofold and supplementing with 2.3 m*M* sodium pyruvate (not a normal component) enhanced parasite development through the entire life cycle. Insulin is the only hormone present in the medium (8 mg/l) and its importance to parasite development *in vitro* was confirmed by Kyle and McDougald.[59] Removal of insulin from the culture medium resulted in cessation of host cell replication; and although the 2× level did not enhance cell proliferation over the standard level, it significantly enhanced the development of all stages of the parasite.

I. ENVIRONMENTAL CONDITIONS

Although the basal temperature of the chicken is 41°C and *E. tenella* grows optimally in cell culture at this temperature, it is recommended that all culturing be done at 40.5°C. This half degree of difference acts as a good safety margin for temperature fluctuations in the incubator since temperatures above 41°C are detrimental to parasite growth.[60] At temperatures lower than 40.5°C parasite development will occur at a slower rate. Below 36°C parasite development is minimal. The asexual development of *E. bovis* has been achieved at 39°C.[61] Although *T. gondii* is usually cultured between 37° and 39°C, the tachyzoites can be maintained at lower temperatures such as 22°C or 4°C and passed every few weeks to reduce the number of passages required to maintain the culture.[62]

Another important factor in cultivating the coccidia is the relative humidity of the incubator, although this is not an issue when a closed vessel is used. In open systems, evaporation can occur quickly and just a 10% loss of volume will alter the osmolarity of the medium to a point whre cultured conditions are dramatically changed and the host cells are damaged. Unless the incubator has an automatic humidity control, water must be supplemented in a pan or glass dish to achieve saturating levels of humidity; unfortunately this process enhances the opportunity for fungal contamination. The oxygen level is also a critical variable in maintaining the proper environment for cells in culture. It is essential for growth and metabolism of cells. Excessive levels of oxygen are detrimental to the host cells. The normal atmospheric levels of oxygen are 18.0 to 21.0% (131 ± 15 mmHg) depending on altitude. *In vivo* the average oxygen tensions are 30 to 50 mmHg in mixed venous blood, 95 mmHg (12.0%) in arterial blood, 20 to 40 mmHg in many tissues (as connective tissue), and 20 mmHg in gut mucosa.[63-66] In cell culture systems the oxygen concentration in the gas phase is in equilibrium with the oxygen tension in the culture medium. When this equilibrium is reached a gas phase of 18.0% oxygen results in tensions of 125 to 135 mmHg in the medium, a level which is much higher than the *in vivo* correlate. An oxygen level of 2.0 to 4.0% results in 15 to 30 mmHg, which most closely approximates that in the intestinal mucosa (20 mmHg).[63,65,66] It is also important to note that mammalian cells in culture apparently do not exhaust the available oxygen even after extended periods under low oxygen conditions.[65,67]

The influence of oxygen levels in the propagation of coccidia in cell cultures is apparent in the penetration of host cells and subsequent development. Kogut and Patton[68] found nearly twice as many sporozoites invaded host cells in a 4% oxygen environment as compared to atmospheric levels. Tilley and Upton[69] demonstrated the significance of a low oxygen environment in the *in vitro* development of *Eimeria nieschulzi*. *E. nieschulzi* sporozoites under standard culture conditions (5% CO_2, 95% air) developed through first generation schizogony with little development beyond this stage, while cultures incubated in the low oxygen environment of candle jars, developed through third and fourth generation schizogony. There was also a 50-fold increase in the number of developing parasites in the reduced oxygen cultures. In a more detailed study, Chandler-Conroy[70] determined the influence of reduced oxygen on the host cells and the development of *E. tenella* in primary chick kidney cells. The gases were custom mixed and contained 5% CO_2 with 0.0, 0.1, 1.0 or 5.0% oxygen with the balance being nitrogen. Cultures were incubated in modular chambers flushed with the appropriate gas mixture. Reduced oxygen resulted in a decrease in host cell division, although the cells appeared morphologically normal and actually lasted longer in culture compared to cultures incubated at atmospheric levels of oxygen. Total parasitemia and subsequent development increased significantly in cultures with reduced oxygen levels. Schizogony increased by over 60% in the cultures with 0.1% oxygen and development was only slightly retarded (11% of control) in the anaerobic cultures. There was also a twofold increase in gametogony in reduced oxygen cultures. In summary, reduced oxygen levels appear to enhance parasite growth and to reduce host cell division and subsequent peeling due to overgrowth. This environment may more closely mimic the natural environment of the parasite. Not only because the gut is basically anaerobic but, in addition in many phases of the intracellular life of the parasite, the infected cell is starting to die from the infection and subsequently becomes less oxygen rich.

This area of cell culture needs much more exploration since previous to the studies mentioned here, the models used for propagating parasites within cultured cells have been biased by using methods developed for culturing the host cells with little regard for the atmospheric conditions required for the parasites.

IV. QUANTITATION OF PARASITE GROWTH

Anyone who has worked with intracellular parasites for a number of years can relate to the tedium associated with quantitating parasite growth. Millions of hours have been spent by parasitologists around the world determining the amount of parasite growth in various culture systems under various conditions. In the past few years many laboratories have been replacing visual assessment of parasite growth with the incorporation of radiolabeled nucleic acid precursors by the intracellular parasite for quantitating growth. The most common labeled precursor for the coccidia is ^{3}H-uracil. Both the *T. gondii* and the *Eimeria* have the enzyme uracil phosphoribosyl transferase and are capable of salvaging this pyrimidine. The host cells lack this enzyme and therefore only incorporate background levels of ^{3}H-uracil. This method was originally introduced in studies by Pfefferkorn with *T. gondii*[71] and has more recently been used to assess the intracellular growth of *Eimeria tenella*[43] and *Eimeria bovis*.[26] Quantitation via the incorporation of radiolabeled uracil not only eliminates the need for visual assessment but also gives a true measurement of growth in cases of asynchronous development. Since the *Eimeria* develop asynchronously, developing schizonts are variable in size, making it difficult to quantitatively evaluate drug effects. In contrast, the amount of labeled uracil incorporated directly correlates to the rate at which the overall infection is developing relative to labeled control cultures.

Uracil incorporation can be detected within the first few hours after the parasites become intracellular while significant morphological change is not apparent until 36 to 48 h later.[43] Therefore, parasite growth can be measured without waiting for a morphological change. The lack of incorporation of labeled uracil by extracellular sporozoites followed by incorporation immediately after penetrating the host cell suggests that the parasite has been stimulated to produce nucleic acids or has become more permeable to uracil once intracellular. For large-scale screening purposes the uracil assay can be run in microtiter plates and toxic effects of drug treatments on host cells can be determined using ^{3}H-uridine incorporation to measure the growth of normal host cell cultures.[43] Although the use of radiolabeled precursors eliminates the need to visually inspect each experimental group, it is always important to confirm interesting results in a visual assay since there is the remote possibility that an observed effect is a result of altered incorporation of the precursor and is unrelated to growth.

REFERENCES

1. **Doran, D. J.,** Cultivation of Coccidia in avian embryos and cell cultures, in *The Coccidia: Eimeria, Isospora, Toxoplasma and Related Genera,* Hammond, D. M. and Long, P. L., Eds., University Park Press, Baltimore, 1973, 183.
2. **Doran, D. J.,** Behavior of Coccidia in vitro, in *The Biology of the Coccidia,* Long, P. L., Ed., University Park Press, Baltimore, 1982, 229.
3. **Long, P. L.,** The growth of embryo-adapted strains of *Eimeria* in cell cultures, in *Proc. 3rd Int. Congr. of Parasitol.,* Munich, 1, 121, 1974.
4. **Ryley, J. F. and Wilson, R. G.,** The development of *Eimeria* brunetti in tissue culture, *J. Parasitol.,* 58, 660, 1972.
5. **Doran, D. J.,** Survival and development of 5 species of chicken coccidia in primary chicken kidney cell cultures, *J. Parasitol.,* 57, 1135, 1971.
6. **Naciri-Bontemps, M.,** Reproduction of the cycle of coccidia *Eimeria acervulina* (Tyzzer, 1929) in cell cultures of chicken kidneys, *Ann. Rech.,* 7, 223, 1976.
7. **Doran, D. J.,** Cultivation and freezing of poultry coccidia, *Acta Vet. (Brno.),* 38, 25, 1969.
8. **Doran, D. J. and Augustine, P. C.,** *Eimeria dispersa* and *Eimeria gallopavonis:* Infectivity, survival, and development in primary chicken and turkey kidney cell cultures, *J. Protozool.,* 24, 172, 1977.
9. **Augustine, P. C. and Doran, D. J.,** Development of *Eimeria meleagrimitis* Tyzzer from sporozoites and merozoites in turkey kidney cell cultures, *J. Protozool.,* 25, 82, 1978.
10. **Patton, W. H.,** *Eimeria tenella:* cultivation of the asexual stages in cultured animal cells, *Science,* 150, 767, 1965.
11. **Fayer, R.,** Gametogony of *Sarcocystis sp.* in cell culture, *Science,* 175, 65, 1972.
12. **McDougald, L. R. and Jeffers, T. K.,** *Eimeria tenella* (Sporozoa, Coccidia): gametogony following a single asexual generation, *Science,* 192, 258, 1976.
13. **Fayer, R. and McLoughlin, D. K.,** Effect of amprolium on *Eimeria adenoides* in cell culture, *J. Parasitol.,* 56, 388, 1970.
14. **Strout, R. G. and Ouellette, C. A.,** *Eimeria tenella:* screening of chemotherapeutic compounds in cell cultures, *Exp. Parasitol.,* 33, 477, 1973.
15. **Ryley, J. F. and Wilson, R. G.,** Studies on the mode of action of the coccidiostat robenidine, *Z. Parasitenk.,* 37, 85, 1971.
16. **McDougald, L. R. and Galloway, R. B.,** *Eimeria tenella:* anticoccidial drug activity in cell cultures, *Exp. Parasitol.,* 34, 189, 1973.
17. **Wilson, R. G.,** Studies on the short term application of drugs to *Eimeria tenella* in tissue culture, *Parasitology,* 73, 283, 1976.
18. **Ryley, J. F. and Wilson, R. G.,** Comparative studies with anticoccidials and three species of chicken coccidia in vivo and in vitro, *J. Parasitol.,* 58, 664, 1972.
19. **Ryley, J. F. and Wilson, R. G.,** Drug screening in cell culture for the detection of anticoccidial activity, *Parasitology,* 73, 137, 1976.

20. **Roberts, C. O., Chaparas, S. D., and McLaughlin, D.,** The use of the plaque assay in chemotherapeutic and dermal hypersensitivity studies on *Toxoplasma gondii, Trans. Am. Microsc. Soc.*, 95, 470, 1976.
21. **Pfeiffer, J. R.,** Effects of Monensin on the Ultrastructure of First Generation Development of *Eimeria tenella* (Coccidia) Cultured In Vitro, Master of Science thesis, Department of Zoology, University of New Hampshire, Durham, N. H., 1973.
22. **Boothby, J. M.,** Evaluation of the Ionophorous Antibiotic, Lasalocid, against *Eimeria tenella* (Coccidia) in Cell Culture, Master of Science thesis, Department of Animal Science, University of New Hampshire, Durham, N. H., 1975.
23. **Smith, C. K., II and Strout, R. G.,** *Eimeria tenella:* accumulation and retention of anticoccidial ionophores by extracellular sporozoites, *Exp. Parasitol.*, 48, 325, 1979.
24. **Augustine, P. C. and Danforth, H. D.,** Use of monoclonal antibodies to locate Eimeria sporozoites (Protozoa) in intestinal sections, *Proc. Helminthol. Soc. Wash.*, 51, 361, 1984.
25. **Pfefferkorn, E. R.,** Interferon blocks the growth of *Toxoplasma gondii* in human fibroblasts by inducing the host cell to degrade tryptophan, *Proc. Natl. Acad. Sci. U.S.A.*, 81, 908, 1984.
26. **Hughes, H. P. A., Speer, C. A., Kyle, J. E., and Dubey, J. P.,** Activation of murine macrophages and a bovine monocyte cell line by bovine lymphokines to kill the intracellular pathogens *Eimeria bovis* and *Toxoplasma gondii., Inf. Immun.*, 55, 784, 1987.
27. **Kogut, M. and Lange, C.,** Effect of lymphokine treatment on the invasion of cultured animal cells by *Eimeria tenella, Interferon Res.*, 1, 31, 1988.
28. **Danforth, H. D.,** Use of monoclonal antibodies directed against *Eimeria tenella* sporozoites to determine stage specificity and in vitro effect on parasite penetration and development, *Am. J. Vet. Res.*, 44, 1722, 1983.
29. **Augustine, P. C. and Danforth, H. D.,** Effect of lectins on penetration of cultured cells by *Eimeria meleagrimitis* sporozoites, *Proc. Annu. Meet. Am. Soc. Parasitol.*, 151, 1981.
30. **Crane, M. S. J. and Dvorak, J. A.,** Influence of monosaccharides on the infection of vertebrate cells by *Trypanosoma cruzi* and *Toxoplasma gondii, Mol. Biochem. Parasitol.*, 5, 333, 1982.
31. **Augustine, P. C.,** Effect of polyions, calcium, and enzymes on penetration of cultured cells by *Eimeria meleagrimitis* sporozoites, *J. Parasitol.*, 66, 498, 1980.
32. **Augustine, P. C.,** *Eimeria meleagrimitis* sporozoites: effect of lectins on invasion of cultured cells, *Poult. Sci.*, 64, 2296, 1985.
33. **Saffer, L. D., Long Krug, S. A., and Schwartzman, J. D.,** The role of phospholipase in host cell penetration by *Toxoplasma gondii, Am. J. Trop. Med. Hyg.*, 40, 145, 1989.
34. **Doran, D. J., Vetterling, J. M., and Augustine, P. C.,** *Eimeria tenella:* an *in vivo* and an *in vitro* comparison of the Wisconsin, Weybridge, and Beltsville strains, *Proc. Helminthol. Soc. Wash.*, 41, 77, 1974.
35. **Bedrnik, P.,** The sensitivity of *Eimeria tenella* sporozoites, maintained in tissue culture, to fungizone, *Acta Vet. (Brno.)*, 39, 473, 1970.
36. **Wagenbach, G. E.,** Purification of *Eimeria tenella* sporozoites with glass bead columns, *J. Parasitol.*, 55, 833, 1969.
37. **Bontemps, M. and Yvore, P.,** Technique de purification des suspensions de sporozoites d'*Eimeria* sur colonne des fibres synthetiques, *Ann. Rech. Vet.*, 5, 109, 1974.
38. **Schmatz, D. M., Crane, M. S. J., and Murray, P. K.,** Purification of *Eimeria* sporozoites by DE-52 anion exchange chromatography, *J. Protozool.*, 31, 181, 1984.
39. **Stotish, R. L., Simashkevich, P. M., and Wang, C. C.,** Separation of sporozoites, sporocysts, and oocysts of *Eimeria tenella* by centrifugal elutriation, *J. Parasitol.*, 63, 1124, 1977.
40. **Wisher, M. H. and Rose, M. E.,** The large scale preparation of purified sporozoites of *Eimeria* spp. by metrizamide density gradient centrifugation, *Parasitology*, 88, 515, 1984.
41. **Dulski, P. and Turner, M.,** The purification of sporocysts and sporozoites from *Eimeria tenella* oocysts using Percoll density gradients, *Avian Dis.*, 32, 235, 1988.
42. **Jackson, P. R., Pappes, M. G., and Hansen, B. D.,** Fluorogenic substrate detection of viable intracellular and extracellular pathogenic protozoa, *Science*, 227, 435, 1985.
43. **Schmatz, D. M., Crane, M. S. J., and Murray, P. K.,** *Eimeria tenella:* parasite specific incorporation of ^{3}H-Uracil as a quantitative measure of intracellular development, *J. Protozool.*, 33, 109, 1986.
44. **Matsuoka, T., Callender, M. E., and Shumarel, R. F.,** Embryonic bovine tracheal cell line for *in vitro* cultivation of *Eimeria tenella, Am. J. Vet. Res.*, 30, 1119, 1969.
45. **Doran, D. J.,** Increasing the yield of *Eimeria tenella* oocysts in cell culture, *J. Parasitol.*, 57, 891, 1971.
46. **Doran, D. J. and Augustine, P.C.,** Comparative development of *Eimeria tenella* from sporozoites to oocysts in primary kidney cell cultures from gallinaceous birds, *J. Protozool.*, 20, 658, 1973.
47. **Doran, D. J. and Vetterling, J. M.,** Survival and development of *Eimeria meleagrimitis* Tyzzer, 1929, in bovine kidney and turkey intestine cell cultures, *J. Protozool.*, 15, 796, 1968.
48. **Shibalova, T. A.,** Cultivation of endogenous stages of chicken coccidia in embryos and tissue cultures, *J. Parasitol.*, 56 (4), 315, 1970.

49. **Crane, M. S. J., Schmatz, D. M., Stevens, S., Habbersett, M. C., and Murray, P. K.,** *Eimeria tenella:* in vitro development in irradiated bovine kidney cells, *Parasitology,* 88, 521, 1984.
50. **Schmatz, D. M. and Murray, P. K.,** Cultivation of *Trypanosoma cruzi* in irradiated muscle cells: improved synchronization and enhanced trypomastigote production, *Parasitology,* 85, 115, 1982.
51. **Dvorak, J. A. and Crane, M. S. J.,** Vertebrate cell cycle modulates infection by protozoan parasites, *Science,* 214, 1034, 1981.
52. **Schmatz, D. M.,** In vitro cultivation of coccidia, in *Advances in Cell Culture,* Maramorosch, K., Ed., Academic Press, New York, 5, 241, 1987.
53. **Doran, D. J. and Augustine, P. C.,** *Eimeria tenella:* vitamin requirements for development in primary cultures of chicken kidney cells, *J. Protozool.,* 25, 544, 1978.
54. **Latter, V. S. and Holmes, L. S.,** Identification of some nutrient requirements for the *in vitro* cultivation of *Eimeria tenella, Parasitology,* 81, Proc. XIX, 1980.
55. **Sofield, W. L. and Strout, R.,** Amino acids essential for *in vitro* cultivation of *Eimeria tenella, J. Protozool.,* 21, (Suppl.), 434, 1974.
56. **Freshney, R. I.,** *Culture of Animal Cells — A Manual of Basic Technique,* Freshney, R. I., Ed., Alan R. Liss, New York, 91, 1983.
57. **Sanford, J. C.,** Nutritional Requirements of *Eimeria tenella* (Coccidia) Propagated in Cell Culture Using a Chemically Defined Medium, Master of Science thesis, University of New Hampshire, Durham, 1983.
58. **Waymouth, C.,** Preparation of Medium MAB 87/3 for primary cultures of epithelial cells, *Tissue Culture Association Manual,* 3, 521, Procedure No. 35781, 1976.
59. **Kyle, D. E. and McDougald, L. R.,** Effect of insulin and pancreatic polypeptide on *in vitro* development of *Eimeria tenella, Proc. Annu. Meet. Am. Soc. Parasitol.,* 48, 1985.
60. **Strout, R. G., Ouellette, C. A., and Gangi, D. P.,** *Eimeria tenella:* temperature and asexual development in cell culture, *Exp. Parasitol.,* 25, 324, 1969.
61. **Fayer, R. and D. M. Hammond,** Development of first-generation schizonts of *Eimeria bovis* in cultured bovine cells, *J. Protozool.,* 14, 764, 1967.
62. **Akinsha, G. T.,** Prolonged preservation of *Toxoplasma* in tissue cultures, *Byull. Eksp. Biol. Med.,* 58, 98, 1964.
63. **Richter, A., Sanford, K., and Evans, U.,** Influence of oxygen and culture media on plating efficiency of some mammalian tissue cells, *J. Natl. Cancer Inst.,* 49, 1075, 1972.
64. **Balin, A. K., Goodman, D., Rasmussen, H., and Cristofalo, V. J.,** Atmospheric stability in cell culture vessels, *In Vitro,* 12, 687, 1976.
65. **Taylor, W. G., Camalier, R., and Sanford, K.,** Density dependent effects of oxygen on the growth of mammalian fibroblasts in culture, *J. Cell Physiol.,* 95, 33, 1978.
66. **Taylor, W. G. and Camalier, R.,** Modulation of epithelial cell proliferation in culture by dissolved oxygen, *J. Cell Physiol.,* 111, 21, 1982.
67. **Balin, A. K., Goodman, D., Rasmussen, H., and Cristofalo, V. J.,** Effect of oxygen tension on the growth and metabolism of WI-38 cells, *J. Cell Physiol.,* 89, 235, 1976.
68. **Kogut, M. and Patton, W.,** Effects of a low oxygen atmosphere on the asexual development of *Eimeria tenella* in a mammalian cell line, *Am. Soc. of Parasitol. 54th Annu. Meet.,* abstr. #113, 1979.
69. **Tilley, M. and Upton, S. J.,** A comparative study on the development of *Eimeria nieshulzi in vitro* under aerobic and reducing conditions, *J. Parasitol.,* 74, 1042, 1988.
70. **Chandler-Conroy, N.,** *Eimeria tenella:* Effect of Oxygen Concentration on Parasite Development in Cell Culture, Master of Science thesis, University of New Hampshire, Durham, 1987.
71. **Pfefferkorn, E. R. and Pfefferkorn, L. C.,** Specific labeling of intracellular *Toxoplasma gondii* with uracil, *J. Protozool.,* 24, 449, 1977.

Chapter 12

PATHOLOGY OF COCCIDIAL INFECTIONS

Martin W. Gregory

TABLE OF CONTENTS

I. OVERVIEW

Coccidiosis usually occurs only if an animal is subjected to heavy infection when its resistance is low. Low resistance can be related to age, immune status, nutrition, or stress. Signs of intestinal coccidiosis usually include diarrhea, with or without blood loss. Lesions are caused primarily by inflammation, epithelial hyperplasia, and epithelial loss. They include villus atrophy, mucosal thickening, polyps, crypt atrophy, denudation, and hemorrhage. The most pathogenic species tend to be those that destroy crypt stem cells in the large intestine, causing denudation and hemorrhage.

II. INTRODUCTION

A successful parasite is one that infects every available host while causing minimal harm. The success of coccidia is spectacular: they probably infect every wild individual in every vertebrate species, and under "natural" conditions they cause no disease. It seems like a perfect example of peaceful co-existence.

But this harmony is precarious: a slight change can upset the balance, such that the coccidia multiply too much, and cause too much damage to the host. In other words, they cause coccidiosis.* I shall deal first with the factors that tip the balance, then with the lesions and their pathogenesis, using the sheep as a basis for comparison. The last section will summarize the salient points of the most important coccidioses.

III. FACTORS THAT DETERMINE THE OCCURRENCE OF COCCIDIOSIS

In many species (the horse, for example) coccidiosis occurs rarely, if ever. In some species (e.g., man) it occurs mainly in association with immune deficiency due to disease or stress, while in others (e.g., poultry, sheep) it occurs mainly in association with intensive

* In this chapter "coccidiosis" means "clinical coccidiosis". Generally speaking, if an infectious organism is present without causing clinical signs, the infection is said to be "subclinical". This implies that the organism's presence is abnormal. In the case of coccidia, the "normal" situation, (except perhaps in man) is that virtually every host is infected. Most "healthy" animals can therefore be said to have "subclinical coccidiosis", so this term is not very useful.

husbandry. Many factors probably contribute in every case, and I shall try to simplify the picture as far as current knowledge allows.

A. PARASITE SPECIES

Some species are more pathogenic than others. For example, 1000 oocysts of *Eimeria flavescens* in a susceptible rabbit will cause more severe weight loss than 1,000,000 *E. irresidua.*[1] The reasons for the differences will be partly explained under Pathogenesis.

B. INFECTING DOSE

In most coccidia/host relationships, the pathogenic effect is proportional to the number of infective oocysts ingested.[1]** As the infecting dose increases, the pathogenic effect continues to increase, unless immune mechanisms come into play. With *E. crandallis* the pathogenic effect was found to be proportional to the infecting dose with very large or with small inocula but not in the middle range of doses.[2] This may be related to the unusual life-cycle of *E. crandallis.*[3]

C. RATE OF INFECTION

The pathogenic effect of a given number of oocysts can be much less when given in small doses over several days or weeks than when given as a single dose.[2]

D. INTERACTION BETWEEN SPECIES OF COCCIDIUM

Little work has been done on this subject — a fact lamented by Marquardt.[4] It is clear, however, that interaction can be very marked. If two coccidial species affect the same part of the gut, (e.g., *E. brunetti* and *E. mitis*)[5] they appear to compete, and their combined effect is not more than the effect of only one species. If two species of coccidium affect different parts of the gut, (e.g., *E. mitis* and *E. acervulina*),[5] their combined effect can be more than that of a double dose of either species. *E. piriformis* is the only species that affects the colon and rectum of the rabbit, and it aggravates the effects of any other rabbit coccidium.[6]

E. INTERACTION WITH OTHER PATHOGENS

Coccidiosis is a disease of young animals, and it therefore tends to coincide with stress-related diseases. Such diseases are multifactorial, involving whatever opportunist pathogens happen to be around at the time, including viruses, bacteria, and fungi. The pathogenic roles of enteric bacteria are very complex even in the absence of coccidia.[7] Studies on interactions have been made in chickens[8-10] and lambs.[11] Parvovirus is thought to aggravate coccidiosis in calves, and/or vice versa.[12,13] This is probably partly because parvovirus attacks the same cells as the most pathogenic bovine coccidia: the stem cells of the crypts. Immunosuppression may also be involved. Helminths apparently interact as well.[14]

F. AGE

Newborn rabbits are resistant to coccidial infection.[15] In sheep, susceptibility to a combination of *E. ovinoidalis* and *E. crandallis* infection was found to increase progressively up to at least 4 weeks of age.[16] Resistance correlated well with antibodies acquired via the colostrum. However, Catchpole and Devonshire[17] showed that even colostrum-deprived lambs were resistant to the pathogenic effects of 10,000 *E. ovinoidalis* oocysts for the first week of life. Gouet et al.[11] managed to infect newborn lambs using heavy doses of this organism (150,000), but only in lambs that had been inoculated with normal gut flora.

** In very heavy infections there may be a "crowding" effect, which will result in a relative reduction in *oocyst output*. But here we are considering the *pathogenic effect*, to which the oocyst output is irrelevant.

Axenic lambs were relatively resistant. In lambs there is therefore an increase in susceptibility with age that is affected by microflora.

G. IMMUNE STATUS

1. Passive Immunity

As mentioned in the preceding paragraph, lambs acquire anticoccidial antibodies via the colostrum, and the resistance of field lambs to coccidiosis correlates well with such antibodies.[16] However, there is no direct evidence that these circulating antibodies are involved in parasite rejection.[17]

2. Active Immunity

Adult animals are normally highly resistant to coccidiosis, as a result of active immunity. Some coccidia stimulate the rapid development of solid immunity, while others require repeated infections. Coccidia can cause severe intractable diarrhea in humans with acquired immune deficiency syndrome (AIDS).[18,19] Coccidiosis occurs in other species in which the cell-mediated immune response is impaired for any reason. One such reason can be stress.

H. STRESS

An animal's resistance to coccidial infection can be reduced by adverse conditions[20,21] such as dietary changes,[12,13,22-24] prolonged travel,[25,26] cold weather,[27] hot dry weather,[12,28] changes in social environment[29] or severe concurrent infection.[30] Dubey[31] found that *Sarcocystis* dramatically increased the susceptibility of goats to intestinal coccidiosis caused by various *Eimeria* species. In calves, coccidiosis is a problem in very cold weather.[27,32] It also occurs in feedlots where calves or lambs arrive after a long journey in which they may have suffered from heat, cold, hunger, or thirst.[26,33] In Africa[34] and Australia[12,22] bovine coccidiosis is sometimes seen in hot dry weather. Dogs may develop coccidiosis due to *Isospora canis* when subjected to weaning stress.[24] Even adult sheep and goats have been known to suffer severe outbreaks of coccidiosis, with high mortality.[25,36-38]

I. NUTRITION

In Great Britain, twin lambs suffer more from coccidiosis than do singles, and triplets suffer even more than twins. This is probably a reflection of the nourishment received by the lamb, and may in turn be a reflection of the ability of the gut to repair damage. One of the effects of monensin was found to be an elevation of selenium levels[39] which could also be a factor in resistance to coccidial infection. Selenium helps the immune response to *E. tenella* of chickens.[40]

IV. CLINICAL SIGNS AND SYMPTOMS

Humans infected with *Isospora belli* have complained of fever, headache, abdominal pain,and diarrhea.[41] In other animals the clinical signs of intestinal coccidiosis are very variable. They may be barely detectable, or there may be acute bloody diarrhea, chronic diarrhea and emaciation, or even sudden death. The severity depends on many of the factors discussed. When partially immune animals are reinfected, they may show little improvement in terms of diarrhea and lesions, but a considerable improvement in terms of body weight and mortality, in comparison with nonimmune controls.[16,42]

Diarrhea is caused by failure to absorb water and electrolytes. It does not always occur,[6] and it tends to be more severe if the large intestine is affected. Blood in the feces may be fresh and clotted (if the hemorrhage is in the large intestine), or dark and partly digested (if it is from the small intestine). There may also be loss of fibrin and other proteins into the gut lumen, and these may appear in the form of whitish streaks or as gelatinous casts.

The changes in the feces are accompanied by soiling of the perineum, and various degrees of inappetence, abdominal discomfort, and weight loss, sometimes death. Nervous signs sometimes occur in calves.[43] Hepatic coccidiosis in rabbits causes inappetence, emaciation, jaundice, and constipation.[44] Renal coccidiosis in geese causes diarrhea, weakness, ataxia, and emaciation.[45]

V. PATHOGENESIS

In 1929, Jeannette Chapman[46] wrote "In spite of the great amount of work that has been done on the subject . . . the mechanism of the pathologic effect of the coccidia is still not entirely clear". This is still true, but it is now clear that several mechanisms are involved.

When the balance is upset, too many parasites multiply and the host is unable to control them. Two damaging factors come into play: the parasites themselves, and the host reaction to them. A severe host reaction may do more damage than the parasites themselves, and it is not always easy to say what is caused by which. Both cause structural changes, which impair function, leading to physiological changes. Ruff and Allen (Chapter 13) will deal with the physiological changes; I shall deal with the structural changes themselves.

Some of the host reactions are general responses to interference with host tissues, others are specific responses to coccidia. Again, it is not always clear which is which. I shall describe the structural changes that can occur; then I shall relate them where possible to particular parasite stages, and finally, I shall relate them to particular coccidial infections. First some words about the site of lesions and their extent. I shall view these from the point of view of the parasite, as if it were making rational decisions. However unrealistic this approach is, it can help us to understand the reasons for the adaptations we see.

A. THE SITE OF INFECTION

The small intestine is the most popular site for coccidia. In ruminants the organ is very long, providing a large number of enterocytes and allowing enormous multiplication of the parasite, usually without too much damage being caused. If absorption is impaired, the large intestine can compensate to some extent. Coccidia that infect the large intestine tend to be more pathogenic than those that infect the small intestine. This may be partly because the rate of cell turnover is lower, partly because if the large intestine is damaged, there is no absorptive gut lower down to compensate, and partly because in the large intestine there tend to be more opportunist organisms ready to exploit a damaged mucosa. *E. stiedai* chooses to infect the bile duct epithelium of rabbits, where there are no other organisms, and no danger of loss of absorptive function. The disadvantage is that the cell turnover rate is very low. But this rises dramatically in response to the parasite, so there is no shortage of host cells, no ulceration occurs, and pathogenic effects are minor unless the infection is very heavy.[44] *E. truncata* affects the kidney tubules of geese. There appears to be no proliferation of host cells, so the damage caused can be severe.[45] Giant meronts (referred to as *E. gilruthi*) are often found in the abomasum of ruminants. They seem to cause little harm. Occasionally giant meronts are seen in the mesenteric lymph nodes of goats[47,48] and lambs,[48,49] but their pathological significance is slight. No doubt they arise from sporozoites that arrive there accidentally. Oocysts have also been seen in lymph nodes.[50]

B. THE CELLS INFECTED

The disadvantage of enterocytes, from the coccidium's point of view, is that the turnover is so rapid that the parasite has little time to exploit the cells before their lifespan is completed. Most coccidia therefore aim to infect cells low down in the crypts. Only those coccidia that produce very small meronts can afford to infect cells that are already on the villus — e.g., *E. acervulina*, *E. bakuensis* (second meronts), and *Isospora suis*. Such species produce lesions that heal quickly because the stem cells in the crypts are undamaged. Coccidia that infect the crypts run the risk of damaging the stem cells. Some infect the stem cells without killing them (e.g., *E. bakuensis*,[51] *E. crandallis*[3]). Some kill the stem cells (e.g., *E. ovinoidalis*,[50] *E. flavescens*[52]). If all the stem cells in an area are killed, the epithelium cannot be replaced except by encroachment from undamaged areas. Some coccidial stages avoid the problem by causing their host cell to migrate from the epithelium into the lamina propria (e.g., large meronts of *E. tenella*,[53] *E. necatrix*,[53] and *E. crandallis*).[3]

C. THE EXTENT OF THE LESIONS

Some stages are released into the gut lumen, in which they tend to spread widely, leading to diffuse lesions (e.g., *E. crandallis* first generation merozoites).[3] Other stages tend to invade locally when released — these give rise to "colonies" as mentioned by Tyzzer et al.[54] (e.g., *E. bakuensis, E. christenseni, E. tenella, E. necatrix, E. acervulina* merozoites, and *E. flavescens* third stage merozoites[1]). In these infections each colony represents the site of a first generation meront, which in turn represents one sporozoite, or (in the case of *E. flavescens*) one merozoite.

D. HISTOPATHOLOGY

I shall first offer an overview of the main processes involved, then I shall relate these to particular stages and, in the last section, to particular coccidial species.

1. Vascular Changes

Hyperemia, edema, and hemorrhage are visible grossly, and give a good indication of the severity of the damage. They may be part of a general inflammatory response, which may include a response to opportunist bacteria. Products of mucosal mast cells may increase permeability of endothelia and epithelia, leading to edema and increased fluid loss. In chickens, the most severe intestinal lesions are associated with hemorrhage caused by meronts in the lamina propria. How they cause the hemorrhage is still unclear.[55,56] In mammals, hemorrhage can also be associated with meronts,[50] but is usually associated with gamonts and denudation of the large intestine.[50,57]

2. Cellular Infiltration

Neutrophils, eosinophils, heterophils, lymphocytes, and macrophages tend to accumulate in any damaged tissues as part of the inflammatory response. Mucosal mast cells and globule leukocytes[58] may also increase in number.[59,60] Neutrophils can be very numerous where the epithelial covering is broken, and particularly if there is invasion by bacteria. They may also contribute to the cell destruction.[61] Macrophages play a part in the destruction of first generation meronts,[50] (Figure 1) as do neutrophils and eosinophils. Multinucleate giant cells engulf oocysts in the tissues.[52] Macrophage-like cells are sometimes seen at the bases of coccidial polyps, where they are associated with degeneration of infected crypts (Figure 2).[51] They are also associated with crypt degeneration in the absence of polyp formation.[50]

3. Epithelial Hyperplasia

All coccidial infections incur damage to epithelia, to which the normal response is hyperplasia (cell proliferation). In most cases it is only the uninfected cells that proliferate, but in some coccidial infections the infected cells continue to divide. With *E. crandallis* and with polyp-forming coccidia, the pro-gamont stage may divide in synchrony with the host cell that it occupies.[62] With *E. stiedai*, infected cells may divide,[63] but synchronous division has not been reported.

Intestinal epithelial cells (enterocytes) arise from stem cells near the bases of the crypts of Lieberkühn. Their increased proliferation results in lengthening of the crypts (Figure 7)

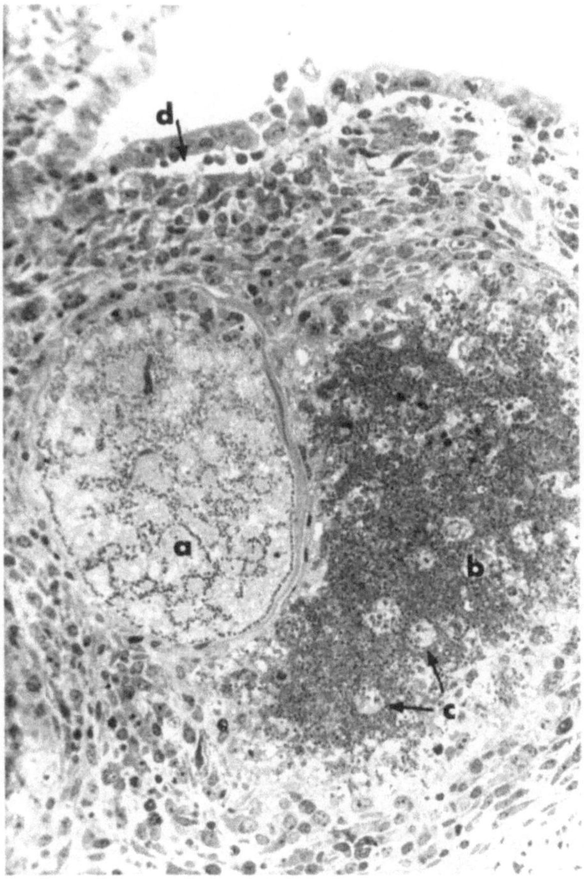

FIGURE 1. Two first generation meronts of *Eimeria crandallis* in the small intestinal mucosa of a lamb. The one on the left (a) is immature and apparently intact. The one on the right (b) is almost mature; the host cell has disintegrated, and the meront has been invaded by macrophages (c) that are mopping up merozoites. The surrounding lamina propria is heavily infiltrated by inflammatory cells, particularly macrophages. The surface epithelium is separated from the lamina propria by a bleb (d) of fluid, containing lymphocytes. Above the bleb, enterocytes are separating from one another and sloughing away. (1 μm methacrylate, toluidine blue (TB).

and (in the small intestine) a decrease in the villus/crypt ratio.[64] At first it was thought that crypt hyperplasia was purely a response to epithelial cell loss and villus atrophy, but Ferguson[65] showed that, in some cases at least, it preceded villus atrophy. Ferguson and Jarrett[66] also showed that these mucosal changes could be thymus dependent. Crypt hyperplasia is usually accompanied by a rise in the rate of cell-turnover,[64] which in turn means immaturity of surface epithelial cells, and impairment of their absorptive function. This can be a cause of diarrhea (Figure 19).

Hyperplasia is either focal or diffuse. Focal hyperplasia causes the formation of "oocyst patches" (Figures 3 and 4) and polyps (Figures 5 and 6), as in *E. bakuensis, E. ahsata* (sheep), *E. christenseni,* and *E. arloingi* (goats). Such focal lesions are thought to result from release of merozoites within the lamina propria rather than into the intestinal lumen.[51] The mechanism of polyp formation is not well understood. They have been found in natural infections of *E. bakuensis,* but in the laboratory, this organism caused only oocyst patches.[51]

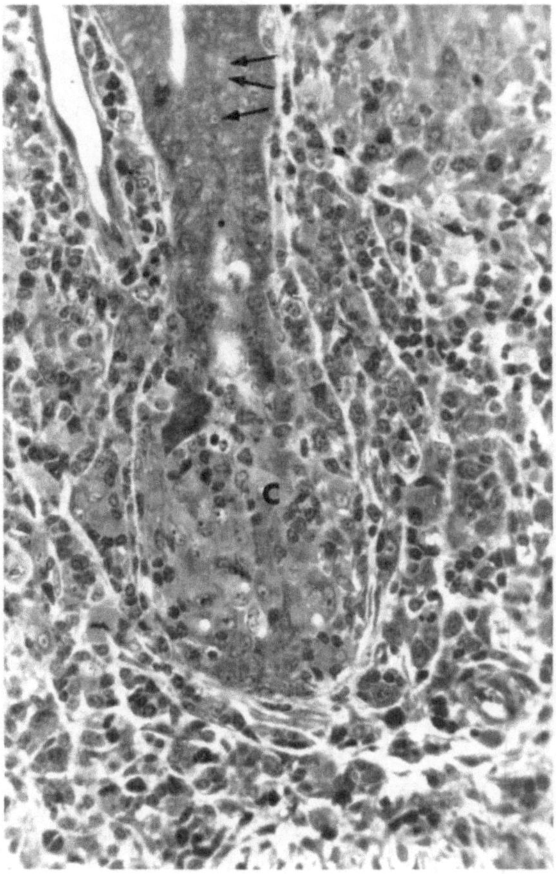

FIGURE 2. A crypt at the base of a coccidial polyp in the small intestine of a lamb. The crypt enterocytes are infected with progamonts (arrows), probably of *E. bakuensis*, and are apparently being destroyed by a cluster (C) of macrophage-like cells. (3 μm methacrylate, TB.) (See also Figures 5 and 6).

Polyps are formed by enormous lengthening of the villi, which may be due to crypt hyperplasia, or to a lengthening of the lifespan of each enterocyte. Probably it is due to both. Diffuse hyperplasia occurs for example with *E. crandallis*,[3,62] (Figure 7) *E. weybridgensis*[63] (sheep), and *E. stiedai*[44] (rabbits) (Figures 8 and 9). It is thought to result when the merozoites are released into the lumen. Filamentous organisms are sometimes seen attached to infected enterocytes. Their identity and significance are unknown.[67]

4. Epithelial Loss

Enterocytes normally complete their lifespan in a few days. When they reach the mucosal surface, they are shed into the lumen or absorbed by neighboring enterocytes. In coccidial infections, enterocytes may be lost prematurely. The causes of premature cell loss are not entirely clear. One cause may be anoxia, which can occur in ruminants when giant meronts separate the epithelium from its blood supply. Another is a hypersensitivity reaction, in which mast cell products cause the accumulation of fluid beneath epithelia in the form of blebs, which separate sheets of epithelium from the underlying stroma (Figure 1). The mechanism of cell death in these cases is necrosis.

Another mechanism is apoptosis — programmed cell death. The program for the cell's lifespan may be changed by events occurring during that lifespan, including the effects of

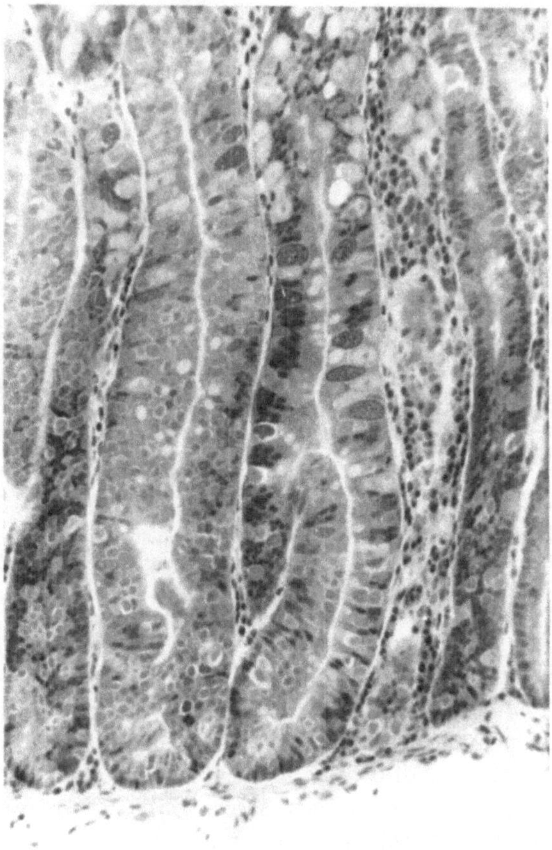

FIGURE 3. The bases of crypts in an "oocyst patch" in the small intestine of a lamb infected with *E. bakuensis*. Every cell appears to be infected with gamonts, and yet there is no tendency for infected cells to separate as they do in some other coccidial infections (see Figures 10 and 11). (3 μm methacrylate, hematoxylin & eosin (H & E). (Reprinted with permission from Gregory, M. W. et al., *Int. J. Parasitol.*, 17, 1113, 1987. Copyright 1987, Pergamon Press PLC.)

cytotoxins[68] and of host immune cells.[69] In coccidial infections we may see apoptosis of epithelial cells in large numbers, both on the surface and in the crypts (Figures 10 and 11). Infected and uninfected cells may be involved, and the immediate cause is not always apparent. In some cases macrophage-like cells are seen very close to them (Figure 2); these cells may be involved in the cause of death, or simply in clearing up the mess.[50] Fernando and Pasternak[70] have shown that crypt cells infected with *E. necatrix* are more sensitive to proteolytic enzymes than uninfected cells. Pasternak and Fernando[71] surveyed the host cell response to coccidial infection.

Epithelial loss in the small intestine results immediately in villus atrophy.[72,73] If the underlying crypts are intact, the lesion will be transient: crypt hyperplasia will cause new cells to pour out of the crypts and restore the villous architecture. A similar pathological picture is produced by other agents that destroy villus epithelial cells while leaving crypts intact, e.g., transmissible gastroenteritis virus of piglets,[74] and food allergy.[65] Such conditions result in reduced surface area of absorptive epithelium, increased cell turnover rate, increased immaturity of enterocytes, and consequently reduced absorption of nutrients (Figure 19).[75]

If the cell loss extends to the crypts, it can result in crypt atrophy. If the stem cells in

FIGURE 4. Stereo pair scanning electron micrograph (SEM) of part of an "oocyst patch" in the small intestine of a lamb infected with *E. bakuensis*. The small villi towards the bottom left are normal and uninfected, while those filling the rest of the picture are infected with gamonts and oocysts. The specimen has been freeze-cracked in ethanol. (Reprinted with permission from Gregory, M. W. et al., *Int. J. Parasitol.*, 17, 1113, 1987. Copyright 1987, Pergamon Press, PLC.)

the crypts are lost over a wide area, the epithelium cannot regenerate, and the area becomes denuded (Figures 12 to 14). Such a lesion can heal only by encroachment from the edge. A similar pathological picture is produced by other agents that destroy crypt stem cells, e.g., feline parvovirus (the cause of cat enteritis) and ionizing radiation. Denudation means two things: loss of absorptive epithelium, and access to the tissues by opportunist bacteria and fungi. Loss of absorptive epithelium can mean prolonged diarrhea; invasion by opportunist bacteria can mean tissue necrosis and death of the host. One of the causes of death could be absorption of toxins resulting from excessive proteolysis.[46,76]

5. Depletion of Lymphocytes in Peyer's Patches

This change (Figure 15) appears to be related to the severity of the disease, and is probably a manifestation of the mobilization of forces in the immune system. Most of the cell loss is from the follicles themselves, most of which are of the B-cell line.[13,50,52]

E. LESIONS ASSOCIATED WITH SPOROZOITES

Numbers of sporozoites are always low in relation to later stages, and their presence is usually transient, so the damage they cause to the tissues is minimal. On the other hand, they can cause dramatic changes to the individual cells that they infect. These include varying degrees of hypertrophy of nucleoli (Figure 16),[77] nuclei and cytoplasm. They may also make the host cell invasive (Figure 16).[3,53,71]

F. LESIONS ASSOCIATED WITH MERONTS AND MEROZOITES

Massive numbers of meronts cause whitening and thickening of the mucosa. The giant meronts of ruminant coccidia are visible to the naked eye as pinpoint white spots. Meronts in large numbers can cause edema and hemorrhage in chickens[55] and in lambs.[78] Cells infected with small meronts tend to remain in the epithelium. Cells destined to contain large meronts (e.g., *E. necatrix*,[53] *E. tenella*,[53] *E. crandallis*[3]) migrate into the lamina propria. Even large meronts may elicit little or no inflammatory reaction and cause little harm.[79] In large numbers they can cause swelling of the mucosa, with a degree of epithelial loss that can be considerable if it extends to the crypts (Figure 1).[50] In *E. tenella* and *E. necatrix* infections, they cause severe hemorrhage.[55,56] Even in sheep, very large numbers of first

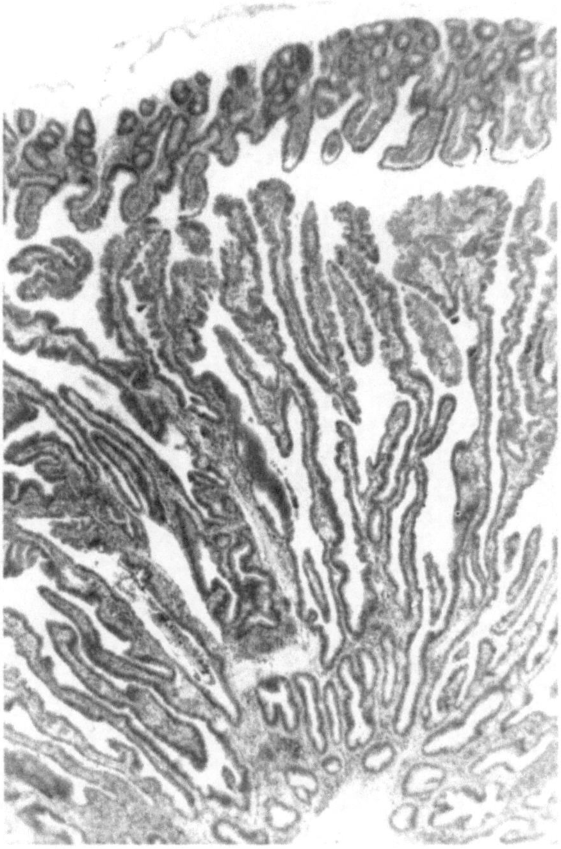

FIGURE 5. Part of a coccidial polyp in the small intestine of a lamb, caused presumably by *E. bakuensis*. At the top is the intestinal wall, with normal uninfected villi. The base of the polyp is at the foot of the picture, and the rest is filled with the enormously enlarged villi of the polyp. Virtually every epithelial cell on the polyp is infected: with pro-gamonts at the base, and gamonts or oocysts at the periphery. (4 μm paraffin, H & E.)

generation meronts can have devastating effects.[36,78] Giant meronts in ruminants are often surrounded by inflammatory cells, particularly neutrophils. Host cells may enter the meront and engulf merozoites (Figure 1). Individual meronts may elicit no apparent host reaction, even though neighboring meronts are surrounded and infiltrated by host cells.[50] Meronts that have been destroyed are seen as ghosts: agglomerations of macrophages and cell debris. In *E. bakuensis* infection, the refractile remains of the meront wall may still be visible.[63]

G. LESIONS ASSOCIATED WITH GAMONTS

In most coccidia/host relationships, this stage is the most pathogenic, presumably because it is the most numerous. Aggregations of gamonts cause whitening and thickening of the mucosa, which may be focal (as in *E. bakuensis* and *E. acervulina*) or diffuse (as in *E. crandallis* and *E. maxima*). Swelling, edema, and petechial or gross hemorrhages are sometimes seen. Where there is denudation, there may be a diphtheritic membrane of necrotic material.[52] Host cells infected with gamonts react in various ways. They may continue to divide and produce hyperplastic lesions that are diffuse (*E. stiedai* — Figure 9, *E. crandallis* — Figure 7), or focal (*E. bakuensis* polyps — Figure 5). They may stop dividing but remain

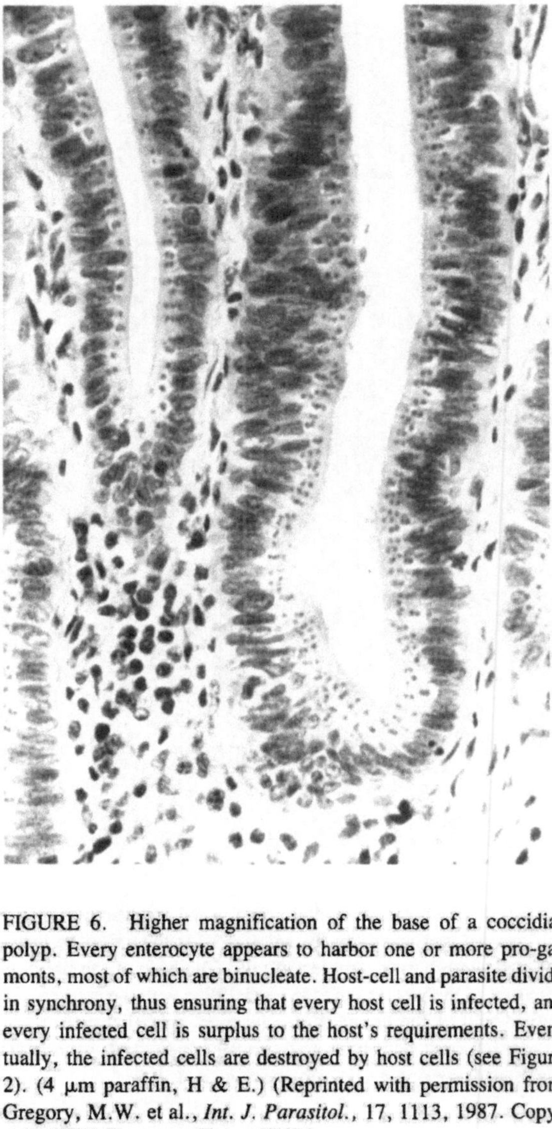

FIGURE 6. Higher magnification of the base of a coccidial polyp. Every enterocyte appears to harbor one or more pro-gamonts, most of which are binucleate. Host-cell and parasite divide in synchrony, thus ensuring that every host cell is infected, and every infected cell is surplus to the host's requirements. Eventually, the infected cells are destroyed by host cells (see Figure 2). (4 μm paraffin, H & E.) (Reprinted with permission from Gregory, M.W. et al., *Int. J. Parasitol.*, 17, 1113, 1987. Copyright 1987, Pergamon Press, PLC.)

in place until the parasite matures (as in *E. acervulina* and in *E. bakuensis* "oocyst patches" — Figures 3 and 4) or they may stop dividing and separate from their neighbors (as with *E. ovinoidalis*[50] — Figures 10 and 11, and *E. flavescens* — Figures 13 and 14.[52]) Where the cells stop dividing and separate from their neighbors there is marked epithelial loss: many cells (uninfected as well as infected) detach from the basement membrane and are lost in the gut lumen. Many of these cells do not remain attached long enough for the gamonts to mature, so the detachment is clearly of no benefit to the parasite. It must be assumed to be a host reaction, but in heavy infections it results in denudation of large areas of the mucosa, which may take weeks to heal (Figures 17 and 18).

H. LESIONS IN UNINFECTED TISSUES

Most pathological studies have been done by parasitologists rather than pathologists, and most workers have ignored changes in parts of the body in which the organism was not found. Lozanov and Koinarski[76] reported degenerative changes in the liver, kidney, and

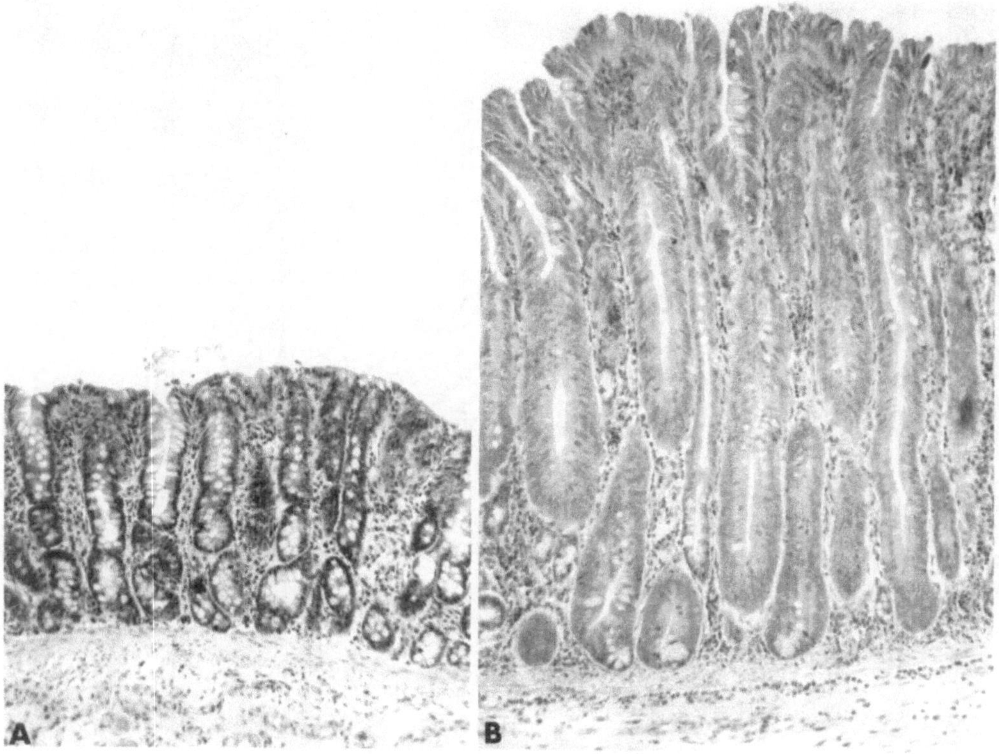

FIGURE 7. (A) Cecal mucosa of coccidia-free lamb. (B) Cecal mucosa of a lamb 16 days after infection with *E. crandallis*, at the same magnification as Figure 7(A). Note the crypt hyperplasia, the lack of goblet-cell differentiation, and the relative lack of inflammatory reaction. This hyperplastic mucosa can be looked upon as a spread-out polyp, with pro-gamonts at the base, and gamonts and oocysts at the surface. The pro-gamonts divide in synchrony with their host cells in the same way as those of *E. bakuensis* in polyps (see Figures 5 and 6). (3 μm methacrylate, H & E.)

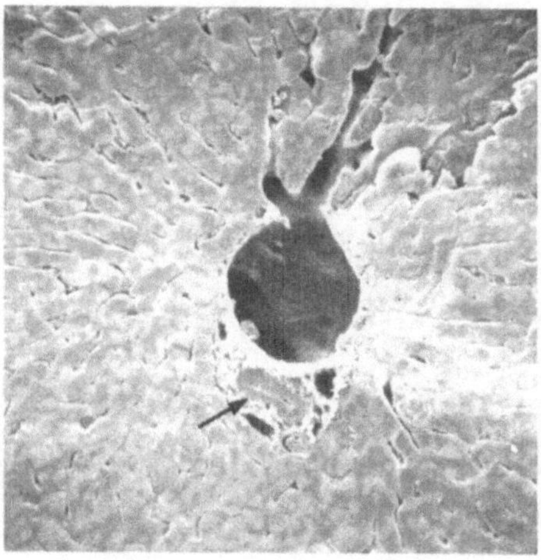

FIGURE 8. SEM of freeze-cracked liver of a coccidia-free rabbit, to compare with Figure 9. Immediately below the portal venule in the centre is a bile duct (arrow).

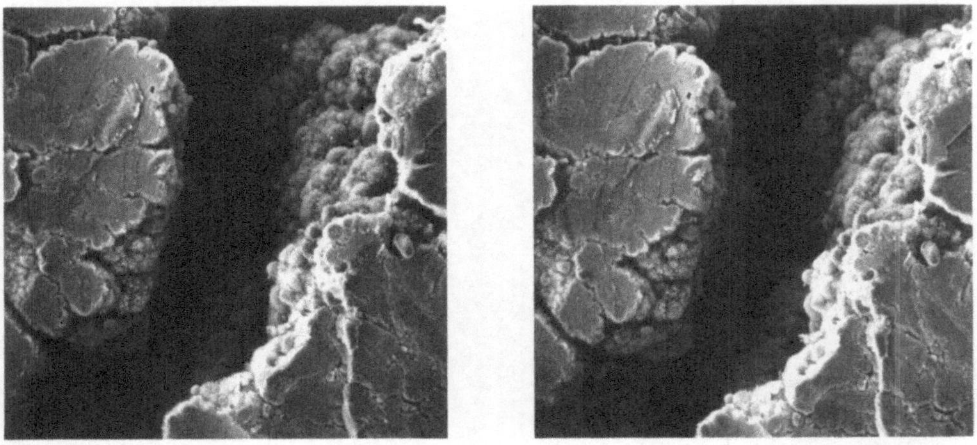

FIGURE 9. Stereo-pair SEM of freeze-cracked liver of a rabbit infected with *E. stiedai* 20 days previously. The magnification is the same as Figure 8; the whole field is filled with bile-duct epithelium that is shedding oocysts into the lumen. Host cells infected with gamonts continue to divide, but synchronous division has not been reported.

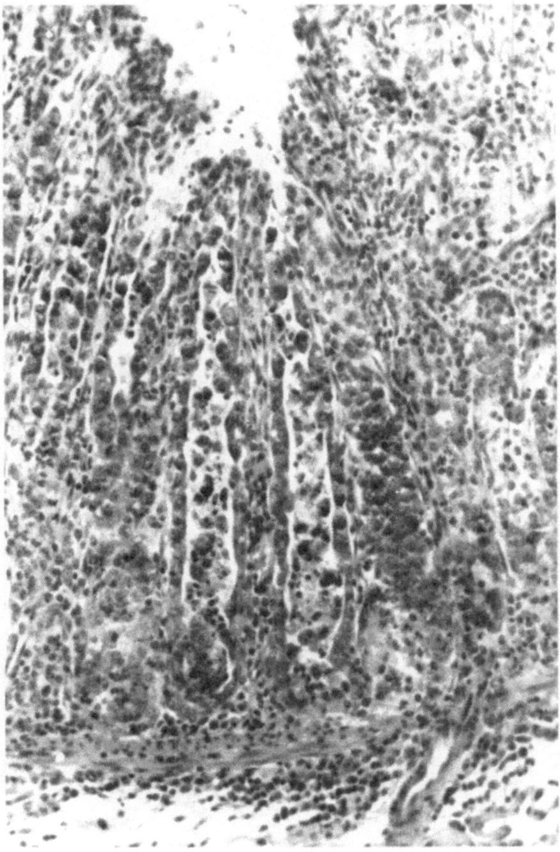

FIGURE 10. Cecal mucosa of a lamb 13 days after infection with *E. ovinoidalis*. The crypt epithelial cells are infected with gamonts, and are separating and sloughing away into the gut lumen. Compare with Figures 3 and 7, to see the differing reactions to infection with gamonts. Figures 3 and 10 are at the same magnification. (3 μm paraffin, H & E.)

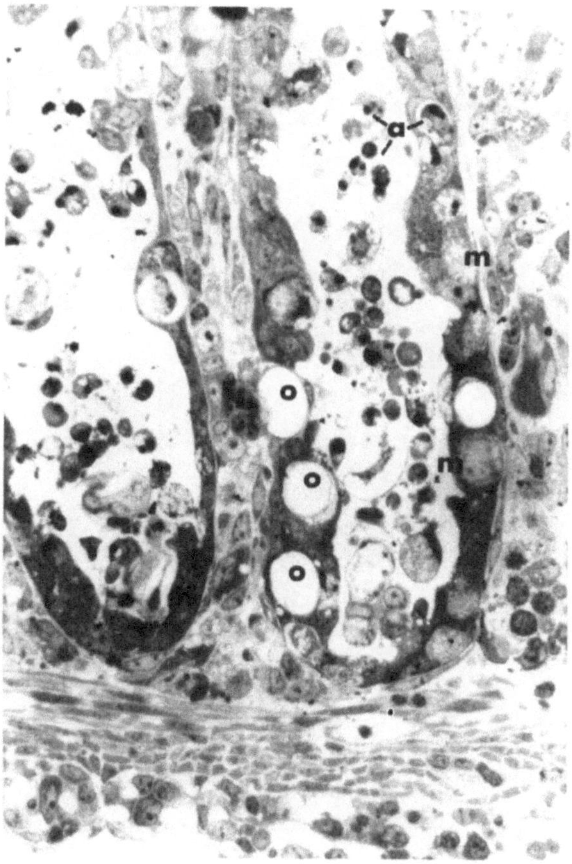

FIGURE 11. Higher magnification of a similar specimen to that seen in Figure 10, showing apoptosis of enterocytes in cecal crypts infected with *E. ovinoidalis*. Several oocysts (o) and macrogamonts (m) are seen. Apoptotic bodies (a) are seen in epithelial cells and in the lumen. (1 μm methacrylate, methylene blue, basic fuchsin (MB/BF).)

heart of turkeys infected with *E. adenoeides*, which they attributed to the effects of toxic cell debris in the intestine. Majid and Winter[80] reported membranoproliferative glomerulonephritis in lambs said to be suffering from coccidiosis. It was characteristic of the lesions caused by antigen/antibody complexes being trapped in glomeruli. There were no pathological data from the site of coccidial infection. Radostits and Stockdale[32] found no lesions in the central nervous system in cases of nervous coccidiosis in calves.

I. CONCLUSION

Figure 19 summarizes the pathogenetic pathways that can lead to clinical disease and death. The pathogenicity of each coccidial species depends partly on its biological characteristics — in other words, on which cells it infects, in which part of the gut, and whether this infection results in the destruction of the host cells. In Table 1, I attempt to arrange these characteristics in order of increasing importance: the first three can be regarded as factors that reduce pathogenicity, while the last three increase it. We know too little about most coccidial species to classify them on this basis, but in Table 1 I have tried to put the better known species in logical order to show that their pathogenicity accords roughly with the characteristics listed.

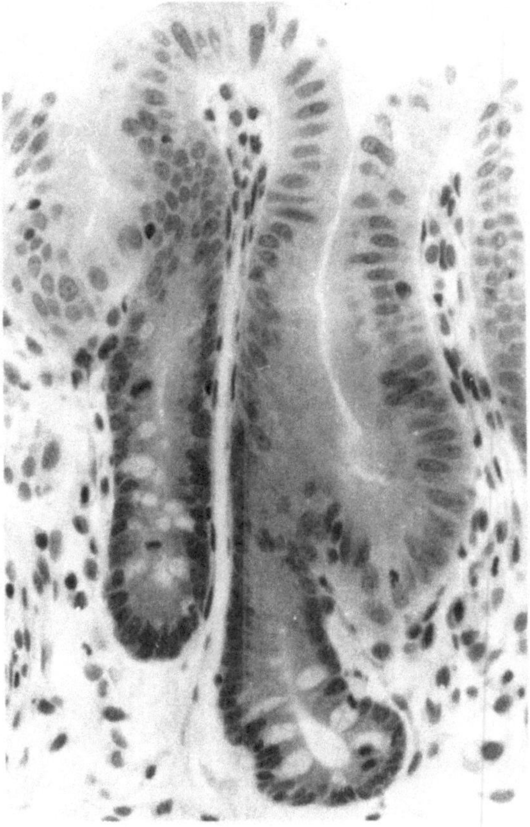

FIGURE 12. Cecum of a coccidia-free rabbit, for comparison with Figures 13 and 14. (3 μm methacrylate, H & E.) (Reprinted with permission from Gregory, M. W. and Catchpole, J., *Int. J. Parasitol.*, 16, 131, 1986. Copyright 1986, Pergamon Press PLC.)

VI. THE MAJOR COCCIDIOSES

A. MAMMALS

1. Man

Coccidiosis in man is rare. *Isospora belli* is the species most commonly associated with disease. It affects the small intestine, causing villus atrophy in a manner similar to that of *I. suis* in pigs (see below). The finding of oocysts in feces or of coccidial stages in intestinal biopsies is regarded as diagnostic.[41] Experimental infections have been reported, using *I. belli* and *Sarcocystis* species (then thought to be "*I. hominis*"[81,82]). *Sarcocystis* species are less pathogenic.[81] The disease caused by *I. belli* tends to be acute in onset, with fever, headache, diarrhea, abdominal pain, debility, and weight loss, and it progresses to a phase of chronic diarrhea. It was said to be self-limited, "usually of 6 weeks to 6 months duration".[41] By a veterinary yardstick, even 6 weeks is long. Brandborg et al.[41] reported cases that lasted years, in at least some of which the coccidiosis was a complication of another condition such as Hodgkins disease. Lesions included villus atrophy, crypt hyperplasia, and flat mucosa. The chronicity was explained in at least one case by the constant finding of sporulated oocysts in the duodenum but not in the feces. The disease occurs mainly in association with impairment of the immune system, and is one of the problems associated with AIDS.[18] It can be intractable.[19]

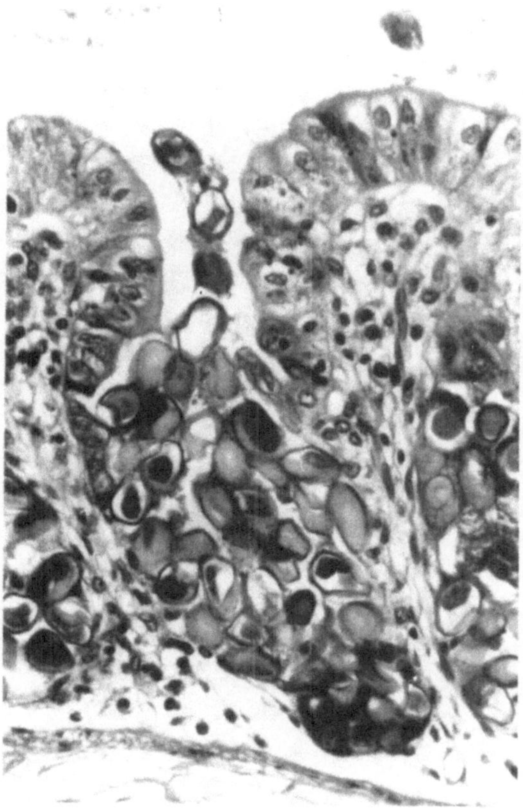

FIGURE 13. Cecum of rabbit 9 days after infection with *E. flavescens*. The crypts are saturated with coccidial oocysts, some of which are seen flowing out into the gut lumen. The surface epithelium is nearing the end of its lifespan, and would soon have sloughed away leaving no crypt stem cells to regenerate it. (4 μm paraffin, H & E.)

2. Cattle (*Bos*)

E. bovis and *E. zuernii* are the principal pathogens. Lesions are due to the gamonts attacking crypt cells in the large intestine, causing diarrhea that is often bloody.[20,83-85] Calves may be affected from about 3 weeks of age up to a year or more. The disease is commonest in very cold or very hot weather,[12,32] or when calves are moved to feedlots[86] or are infected with parvovirus.[13] Nervous signs may also occur due to a neurotoxin found in the serum.[43,87] No gross or histological lesions were found to explain the nervous signs.[32]

3. Sheep (*Ovis*)

E. ovinoidalis is the main pathogen. In Great Britain, the disease affects unweaned lambs around 6 weeks of age,[88] whereas in the U.S. it tends to appear when they are moved to feedlots.[33] Newborn lambs are relatively resistant to infection, and their susceptibility increases up to about 4 weeks of age.[16] The gamonts are the pathogenic stage, and they damage the crypts in the caecum and colon (Figures 10 and 11), causing diarrhea which is often bloody. The resulting denudation may take several weeks to heal (Figure 17), during which time the feces are soft or liquid and the weight gain is poor. *E. crandallis* is often present in large numbers in cases of coccidiosis,[88] and has been shown experimentally to cause severe disease in coccidia-free lambs[2] and to exacerbate infections with *E. ovinoidalis* in field lambs.[89] It tends also to be present in high numbers in healthy lambs,[90] so its

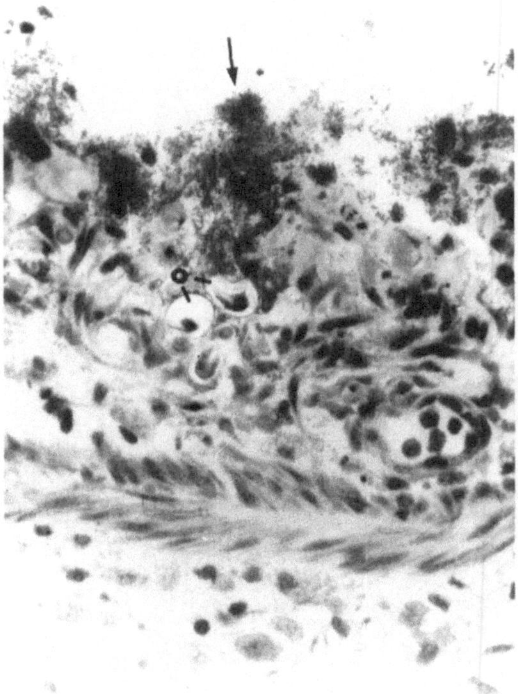

FIGURE 14. Cecum of rabbit 12 days after a heavy infection with *E. flavescens*, same magnification as Figures 12 and 13. The crypts have been destroyed, so the epithelium cannot regenerate except by spread from isolated pockets of surviving stem cells. In this case, the mucosa was invaded by bacteria (arrow), and the rabbit would have died. Some oocysts (o) have been trapped in the mucosa. (3 μm methacrylate, Giemsa.) (Reprinted with permission from Gregory, M. W. and Catchpole, J., *Int. J. Parasitol.*, 16, 131, 1986. Copyright 1986, Pergamon Press PLC.)

pathogenic role in the field is hard to assess. The most pathogenic stage of *E. crandallis* is the gamont, which affects crypts in the small and sometimes also in the large intestine, causing diffuse hyperplasia (Figure 7); it does not typically cause denudation because it multiplies in synchrony with its host cells, thus ensuring that sufficient host cells are always available.[62] *E. bakuensis* is among the most prevalent species in sheep. Its gamonts cause oocyst patches (Figures 3 and 4) and, apparently, polyps (Figures 5 and 6), but there is little evidence that much harm results.[51] *E. ahsata* causes similar lesions, and has been reported as highly pathogenic.[26,91] Increasing daily doses of *E. ahsata* in lambs reared coccidia-free produced no signs of disease.[92]

4. Goats (*Capra*)

E. arloingi is the most commonly reported coccidium in goats.[93] It causes oocysts patches and polyps in the small intestine,[94] as does *E. christenseni.*[95] Apparently they are much more pathogenic than the corresponding lesions in sheep. In the laboratory, *E. ninakohlyakimovae*[95,96] and *E. caprina*[95] produce diseases similar to that caused by *E. ovinoidalis* in sheep.

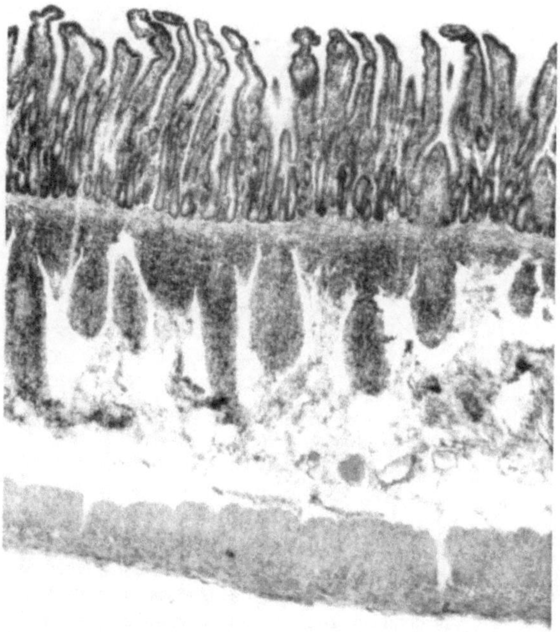

FIGURE 15. Ileum of a lamb, 17 days after *E. ovinoidalis* infection. The submucosa is hyperemic and edematous, the muscularis is hypertrophic, and the Peyer's patch follicles are somewhat depleted of lymphocytes. This depletion appears to be a response to the damage in the large intestine caused by the later stages of this parasite. There is some crypt hyperplasia, but the villous architecture is undisturbed. (5 μm paraffin, H & E.)

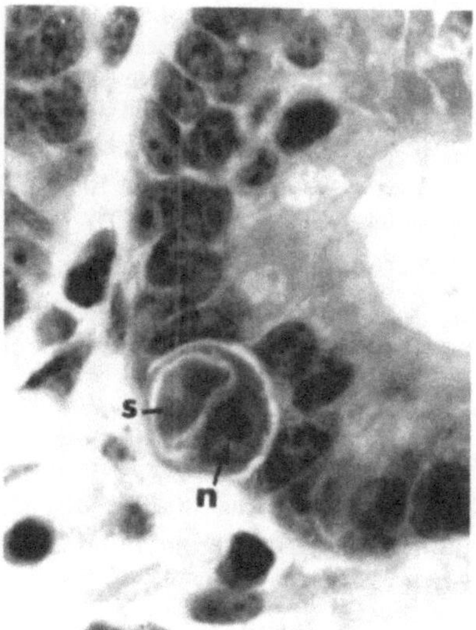

FIGURE 16. Hypertrophic nucleolus (n) in a jejunal crypt epithelial cell infected with a sporozoite (s) 3 days after inoculation of a lamb with *E. crandallis*. The infected host cell is migrating into the lamina propria. (3 μm methacrylate, H & E.)

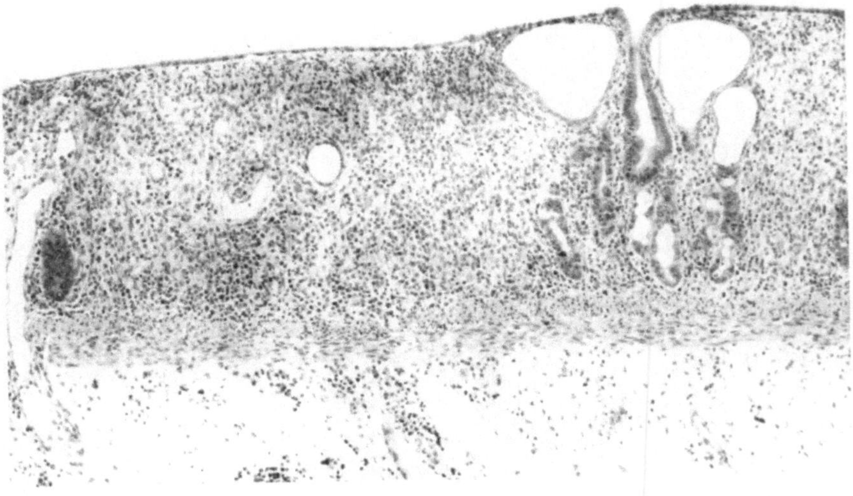

FIGURE 17. Cecum of a lamb 17 days after receiving 100,000 oocysts of *E. ovinoidalis*. Note the severe destruction of crypt tissue, despite the fact that this lamb had received an "immunizing" inoculum of 100 oocysts 28 days before the challenge. The mucosa is occupied instead by granulation tissue, and some areas were still denuded. Here an isolated residuum of stem cells has given rise to a group of crypts. Some of the regenerating crypts have failed to break through to the surface, and have formed cysts; others are producing a spreading sheet of epithelial cells, the edge of which can be seen at the extreme left. The mucosa is hyperaemic and the submucosa edematous. (3 μm methacrylate, H & E.) (Reprinted with permission from Gregory, M. W. and Catchpole, J., *Int. J. Parasitol.*, 17, 1099, 1987. Copyright 1987 Pergamon Press PLC.)

5. Camels (*Camelus*)

Hussein et al.[97] found that *E. dromedarii, E. rajasthani,* and *E. cameli* were all pathogenic, causing diarrhea, sometimes bloody, in camel calves. They affected the jejunum and ileum, causing villus atrophy and denudation.

6. Pigs (*Sus*)

Pig coccidioses have been reviewed by Stuart and Lindsay.[98] The most troublesome species is *Isospora suis,* which multiplies very rapidly in the small intestine of piglets less than 2 weeks old.[93,100] It affects the villus epithelium of the small intestine, causing sudden severe villus atrophy and diarrhea.[99,101-103] It does not affect the crypts, so if the piglet can survive the initial onslaught it recovers rapidly. *E. scabra* occasionally causes trouble in older pigs[104] and its pathology was investigated experimentally by Vitovec et al.[105] The pathogenic mechanism appeared similar to that of *I. suis,* lesions and symptoms resolving rapidly.

7. Rabbits (*Oryctolagus*)

Unlike most other species, *E. stiedai* causes chronic disease, lasting up to 7 weeks.[106] It affects the bile duct epithelium, causing enormous biliary hyperplasia (Figures 8 and 9), and, in heavy infections, cirrhosis. An excellent description has been given by Smetana.[4] Of the intestinal coccidia in rabbits, *E. flavescens* and *E. intestinalis* are the most pathogenic.[6,52] The gamonts of *E. flavescens* destroy the crypts of the cecum, causing denudation and severe diarrhea (Figures 12 to 14 and 18). *E. intestinalis* affects the crypts of the small intestine, and the pathology has been studied by Peeters et al.[107] *E. piriformis* affects the crypts of the rectum in its gamont stage, and its pathogenic effect can be additive.[6]

8. Dogs (*Canis*)

Isospora canis can cause coccidiosis in puppies subjected to weaning stress.[24]

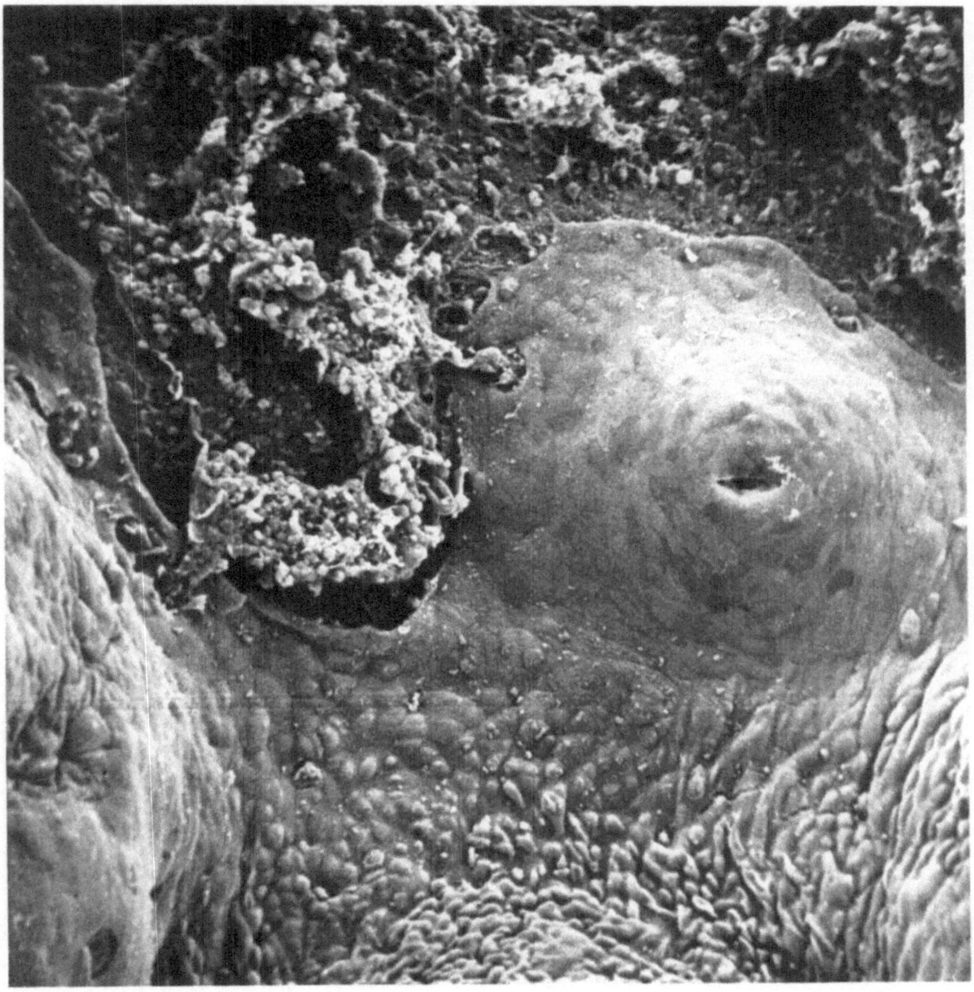

FIGURE 18. SEM of rabbit cecum 12 days after infection with *E. flavescens*. At the top of the picture the mucosa is denuded, and covered with fibrinous exudate and cell debris. Lower down, surviving crypts are pouring epithelial cells onto the denuded surface. Unlike the case shown in Figure 14, this rabbit would probably have survived. (Reprinted with permission from Gregory, M. W. and Catchpole, J., *Int. J. Parasitol.*, 16, 131, 1986. Copyright 1986, Pergamon Press PLC.)

B. BIRDS

1. Chickens (*Gallus*)

The following descriptions are based mainly on Long[108] and Fernando.[56] In *E. tenella* infection, hemorrhagic or whitish lesions 1 to 5 mm in diameter appear in the ceca on day 5. If the bird survives longer, the lesion starts to resolve, but the contents often form a solid or semisolid core that may stay for several days before being passed. *E. necatrix* causes hemorrhagic lesions throughout the small intestine, particularly the mid-jejunum. In both species the lesions are caused by the large second generation meronts. Infiltration with granulocytes increases rapidly as the meronts develop. Rupture of blood capillaries and of the epithelium precedes the release of merozoites.[55,56] In *E. tenella,* remaining epithelium is then invaded by merozoites to form gamonts, which are less damaging than the meronts. New epithelium forms rapidly, the inflammatory cells diminish, and discrete lymphoid foci increase in number. In *E. necatrix* infection, the crypt architecture is often maintained and a layer of uninfected cells often lines the sac of infected cells, allowing regeneration of a

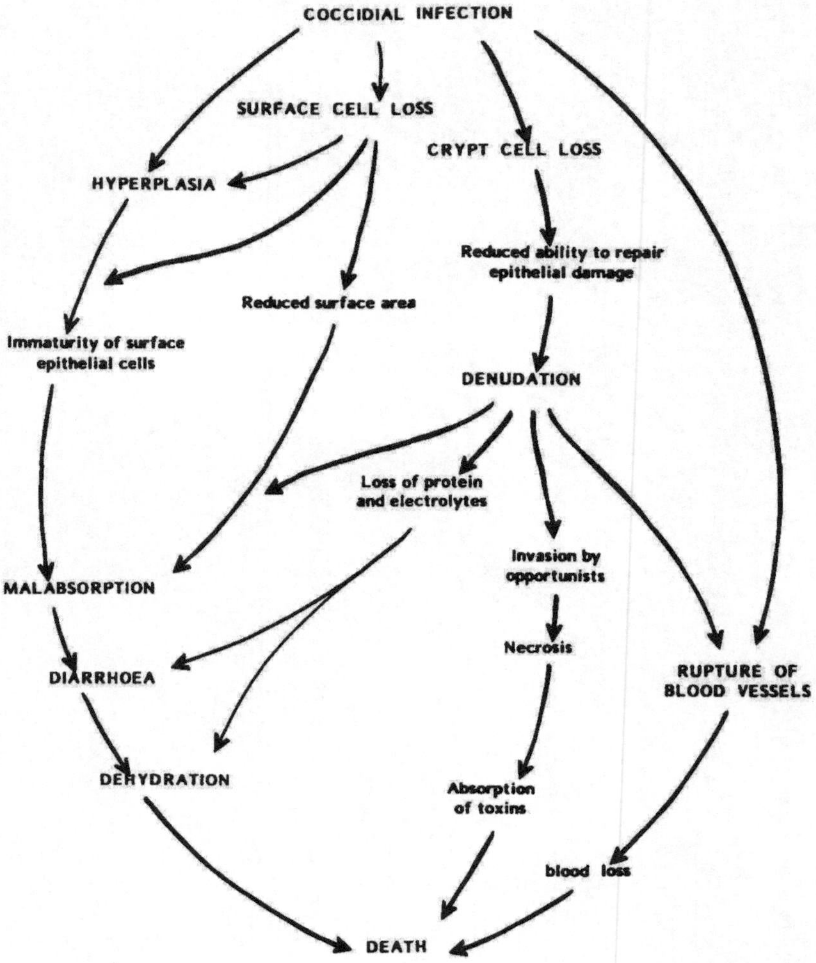

FIGURE 19. Pathogenesis of intestinal coccidiosis. This diagram shows the possible causal pathways of the changes that may take place. With the less pathogenic coccidia, effects are restricted to the left-hand pathways, via hyperplasia and surface cell loss. The more pathogenic coccidia cause changes along all four pathways, but particularly those leading to crypt cell loss and rupture of blood vessels.

crypt.[54,109,110] There is some destruction of muscle layers, and a very variable leukocyte infiltration around or within crypts. In mature birds, many mononuclear cells are seen in the lamina propria.[54] Ghosts of meronts are seen on day 6. Gamonts develop in the ceca and are relatively harmless.

In other species of *Eimeria* all stages develop in the epithelium, the pathogenic stages are the gamonts, there is little or no hemorrhage and (except in *E. brunetti*) no denudation. *E. acervulina* affects the villus epithelium of the duodenum, causing discrete whitish lesions which involve a degree of villus atrophy and increased cellularity of the lamina propria but no hemorrhage.[64,72,111] *E. mitis* causes damage to the lower intestine. *E. maxima* causes petechial hemorrhages in the upper two thirds of the intestine, with thick mucus flecked with blood. Villus atrophy occurs, and the tips of villi may slough away. *E. brunetti* affects the lower third of the intestine, causing pinpoint hemorrhages, and mucoid enteritis, sometimes with blood. There is loss of villus epithelium and of fluid and electrolytes, sometimes leading to denudation, and necrotic enteritis. *E. praecox* is relatively nonpathogenic: there is little or no inflammatory reaction, even in heavy infections.[54,112]

TABLE 1
Factors Contributing to the Pathogenicity of Coccidial Species

Increasing pathogenicity →

	sd	dv	lc	si	vi	li	cr	bl
E. ahsata	+?	+?	+	+				
E. arloingi	+?	+?	+	+				
E. christenseni	+	+	+	+				
E. bakuensis	+	+	+	+				
E. weybridgensis	+	+		+				
E. crandallis	+	+		+		+		
E. stiedai		+						
E. acervulina			+	+	+			
I. belli				+	+			
I. suis				+	+			
E. intestinalis				+	+			
E. maxima				+	+			
E. brunetti					+	+		
E. necatrix				+				+
E. tenella						+		+
E. ovinoidalis				+	+	+	+	+
E. ninakohlyakimovae				+	+	+	+	+
E. caprina				?	+	+	+	+
E. bovis				+	+	+	+	+
E. zuernii				+	+	+	+	+
E. flavescens					+	+	+	+

Increasing pathogenicity ↓

Note: Key: sd = synchronous division of pro-gamonts and their host cells, dv = division of infected host cells occurs, lc = localized lesions, si = damage to small intestine, vi = villus epithelial cells destroyed, li = damage to large intestine, cr = crypt stem cells destroyed, bl = blood vessels damaged.

2. Turkeys (*Meleagris*)

E. adenoeides causes very severe lesions in the lower intestine and ceca, producing white intestinal casts. *E. meleagrimitis* may cause hemorrhagic enteritis in the duodenum and ileum.[113,114]

3. Ducks (*Anas*) Geese (*Anser*)

Coccidioses of the Anseriformes have been reviewed by Gajadhar et al.[115] *Tyzzeria perniciosa* is the most pathogenic species in ducks, causing lesions like those of *E. necatrix* in chickens. *E. truncata* affects the tubule epithelium of the goose kidney, and can cause serious losses.[45,116] Affected birds show weakness, diarrhea, and ataxia; the kidneys appear pale, swollen, and covered with small whitish nodules; the tubules are distended with urates and oocysts. *E. anseris* and *E. nocens* can also be pathogenic; the former is associated with papilliform lesions throughout the small intestine.[117,118]

4. Pheasant (*Phasianus*), Partridge (*Perdix, Alectoris*); Quail (*Coturnix, Colinus*)

Coccidia of game birds have been reviewed by Norton.[119] Acute infections with heavy losses are usually associated with species which develop in the ceca such as *E. colchici* (pheasant), *E. legionensis* (red-legged partridge), and *E. tsunodai* (Japanese quail).

REFERENCES

1. **Norton, C.C., Catchpole, J., and Joyner, L. P.,** Redescriptions of *Eimeria irresidua* Kessel & Jankiewicz, 1931 and *E. flavescens* Marotel & Guilhon, 1941 from the domestic rabbit, *Parasitology,* 79, 231, 1979.
2. **Catchpole, J. and Gregory, M. W.,** Pathogenicity of the coccidium *Eimeria crandallis* in laboratory lambs, *Parasitology,* 91, 45, 1985.
3. **Gregory, M. W., Catchpole, J., and Norton, C. C.,** Observations on the endogenous stages of *Eimeria crandallis* in domestic lambs *(Ovis aries), Int. J. Parasitol.,* 19, 907, 1989.
4. **Marquardt, W. C.,** Some problems of host and parasite interactions in the coccidia, *J. Protozool.,* 23, 287, 1976.
5. **Joyner, L. P. and Norton, C. C.,** *Eimeria mitis* in mixed infections with *E. acervulina* and *E. brunetti* in the fowl, *Parasitology,* 86, 381, 1983.
6. **Coudert, P.,** Some peculiarities of rabbit coccidiosis, *Proc. 5th Int. Conf. Coccidiosis,* Tours, France, 17—20 October 1989 INRA Publ., 1989.
7. **Prescott, J. F.,** Intestinal disorders and diarrhoea in the rabbit, *Vet. Bull.,* 48, 475, 1978.
8. **Bradley, R. E. and Radhakrishnan, C. V.,** Coccidiosis in chickens: obligate relationship between *Eimeria tenella* and certain species of cecal microflora in the pathogenesis of the disease, *Avian Dis.,* 17, 461, 1973.
9. **Dykstra, D. D.,** Relationships of anaerobic bacteria to coccidiosis in the chicken, *Diss. Abstr. Int.,* 38, 2962-B, 1978.
10. **Kageyama, A., Fukuta, T., Baba, E., and Arakwa, A.,** The influence of various bacteria on the cecal mucosa of monoflora chickens infected with *Eimeria tenella.* A scanning electron microscope study, *Zentralbl. Bakteriol. Hyg. A.,* 265, 353, 1987.
11. **Gouet, P., Yvoré, P., Naciri, M., and Contrepois, M.,** Influence of digestive microflora on parasite development and the pathogenic effect of *Eimeria ovinoidalis* in the axenic, gnotoxenic and conventional lamb, *Res. Vet. Sci.,* 36, 21, 1984.
12. **Parker, R. J., Boothby, K., Polkinghorne, I., and Holroyd, R. G.,** Coccidiosis associated with post-weaning diarrhoea in beef calves in a dry tropical region, *Aust. Vet. J.,* 61, 181, 1984.
13. **Durham, P. J. K., Johnson, R. H., and Parker, R. J.,** Exacerbation of experimental parvoviral enteritis in calves by coccidia and weaning stress, *Res. Vet. Sci.,* 39, 16, 1985.
14. **Catchpole, J. and Harris, T. J.,** Interaction between coccidia and *Nematodirus battus* in lambs on pasture, *Vet. Rec.,* 124, 603, 1989.
15. **Dürr, U. and Pellérdy, L.,** The susceptibility of suckling rabbits to infection with coccidia, *Acta Vet. Hung.,* 19, 453, 1969.
16. **Gregory, M. W. and Catchpole, J.,** Ovine coccidiosis: heavy infection in young lambs increases resistance without causing disease, *Vet. Rec.,* 124, 458, 1989.
17. **Catchpole, J. and Devonshire, R.,** The effect of colostrum on the susceptibility of lambs to coccidiosis, *Proc. 5th Int. Conf. Coccidiosis,* Tours, France, 17-20 October 1989, INRA Publ., 1989.
18. **Dehovitz, J. A., Pape, J. W., Boncy, M., and Johnson, W. D.,** Clinical manifestations and therapy of *Isospora belli* infection in patients with the acquired immunodeficiency syndrome, *N. Engl. J. Med.,* 315, 87, 1986.
19. **Restrepo, C., Macher, A. M., and Radany, E. H.,** Disseminated extraintestinal isoporiasis in a patient with acquired immune deficiency syndrome, *Am. J. Clin. Pathol.,* 87, 536, 1987.
20. **Yvoré, P., Naciri, M., and Esnault, A.,** La coccidiose du veau, *Rec. Méd. Vét.,* 159, 217, 1983.
21. **Gräfner, G. and Graubmann, H. D.,** Betrachtungen zur Pathogenität von *Eimeria*-Arten am Beispiel der Rinderkokzidiose, *Angew. Parasitol.,* 20, 209, 1979.
22. **Parker, R. J., Jones, G. W., Ellis, K. J., Heater, K. M., Schroter, K. L., Tyler, R., and Holroyd, R. G.,** Post-weaning coccidiosis in beef calves in the dry tropics: experimental control with continuous monensin supplementation via intra-ruminal devices and concurrent epidemiological observations, *Trop. Anim. Health Prod.,* 18, 198, 1986.
23. **Pout, D. D.,** Coccidiosis of lambs. IV. The clinical response to infections of *E. arloingi* "B" and *E. crandallis* in laboratory lambs, *Br. Vet. J.,* 130, 54, 1974.
24. **Todd, K. S.,** Discussion of work on *Isospora* species, *J. Protozool.,* 28, 247, 1981.
25. **Dent, C. H. R.,** Coccidiosis could be a problem in your livestock, *Agric. Gaz. N. S. W.,* 91, 34, 1980.
26. **Mahrt, J. L. and Sherrick, G. W.,** Coccidiosis due to *Eimeria ahsata* in feedlot lambs in Illinois, *J. Am. Vet. Med. Assoc.,* 146, 1415, 1965.
27. **Niilo, L.,** Experimental winter coccidiosis in sheltered and unsheltered calves, *Can. J. Comp. Med.,* 34, 20, 1970.
28. **Jolly, W. R. and Bergstrom, R. C.,** Summer coccidiosis, *Vet. Med. Small Anim. Clin.,* 72, 218, 1977.
29. **Gross, W. B.,** Effect of social environment and oocyst dose on resistance and immunity to *Eimeria tenella* challenge, *Avian. Dis.,* 29, 1018, 1985.

30. **Hateren, A. K., Cuperus, T., and Walgemoed, J.,** Is coccidiosis in adult cattle an exception ? (in Dutch, summary in English), *Tijdschr. Diergeneeskd.*, 110, 558, 1985.
31. **Dubey, J. P.,** Immunity to sarcocystosis: modification of intestinal coccidiosis, and disappearance of sarcocysts in dairy goats, *Vet. Parasitol.*, 13, 23, 1983.
32. **Radostits, O. M. and Stockdale, P. H. G.,** A brief review of bovine coccidiosis in Western Canada, *Can. Vet. J.*, 21, 227, 1980.
33. **Newsom, I. E. and Cross, F.,** An outbreak of coccidiosis in lambs, *J. Am. Vet. Med. Assoc.*, 77, 232, 1930.
34. **Gregory, M. W.,** unpublished data, 1970.
35. **Chineme, C. N. and Njoku, C. O.,** Gastrointestinal globidiosis of Yankasa sheep in Nigeria, *Vet. Rec.*, 103, 418, 1978.
36. **Marsh, H. and Tunnicliff, E. A.,** Enteritis in sheep caused by infection with the protozoan parasite *Globidium gilruthi*, *Am. J. Vet. Res.*, 2, 174, 1941.
37. **Rac, R. and Willson, R. L.,** Globidiosis in sheep, *Aust. Vet. J.*, 35, 455, 1959.
38. **Mugera, G. M. and Bitakaramire, P.,** Gastro-enteritis in Kenya goats caused by *Globidium gilruthi*, *Vet. Rec.*, 82, 595, 1968.
39. **Anderson, P. H., Berrett, S., Catchpole, J., Gregory, M. W., and Brown, D. C.,** Effect of monensin on the selenium status of sheep, *Vet. Rec.*, 113, 498, 1983.
40. **Colnago, G. L., Jensen, L. S., and Long, P. L.,** Effects of selenium and vitamin E on development of immunity to *E. tenella* (abstr.), *Poult. Sci.*, 61, 1374, 1982.
41. **Brandborg, L. L., Goldberg, S. B., and Breidenbach, W. C.,** Human coccidiosis — a possible cause of malabsorption, *N. Engl. J. Med.*, 283, 1306, 1970.
42. **Long, P. L., Johnson, J., and Wyatt, R. D.,** Pathological and clinical effects of *Eimeria tenella* in partially immune chicks, *J. Comp. Pathol.*, 91, 581, 1981.
43. **Isler, C. M., Bellamy, J. E. C., and Wobeser, G. A.,** Labile neurotoxin in serum of calves with "nervous" coccidiosis, *Can. J. Vet. Res.*, 51, 253, 1987.
44. **Smetana, H.,** coccidiosis of the liver in rabbits. III. Experimental study of the histogenesis of coccidiosis of the liver, *Arch. Pathol.*, 15, 516, 1933.
45. **Sezen, I. Y., Rudolph, R., and Greuel, E.,** Zur Pathologie der experimentellen Nierenkokzidiose *(E. truncata)* bei der Hausgans *(Anser anser domesticus)*, *Dtsch. Tierärztl. Wochenschr.*, 88, 502, 1981.
46. **Chapman, J.,** A study of coccidiosis in an isolated rabbit colony. The clinical symptoms, pathology, immunology, and attempted therapy of the disease, *Am. J. Hyg.*, 9, 389, 1929.
47. **Lima, J. D.,** Development of *Eimeria* species in mesenteric lymph nodes of goats, *J. Parasitol.*, 65, 976, 1979.
48. **Lotze, J. C., Shalkop, W. T., Leek, R. G., and Behin, R.,** Coccidial schizonts in mesenteric lymph nodes of sheep and goats, *J. Parasitol.*, 50, 205, 1964.
49. **Gregory, M. W. and Catchpole, J.,** unpublished data, 1987.
50. **Gregory M. W. and Catchpole, J.,** Ovine coccidiosis: pathology of *Eimeria ovinoidalis* infection, *Int. J. Parasitol.*, 17, 1099, 1987.
51. **Gregory, M. W., Catchpole, J., Pittilo, R. M., and Norton, C. C.,** Ovine coccidiosis: observations on "oocyst patches" and polyps in naturally-acquired infections, *Int. J. Parasitol.*, 17, 1113, 1987.
52. **Gregory, M. W. and Catchpole, J.,** Coccidiosis in rabbits: the pathology of *Eimeria flavescens* infection, *Int. J. Parasitol.*, 16, 131, 1986.
53. **Fernando, M. A., Lawn, A. M., Rose, M. E., and Al-Attar, M. A.,** Invasion of chicken caecal and intestinal lamina propria by crypt epithelial cells infected with coccidia, *Parasitology*, 86, 391, 1983.
54. **Tyzzer, E. E., Theiler, H., and Jones, E.,** Coccidiosis in gallinaceous birds. II. A comparative study of species of *Eimeria* of the chicken, *Am. J. Hyg.*, 15, 319, 1932.
55. **Tyzzer, E. E.,** Coccidiosis in gallinaceous birds, *Am. J. Hyg.*, 10, 269, 1929.
56. **Fernando, M. A.,** Pathology and pathogenicity, in *The Biology of the Coccidia*, Long, P. L., Ed., Edward Arnold, London, 1982, chap. 7.
57. **Friend, S. C. E. and Stockdale, P. H. G.,** Experimental *Eimeria bovis* infection in calves: a histopathological study, *Can. J. Comp. Med.*, 44, 129, 1980.
58. **Gregory, M. W.,** The globule leukocyte and parasitic infection — a brief history, *Vet. Bull.*, 49, 821, 1979.
59. **Pierce, A. E., Long, P. L., and Horton-Smith, C.,** Immunity to *Eimeria tenella* in young fowls, *Immunology*, 5, 129, 1962.
60. **Gregory, M. W. and Nolan, A.,** Globule leucocytes and Peyer's patches in lambs infected with coccidia, *Res. Vet. Sci.*, 30, 385, 1981.
61. **Weiss, S. J.,** Tissue destruction by neutrophils, *N. Engl. J. Med.*, 320, 365, 1989.
62. **Gregory, M. W., Catchpole, J., Norton, C. C., and Pittilo, R. M.,** Synchronised division of coccidia and their host cells in the ovine intestine, *Parasitol. Res.*, 73, 384, 1987.
63. **Gregory, M. W.,** unpublished data, 1987.

64. **Fernando, M. A. and McCraw, B. M.,** Mucosal morphology and cellular renewal in the intestine of chickens following a single infection of *Eimeria acervulina, J. Parasitol.*, 59, 493, 1973.
65. **Ferguson, A.,** Pathogenesis and mechanisms in the gastrointestinal tract, in *Proc. 1st Food Allergy Workshop,* Medical Education Services, Oxford, 1979, 28.
66. **Ferguson, A. and Jarrett, E. E. E.,** Hypersensitivity reactions in the small intestine. I. Thymus dependence of experimental 'partial villous atrophy', *Gut,* 16, 114, 1975.
67. **Gregory, M. W., Pittilo, R. M., Ball, S. J., and Hutchison, W. M.,** Scanning electron microscopy of filamentous organisms associated with coccidial infections in cats and sheep, *Ann. Trop. Med. Parasitol.*, 79, 473, 1985.
68. **Bennett, R. E., Harrison, M. W., Bishop, C. J., Searle, J., and Kerr, J. F. R.,** The role of apoptosis in atrophy of the small gut mucosa produced by repeated administration of cytosine arabinoside, *J. Pathol.*, 142, 259, 1984.
69. **Duvall, E. and Wyllie, A. H.,** Death and the cell, *Immunol. Today,* 7, 115, 1986.
70. **Fernando, M. A. and Pasternak, J.,** *Eimeria necatrix:* induced proteolytic sensitivity of infected chick crypt cells, *Exp. Parasitol.*, 56, 20, 1983.
71. **Pasternak, J. and Fernando, M. A.,** Host cell response to coccidian infection: an introspective survey, *Parasitology,* 88, 555, 1984.
72. **Pout, D. D.,** Villous atrophy and coccidiosis, *Nature London,* 213, 306, 1967.
73. **Pout, D. D.,** Coccidiosis of lambs. III. The reaction of the small intestinal mucosa to experimental infections with *E. arloingi* "B" and *E. crandallis, Br. Vet. J.*, 130, 45, 1974.
74. **Jubb, K. V. F. and Kennedy, P. C.,** *Pathology of Domestic Animals,* Vol. 2, 2nd ed., 1972, 134.
75. **Turk, D. E.,** Intestinal parasitism and nutrient absorption, *Fed. Proc.*, 33, 106, 1974.
76. **Lozanov, L. and Koinarski, V.,** Morphology and histokinesis of the changes in turkey poults caused by an experimental *Eimeria adenoeides* infection, (in Bulgarian, summary in English), *Vet. Sci. Sofia,* 22, 43, 1985.
77. **Marquardt, W. C., Bryan, J. H. D., and Long, P. L.,** Nucleolar hypertrophy as an indicator of transcription in cells infected with second generation meronts of *Eimeria tenella, J. Protozool.*, 31, 569, 1984.
78. **Joyner, L. P., Norton, C. C., and Catchpole, J.,** unpublished data, 1976.
79. **Hammond, D. M., Bowman, G. W., Davis, L. R., and Simms, B. T.,** The endogenous phase of the life cycle of *Eimeria bovis, J. Parasitol.*, 32, 409, 1946.
80. **Majid, H. N. and Winter, H.,** Glomerulonephritis in lambs with coccidiosis, *Aust. Vet. J.*, 63, 314, 1986.
81. **Levine, N. D.,** *Veterinary Protozoology,* Iowa State University Press, Ames, 1985, chap. 8.
82. **Hermier, M., Mojon, M., Piens, M. A., Louis, J. J., and Descos, B.,** Forme grave d'isosporose à *Isospora hominis, Pédiatrie,* 36, 211, 1981.
83. **Hammond, D. M., Davis, L. R., and Bowman, G. W.,** Experimental infections with *Eimeria bovis* in calves, *Am. J. Vet. Res.*, 5, 303, 1944.
84. **Hammond, D. M.,** Cellular interaction between bovine coccidia and their hosts, *Prog. Protozool.*, 1965, 53.
85. **Stockdale, P. H. G.,** The pathogenesis of the lesions produced by *Eimeria zuernii* in calves, *Can. J. Comp. Med.*, 41, 338, 1977.
86. **Fox, J. E.,** Bovine coccidiosis, *Modern Vet Prac.*, 59, 599, 1978.
87. **Fanelli, H. H.,** Observations on "nervous coccidiosis" in calves, *Bovine Practitioner,* No. 18, 50, 1983.
88. **Gregory, M. W., Joyner, L. P., Catchpole, J., and Norton, C. C.,** Ovine coccidiosis in England and Wales 1978—1979, *Vet. Rec.*, 106, 461, 1980.
89. **Gregory, M. W., Catchpole, J., Nolan, A., and Hebert, C. N.,** Ovine coccidiosis: studies on the pathogenicity of *Eimeria ovinoidalis* and *E., crandallis* in conventionally-reared lambs, including possible effects of passive immunity, *Dtsch. Tierärztl. Wochenschr.*, 96, 287, 1989.
90. **Gregory, M. W. and Catchpole, J.,** Output of coccidial oocysts (particularly *Eimeria crandallis*) by naturally-infected lambs: daily and hourly patterns and clinical significance, *Dtsch. Tieraerztl. Wochenschr.*, 94, 521, 1987.
91. **Smith, W. N., Davis, L. R., and Bowman, G. W.,** The pathogenicity of *Eimeria ah-sa-ta,* a coccidium of sheep, *J. Protozool.*, 7(Suppl), 8, 1960.
92. **Catchpole, J., Norton, C. C., and Joyner, L. P.,** Experiments with defined multispecific coccidial infections in lambs, *Parasitology,* 72, 137, 1976.
93. **Norton, C. C.,** Coccidia of the domestic goat *Capra hircus,* with notes on *Eimeria ovinoidalis* and *E. bakuensis* (syn. *E. ovina*) from the sheep *Ovis aries, Parasitology,* 92, 279, 1986.
94. **Sayin, F., Dincer, S., and Milli, U.,** The life cycle and pathogenicity of *Eimeria arloingi* (Marotel, 1905) Martin, 1909, in angora kids and an attempt at its transmission to lambs, *Zentralbl. Vet. Med. B,* 27, 382, 1980.
95. **Norton, C. C. and Gregory, M. W.,** unpublished data, 1985.

96. **Yvoré, P., Esnault, A., and Naciri, M.,** La coccidiose caprine: effet de contaminations mono ou multispécifiques, *Rec. Méd. Vét.*, 161, 347, 1985.
97. **Hussein, H. S., Kasim, A. A., and Shawa, Y. R.,** The prevalence and pathogenicity of *Eimeria* infections in camels in Saudi Arabia, *J. Comp. Pathol.*, 97, 293, 1987.
98. **Stuart, B. P. and Lindsay, D. S.,** Coccidiosis in swine, *Vet. Clin. North Am. Food Anim. Prac.*, 2, 455, 1986.
99. **Stuart, B. P., Lindsay, D. S., Ernst, J. V., and Gosser, H. S.,** *Isospora suis* enteritis in piglets, *Vet. Pathol.*, 17, 84, 1980.
100. **Matuschka, F. R. and Heydorn, A. O.,** Die Entwicklung von *Isospora suis* Biester und Murray 1934 (Sporozoa: Coccidia: Eimeriidae) im Schwein, *Zool. Beitr.*, 26, 405, 1981.
101. **Robinson, Y., Morin, M., Girard, C., and Higgins, R.,** Experimental transmission of intestinal coccidiosis to piglets: clinical, parasitological and pathological findings, *Can. J. Comp. Med.*, 47, 401, 1983.
102. **Stuart, B. P., Gosser, H. S., Allen, C. B., and Bedell, D. M.,** Coccidiosis in swine; dose and age response to *Isospora suis*, *Can. J. Comp. Med.*, 46, 317, 1982.
103. **Vitovec, J. and Koudela, B.,** Pathology of natural isosporosis in nursing piglets, *Folia Parasitol.*, 34, 199, 1987.
104. **Hill, J. E., Lomax, L. G., Lindsay, D. S., and Lynn, B. S.,** Coccidiosis caused by *Eimeria scabra* in a finishing hog, *J. Am. Vet. Med. Assoc.*, 186, 981, 1985.
105. **Vitovec, J., Koudela, B., and Sterba, J.,** Pathology and pathogenicity of *Eimeria scabra* (Henry, 1931) in experimentally infected pigs, *Folia Parasitol.*, 34, 299, 1987.
106. **Barriga, O. O. and Arnoni, J. V.,** *Eimeria stiedae:* weight, oocyst output, and hepatic function of rabbits with graded infections, *Exp. Parasitol.*, 48, 407, 1979.
107. **Peeters, J. E., Charlier, G., Antoine, O., and Mammerickx, M.,** Clinical and pathological changes after *Eimeria intestinalis* infection in rabbits, *Zentralbl. Vet. Med. B*, 31, 9, 1984.
108. **Long, P. L.,** Pathology and pathogenicity of coccidial infections, in *The Coccidia*, Hammond, D. M. and Long, P. L., Eds., University Park Press, Baltimore, 1973, chap. 7.
109. **Stockdale, P. H. G. and Fernando, M. A.,** The development of the lesions caused by second generation schizonts of *Eimeria necatrix*, *Res. Vet. Sci.*, 19, 204, 1975.
110. **Smith, R. R., Ruff, M. D., and Witlock, D. R.,** *Eimeria necatrix* in the chicken: response of the host jejunum to infection and subsequent absorption of methionine and glucose, *Proc. Helminthol. Soc. Wash.*, 47, 235, 1980.
111. **Michael, E. and Hodges, R. D.,** Scanning electron microscopy of the duodenal mucosa of chickens infected with *Eimeria acervulina*, *Parasitology*, 71, 229, 1975.
112. **Long, P. L.,** Studies on *Eimeria praecox* Johnson, 1930, in the chicken, *Parasitology*, 57, 351, 1967.
113. **Joyner, L. P.,** Coccidiosis in turkeys and its control, *Folia Vet. Lat.*, 3, 110, 1973.
114. **Ruff, M. D., Augustine, P. C., and Madden, P. A.,** *Eimeria meleagrimitis*, *E. adenoeides*, and *E. dispersa:* severity of infection and changes in the intestinal mucosa of the turkey, *Exp. Parasitol.*, 51, 87, 1981.
115. **Gajadhar, A. A., Wobeser, G., and Stockdale, P. H. G.,** Coccidia of domestic and wild waterfowl (Anseriformes), *Can. J. Zool.*, 61, 1, 1983.
116. **Farr, M. M. and Wehr, E. E.,** *Eimeria truncata* associated with morbidity and death of domestic goslings, *Cornell Vet.*, 42, 185, 1952.
117. **Randall, C. J. and Norton, C.C.,** Acute intestinal coccidiosis in geese, *Vet. Rec.*, 93, 46, 1973.
118. **Versényi, L. and Pellérdy, L.,** Pathological and immunological investigations of the anseris-coccidiosis of the domestic goose *(Anser anser dom.)*, *Acta. Vet. Hung.*, 20, 103, 1970.
119. **Norton, C. C.,** Coccidiosis in gamebirds: a review, in *Research in Avian Coccidiosis*, McDougald, L. R., Long, P. L., and Joyner, L. P., Eds., University of Georgia, Athens, 1986, 85.

Chapter 13

PATHOPHYSIOLOGY

M. D. Ruff and P. C. Allen

TABLE OF CONTENTS

I. INTRODUCTION

In several recent reviews the physiological reasons for the pathological changes seen in birds with avian coccidiosis have been discussed.[1-6] This paper will not repeat information in those reviews unless it is pertinent to new conclusions or overall hypotheses on how coccidiosis induces the observed pathological changes. For example, the effect of anorexia, reduced feed intake, and maldigestion are discussed in several of those reviews but are not covered here because no new information is available. Similarly, many of the references given in those articles will not be given here as it is not the intent of this paper to be a complete review on the subject of pathophysiology. Rather, the emphasis will be on aspects not previously covered or on results obtained.

II. PATHOPHYSIOLOGY IN THE INTESTINE

A. STRUCTURAL CHANGES

1. Physical Changes

A large number of investigations have reported descriptive pathology of enteric coccidia infections in birds[7-10] and mammals.[11-15] The common theme, with some variations, has been the observations of villus flattening or atrophy, crypt hyperplasia and decreases in villus/crypt ratios accompanied by varying degrees of malaise, diarrhea, malabsorption, and reduction in growth rate.

Coccidia which infect and develop in crypt stem cells, particularly of the lower small intestine, cecum, and colon are highly pathogenic because they tend to destroy the crypt stem cells that they invade. Examples include *E. vermiformis* and *E. falciformis* in mice,[11,12] *E. caviae* in guinea pigs,[16] *E. intestinalis* and *E. flaviscens* in rabbits,[17-19] *E. crandallis* and *E. ovinoidalis* in sheep,[20-21] *E. caprina* and *E. ninakohlyakimova* in goats,[22] and *E. scabra* in pigs.[23] This destruction prevents renewal of villus epithelium and results in denuding of villi with attendant fluid loss, hemorrhage,susceptibility to bacterial invasion (particularly by coliform bacteria), and subsequent formation of necrotic lesions. Severe infections are frequently accompanied by watery diarrhea and death.

The pathogenic effects of coccidia that primarily inhabit the villus epithelium are more variable. Some such as *E. wyomingensis*[24,25] in cattle produce no overt disease even at high infecting doses. *Eimeria mitis*[26] in chickens and *Eimeria lettyae*[27] in bobwhite quail, produce no gross lesions even in heavy infections, but do cause depigmentation and retarded weight gain. At high doses *E. lettyae* causes death in young, but not mature bobwhites. The turkey coccidia, *E. meleagrimitis, E. adenoeides,* and *E. dispersa,* also produce few gross lesions but cause malabsorption and reduced weight gain.[9,28]

In most of these studies, observations of structural changes have been confined to the area of parasite development. However, it is now clear that the physiology and biochemistry of the entire intestine is affected. For example, in chicks infected with *E. acervulina* or *E. mitis,* segments remote from primary sites of infection are increased in length and weight, exhibit mucosal hyperplasia and metabolic changes.[29,30] Intestinal distension, dilation,and flaccidity accompanied by decreased motility has been noted in rabbits infected with *E.*

intestinalis, *E. magna*, and *E. perforans* infections.[31] This condition promotes intussusception.[32]

2. Cellular Changes in the Mucosal Epithelium

The gross appearance of the intestinal mucosa is generally similar among humans, cats, dogs, pigs, rabbits, chickens, and turkeys.[33] From the duodenum through the ileum, villi gradually change shape, going from broad ridge to leaf shape in the duodenum to finger and tongue shape in the ileum. They also become shorter and narrower. It is generally accepted that these changes are gradations of a single process related to the rate of cell turnover,[34] and are influenced by the degree of exposure to physical and chemical damage from gut lumen contents.[35] Increased mitotic activity leading to increased cell turnover is the most common change reported in coccidia infections. There is also an increased rate of stem cell differentiation so that goblet cell hyperplasia[35] and increased numbers of enteroendocrine cells[36] are detected. Thus, not only do infections cause loss of absorptive area due to villar flattening, but the mucosal population also has a decreased percentage of functionally absorptive cells.

Nuclear and nucleolar hyperplasia of infected enterocytes has been reported.[37] Endogenous life stages of *Eimeria stigmosa* in snow geese occur in the nucleoplasm of enterocytes, principally in the apical halves of villi, causing nuclear hypertrophy. Nuclei containing gamonts are up to four times normal size.[10]

Infection of chicks with *E. necatrix* drastically alters invaded host crypt cells.[38] Changes include increased cell surface area, increased synthesis of genomic DNA, increased resistance of the plasma membrane to shear, hypotonic shock, and ultrasonication, increased accumulation of gel phase lipid in the plasma membrane and the ability to become migratory. The host cells ultimately become more suscepible to lysis by low concentrations of trypsin. The major pathology of *E. necatrix* infections is associated with the rupture of these cells.

However, in some instances, diminished mitotic activity is seen in the crypts, as reported in the early stages of *E. intestinalis* infections.[17] Actual crypt atrophy in the small intestine is associated with the release of first stage giant meronts of *E. ovinoidalis* in sheep.[21] It has been suggested that this is caused by a release of a toxic substance from the giant meronts, since the overall process consisting of increased apoptosis and decreased mitosis is similar to that elicited by cytotoxic drugs.

The diarrhea seen in *E. ovinoidalis* infections is associated with release of second generation meronts and gamonts. Crypt cells fail to multiply and infected as well as uninfected cells are shed when the parasite matures. Tissue culture studies suggest that separation of sheets of cells containing parasites could be due to a diffusible product from the infected cells.[39]

B. METABOLIC CHANGES

Metabolic changes accompany the structural changes in the intestinal mucosa that arise during infection by intestinal coccidia such as *E. acervulina* in chickens.[40] Measurements of CO_2 evolved from C-1 and C-6 ^{14}C-labeled glucose showed that the C-1/C-6 ratio of infected duodenal tissue is increased, reaching a maximum coinciding with the presence of large numbers of gamonts. This increase in ratio is indicative of a metabolic shift away from mitochondrial oxidation towards oxidation through the pentose phosphate pathway. Mitochondria isolated from infected tissues oxidize both octanoic acid and α-ketoglutaric acids at lower rates than mitochondria from control tissues. These results indicate damaged mitochondrial function. Electron micrographs reveal mitochondria in large patches of uninfected cells from infected epithelium that are vacuolated and contain degenerate cristae. Feed withdrawal for 48 h does not mimic the ultrastructural or metabolic effects of infection. The fact that the ultrastructual changes are seen primarilyin uninfected cells suggests that they are not caused by the parasite directly, but by some product of host-parasite interaction. These findings suggest that coccidial infections cause decreased energy output in the infected tissues that should decrease absorption of actively transported nutrients and further aggravate the malabsorption brought about by loss of mucosal absorptive surface.[40]

Metabolic changes also occur in portions of the intestine not primarily infected with the parasite as illustrated in the studies of *E. acervulina* infection.[29] Lengthening of the distal third of the chick small intestine is the first initial gross response. This is followed by increased mucosal mass and increased oxidative activity of the mucosal mitochondria. These reactions however, may be basically nonspecific, and initiated by products elaborated by immune interactions as suggested in a recent review.[41]

Although the mucosa of the chick distal small intestine becomes hyperplastic during *E. acervulina* infections its ability to absorb carotenoids is diminished. Conversely, activities of the brush border enzymes sucrase and alkaline phosphatase are increased.[30] This finding is consistent with observations of compensatory absorption of nutrients such as glucose and methionine in this tissue.[2,3]

C. CHANGES IN NUTRIENT ABSORPTION

This subject has been adequately reviewed by Turk[1] and Ruff.[2,3] Little research has been done in this area since 1985. Basically, coccidiosis produces a major malabsorption of most nutrients in the region of the intestine where infection occurs. No effect on nutrient absorption has been demonstrated with cecal coccidiosis (*E. tenella*). Malabsorption can be related directly to changes in the mucosal architecture with some species (*E. acervulina, E. adenoeides*) or can occur in the absence of any visible change or only slight change in the mucosa (*E. mitis, E. meleagrimitis*). Compensatory absorption can occur in noninfected regions.

III. PATHOPHYSIOLOGY IN EXTRAINTESTINAL TISSUE

As will be elaborated below, the pathological effects of enteric coccidial infections in chickens are not restricted to effects on the gut nor do they simply cause reduced weight gain because of anorexia or malabsorption. There is also a series of compositional and metabolic changes in infected tissues and organs.

A. TISSUE COMPOSITION

The levels of moisture, ash, protein, and DNA in the breast, thigh, heart, and liver do not differ among uninoculated broilers and broilers inoculated with *E. acervulina, E. maxima, E. brunetti,* or *E. tenella.*[42] In contrast, RNA decreases in the breast, thigh, and liver of broilers infected with *E. acervulina* and in the liver of broilers infected with *E. brunetti.* Lipid content decreases only in the liver with *E. acervulina* and *E. brunetti* and only in the breast with *E. tenella* infections.

Heart weights are reduced in turkeys by infection with *E. meleagrimitis*[43] or *E. adenoeides*[44] and in chicks infected with *E. acervulina.*[45] However, weights expressed as percentages of body weights do not differ significantly from control weights. This finding suggests that heart sizes and weights are affected by the general growth depression that accompanies infection. Heart lipid and glycogen levels in turkeys and glycogen levels in chicks are reduced at the acute stages of infection.[28,45]

In hearts from infected chicks, glycogen is increased 1.6-fold compared to controls at the beginning of recovery from *E. acervulina* infection. This increase is not as great as normally occurs upon feed withdrawal in chicks, but again suggests effects of nutrient deprivation. Paradoxically this effect occurs at a time when chicks are beginning to refeed.

The wet and dry weights of ceca of chickens infected with *E. tenella* increase compared with the weights in uninoculated controls.[46] Moisture and lipid levels do not change or increase only slightly. Cecal protein and DNA increase indicating an increase in cecal tissue. The changes in structure stimulated by the coccidial infection may also result in a change in function. The cecal muscle from chickens infected with *E. tenella* has an increased amplitude of contraction that is directly related to the severity of infection.[47] The filling and emptying of the ceca is also inhibited by infection.[48]

B. METABOLIC CHANGES

Eimeria maxima infection in chicks causes a decrease in liver glycogen during the acute phase followed by an overshoot in liver glycogen concentrations at about 8 days PI, and another at 10 to 11 days PI.[49] These changes are similar to those described for *E. acervulina* infections,[42,50] including elevation in the normal range of liver glycogen values seen about 7 days PI. Infection with the cecal coccidium, *E. tenella* caused moderate decreases in liver glycogen, but no increase of normal values during recovery from infection.

Analysis of activities of the two enzymes that control the synthesis and breakdown of liver glycogen, glycogen phosphorylase and glycogen synthase, in livers of chicks infected with either *E. acervulina* or *E. tenella* failed to reveal any correlations between total activities of these enzymes (or the ratios of their activities) and the changes in liver glycogen concentration with time after infection.[51] These findings suggest that allosteric modulators of these enzymes elaborated during infection may affect their activities and thus control the levels of liver glycogen.

Infection of chicks with *E. acervulina* cause metabolic changes in heart tissue.[44] At the acute stage of infection oxidation of octanoic acid by heart mitochondria is depressed. At 7 days PI, when chicks are recovering, oxidation of octanoic acid is increased threefold indicating an increase in mitochondrial enzymes capable of oxidizing fatty acids. This finding is consistent with the concurrent increase in heart glycogen indicating that the heart tissue shifts from oxidizing carbohydrates to fat. The stimulus for this shift is not known, but may be linked to hormonal changes that occur with infection.[36]

Infected turkeys have lower heart rates and mean blood pressure than uninoculated controls. When stressed by immersing the birds feet in ice water or by injection of epinephrine, the increase in blood pressure and decrease in heart rate are greater and more prolonged in infected turkeys than in the controls.[28]

IV. CLINICAL CHEMISTRY

Coccidiosis causes several changes in the blood of infected birds besides decreases in the levels of plasma pigment and other nutrients. Total serum protein decreases during infection with most avian species.[49,52] The changes in the different serum protein fractions, however, differs with the coccidial species studied. In *E. acervulina*-infected chickens, all protein components are decreased; the greatest decrease being in the alpha globulins. In *E. tenella*-infected chickens, the level of albumin decreases but the alpha-1, beta, and gamma-1 globulins increased. In *E. maxima*-infected chickens, only the albumin and alpha-1 globulin changed (decrease). In turkeys, infection with *E. meleagrimitis* or *E. adenoeides* increases the levels of alpha-2 and gamma globulins and decreases albumin.[53] The increase in the alpha-2 globulin is not due to pre-beta-lipoprotein, nor is the increase due to ceruloplasmin. The portomicrons are sharply decreased. Other changes in blood composition are discussed in the section on plasma pigment.

The hematopoetic balance can also be upset by coccidiosis a decrease in packed cell volume with species such as *E. tenella*. Infection of turkeys with *E. adenoeides* increases the number of immature RBCs from 0.4 to 3.5%.[54] At the same time the RBCs from infected turkeys are 50 to 65% less fragile to osmotic shock.

Eimeria acervulina infection in chicks causes significant decreases in total plasma lipids from 4 to 8 days PI, the largest decrease being at 5 days PI.[55] All classes of lipids decrease and the lipoprotein banding patterns after ultracentrifugation is altered. The decreases in lipids seen between days 4 and 6 PI, particularly in the portimicron + very low density lipoprotein fraction (highest in triglycerides) are similar to those evoked by 48 h of feed withdrawal, and so might reflect effects of anorexia and malabsorption occurring at that time. Between 7 and 8 days PI, however, infected chicks are beginning to refeed, and their plasma lipids are still depressed. During this time the low density lipoprotein (LDL) band almost disappears. Conversely, 48 h feed withdrawal causes a marked increase in this band. The continued decrease inplasma lipids on days 7 through 9 could be due to some interference with lipoprotein assembly or synthesis of the apoproteins in the intestine and liver. The effect of infection of LDL lasts up to 15 days PI.

Plasma protein concentrations are decreased during the acute phase of *E. maxima* infections,[49] but plasma glucose, FFA and uric acid levels are unaffected. The findings are similar to those where feed had been withheld. In general, plasma glucose levels appear to be maintained during coccidial infections at the expense of carbohydrates and fat reserves.

The effect of coccidial infections on the levels of microelemnts in the blood is dependent on the mineral, species of coccidia, and stage of the infection.[56] Copper levels in the plasma are increased by *E. brunetti* and *E. tenella.* Zinc levels are decreased by *E. acervulina* and *E. necatrix* on the 7th day but increased by *E. acervulina, E. brunetti,* and *E. tenella* on the 9th day PI. The same three species decreased iron levels on days 5 to 7 PI.

Plasma glutamic oxaloacetic transaminase (GOT) increases and carotenoid levels decreases in turkeys inoculated with *E. meleagrimitis.*[43] These levels are not affected by feed or water deprivation in the absence of infection.[57] Plasma carotenoid levels are lower and plasma glucose levels higher in turkeys with *E. adenoeides* infections.[41] The first species decreases plasma and liver retinol and liver retinylpalmitate levels whereas the second species did not.[58]

Rabbits infected with *E. intestinalis,* exhibit hemodilution at the peak of infection as shown by a 5.25% decrease in packed cell volume and a fall in plasma potassium, sodium, and chloride.[17] This is accompanied by a marked decrease in feed intake. Through this shortage of energy intake, plasma glucose levels are maintained at 52.1% of controls. This maintenance takes place at the expense of liver glycogen which is decreased significantly. The need for energy is met by increased catabolism of proteins and rapid mobilization of lipids to support gluconeogenesis, since there is significant uremia and lipemia. These findings also explain the rapid weight loss in infected animals. Clinical signs and symptoms are most severe 8 to 12 days PI and coincide with gametogenesis. There is no significant change in plasma SGOT activity, although the trend is to higher mean value.

At onset of diarrhea and weight loss and oocyst output, there is a reduction in serum albumin levels and plasma alkaline phosphatase activity in lambs infected with *E. crandallis.*[20] Globulins (determined by difference between albumin and total protein) do not change significantly. There are no consistent trends in plasma Na^+, K^+, or Cl^-. Over the course of infections (30 days) there are gradual, but not significant rises in PCV, HCT, and Hb. Clinical and biochemical findings are consistent with loss of absorptive intestinal epithelium that results in water loss (diarrhea). Lowered alkaline phosphatase is presumed to be a sign of loss of absorptive epithelium. Infections with *E. ahsata, E. ovina,* and *E. weybridgensis* product no significant clinical signs.

V. PIGMENTATION

A sensitive pathophysiological manifestation of all coccidia infections in domestic birds is the decreased pigmentation of plasma and tissues by dietary carotenoids. The severity of the effect, however, depends upon the parasite and the portion of the intestine with which it primarily interacts.[3,59] The cecal parasite, *E. tenella*, produces moderate and transient decreases in plasma carotenoids which can be accounted for by blood loss[3,6,60] and are not due to malabsorption. A similar decrease is found with *E. adenoeides* infections in turkeys but the mechanism of the decrease is malabsorption.[44] On the other hand, species such as *E. acervulina* and *E. maxima* parasitizing the small intestine elicit marked, long lasting decreases in plasma carotenoids. This effect can be accompanied by significant and sustained reduction in weight gain and changes in body metabolism. Such carotenoid decreases are considered to be partially the result of malabsorption and partially the result of change in blood transport and dposition of carotenoids. Carotenoid malabsorption during coccidiosis can occur with or without gross architectural changes in the gut.[9,26,27] Therefore, a general mechanism is needed to explain this process.

A. PLASMA CAROTENOIDS

In the chick, plasma carotenoids are concentrated primarily in high density lipoprotein (HDL) particles.[61] They appear to be bound by the main HDL protein, apolipoprotein A-1 (apo A-1). Apo A-1 and carotenoids diminish proportionally during coccidia infection. Importantly, the small intestine is a major site for the biosynthesis of apo A-I[62] where it is known to be synthesized in response to eating and to be incorporated into nascent lipoproteins and exported from the gut via Golgi.[63] Whether carotenoids are bound directly to this protein as they cross the intestinal brush border, or whether they are bound to another receptor before being packaged into lipoproteins is not known.

The concept of a protein carrier for carotenoid pigments is not new. However, from the facts that not all species absorb carotenoids to the same degree,[64] and that all carotenoids are not absorbed at the same rate within a species,[65] it would appear that the carrier is both species specific and stereospecific. It is suggested that the high helical content of chick apo A-1 (90% compared to 70% in humans and 60% in rat) confers these specificities.[66]

Results of studies of *E. acervulina* infected chicks on diets supplemented with varying levels of the carotenoid lutein are consistent with a carrier model for carotenoid absorption, show plasma saturation at lower supplementation levels (indicating decreased availability of a carrier)[66] and indicate that carotenoid malabsorption due to coccidiosis cannot be reversed by increasing dietary carotenoid supplementation.

B. MUCOSAL CAROTENOIDS

Recent work[29,30,67] has reaffirmed the important concept[59] that carotenoid absorption over the entire small intestine is affected by coccidial infections regardless of the primary site of infection. Malabsorption is coexistent with a general hyperplasia that is a result of increased mitotic activity in the crypts and altered metabolic states. Mucosal carotenoids apparently concentrate within Golgi fractions[68] where they are packaged, presumably with their carrier apo A-1, for export from the intestine in lipoprotein particles. If the biosynthesis of this carrier is reduced, malabsorption should result. Incorporation of ^{3}H-leucine into total immunoprecipitable apo A-I extracted from discs from infected mucosa and liver is significantly decreased compared to control at 7 days PI.[69] On the other hand incorporation of label into total protein was increased to 175% of control. This latter finding is consistent with the assumption that the mucosa is actively secreting antibody into the gut lumen during infection.[70] In order to do this, mucosal cells must synthesize secretory component (SC), the receptor and transporter of secretory Igs.[71] Posttranslational processing of SC takes place in the Golgi.

The observations above lead to the hypothesis that in the altered population of mucosal cells elicited by coccidia infections there may be a shift or redirection of protein synthesis which supports transport of secretory antibodies towards the gut lumen and subverts the synthesis of binding proteins or carriers that transport nutrients from the gut lumen into the blood.

C. TISSUE CAROTENOIDS

Coccidiosis has severe effects on the form of the individual carotenoid in the blood, liver, and tissue.[72] Serum concentrations of beta-carotene, cryptoxanthin, oxolutein, and lutein are reduced by infection with *E. acervulina*. Monoester C is not affected. In the liver, the diester, beta-carotene, several monoesters, cryptoxanthin, and lutein are all reduced. In the toe web, free lutein is reduced in a stepwise fashion with time. Monoester is not affected while diester and cryptoxanthin are not reduced until day 10 PI.

VI. HORMONES AND TOXINS

A. HORMONES

1. Peptide Hormones

Prolactin (PRL) is present in all vertebrates. In addition to its role in reproduction, it apparently can also function in osmoregulation, growth and development[73] and immunoregulation.[74,75] It, along with the other pituitary hormones, has been classified as a stress hormone.[73] Infections of chickens with both *E. tenella* and *E. maxima* elicit rises in plasma levels of PRL. It is believed that these increases are homeostatic responses to the osmotic stress of water loss from infection-associated diarrhea.[76,77] However, in the case of *E. tenella* infections, PRL levels are still increased at a time when body water balances are returning to normal. This observation suggests that other factors may also influence PRL increases in coccidia infections. Interestingly, PRL levels are also generally raised in patients suffering from celiac disease,[78] a condition which causes enteric pathology similar to that seen in coccidial infections.

Because one of the salient pathophysiological manifestations of coccidial infections is decreased growth rate, it might be expected that growth hormone (GH) levels would be reduced. However, no significant changes in plasma levels of GH are seen in *E. maxima* infections.[77] The investigators suggest that GH probably does not influence growth depression during coccidiosis.

Although a role for somatomedin C has been implicated in the pathophysiology of sarcosporidiosis in cattle,[79] there have been no investigations to determine the involvement of it or other insulin-like growth factors in coccidiosis.

Infection of chickens with *Eimeria* sp. that parasitize the upper and middle portions of the small intestine cause increases in plasma levels of glucagon-like immunoreactants (GLI), but not necessarily pancreatic glucagon (PG). Variations of PG and insulin (PI) likely reflect changes in feed intake during infection.[80] The increases in GLI reflect changes in the intestine, particularly an increased rate of stem cell differentiation that results in production of enteroendocrine cells containing GLI. Numbers of these cells have been found to increase significantly in the middle and lower two thirds of the chick small intestine. Coccidia that infect the anterior portions of the small intestine (*E. acervulina* and *E. maxima*) appear to stimulate the greatest production of GLI containing cells.[36] Interestingly, plasma levels of GLI are also increased in human patients with active celiac disease.[81]

2. Thyroid Hormones

Both T_4 and T_3 decrease during *E. maxima* infections.[77] These decreases apparently follow trends in feed consumption. It is suggested that reduction in T_3 is therefore linked to reduction in metabolic rate resulting from anorexia which is seen in coccidial infections in general.[82] Reduction may also be related to hypoproteinemia seen during infection, since T_3 and T_4 are bound by plasma proteins. Plasma proteins may be lost through the gut during infection.[83]

3. Steroid Hormones

Plasma levels of adrenal steroids (corticosterone, cortisol) generally are increased in response to stress and to infections.[73] However, it is questionable whether animals perceive parasitic infections as stressful.[84] For instance, corticosterone levels are not significantly increased in *E. maxima* infections.[3] The investigators suggest that this is because birds voluntarily reduced feed intake, and therefore did not feel stress of feed deprivation.

Adrenal concentrations of corticosterone are higher in resistant chicks than in chicks susceptible to *E. tenella*.[85] Investigators suggest that differences might be due to different rates of release, meaning susceptible chicks would release corticosterone more abundantly.

It is well known that site specificity and host specificity of coccidial infections can be compromised by immunosuppression with pharmacological doses of steroids.[86]

In a recent review[87] it was stated that with some exceptions, males are more susceptible to parasitic infection than females, and susceptibility to many parasites increases during pregnancy. As an example, male bobwhite quail appear to be more susceptible than females to *E. lettyae*.[27] Infection also delays sexual maturity in males, and adversely affects egg laying. Infection with *E. uzura* increases the levels of circulating androgens in male japanese quail.[88] However, no differences between sexes are found in the oocyst output from Wood mice infected with *E. hungaryensis*.[89]

It is thought that in general, estrogens enhance humoral immunity while inhibiting cell-mediated immunity, whereas male hormones tend to suppress both types of response.[87] Administration of testosterone to chicks *in ovo* can effectively bursectomize them. But this did not inhibit the development of immunity to *E. tenella* post-hatching.[90]

4. Catecholamines

Little work has been done to determine if coccidial infections cause stress-related changes in host catecholamine metabolism. It has been noted, however, that infection of turkeys with *E. adenoeides* caused more pronounced responses to injection with pharmacological levels of epinephrine, in that mean blood pressure was increased and mean heart rate was decreased to greater extent than in control turkeys.[28]

Catecholamines are synthesized in the brain and at sympathetic nerve endings, in addition to the adrenal medulla, and can have profound effects on smooth muscle, adipose tissue, myocardium, liver, brain, and blood cells.[73] Catecholamines are known to have a major influence on lymphocyte recirculation and distribution. Further, β-adrenergic stimulants seem to have an inhibitory effect on the immune system in general. No investigations of the effects of coccidiosis on these physiological functions of catecholamines has been reported.

B. TOXINS

Toxins have been suggested to have a possible role in the pathophysiological changes of coccidiosis. Cecal extracts of chickens produce death from intervascular coagulation in 20 to 60 sec after intravenous injection into the wing vein of control chickens. This activity is 2 to 5 times as great in an extract from *E. tenella*-infected chickens as that from uninoculated controls.[91] The extract demonstrated a thromboplastin-like activity. The unidentified toxic substance is localized in the microsomal fraction of both the mucosal and muscular layers of the ceca.[92] It is hypothesized that the release of this toxic activity in infected chickens could deplete clotting factors and lead to a hypocoagulative state that results in increased hemorrhage, blood loss, and ultimately death.

Nervous signs and symptoms have been associated with coccidia infections of sheep, goats, dogs, chicks, rabbits, and cattle.[93] In cattle infected with *E. bovis* or *E. zuernii,* these include seizures, and opisthotonus with extensor rigidity.[94] Calves with neurological signs have low plasma levels of copper and high levels of glucose. These blood values may be signs of but not cause of neurological symptoms. The following are not believed to be involved in pathogenesis of nervous coccidial infections: disturbances of serum sodium, potassium, phosphorus, calcium or magnesium, vitamin A deficiency, anemia, lead toxicity, uremia, *Haemophilus* infections, meningoencephalitis, gross alterations in intestinal bacterial flora, or hepatopathy. The syndrome is hard to reproduce experimentally. The investigators suggest that clinical signs could possibly be induced by peptide hormones common to CNS and gastrointestinal tract such as beta-endorphin, leucine enkephalin methionine, enkephalin, somatostatin, and ACTH. However, they did isolate from plasma a substance with a molecular weight higher than 300,000 that had neurotoxic activity. It was found only in plasma, not cerebrospinal fluid, was precipitable from serum with 30% (w/v) $(NH_4)_2SO_4$ and was labile to heat (60°/30 min) and exposure to acidic pH (pH 5.5) or cysteine (1.75 g/100 ml). Ten milliliters of infected calf serum per kilogram produced nervous signs and symptoms in mice.[94-96]

It is suggested that the atrophy of small intestinal crypts seen in *E. crandallis* infections is caused by the release of a toxic substance from the first stage giant meronts, since the process appears similar to that produced by cytotoxic drugs.[21]

C. OTHER BIOCHEMICAL MESSENGERS

There are many host hormones and effecter molecules other than those mentioned above, whose altered productions are likely important in the pathophysiology of coccidia infections. In helminth infections, for example, prominent roles for 5-hydroxytryptamine and prostaglandins in the enteric changes seen have just recently been summarized.[41] However, in coccidial infections nothing is known about the function of these or other local chemical mediators, neurotransmitters, and other peptide hormones.

VII. IMMUNOPATHOLOGY

The immunosuppression effects of *E. neischulzi* infections in rats are well documented.[97] Recent investigations[98] show that the most marked effect of *E. neischulzi* infection is inhibition of peripheral eosinophilia. In concurrent infections with *N. braziliensis,* eosinophils were produced in the bone marrow, as in helminth infectins alone, but were somehow prevented from infiltrating the gut and aiding in the expulsion of the worm. *Eimeria neischulzi* infections also cause a decrease in IEL at 8 days PI, also suggesting immunosuppression.[99]

Inflammation and infiltration of infected gut tissues with mononuclear cells is a common finding of *Eimeria* infections. However, Castro and Duszynski[100] found that there is an initial inflammatory response followed by reduced inflammation in *E. neischulzi* infection in rats that is associated with gamont formation and the release of oocysts. Thereafter inflammation reappears. Reduced inflammation occurred not only in the gut but other tissues, since granulation around sterile strings implanted under abdominal wall was reduced also. However, peripheral leukocyte numbers did not change significantly. The mechanism of immunosuppression is unknown. It may, however, be due to local production of corticosteroids.[101]

Inflammatory reactions are apparently not completely effective in removing *E. neischulzi* because the parasites (sporozoites) are retained in the intestine for up to 21 days PI as determined by passing infections to clean hosts upon feeding infected intestinal tissues. The specific cells in which it resides during this time are unknown.

There is no evidence of episodes of marked immunosuppression during *E. bovis* infections as determined by lack of effect of con-A stimulation of mixed lymphocyte cultures.[102]

VIII. INTERACTIONS

During the early years of commercial poultry production coccidiosis often caused severe problems, and the importance of subacute disease was not established. Later we began to differentiate between clinical coccidiosis, causing adverse economically important effects, and "subclinical" coccidiosis sometime referred to as coccidiasis. It was generally accepted that low levels of coccidiosis in the field had no economic importance. Over the past 10 to 15 years, evidence has accumulated on disease interactions leading to a reappraisal of earlier conclusions. Data clearly show that the effects of coccidiosis and other diseases are often more severe when two or more are present together, even if neither condition is severe enough to produce adverse effects alone. Such interactions are usually difficult to diagnose in the field.

The interactions that occur between coccidia and other disease conditions are of four types. The first is where the effects of one disease on a particular parameter are so great that one condition essentially produces most of the effect produced in combination. The second, and most common interaction, is an additive effect approximately equal to the sum of the effects from the two diseases individually. The third is a synergism where the total effect is far greater than the sum of the effects of the individual conditions. Finally, in a few cases, one condition will ameliorate the effects of the other.

A. NUTRITION

Workers in coccidiosis have observed for many years that severe coccidiosis in floor pens is often associated with increased frequency of leg problems and nutrient deficiency diseases such as exudative diathesis. Infection with any of several species of coccidia potentiated selenium deficiency in chicks fed critical levels of that mineral.[103] The symptoms of copper toxicity are exacerbated in chicks given multiple infections of *E. acervulina* or *E. tenella*.[104] The first species also increases bone mineral retention and exacerbates cobalt toxicity.[105] Bone magnesium uptake and calcium loss are greater following *E. acervulina* infection compared with chicks not receiving coccidia.[106] Infection with the same species also increases the concentration of liver copper and kidney zinc.[107] Coccidiosis does not always exacerbate the effects of mineral toxicity. The growth depression effect of zinc is ameliorated by infection with *E. acervulina*.[107,108]

B. STRESS

Increasing levels of social stress causes an increased resistance to infection with *E. necatrix*[109] or *E. tenella*.[110] Chickens are also more susceptible to primary coccidial infection under conditions of low stress than under conditions of high stress.

C. DISEASES

1. Bacteria

The interactions between coccidiosis and bacteria can be viewed from two perspectives. Interactions where the bacteria has an influence on the severity of the coccidial infection and those where coccidiosis affects the bacterial infection. Examples of the former come from early work in gnotobiotic chickens that showed that the gastrointestinal flora is necessary for the development of typical cecal coccidiosis.[111] Effects of cecal coccidiosis such as mortality, weight loss, and cecal lesion scores were less in gnotobiotic chickens than in conventional ones.

Coccidiosis affects the bacterial populations in the ceca of gnotobiotic chickens. For example, *E. tenella* stimulated an eightfold increase in the numbers of *C. perfringens* in the ceca of monoflora chickens.[112] Such an increase can increase the pathological effects of this bacteria in conventional chickens. Mortality due to necrotic enteritis was highest (53%) in chickens inocualted with *E. acervulina* before infection with clostridia.[113] Under field conditions, coccidia can play a significant role in the occurrence of ulcerative enteritis. An interaction between *Escherichia coli* and *E. brunetti* has also been reported.[114]

There are numerous examples of coccidial interactions with *Salmonella*. Concurrent infections of *S. typhimurium* and *E. tenella* given at the same time or 2, 4, or 6 days after inoculation with the coccidia, did not affect the number of salmonella in the ceca.[115] When salmonella were given for 5 consecutive days after inoculation with coccidia, however, the number of salmonella recovered 7, 10, 14, or 28 days later increased in the ceca and liver. Scanning electron microscopic studies confirmed that damage to the cecal mucosa was greater in chickens with simultaneous infections.[116]

The increase in the number of salmonella isolated is not merely a condition of oral inoculation with a bacterial culture. Contaminated feed[117] or contaminated cages[118] will produce the same effect. Several species of salmonella will increase after cecal coccidiosis[119] and infection with *E. acervulina* and *E. maxima* will also enhance salmonella infection.[120] The mechanism by which coccidiosis induces this increase is not known but Baba et al.[119] found that the increased multiplication of salmonella in the ceca is associated with decreased concentrations of volatile fatty acids.

2. Viruses

One of the first coccidiosis/virus interactions described was the interference with the development of immunity to coccidia in chicks exposed to Marek's disease virus.[121] This effect was not seen when chicks were exposed to Marek's disease virus several weeks after exposure to coccidiosis.

Infectious bursal disease exposure prior to infection with *E. tenalla* produced several significant interactions including earlier cecal hemorrhage, higher mortality, and greater lesion scores.[122] The effect on weight gain was additive but the effect on mortality was greater than additive. The same virus also reduced the degree of protection to subsequent challenge with cecal coccidiosis.[123] There was a significant interaction between cecal coccidiosis and reticuloendothelial virus.[124]

The depression of weight gain was greater when reovirus and *E. acervulina* or *E. mitis* were present together than when either disease organism was present alone.[125] The most obvious interaction was a marked increase in swollen hocks when both coccidia and reovirus were present together (20 to 27%) compared with the frequency when either was present alone (0 to 10%). There was some indication that early coccidiosis could increase the ability of some of the virus isolates to infect various tissues of the chicken host. A second reovirus that caused a stunting rate of 28% when present alone, gave stunting rates of 50 and 57% when *E. acervulina* or *E. maxima* were present concurrently with the virus, even though the coccidia alone produced little stunting at the low infection levels used.[126]

3. Mycotoxins

An interaction between coccidia and mycotoxins was first reported by Edds et al.[127] who found that dietary aflatoxin at 2 ppm increased mortality from cecal coccidiosis. Since then, several studies have shown interactions among other species of coccidia and other mycotoxins.[128-130] The adverse effects on weight gain and plasma pigment are greater when both coccidia and mycotoxins are present together than when either is present alone. The combined effect is generally additive.

The interaction of aflatoxin and *E. tenella* on mortality is, however, greater than additive. With severe cecal coccidiosis, mortality was 60% when aflatoxin is present even though the mortality caused by coccidiosis or aflatoxin alone was 28 and 10%, respectively. The mechanism of this synergistic action might be the amount and speed of cecal hemorrhage when both conditions were present together.[128]

Coccidia/mycotoxin interactions can affect anticoccidial efficacy. Aflatoxin reduced the efficacy of monensin against *E. tenella.*[128] Conversely, there was no reduction in the control of *E. acervulina* by monensin[131] or amprolium[127] when aflatoxin was present. Mortality with DON was greater (32%) when nicarbazin was present than mortality when nicarbazin was absent (9%; Ruff, Doerr, and Huff, 1989, unpublished data).

IX. INTEGRATION OF PATHOPHYSIOLOGICAL CHANGES

A. MAJOR PHYSIOLOGICAL STRESSES CAUSED BY COCCIDIOSIS

The various physical, physiological, and metabolic effects described above interact to produce the end effects of the coccidial infection. For example, mortality during *E. tenella* infection results from four major physiological stresses that act to cause death.[132] Those are (1) hypothermia, (2) depletion of carbohydrate stores, (3) metabolic acidosis, and (4) renal tubule-cell dysfunction. These effects are not related to reduced feed intake. Hemorrhage accentuates the lowered body temperature. The lowered body temperature may also be due to effects of the infection on the hypothalmus which in turn would place an added demand on the energy reserves to maintain body temperature. In birds, the level of blood glucose is relatively stable, yet before death in moribund birds with cecal coccidiosis glucose levels were only 40% of the controls. Metabolic acidosis exacerbates other pathological responses. For example, it stimulates respiratory compensation (panting) that further aggravates the cooling of the bird. The decrease in hemoglobin and protein seen with infection would make the acidosis worse because the normal buffering capacity of the blood would be reduced. Diarrhea also contributes to the electrolyte shifts. The change in blood pH also affects the function of the kidneys.

The severity of pathological changes during coccidiosis are not always apparent. For example, in severe infections with *E. tenella*, change in PCV would indicate up to a 50% loss of blood. Actually, the blood loss over a two day period may exceed the total blood volume of the bird.[133] There may also be an inverse relation between cecal lesion score and total blood loss leading to erroneous conclusions about which infection is more severe. A strain of cecal coccidiosis that produces severe lesions actually loses less blood than one with less severe lesions. This is because blood coagulates in the ceca more rapidly in chickens given the pathogenic strain and is not expelled in the feces.

B. CYTOKINES

It has been shown recently that interleukin-1 (IL-1) plays a major role in the pathophysiology of acute phase responses to infection.[101,134] Intraperitoneal injection of chicks with Sephadex (a type of sterile immunization that stimulates macrophage/monocyte infiltration) causes significant increases in plasma levels if IL-1 and cortisone.[134] Both the immunization with Sephadex, and injection of IL-1 produce anorexia, followed by reduced feed efficiency, weight gain depression and a variety of metabolic effects including increased serum copper, decreased zinc, and decreased accretion of muscle protein due to increased rate of degradation.[134] The physiological changes produced by IL-1 in chicks are very similar to those occurring in coccidia infections.

REFERENCES

1. **Turk, D. E.,** The effects of coccidiosis on intestinal function and gut microflora, in *Avian Coccidiosis,* Long, P. L., Boorman, K. N., and Freeman, B. M., Eds., Br. Poultry Sci., Edinburgh, 227, 1978.
2. **Ruff, M. D.,** Malabsorption from the intestine of birds with coccidiosis, in *Avian Coccidiosis,* Long, P. L., Boorman, K. N., and Freeman, B. M., Eds., Br. Poultry Sci., Edinburgh, 281, 1978.
3. **Ruff, M. D.,** Reasons for inadequate nutrient utilization during avian coccidiosis: a review in *Research in Avian Coccidiosis,* McDougald, L. R., Joyner, L. P., and Long, P. L., Eds., University of Georgia, Athens, 169, 1986.
4. **Ruff, M. D.,** Interaction of avian coccidiosis with other diseases: a review, *Proc. 5th Int. Coccidiosis Conf.,* Tours, France, Yvore, P., Ed., in press.
5. **Fernando, M. A.,** Pathology and pathogenicity, in *The Biology of the Coccidia,* Long, P. L., Ed., University Park Press, Baltimore, 287, 1982.

6. **Yvore, P.,** Physiology of the host response to coccidial infections, in *Research in Avian Coccidiosis,* McDougald, L. R., Joyner, L. P., and Long, P. L., Eds., University of Georgia, Athens, 148, 1986.
7. **Witlock, D. R., Lushbaugh, W. B., Danforth, H. D., and Ruff, M. D.,** Scanning electron microscopy of the cecal mucosa in *Eimeria-tenella*-infected an uninfected chickens, *Avian Dis.*, 19, 293, 1975.
8. **Witlock, D. R. and Ruff, M. D.,** Comparison of the intestinal surface damage caused by *Eimeria mivati, E. necatrix, E. maxima, E. brunetti,* and *E. acervulina* by scanning electron microscopy, *J. Parasitol.*, 63, 194, 1977.
9. **Ruff, M. D., Augustine, P. C., and Madden, P. A.,** *Eimeria meleagrimitis, E. adenoides* and *E. dispersa:* severity of infection and changes in intestinal mucosa of the turkey, *Exp. Parasitol.*, 51, 87, 1981.
10. **Gajadhar, A. A., Rainnie, D. J., and Cawthorn, R. J.,** Description of the goose coccidium *Eimeria stigmosa* (Klimes, 1963), with evidence of intranuclear development, *J. Parasitol.*, 72, 588, 1986.
11. **Blagburn, B. L. and Todd, K. S., Jr.,** Pathological changes and immunity associated with experimental *Eimeria vermiformia* infections in *Mus musculus, J. Protozool.*, 31, 556, 1984.
12. **Mesfin, G. M., Bellamy, J. E., and Stockdale, P. H. G.,** The pathological changes caused by *Eimeria falciformis* var. *pragensis* in mice, *Can. J. Comp. Med.*, 42, 496, 1978.
13. **Peeters, J. E., Pohl, P., and Charlier, G.,** Infectious agents associated with diarrhoea in commercial rabbits: a field study, *Ann. Rech. Vet.*, 15, 335, 1984b
14. **Gregory, M. W., Joyner, L. P., Catchpole, J., and Norton, C. C.,** Ovine coccidiosis in England and Wales, 1978—1979, *Vet. Rec.*, 106, 461, 1980.
15. **Higgs, S. and Nowell, F.,** Laboratory studies with clones of *Eimeria hungaryensis,* a parasite of the wood mouse, *Apodemus sylvaticus, Parasitology,* 92, 231, 1986.
16. **Muto, M., Sugisaki, M., Yusa, T., and Noguchi, Y.,** Studies on guinea pigs. I. Clinico-pathological observation, *Exp. Anim. (Tokyo),* 34, 23, 1985.
17. **Peeters, J. E., Charlier, G., Antoine, O., and Mammerickx, M.,** Clinical and pathological changes after *Eimeria intestinalis* infection in rabbits, *Zentralbl. Vet. Med. B,* 31, 9, 1984.
18. **Gregory, M. W. and Catchpole, J.,** coccidiosis in rabbits: the pathology of *Eimeria flavescens* infection, *Int. J. Parasitol.*, 16, 131, 1986.
19. **Licois, D. and Guillot, J. F.,** Evolution du nombre de colibacilles chez des lapereaux atteints de coccidiose intestinale, *Rec. Med. Vet.*, 156, 555, 1980.
20. **Catchpole, J. and Gregory, M. W.,** Pathogenicity of the coccidium *Eimeria crandallis* in laboratory lambs, *Parasitology,* 91, 45, 1985.
21. **Gregory, M. W. and Catchpole, J.,** Ovine coccidiosis: pathology of *Eimeria ovinoidalis* infection, *Int. J. Parasitol.*, 17, 1099, 1987.
22. **Norton, C. C.,** Coccidia of the domestic goat, *Capra hercus* with notes on *Eimeria ovinoidalis* and *E. bakuensis* (syn. *E. ovina)* from the sheep *Ovis aries, Parasitology,* 92, 279, 1986.
23. **Vitovec, J., Koudela, B., and Sterba, J.,** Pathology and pathogenicity of *Eimeria scabra* (Henry, 1931) in experimentally infected pigs, *Folia Parasitol. (Praha),* 34, 299, 1987.
24. **Lindsay, D. S., Ernst, J. V., Benz, G. W., and Current, W. L.,** The sexual stages of *Eimeria wyomingensis* Huizina and Winger (1942) in experimentally infected calves, *J. Parasitol.*, 74, 833, 1988.
25. **Ernst, J. V. and Benz, G. W.,** Attempts to produce experimental *Eimeria wyomingensis* infections in calves, *J. Parasitol.*, 66, 625, 1980.
26. **Ruff, M. D. and Edgar, S. A.,** Reduced intestinal absorption in broilers during *Eimeria mitis* infections, *Am. J. Vet. Res.*, 43, 507, 1982.
27. **Ruff, M. D. and Wilkins, G. C.,** Pathogenicity of *Eimeria lettyae* Ruff, 1985 in the northern bobwhite *(Colinus virginanus), J. Wildl. Dis.*, 23, 121, 1987.
28. **Augustine, P. C., Kuenzel, W. J., and Thomas, O. P.,** Effect of coccidiosis on heart composition and function in young turkeys, *Avian Dis.*, 26, 346, 1982.
29. **Allen, P. C.** Physiological responses of chicken gut tissue to infection with *Eimeria acervulina, Avian Dis.*, 28, 868, 1984.
30. **Allen, P. C.,** Physiological responses of chicken gut tissue to coccidial infection: comparative effects of *Eimeria acervulina* and *Eimeria mitis* on mucosal mass, carotenoid content and brush border enzyme activity, *Poult. Sci.*, 66, 1306, 1987.
31. **Fioramonti, J., Sorraing, J. M., Licois, D., and Bueno, L.,** Intestinal motor transit disturbances associated with experimental coccidiosis *(Eimeria magna)* in the rabbit, *Ann. Rech. Vet.*, 12, 413, 1981.
32. **Weisbroth, S. H. and Scherr, S.,** Fatal intussuseption associated with intestinal coccidiosis *(Eimeria perforans)* in a rabbit, *Lab. Anim. Sci.*, 5, 79, 1975.
33. **Loehry, C. A. and Creamer, B.,** Three-dimensional structure of the human small intestinal mucosa in health and disease, *Gut,* 10, 6, 1969.
34. **Townley, R. R. W., Cass, M. H., and Anderson, C. M.,** Small intestinal pattern of coeliac disease and idiopathic steatorrhoea seen in other situations, *Gut,* 5, 51, 1964.
35. **Smith, R. R., Ruff, M. D., and Witlock, D. R.,** *Eimeria necatrix* in the chicken: response of the host jejunum to infection and subsequent absorption of methionine and glucose, *Proc. Helminthol. Soc. Wash.*, 47, 235, 1980.

36. **Allen, P. C. and McMurtry, J. P.**, Changes in pancreatic hormones associated with coccidiosis: relationship with changes in the small intestine, *Vet. Parasitol.*, in press.
37. **Marquardt, W. C., Bryan, J. H. D., and Long, P. L.**, Nucleolar hypertrophy as an indicator of transcription in cells infected with second genration meronts of *Eimeria tenella, J. Protozool.*, 31, 569, 1984.
38. **Fernando, M. A. and Pasternak, J.**, *Eimeria necatrix*: induced proteolytic sensitivity of infected crypt cells, *Exp. Parasitol.*, 56, 20, 1983.
39. **Urquhart, C.**, Morphological and adhesive changes to cultured chick kidney cells following parasitization with *Eimeria tenella* (Protozoa: Coccidia), *Parasitology*, 82, 175, 1981.
40. **Allen, P. C. and Danforth, H. D.**, The effects of *Eimeria acervulina* infection on the metabolism of chick duodenal tissue, *Vet. Parasitol.*, 14, 105, 1984.
41. **Castro, G. A.**, Immunophysiology of enteric parasitism, *Parasitol. Today*, 5, 13, 1989.
42. **Ruff, M. D., Allen, P. C., and Chute, M. B.**, Composition of heart, liver, and skeletal muscle from broilers with coccidiosis, *Poult. Sci.*, 60, 1807, 1981.
43. **Augustine, P. C. and Thomas, O. P.**, *Eimeria meleagrimitis* in young turkeys: effects on weight, blood, and organ parameters, *Avian Dis.*, 23, 854, 1979.
44. **Augustine, P. C. and Thomas, O. P.**, Effect of time on response to *Eimeria adenoeides* and *Eimeria meleagrimitis* infection in young turkeys, *Avian Dis.*, 25, 366, 1981.
45. **Allen, P.C.**, The effect of *Eimeria acervulina* infection on hearts of young broiler chicks, *Avian Dis.*, 31, 637, 1987.
46. **Witlock, D. R.**, Changes in cecal composition with *Eimeria tenella* infection, *Poult. Sci.*, 61, 57, 1982.
47. **Witlock, D. R. and Fetterer, R. H.**, *Eimeria tenella*: cecal composition in infected chickens, *Exp. Parasitol.*, 56, 186, 1983.
48. **Arakawa, A., Sayama, M., Baba, E., and Fukata, T.**, Cecal filling and defecation of chickens infected with *Eimeria tenella, Zentralbl. Bakteriol. Milrobiol. Hyg. A*, 264, 337, 1987.
49. **Chapman, H. D., Fernandes, D. L., and Davison, T. F.**, A comparison of the effects of infection with *Eimeria maxima* and dietary restriction on weight gain, plasma metabolites and liver glycogen in the immature folw, *Gallus domesticus, Parasitology*, 84, 205, 1982.
50. **Ruff, M. D. and Allen, P. C.**, Changes in liver glycogen in broilers during coccidiosis, *Vet. Parasitol.*, 10, 1447, 1982.
51. **Allen, P. C.**, Effects of coccidiosis on activities of glycogen phosphorylase and glycogen synthase in livers from 2- to 3-week-old broiler chicks, *Proc. Helminthol. Soc. Wash.*, 55, 50, 1988.
52. **Ruff, M.D. and Augustine, P. C.**, Effects of coccidiosis on the electrophoretic pattern of serum proteins in chickens, *J. Parasitol.*, 68, 107, 1982.
53. **Augustine, P. C.**, Electrophoretic separation of serum proteins and lipoproteins of young turkeys infected with *Eimeria meleagrimitis* or *Eimeria adenoeides, Poult. Sci.*, 64, 1644, 1985.
54. **Augustine, P. C. and Witlock, D. R.**, Decreased osmotic fragility of red blood cells of *Eimeria adenoeides*-infected turkeys, *Avian Dis.*, 28, 343, 1984.
55. **Allen, P. C.**, The effect of *Eimeria acervulina* infection of plasma lipids and lipoproteins in young broiler chicks, *Vet. Parasitol.*, 30, 17, 1988.
56. **Turk, D. E.**, Microelements in the circulation of coccidiosis-infected chicks, *Poult. Sci.*, 65, 2098, 1986.
57. **Augustine, P. C.**, Effect of feed and water deprivation on organ and blood characteristics of young turkeys, *Poult. Sci.*, 61, 796, 1982.
58. **Augustine, P. C. and Ruff, M. D.**, Changes in carotenoid and vitamin A levels in young turkeys infected with *Eimeria meleagrimitis* of *E. adenoides, Avian Dis.*, 27, 963, 1983.
59. **Yvore, P.**, Comparison de l'incidance nutritionelle d'une coccidiose intestinale *(E. acervulina)* et d'une coccidiose caecale *(E. tenella), Folia Vet. Lat.*, IV, 408, 1974.
60. **Ruff, M. D. and Fuller, H. R.**, Some mechanisms of reduction of carotenoid levels in chickens infected with *Eimeria acervulina* or *E. tenella, J. Nutr.*, 105, 1447, 1975.
61. **Allen, P. C.**, Effect of *Eimeria acervulina* infection on chick *(Gallus domesticus)* high density lipoprotein composition, *Comp. Biochem. Physiol.*, 87B, 313, 1987.
62. **Rajavashisth, T. P., Dawson, P. A., Williams, D. L., Shakelford, J. E., Lebherz, H., and Lusis, A. S.**, Structure, evolution and regulation of chicken apolipoprotein A-I, *J. Biol. Chem.*, 262, 7058, 1987.
63. **Glickman, R. M. and Green, P. H. R.**, The intestine as a source of apolipoprotein A-I, *Proc. Natl. Acad. Sci., U.S.A.*, 74, 2569, 1977.
64. **Ganguly, J., Krishnamurthy, S., and Mahadevan, S.**, The transport of carotenoids, vitamin A and cholesterol across the intestine of rats and chickens, *Biochem. J.*, 71, 756, 1959.
65. **Quackenbush, F. W., Kvakovszky, S., and Hoover, T.**, Deposition of individual carotenoids in the avian skin, *J.A.O.A.C.*, 48, 1241, 1965.
66. **Allen, P. C. and Brown, E. M.**, Binding of lutein in chick high density lipoprotein particles, *Proc. F.A.S.E.B.*, 72nd Annu. Meet. Las Vegas, Abstr. 2823, 1988.
67. **Allen, P. C. and Lillehoj, H. S.**, Pathophysiology of *Eimeria maxima* infections in white leghorn chickens, *Poult. Sci.*, 66 (Suppl. 1), 55, 1987.

68. **Allen, P. C.,** Biochemical changes in the intestinal mucosa associated with coccidiosis, in *Research in Avian Coccidiosis, Proc. Georgia Coccidiosis Conf.*, McDougald, L. R., Joyner, L. P., and Long, P., Eds., University of Georgia, Athens, 1986, 194.
69. **Allen, P. C.,** The effect of *Eimeria acervulina* infection on apolipoprotein A-I synthesis in the chick intestine, *Poult. Sci.*, 67, (Suppl. 1), 45, 1988.
70. **Rose, E. M.,** Immune responses to Eimeria infections, in *Research in Avian Coccidiosis, Proc. Georgia Coccidiosis Conf.*, McDougald, L.R., Joyner, L. P. and Long, P. L Eds., University of Georgia, Athens, 1986, 449.
71. **Kloppel, T. M., Hoops, T. C., and Brown, W. R.,** Advances in cellular mechanisms of secretory component synthesis and secretion and the transport of IgA, in *Adv. Exp. Med. Biol.*, McGhee, J. R., Mestecky, J., Ogra, P. L., and Bienenstock, L., Eds., Plenum Press, New York, 216B, 1061, 1987.
72. **Tyczkowski, J. K., Ruff, M. D., and Hamilton, P. B.,** Altered metabolism of oxycarotenoids in chickens during infection with *Eimeria acervulina, Poult. Sci.*, in press.
73. **Berczi, I.,** *Pituitary Function and Immunity,* CRC Press, Boca Raton, FL, 1986.
74. **Singh, U. and Owen, J. J. T.,** Studies on the maturation of thymus stem cells. The effects of catecholamines, histamine and peptide hormones on the expression of T cell alloantigens, *Eur. J. Immunol.*, 6, 59, 1976.
75. **Bhat, G., Gupta, S. K., and Maiti, B. R.,** Influence of prolactin on mitotic activity of the bursa of Fabricius of the chick, *Gen. Comp. Endocrinol.*, 52, 452, 1983.
76. **Chadwick, A., Rapson, E. B., Carlos, G. M., and Lee, D. L.,** Circulating prolactin concentrations in chickens infected with *Eimeria tenella, Br. Poult. Sci.*, 26, 17, 1985.
77. **Davison, T. F., Chapman, H. D., and Harvey, S.,** Endocrine changes in the fowl during infection with *Eimeria maxima, Res. Vet. Sci.*, 38, 296, 1985.
78. **Stevens, F. M., McCarthy, C. F., and Craig, A.,** Is prolactin trophic to the intestine in coeliac disease, *Gut,* 19, a992, 1978.
79. **Elsasser, T. H., Rumsey, T. S., Hammond, A. C., and Fayer, R.,** Influence of parasitism on plasma concentrations of growth hormone, somatomedin-C and somatomedin-binding proteins in calves, *J. Endocrinol.*, 116, 191, 1988.
80. **Allen, P. C. and McMurtry, J. P.,** Changes in pancreatic hormones associated with coccidiosis. *Poult. Sci.*, 63, 1129, 1984.
81. **Bloom, S. R. and Polak, J. M.,** The effect of gastrointestinal disease on gut hormone release, in *Gut Peptides, Secretion, Function and Clinical Aspects,* Miyoshi, A. and Grossman, M. I., Eds., Elsevier, New York, 1979, 356.
82. **Sykes, A. H.,** Resting oxygen consumption in chicks infected with intestinal coccidiosis, *Res. Vet. Sci.*, 11, 489, 1970.
83. **Preston-Mafham, R. A. and Sykes, A. H.,** Changes in the permeability of the mucosa during intestinal coccidiosis infections of the fowl, *Experientia,* 23, 972, 1967.
84. **Riley, V.,** Psychoneuroendocrine influences on immunocompetence and neoplasia, *Science,* 212, 1100, 1981.
85. **Challey, J. R.,** Changes in adrenal constituents and their relationship to corticosterone secretion in chickens selected for genetic resistance and susceptibility to cecal coccidiosis, *J. Parasitol.*, 52, 967, 1966.
86. **Long, P.,** Development (schizogony) of *Eimeria tenella* in the liver of chickens treated with corticosteroids, *Nature*, 225, 290, 1970.
87. **Alexander, J. and Stimson, W. H.,** Sex hormones and the course of parasitic infection, *Parasitol. Today,* 4, 189, 1988.
88. **Ruff, M. D., Abdel Nabi, M. A., Clarke, R. N., Morarak, M., and Ottinger, M. A.,** Effect of coccidiosis on reproductive maturation of male japanese quail, *Avian Dis.*, 32, 41, 1988.
89. **Hiss, S. and Nowell, F.,** Laboratory studies with clones of *Eimeria hyngaryensis,* a parasite of the wood mouse, *Apodemus sylvaticus, Parasitology,* 92, 231, 1986.
90. **Lillehoj, H. S.,** Effects of immunosuppression on avian coccidiosis: Cyclosporin A but not hormonal bursectomy abrogates host protective immunity, *Infec. Immun.*, 55, 1616, 1987.
91. **Witlock, D. R.,** Demonstration of thromboplastin-like activity in extracts from *Eimeria tenella*-infected and noninfected chicken ceca, *Proc. Helminthol. Soc. Wash.*, 47, 122, 1980.
92. **Witlock, D. R. and Brimfield, A. A.,** Further studies on toxic extracts from *Eimeria tenella*-infected chicken ceca, *Proc. Helminthol. Soc. Wash.*, 49, 305, 1982.
93. **Hickox, R. F.,** Central nervous system coccidiosis: more common than veterinarians think, *DVM News Mag.*, 12, 54, 1981.
94. **Isler, C. M., Bellamy, J. E. C., and Wobeser, G. A.,** Labile neurotoxin in serum of calves with nervous coccidiosis, *Can. J. Vet. Res.*, 51, 253, 1987.
95. **Isler, C. M., Bellamy, J. E., and Wobeser, G. A.,** Pathogénesis of neurological signs associated with bovine enteric coccidiosis: a prospective study and review, *Can. J. Vet. Res.*, 51, 261, 1987.
96. **Isler, C. M., Bellamy, J. E., and Wobeser, G. A.,** Characteristics of the labile neurotoxin associated with nervous coccidiosis, *Can. J. Vet. Res.*, 51, 271, 1987.

97. **Bristol, J. R., Pinon, A. J., and Mayberry, L. F.,** Interspecific interactions between *Nippostrongylus braziliensis* and *Eimeria neschulzi* in the rat, *J. Parasitol.*, 69, 372, 1983.
98. **Upton, S. J., Mayberry, L. F., Bristol, J. R., Favela, S. H., and Sambrano, G. R.,** Suppression of peripheral eosinophilia by the coccidium *Eimeria neischulzi* in experimentally infected rats, *J. Parasitol.*, 73, 300, 1987.
99. **Rose, E. and Hesketh, P.,** Coccidiosis: T-lymphocyte dependent effects of infection with *Eimeria neischulzi* in rats, *Vet. Immunol. Immunopathol.*, 3, 499, 1982.
100. **Castro, G. A. and Duszynski, D. W.,** Local and systemic effects on inflammation during *Eimeria nieschulzi* infection, *J. Protozool.*, 31, 283, 1984.
101. **Klasing, K. C., Laurin, D. E., Peng, R. K., and Fry, D. M.,** Immunologically mediated growth depression in chicks: influence of feed intake, corticosterone and interleukin-1, *J. Nutr.*, 117, 1637, 1987.
102. **Hughes, H. P. A., Whitmere, W. M., and Speer, C. A.,** Immunity patterns during acute infection by *Eimeria bovis*, *J. Parasitol.*, 75, 86, 1989.
103. **Pesti, G. M. and Combs, G. F., Jr.,** Studies on the enteric absorption of selenium in the chick using localized coccidial infections, *Poult. Sci.*, 55, 2265, 1976.
104. **Czarnecki, G. L. and Baker, D. H.,** Feed additive interactions in the chicken: Reduction of tissue copper deposition by dietary roxarsone in healthy and in *Eimeria acervulina*-infected or *Eimeria tenella*-infected chicks, *Poult. Sci.*, 63, 1412, 1984.
105. **Brown, D. R. and Southern, L. L.,** Effect of *Eimeria acervulina* infection in chicks fed excess dietary cobalt and/or manganese, *J. Nutr.*, 115, 347, 1985.
106. **Giraldo, C. A., Brown, D. R., Watkins, K. L., and Southern, L. L.,** Responses to excess dietary magnesium as affected by experimental *Eimeria acervulina* infection or by dietary ammonium chloride ingestion in the chick, *J. Nutr.*, 117, 1053, 1987.
107. **Bafundo, K. W., Baker, D. H., and Fitzgerald, P. R.,** Lead toxicity in the chick as affected by excess copper and zinc and by *Eimeria acervulina* infection, *Poult. Sci.*, 63, 1594, 1984.
108. **Bafundo, K. W., Baker, D. H., and Fitzgerald, P. R.,** The iron-zinc interrelationship in the chick as influenced by *Eimeria acervulina* infection, *J. Nutr.*, 114, 1306, 1984.
109. **Gross, W. B.,** Plasma steroid tendency, social environment and *Eimeria necatrix* infection, *Poult. Sci.*, 55, 1508, 1976.
110. **Gross, W. B.,** Effect of social environment and oocyst dose on resistance and immunity to *Eimeria tenella* challenge, *Avian Dis.*, 29, 1018, 1985.
111. **Visco, R. J.,** *Eimeria tenella* in gnotobiotic chickens: hematocrit, weight change, cecal pathology, and mortality, *J. Parasitol.*, 61, 194, 1975.
112. **Dykstra, D. D. and Reid, W. M.,** Monensin, *Eimeria tenella* infection, and effects on the bacterial populatins in the ceca of gnotobiotic chickens, *Poult. Sci.*, 57, 398, 1978.
113. **Al-Shekhly, F. and Al-Saieg, A.,** The role of coccidia in the occurrence of necrotic enteritis of chickens, *Avian Dis.*, 24, 324, 1980.
114. **Nagi, M. S. and Mathey, W. J.,** Interaction of *Escherichia coli* and *Eimeria brunetti* in chickens, *Avian Dis.*, 16, 864, 1972.
115. **Arakawa, A., Baba, E., and Fukata, T.,** *Eimeria tenella* infection enhances *Salmonella typhimurium* infection in chickens, *Poult. Sci.*, 60, 2203, 181.
116. **Kim, S., Fukata, T., Baba, E., and Arakawa, A.,** Effects of infection with *Eimeria tenella* upon *Salmonella typhymurium* infection in the gnotobiotic chicks. Scanning electron microscopic study, *Zentrabl. Bakteriol. Milrobiol. Hyg. A*, 260, 238, 1985.
117. **Morishima, H., Baba, E., Fukata, T., and Arakawa, A.,** Effect of *Eimeria tenella* infection in chickens fed the feed artificially contaminated with *Salmonella typhimurium*, *Poult. Sci.*, 63, 1732, 1984.
118. **Kosugi, Y., Baba, E., Fukata, T., and Arakawa, A.,** Effects of cage contamination with coccidia and salmonella on acute salmonellosis in young chickens, *Avian Dis.*, 30, 313, 1986.
119. **Baba, E., Yaono, M., Fukata, T., and Arakawa, A.,** Infection by *Salmonella typhimurium, S. agona, S. enteritidis* or *S. infantis* of chickens with caecal coccidiosis, *Br. Poult. Sci.*, 26, 505, 1985.
120. **Takimoto, H., Baba, E., Fukata, T., and Arakawa, A.,** Effects of infection of *Eimeria tenella, E. acervulina*, and *E. maxima* upon *Salmonella typhimurium* infection in chickens, *Poult. Sci.*, 63, 478, 1984.
121. **Biggs, P. M., Long, P. L., Kenzy, S. G., and Rootes, D. G.,** Relationship between Marek's disease and coccidiosis. II. The effect of Marek's disease on the susceptibility of chickens to coccidial infection, *Vet. Rec.*, 83, 284, 1968.
122. **Giambrone, J. J., Anderson, W. I., Reid, W. M., and Eidson, C. S.,** Effect of infectious bursal disease on the severity of *Eimeria tenella* infections in broiler chicks, *Poult. Sci.*, 56, 243, 1977.
123. **Anderson, W. I., Reid, W. M., Lukert, P. D., and Fletcher, O. J., Jr.,** Influence of infectious bursal disease on the development of immunity to *Eimeria tenella*, *Avian Dis.*, 21, 637, 1977.
124. **Motha, M. X. J. and Egerton, J. R.,** Influence of reticuloendotheliosis on the severity of *Eimeria tenella* infection in broiler chickens, *Vet. Microbiol.*, 9, 121, 1984.
125. **Ruff, M. D. and Rosenberger, J. K.,** Concurrent infections with reoviruses and coccidia in broilers, *Avian Dis.*, 29, 465, 1985.

126. **Ruff, M. D. and Rosenberger, J. K.,** Interaction of low-pathogenicity reoviruses and low levels of infection with several coccidial species, *Avian Dis.*, 29, 1057, 1985.
127. **Edds, G. T., Nair, K. P. C., and Simpson, C. F.,** Effect of aflatoxin B_1 on resistance in poultry against cecal coccidiosis and Marek's disease, *Am. J. Vet. Res.*, 34, 819, 1973.
128. **Wyatt, R. D., Ruff, M. D., and Page, R. K.,** Interaction of alfatoxin with *Eimeria tenella* infection and monensin in young broiler chickens, *Avian Dis.*, 19, 730, 1975.
129. **Ruff, M. D. and Wyatt, R. D.,** Influence of dietary aflatoxin on the severity of *Eimeria acervulina* infection in broiler chickens, *Avian Dis.*, 22, 471, 1978.
130. **Huff, W. E. and Ruff, M. D.,** *Eimeria acervulina* and *Eimeria tenella* infections in ochratoxin A-compromised broiler chickens, *Poult. Sci.*, 61, 685, 1982.
131. **Southern, L. L. and Baker, D. H.,** *Eimeria acervulina* infection during aflatoxicosis in the chick, *Nut. Rep. Int.*, 29, 35, 1984.
132. **Witlock, D. R., Ruff, M. D., and Chute, M. B.,** Physiological basis of *Eimeria tenella*-induced mortality in individual chickens, *J. Parasitol.*, 67, 65, 1981.
133. **Witlock, D. R.,** Physiologic basis of blood loss during *Eimeria tenella* infection, *Avian Dis.*, 27, 1043, 1983.
134. **Dinarello, C. A.,** Interleukin-1 and the pathogenesis of the acute phase response, *N. Engl. J. Med.*, 311, 1413, 1984.

Chapter 14

IMMUNITY TO COCCIDIOSIS

D. Wakelin and M. Elaine Rose

TABLE OF CONTENTS

I. INTRODUCTION

Immunity to infections with the eimeria has been the subject of a number of recent reviews.[1-3] These cover in detail much of the literature upon which our current understanding is based. For this reason, the approach chosen here is deliberately selective. We have decided to concentrate only on those aspects of the host response which can be correlated directly with resistance to infection, and have focused primarily on work published since 1984. Our objective is threefold: (1) to place the topic of immunity to eimeria in the wider context of immunoparasitology, drawing out significant similarities with, as well as differences from, responses to other parasites; (2) to define the capacity of the host to mount protective responses at the site of infection, concentrating upon the intestine, and (3) to draw from recent papers the basis for a model to explain immunity to eimeria which may help to define and focus future research.

II. IMMUNITY TO INFECTION

Immunity to an infectious organism may be expressed, and therefore defined, in several different ways. For the purposes of this review immunity will be equated with *acquired* responses which protect against the organism itself and against the pathologic consequences of infection. Many aspects of host behavior and physiology influence the outcome of contact between potential host and infectious agent and some of these act essentially nonspecifically and independently of prior contact; such expressions of *natural* resistance, or natural immunity, contribute to the phenomenon of host specificity (see Chapter 3). Acquired resistance, or *immunity* in the sense used here, is distinguishable by the fact that it is elicited by specific molecules (antigens) of a particular infectious organism, and therefore is not expressed until the host has had some experience of that organism. As a result of such experience the host develops memory of the antigens concerned, and is able to express an enhanced response on reinfection. Immunity is, therefore, the outcome of a dynamic interplay between two organisms, the host and the parasite, and it is mediated through the activity of a tightly regulated host physiological system (the immune system) which changes adaptively in response to the activity and behavior of the parasite. Any understanding of the basis of immunity to a particular group of parasites therefore requires a detailed knowledge of the parasite as a source of antigenic stimulation and as a target for host-protective immune responses, as well as of the capacity of the host to mount anti-parasite responses.

A. EIMERIA AND THE IMMUNE RESPONSE

In common with all microparasites eimeria are capable of providing the host with a substantial antigenic challenge because of their capacity to undergo repeated cycles of asexual division, during which the numbers of organisms increase exponentially. They differ from other members of the Sporozoa in that the duration of each infection is essentially predetermined and therefore self-limiting (although there are exceptions to this). In consequence, although immunity develops during an initial infection, its effects are often not apparent until a subsequent infection is experienced. An important corollary of this is that only certain facets of the response may be evident. For example, although effector mechanisms capable of affecting schizogonic stages develop, their development may be too late to affect schizogony during the primary infection. If other mechanisms which are effective in preventing development of subsequent infections beyond the initial stages of sporozoite invasion are also induced, then anti-schizogonic stage immunity cannot be expressed, although the potential for this immunity is certainly present. Recognition of this situation is important when artificial stimulation of effective immunity is being considered.

Each stage of the eimerian cycle is a source of antigenic material and, in theory at least, can therefore be the target of stage-specific, host-protective responses. The nature of the antigens released, the ways in which they become available to the host, and the manner in which host responses may be expressed against the parasite depend upon the life cycle stage concerned and its relationship to the cells and tissues of the host. During each developmental cycle eimeria utilize a variety of niches within the host, from the intestinal lumen, to the extracellular spaces of the mucosa, to the intracellular environments of enterocytes and other epithelial cells. The extent to which the host can adversely affect the parasite and control the level and development of infection, and the effector mechanisms that can be employed, are defined by the characteristics of each niche. Thus, in the extracellular (and to some extent the luminal) phases the parasite is, theoretically, susceptible to the action of factors present in tissue fluids, such as antibody, complement, inflammatory mediators and cytokines, as well as potentiated components of natural resistance such as phagocytosis. When inside cells the parasite is inaccessible to many of these factors, and can be affected only by mechanisms operative intracellularly, such as O_2 radicals, lysosomal enzymes, halides, or by the destruction of the host cell itself through some cytotoxic activity.

1. Sporozoites

Sporozoites are released in the intestinal lumen, penetrate the mucus layer covering the mucosa and enter epithelial cells. Antigens are present on the sporozoite surface and some of these molecules may be involved in recognition and penetration of host cells (see Chapter 17). Host responses are likely therefore to be directed against the sporozoite surface and these may influence both survival and development.

2. Schizonts

In development into the schizont the sporozoite undergoes antigenic change, but the intracellular location of the schizont means that antigen is not available to the host unless it becomes expressed at the host cell membrane. Effector mechanisms against this stage may operate indirectly (i.e., against the host cell), or directly by activating intracellular defence mechanisms.

3. Merozoites

The release of merozoites is associated with the mass release of antigenic material from the infected host cell. This antigen, though not directly relevant to the control of this stage may trigger responses capable of affecting later developmental cycles. Merozoites themselves, like sporozoites, have a brief extracellular existence and their surface antigens provide targets for direct anti-parasite responses. As there are differences between the antigenic profiles of each generation, it is possible that these responses may show some stage-specificity (see Section III).

4. Gametocytes

Within the host cell the gametocyte is, like the schizont, protected from many potential effector mechanisms. Antigen may be expressed at the surface of the host cell to function as a target for host responses; intracellular killing mechanisms may also be activated. The male gametes have to pass through an extracellular phase in order to achieve fertilization, and thus their surface antigens may become targets of immunity. Although the wall of the oocyst is antigenic and may be the target of immune responses, its physicochemical nature is such that these responses are unlikely to affect the contents of the cyst.

B. IMMUNITY IN THE ENVIRONMENTS OF *EIMERIA* SPP.

The majority of eimeria are parasites of the intestinal tract, only a few develop in extra-intestinal sites (e.g., *E. stiedai* in the liver, *E. truncata* in the kidney). All are intracellular organisms, capable of living within a number of cell types but adapted primarily for invasion of, and development within, epithelial cells. Although the basic processes involved in the initiation of immune responses against coccidia are the same as those elicited by other intracellular parasites, the way in which these responses can be expressed is determined by the characteristics of the organ parasitized. In the case of the intestine these characteristics are reflected in immunological mechanisms that are often quite distinct from those which operate against parasites in other tissues of the body.[4]

1. Initiation of Immunity

Antigen is made available to the host in at least three different ways: (1) by release from invading and developing stages, (2) by incorporation and expression in host cell membranes, and (3) after phagocytosis and breakdown of extracellular stages. In each case antigen has to be taken up and processed intracellularly before it can be presented to the host immune system. In many cases the cells which fulfill this function of processing and presentation are professional phagocytes (macrophages, dendritic cells) but, given that the essential processes involved (internalization and lysosomal degradation) are common to all cells, antigen presentation can be performed by any cell which shares with phagocytes the expression of cell surface molecules coded for by class II genes of the major histocompatibility complex. These molecules (class II or Ia molecules) combine with processed antigen to create the antigenic determinants which can be recognized by the antigen receptors of T lymphocytes. Recent work has shown that B lymphocytes, which bind antigen via their surface Ig molecules, and then internalize the bound antigen, can act as effective antigen-presenting cells.[5] Enterocytes themselves, in response to lymphokines, may also express class II molecules and thus function as antigen-presenting cells.[6] *In vivo*, antigen-presentation may take place within the intestinal mucosa, in the draining lymph nodes and in the spleen. T cells responding to antigen may proliferate in all three sites, those in the nodes and spleens recirculating to re-enter the intestine to provide an enlarged pool of antigen-specific cells.

2. Development and Expression of Immunity

T lymphocytes of the helper subset play a central role in the initiation and expression of almost all immune responses, whatever the ultimate effector mechanism. T cells that have responded to antigen release a number of soluble factors which regulate the activities of all other cellular elements of the response, they may also interact directly with B lymphocytes, each cell recognizing distinct determinants on the same antigen molecule.

a. Antibody

T-cell-dependent, B-cell-activation can occur within the intestine itself, in the draining lymph nodes and in the spleen. There is a regular traffic of B plasmablasts from lymphoid tissues into the mucosa, primarily of cells which will secrete IgA when mature. Differentiation into plasma cells occurs both within the intestinal wall and within the extraintestinal lymphoid tissues, so that antibody can be released locally, or can enter the mucosa via the bloodstream. IgG molecules are able to cross vessel walls to enter tissue fluids, and this transport is vastly increased when inflammation develops and the permeability of vessel walls changes. IgA and IgM can cross epithelial surfaces directly, being transported by active secretion; IgA can also enter the gut lumen in the bile after transport from the blood stream across the hepatocytes of the liver. IgE also enters the mucosa from the blood, but is normally bound to cells with surface IgE receptors such as mast cells and basophils. The actions of antibodies against eimeria can, in theory, include opsonization for subsequent phagocytosis, participation in cytotoxicity, lysis through activation of complement, prevention of invasion by immobilization in the mucus layer at the epithelial surface or by interference with the process of membrane recognition. All of these activities must be directed against extracellular stages.

Complement will be present within the mucosa in tissue fluids, but occurs in the gut lumen only in very small amounts.

b. T cells

A major effector-activity of T cells is cytotoxicity, normally a property of the $CD8^+$, or $CD4^+$ helper 1, subsets. Cytotoxicity is mediated by T-cell/target-cell interactions and requires recognition of surface antigen bound to class I MHC molecules. Cytotoxicity as such appears to be rarely involved against any intracellular protozoon, although recently it has been shown that it can occur.[7]

A more important expression of direct T cell anti-parasite activity is the release of lymphokines such as IFN-γ, which has the property of activating killing mechanisms within cells harboring parasites. Lymphokines may activate cells directly, after binding to surface receptors, or may do so indirectly by activating other cells, such as macrophages, to release cytokines (e.g., IL-1, TNFα) which can also promote intracellular killing. IL-1 release is also associated with enhanced phagocytic and cytotoxic activity in macrophages.[8]

c. Inflammation

Inflammation in the intestine is a complex phenomenon, with many interacting components. It may be initiated as a result of tissue damage per se, by formation of antigen-antibody complexes, by complement components, by T-cell lymphokines, by activation of amine-containing cells when their surface IgE binds specific antigen, and it is reinforced by the release of factors (enzymes, amines, leukotrienes, prostaglandins, cationic proteins, etc.) from the variety of cells that infiltrates inflamed tissues. Entry and proliferation of several cell types, such as mast cells, basophils and eosinophils, are T-cell-dependent phenomena, amine-containing cell infiltration creates the conditions in which antigen-release leads to increased vascular and epithelial permeability, promoting further infiltration, and facilitating release of immunoglobulins both into the mucosa and into the gut lumen. There are many structural alterations and physiological changes which accompany intestinal inflammation, many of which, together with mediator release, are inimical to parasite development and survival as well as being deleterious to the welfare of the host. One important change concerns the quantity and quality of mucus secreted in the infected intestine. Not only is there hyperplasia of goblet cells, but the biochemical nature of the mucus also changes, increasing its viscosity. Altered mucus secretion is thought to play an important role in resistance to intestinal nematodes and may influence the development of eimeria infections in the immune intestine.[9]

The manifold ways in which effector mechanisms may be generated and operate against eimeria in the intestine are summarized in Figure 1.

C. INVESTIGATION OF IMMUNITY—SYSTEMS UTILIZED

Studies concerned with immunity against coccidiosis have, as their eventual objective, the development of strategies to regulate levels of infection in economically important hosts by immunological means. Ideally, analysis of immune responses should be carried out in the target host, but such work may pose problems. With larger animals, such as cattle, pigs, and sheep, there are logistic constraints (cost, size, housing, difficulty in maintaining coccidia-free) which restrict their use. In addition, although knowledge of their immune systems is progressing rapidly it is, currently, less advanced than that in man, mouse and rat, and specific reagents for analysis of immune responses are of limited availability. Some of these problems, fortunately, do not apply to the domestic fowl, in which the disease is of greatest economic importance,and it has been possible to carry out much useful work directly on this species. Even here, however, there are some limitations, e.g., incomplete knowledge of the immune system, unavailability of commercially produced immunologic reagents and

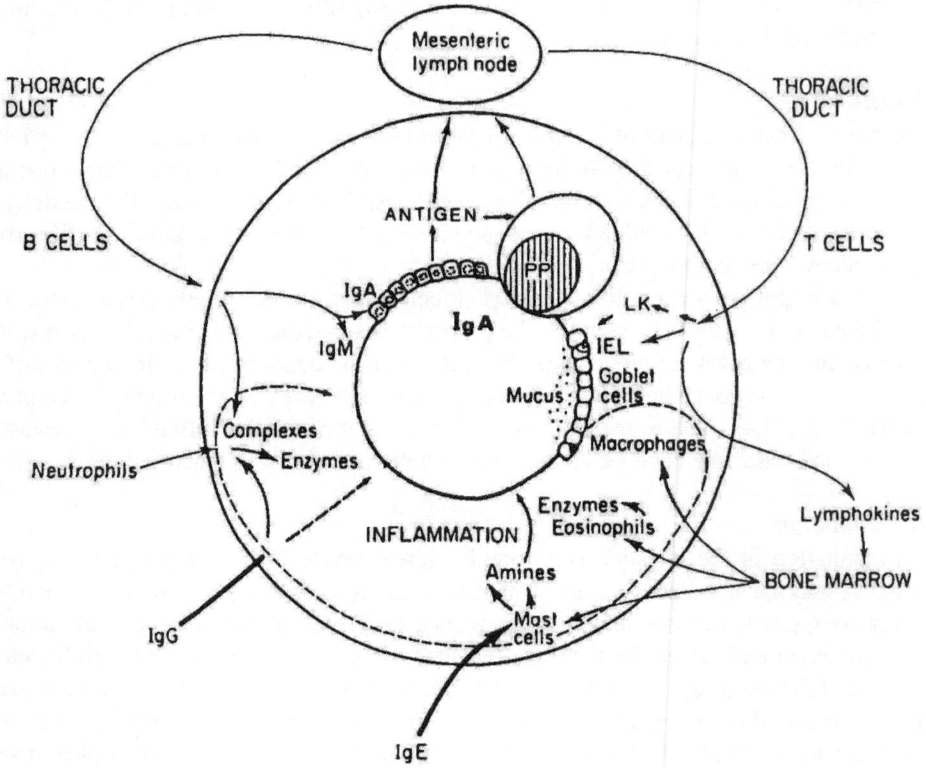

FIGURE 1. Representation of the immune and inflammatory mechanisms available for protective immunity to intestinal infections with eimeria in mammalian hosts. The responses are initiated by recognition of parasite-derived antigen, in the mucosa, in Peyer's patches (PP) and in the draining lymph node. Recirculation of lymphocytes brings B and T cells back into the mucosa. B cells secrete Ig locally as well as systemically; IgA also enters the intestine in bile. Antibodies interact with a variety of myeloid cells to initiate inflammatory changes, and may interact directly with parasites. T cells initiate changes within the mucosa itself, influencing local populations of myeloid-derived cells, and regulate production of these cells in the bone marrow. T cell influences also effect epithelial changes, involving enterocytes, goblet cells and intraepithelial lymphocytes (IEL). In avian hosts mesenteric lymph nodes are absent and true PP are few, but the intestine contains aggregated nodules of lymphoid tissue. (Redrawn from Wakelin, D., *Immunity to Parasites,* Edward Arnold, London, 1984.

a limited supply of inbred lines with defined immunological characteristics. All of these constraints on the use of target species make rodent models, in particular murine systems, a rewarding alternative. The immune system of the mouse is thoroughly investigated, many reagents, including those produced by recombinant DNA technology, are commercially available, and there is a wide range of inbred and genetically defined lines. The use of murine hosts provides the opportunity for detailed and sophisticated analysis of those components of the immune system that are functional in anti-eimeria responses and allows the development of paradigms which can be tested in hosts of economic importance (see Section V).

An important consideration of using eimeria in rodent hosts is the availability of *Eimeria* species which model the biology and characteristics of those which parasitize other hosts. Fortunately there is a good choice of species for work in the mouse (8 or 9) or rat (3 or 4).[10] Among these species there is variation in location in the intestine, pathogenicity and immunogenicity and it is therefore possible to match species of interest in the target animal. For example, *E. maxima* of the fowl, *E. vermiformis* of the mouse, and *E. nieschulzi* of the rat, are all parasitic in the small intestine of their hosts, moderately pathogenic and highly immunogenic; similarly, *E. tenella* of the chicken and *E. pragensis* of the mouse both parasitize the large intestine, are highly pathogenic and moderately immunogenic. To date, comparisons of immune responses of different hosts to *Eimeria* spp. of similar character have revealed no major differences, even between avian and mammalian hosts. For example, antibody responses of rats and chickens to infection with, respectively, *E. nieschulzi* and *E. tenella*, are similar in both isotype and kinetics.[11-14] There is, therefore, no reason to believe that information obtained from model systems could not reasonably be applied to the target animal.

In vivo studies have the merit of being concerned with the response of the whole organism to infection, but the complexity of this response makes analysis of specific components in isolation very difficult, even in well-defined hosts such as the mouse. For this reason *in vitro* systems are an extremely valuable adjunct to *in vivo* studies, allowing individual effectors of immunity to be analysed in detail. However, this approach is limited by the availability of suitable models for *Eimeria* spp. and *Isospora* spp. Although a number of species can be grown in cultured cells and in avian embryonic membranes, few complete their development and many, although capable of initial invasion, do not replicate in these milieu (see Chapter 11). Nevertheless, culture systems have been used, notably for *E. tenella* and *E. bovis,* for investigating mechanisms of immunity. They are most useful in testing antiparasitic effects, e.g., of cytokines capable of acting on intracellular developmental stages. A culture system also enables the invasiveness of sporozoites or merozoites, which have been subjected to various treatments *in vitro*, to be examined more easily than by inoculation into the host. The latter, however, will in most cases be the only measure of ability to continue development. Examples of the use of the culture system in immunologic work include assessment of monoclonal antibodies raised for the identification of important antigens.[15-18] Both *in ovo* and *in vitro* systems have been used to measure the antiparasite effects of lymphoid cell suspensions and antibodies[19-22] and lymphokines[23-25] (see Chapter 11). Although such systems have considerable advantages over the normal host, it must be borne in mind that substances shown to be active *in vitro* or *in ovo* may not necessarily be so *in vivo*. For example, serum samples from animals injected with nonviable parasites, or products derived from them, may be antiparasitic *in vitro* yet the animals themselves remain completely susceptible to infection[26] because a mechanism which would allow suitable contact between the antibodies and the parasite has not been primed by the procedure used for immunization.

III. ANTIGENS OF *EIMERIA* SPP.

Knowledge of the antigens of the eimeria is a prerequisite for the development of vaccines for controlling coccidiosis and important for an understanding of the immune response.

The exquisite species-specificity of immunity is well documented (see previous reviews) and indicates the possession, or expression/presentation, of antigen(s) unique to each species. This has necessitated the inclusion in a live virulent vaccine for coccidiosis in chickens ("Coccivac", Sterwin Inc.) of representatives of all the species of *Eimeria* capable of infecting the domestic fowl, and, more burdensomely, the manipulation of each species for the production of a live attenuated vaccine. Further complexity is provided byvariation within some of the species, notably *E. maxima* and *E. acervulina.* These immunogenic differences between, and within, species are reviewed in detail in Chapter 16.

The life-history of eimerian parasites is intricate, consisting of several (differing) generations of asexual replication, followed by the development of sexual stages. These different life-cycle stages might be expected to differ in immunogenicity and in antigenic composition. Identification of the most immunogenic developmental stage is a starting point for the detailed antigenic anaysis required for work on subunit vaccines and, until recently, work on eimerian antigens was more or less confined to this topic. The findings have been reviewed,[2,3,27] essentially, they indicate that the gametogony stages are not as immunogenic as developing asexual forms of the parasite. Recently published work has demonstrated the immunizing potential of the sporozoite of *E. tenella*[28,29] which, although previously in some doubt, is entirely in keeping with the sporozoite being a target of the immune response.[30-32] Also shown has been the immunogenic superiority of the development of the first generation schizont of *E. tenella* in comparison with the other asexual generations of this species[32-33] and a lack of cross-protection *in vivo* between the first and second asexual generations.[32] This immunological diversity between the stages correlated with antigenic differences in the sporozoite and first and second merozoite, detected by immunofluorescence and immunoblotting. Differences in the antigenic composition of the sporozoite and first generation merozoite of *E. bovis* have also been recorded[34] and it is highly likely that such differences will prove to be a general feature of the eimeria. It remains to be seen whether the location of the most important immunogens will be the same for all species.

During the last 5 years, interest in the development of subunit vaccines for coccidiosis has resulted in the identification and characterization of many antigens, in particular those exposed on the surfaces of the invasive forms, i.e., sporozoites and merozoites. Recent developments include the demonstration of analogies between some of these surface polypeptides and heat shock proteins[35] and, despite their presumed poor immunogenicity, analyses of the antigens of sexual developmental stages.[36-39] The subject is reviewed in Chapter 17.

Attention to date has focused on antigens expressed within, or on the surface of, various developmental stages of the parasite, with little consideration given to any that might be expressed on the surface of infected host cells. That these could be of interest is, however, indicated by the detection by immunofluorescence of specific binding of serum from chickens immunized with *E. tenella* to the surface of cecal crypt enterocytes infected with second generation schizonts of *E. tenella* (Rose and Wisher, unpublished observations). Also, Pasternak and Fernando[40] have cited unpublished observations of a significantly increased amount, in the host-cell membrane, of a protein which is associated with the second generation schizont of *E. necatrix*. Other workers have observed diffuse immunostaining with anti-*E. tenella* serum in cecal tissue from birds infected with *E. tenella* and attributed it to the presence of soluble antigens derived from the parasite.[41]

A. ANTIGENS AS TARGETS

The host's immune defense against eimerian parasites may be directed against both extra- and intracellular developmental stages, with the likelihood that the mechanisms deployed will differ. Extracellular stages include the invasive forms, i.e., the sporozoite and the merozoites of each generation of asexual replication, and the microgametes. It is not known whether the gametocytes are affected by the host response because difficulties in their isolation from host tissue and inability (in most cases) to develop *in vitro* have precluded much experimentation. There have, however, been reports of the production of oocysts of abnormal morphology, and/or sporulation and lowered viability in partially immune hosts,[42-45] which may indicate some deleterious effect on the gametes.

Extracellular sporozoites and merozoites are obvious targets for antibodies and there is abundant literature on the action *in vitro* of humoral and secretory antibodies on these invasive stages, with lethal, neutralizing or agglutinating effects. Reports published before 1983 have been reviewed;[2,3] more recent ones include the action of intestinal contents and monoclonal antibodies on, and electronmicroscopal observations of morphological changes[46-50] in, the parasites. Where appropriate, these have been reviewed in Chapter 17.

In vivo, the peak of the secretory IgA and IgM response to *E. tenella*, as measured in bile, has been shown to coincide with a reduction in the numbers of sporozoites recoverable from the lumen of the small intestine after a challenge inoculum of oocysts given orally.[51]

This suggests that, at certain stages of immunization, sporozoites may be the target of secreted antibodies in the lumen of the intestine or in the apical mucin. Although not demonstrated, it is likely that free merozoites could be similarly affected. Circulating serum antibodies may also gain access to extracellular stages as a result of the increased capillary permeability of the mucosa that occurs during infection[52-53] and on challenge of immunized animals.[54-58] The ability of molecules as large as immune globulins to pass into the intestine as a result of this response has, however, been questioned.[59]

Destruction of the invasive extracellular developmental stages should lead to a decrease in the numbers of subsequent intracellular forms.Comparisons of the numbers of intracellular sporozoites in naive and immune animals have produced conflicting reports, ranging from a reduction of 100% to no effect. Recent work has indicated a reduction, in the immune host of, for *E. tenella*, approximately 50 to 60%,[30-32] for *E. nieschulzi*, approximately 80%,[60] and for *E. adenoeides*, approximately 40%.[31] Invasion by *E. acervulina* or *E. meleagrimitis*, however, was either unaffected, or enhanced in the immunized host.[31] Discrepancies in the reports may be due to the stage of immunization of the host, mechanisms differing at different stages of immunization,[2] or to the species of *Eimeria*, as postulated by Augustine and Danforth[31] in their comparative study. Merozoites are also affected by immunity since smaller numbers are found within the enterocytes of suitably immunized hosts than in naive animals, after intraluminal inoculation and they fail to develop normally.[61-62] Inhibition of development of the invasive stages once within the cells of immunized hosts is a universal finding but it is not known whether this is due to effects exerted on the parasites before or after invasion. The latter is more likely since parasites which are inhibited within immune hosts may resume development on transfer into naive animals.[63-65] Once within the cell, parasites are generally considered to be protected against the effects of antibodies and susceptible only to the actions of T cells. However, enterocytes infected with second generation schizonts of *E. necatrix*[66] or *E. tenella* (Rose and Wisher, unpublished observations) have been shown to become permeable. Until recently this was thought to be confined to small molecules but Danforth and Augustine[67] have now shown that monoclonal antibodies are able to cross both host cell and parasite membranes in *in vitro* cultures of *E. tenella*.

All intracellular developmental stages have been shown to be susceptible to the immune response, the effect being evident in inhibition of development, or degeneration. The antigens eliciting these reactions might be expected to be expressed on the surface of the parasitized cell but little information is available (see earlier).

In an intriguing manifestation of the immune response, the behavior of the sporozoite, rather than the organism itself, appears to be affected, i.e., its migration within mononuclear cells, now considered to be intraepithelial leukocytes (IEL)[65,68-70] rather than macrophages.[71-74] This is a process which normally occurs within approximately the first 24 h of infection in those species in which asexual replication begins in the crypts. Even in those species whose development is confined to the villous epithelium there is movement of at least a proportion of sporozoites from enterocytes to lamina propria and back into enterocytes before growth occurs[70] (See Chapter 4). In all cases it is likely that the migration involves extraintestinal transport via the blood circulation since infections can be induced with blood or homogenates of liver or spleen within a few hours of the inoculation of oocysts.[65,70] The effect of immunity on this early part of the life-cycle has been examined in infections with *E. tenella*[30] and *E. maxima*.[65] Interruptions to the normal pattern were seen in both: in *E. tenella* the major effect was exerted on the final stage, i.e., transfer from transporting IEL to enterocyte within the epithelium of the crypt, whereas in *E. maxima* it occurred earlier since fewer sporozoites arrived at the crypt, being retained in the lamina propria. Nevertheless, in both cases the inhibited sporozoites appeared morphologically normal and those of *E. maxima* were shown to be viable in that they induced infections when transferred to naive animals in blood or homogenates of liver or spleen[65] or intestinal mucosa.[64,65] That

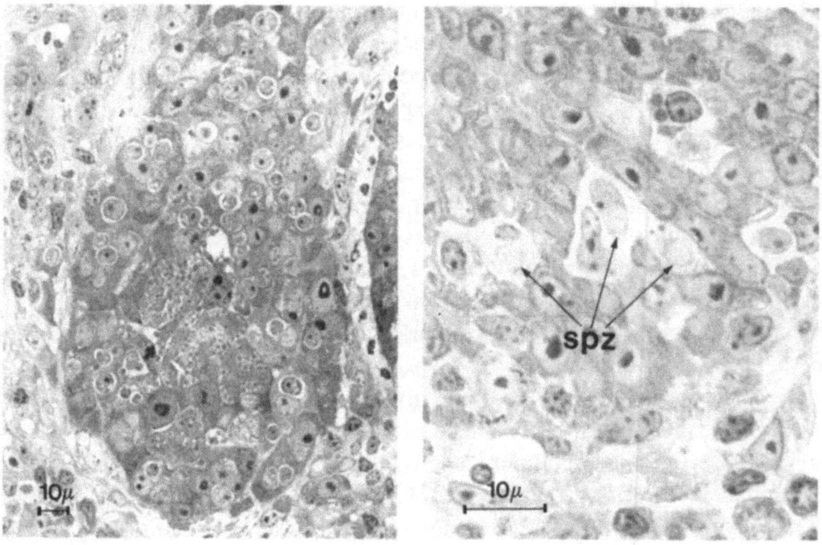

FIGURE 2. *Eimeria tenella:* Sections of ceca from naive (A) or immunized (B) chickens taken 60 h after challenge. (A) contains developing parasites (first generation schizonts and second generation trophozoites) within enterocytes of the crypt whereas (B) has sporozoites (arrowed) within intraepithelial lymphocytes.

these effects on intracellular sporozoites are manifest within 24 h after the introduction of the parasite suggests both a very rapid expression of parasite antigen on the surface of the transporting cell and a very prompt response of the host. In the case of *E. maxima* it seems that the target is the transporting host cells since these do not reach the crypt epithelium whereas in *E. tenella* there is an apparent failure of the morphologically normal sporozoites to transfer to enterocytes (Figure 2).

A proportion of sporozoites escapes the inhibitory effect and immunity is then (apart from the intervening invasive, briefly extracellular merozoites, as discussed earlier) directed against successive trophic stages which are seen degenerating within enterocytes.[62-63,65,75] In *E. vermiformis,* a species in which the primary infection is unusually subject to immune control, similar degeneration of intracellular parasites, coupled with destruction of parasitised crypts is associated with the early termination of infection seen in BALB strains of mouse.[76] Since the enterocyte is now considered to act as an antigen-presenting cell to T lymphocytes[77] and an increase in Ia expression on its surface may be induced by eimerian infections (Mackenzie and Rose, unpublished observations), the enterocyte-Ia-parasite antigen complex is probably an effective target.

B. STIMULATION OF IMMUNITY BY VACCINATION

Immunity to *Eimeria* spp. is readily stimulated by infection with the homologous parasites and animals may be safely vaccinated by controlled dosing with virulent organisms, or the administration of suitably attenuated lines (see Chapter 16). Attempts to stimulate protective immunity in the absence of infection, i.e., with nonviable material have, however, met with little success. In work done before 1980 some procedures have elicited immune responses, both humoral and cell-mediated which were virtually indistinguishable from those detected in infected (immunized) animals, without conferring protection against orally induced infection.[78] Within the current decade, advances made in vaccinating against other protozoan parasites, and some enteric helminths, has stimulated enormous interest in the development of vaccines against the eimerian coccidia, especially those produced by genetic engineering. This has resulted in the publication of several patent applications claiming successful vaccination with extracts of parasites and/or more defined gene products. The data relating to subunit and idiotypic vaccines are reviewed in Chapter 17. Immunization with crude extracts of host tissues or parasites is described below.

There has been little success in vaccinating with extracts of infected host tissue although protection against the weight-depressing effects of (but not the lesions caused by) some, but not all, strains of *E. tenella* has been described. Similar procedures carried out with other species (*E. acervulina, E. brunetti,* and *E. maxima*) were unsuccessful.[79-80]

Several patent applications refer to the vaccinating properties of crude saline or detergent extracts of various developmental stages (e.g., European Patent Application Numbers: 84-108925.3; 85-401246.5; 86-402547.3; 87-111820.4) or a protein prepared from the membrane of the sporocyst (EPA No. 85-200889.5). The data provided do not allow an appraisal of the protective effects claimed. One publication[81] describes the successful vaccination of chickens against *E. tenella* by oral or intramuscular injection of a saline-soluble extract of sporulated oocysts or purified sporozoites. The effects were judged mainly by a reduction in the severity of the lesions induced by a challenge inoculum of oocysts and most of the results described were obtained with the sporozoite antigen. Surprisingly, this was shown to be effective also against challenge with the heterologous species, *E. acervulina* and *E. maxima*. In two separate trials, a similar preparation of sporozoites of *E. nieschulzi,* when injected subcutaneously into rats, caused a highly significant reduction in the oocysts shed after a challenge inoculum of homologous oocysts (Rose and Hesketh, unpublished observations).

Some genetically engineered coccidial antigens have also been shown to protect against infection[82] (see Chapter 17).

Thus, although reports of successful vaccination by nonviable parasite material are distressingly few to date, there is reason to believe that it is possible.

IV. ANTIBODY RESPONSES TO INFECTION

The production of specific antibodies in response to infection with eimeria is well documented. First suggested in 1908[83] results were equivocal until the 1950s, when antibodies were demonstrated in the sera of chickens infected with *E. tenella*[84] and of rabbits infected with *E. stiedai.*[85] This history has been reviewed[86] as have the more recent findings,[87] including the immunoglobulin isotypes involved and the kinetics of the response.[2,3] Subsequent publications relating to serum antibodies have described variations in techniques,[88] detection of antibodies in the sheep,[89,90] demonstration and transfer of immediate type hypersensitivity in the mouse[91] and correlation of the response in chickens with genetic susceptibility to infection.[92]

Evidence for a secretory immune response to eimeria infections was first provided by the demonstration of an increase in the number of IgA-containing cells in the cecal mucosa of chickens infected with *E. tenella* and in the concentration of IgA in the cecal contents.[93] Since then, specific IgA and, in some cases, IgM antibodies have been detected in the bile, intestinal, and fecal mucus and intestinal washes of infected rodents and chickens, and the kinetics of the responses determined (reviewed by References 2, 3, and 94). More recently, the effect of different modes of immunization[95] and of genetic background[92] on the IgA response of the chicken have been described. Also, high concentrations of IgA have been correlated with low recoveries of sporozoites from the intestinal lumina of immunized chickens.[51]

The role of antibodies in the mediation of protective immunity against the eimeria has been the subject of much discussion[1-3,87,94,96] and still remains largely unresolved. The bulk of the evidence, however, points to a contributory, but not fundamental, function for immunoglobulins in the defense of the host against these parasites. The arguments on which this view is based have been presented in detail in the reviews cited above but are summarized here. B-cell-depleted animals (bursectomized and bursa-diseased chickens, Biozzi low-responder, and CBA/N mice) are less resistant to primary infection and slightly more susceptible to challenge than controls. Also, protection can be passively transferred with serum or, maternally, via the yolk of the egg. Repeated treatment with large amounts of antibodies is, however, necessary and protection is not invariably transferable in this way from highly immune animals. In recently published papers there have been contradictory reports of the effect of bursectomy on primary and challenge infections with *E. tenella*, one stating that it had no signficant effect on the magnitude of infections or the acquisition of immunity "although hormonally bursectomized birds had a higher mortality . . . . ",[97] while another reported a more severe intial infection in bursectomized birds and a slower development of immunity to re-infection.[98] A monoclonal antibody prepared against sporozoites of *E. tenella* has been shown to transfer protection against infection (measured by cecal lesion scores) with this species.[98,99] Attempts in our laboratory, however, to transfer immunity to *E. vermiformis* in mice with large amounts of convalescent serum have failed (unpublished observations). Also, the concentrations of antibodies to *E. vermiformis* in the serum of infected mice has not correlated with the magnitude of the infection.[100] The maternal transfer of antibodies to *Eimeria* spp. in the sheep has been reported[89] but no correlation with infections in the lambs was attempted.

These recent findings do not suggest that any revision is warranted of the view that antibodies may participate, albeit in a minor capacity, in immunity to eimerian infections.

V. T-CELL DEPENDENCY OF IMMUNITY

The immune responses which control infections with eimeria are T dependent, i.e., they are initiated and regulated through the activity of T lymphocytes. In this respect, therefore, immunity to eimeria conforms with what is known of immunity against all eukaryotic parasites. The evidence on which this conclusion is based has been obtained from experiments with T-cell-depleted hosts and from studies on mutant athymic rodents (Figure 3a).

Early work on experimentally thymectomized chickens gave rather equivocal results. Pierce and Long[101] found that removal of the thymus resulted in a greater susceptibility to initial infections with *E. tenella*, but no reduction in the development of immunity ultimately. In a series of experiments with *E. maxima* and *E. brunetti* Rose and Long[102] found increased oocyst output in thymectomized birds in only three out of eight infections, and the pattern of increased susceptibility was rather erratic, in that thymectomized irradiated animals were sometimes more susceptible than irradiated or normal controls, sometimes more susceptible only than irradiated controls or only normal controls. All thymectomized inviduals developed complete resistance to reinfection, even when this was given two weeks after initial exposure. Concurrent testing for tuberculin responsiveness in the experimental groups showed that very few thymectomized birds remained tuberculin negative.

It is likely that these data reflect the difficulty of ensuring complete removal of the multilobed thymus in the avian host. In contrast, infections in athymic (nude) mice and rats have shown quite unequivocally that there is an absolute dependence on an intact T lymphocyte population for the development of immunity. In nude mice infected with *E. vermiformis*[103] both the level and duration of primary infections are increased relative to intact controls, the increased length of patency presumably reflecting additional schizogonic cycles. When reinfected, nude mice show no evidence of any resistance, the course of the challenge infection being identical to that in previously uninfected age-matched nudes. Similar data have been reported for infections with *E. ferrisi*,[104] *E. falciformis*,[105-106] and *E. pragensis*[107] in nude mice and with *E. nieschulzi*[108] in nude rats although, with these species, failure to develop immunity during primary infection is reflected only in an increased output of oocysts, the duration of patency being essentially unaltered.[2,107]

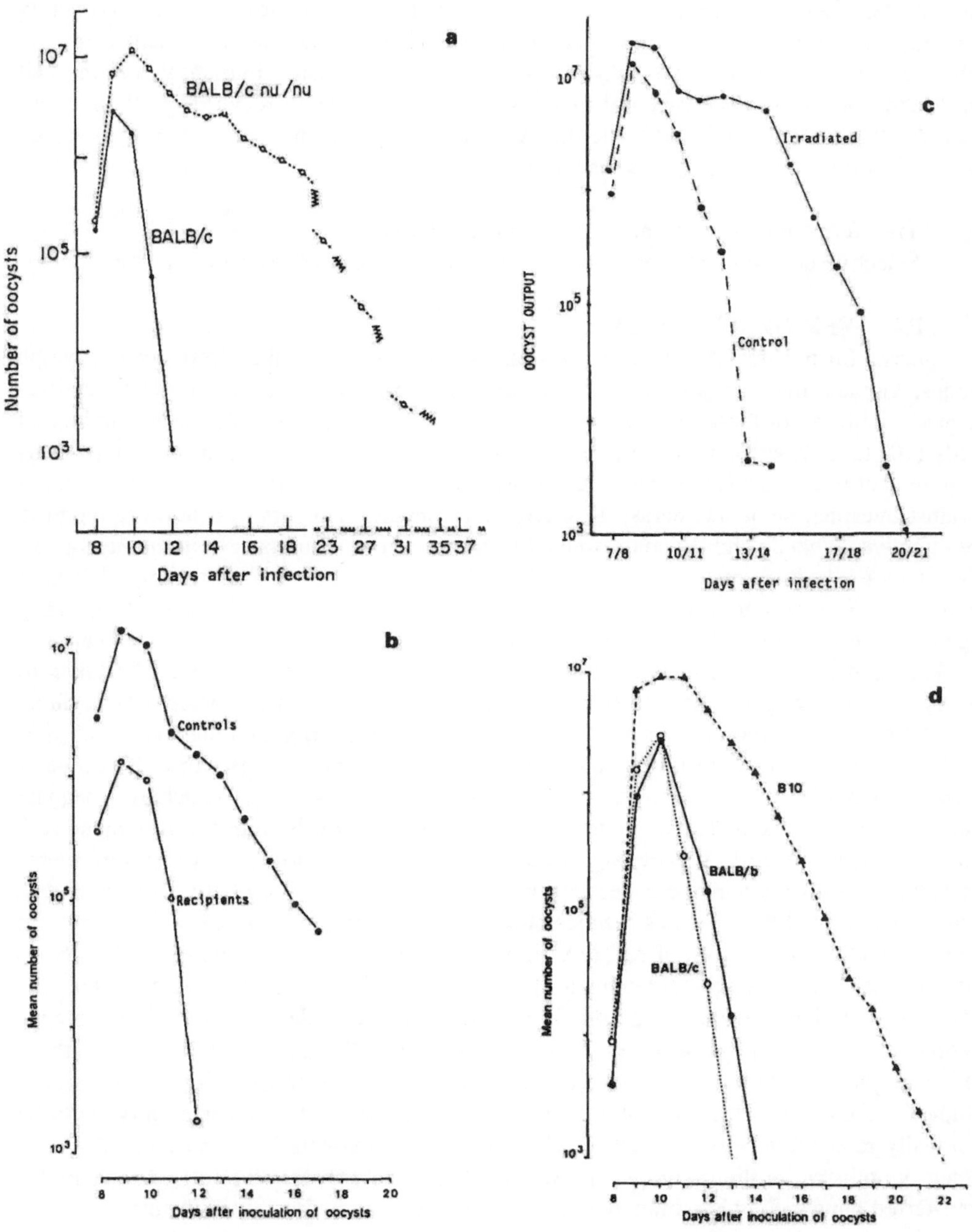

FIGURE 3. The influence of host immunity on the course of infection with *Eimeria vermiformis* in mice. (a) T-cell dependency: infection in BALB/c and BALB/c nude mice. From Rose, M. E., Wakelin, D., and Hesketh, P., *Parasite Immunol.*, 7, 557, 1985. By permission of Blackwell Scientific Publications, Ltd., Oxford.) (b) T-cell dependency: transfer of immunity with immune T cells in NIH mice. (c) Irradiation sensitivity: infection in irradiated and intact NIH mice. (d) Host response phenotype: infection in B10 (H-2)^b), BALB/B (H-2^b) and BALB/c (H-2^d), showing the overriding influence of background genes. (From Joysey, H. S., Wakelin, D., and Rose, M. E., *Infect. Immun.*, 56, 1399, 1988. By permission of the American Society of Microbiologists.)

Demonstration of the T-cell dependence of immunity does not by itself allow any conclusion to be drawn about the mechanisms through which resistance is expressed. The pivotal position of T cells in the development of the immune response means that a variety of mechanisms may function as the ultimate effectors of immunity, as has been amply demonstrated in many other host-parasite systems.[109] Conversely, absence of T cells in thymus-deprived or thymus-deficient hosts will mean that many of the correlates of immunity will be absent or diminished, without necessarily implying that they are functionally essential. Thus, it has been shown that antibody responses, cell-mediated immunity, plasma cell infiltration of the mucosa, mucosal mastocytosis are all severely reduced in athymic animals.[110] Experiments to identify the nature of the T dependent responses have involved manipulation of the host immune system by:

1. Transfer of immunity from immune to naive hosts, or
2. Selective depletion and restoration of immune competence prior to infection.

A. TRANSFER OF IMMUNITY

Successful transfer of immunity with immune lymphocytes taken from spleen, lymph nodes, thoracic duct, or peripheral blood has been described in both chickens[111] and rodents.[112-114] In all of these experiments it was found necessary to transfer large numbers of cells (10^8 to 10^9/recipient) in order to demonstrate significant transfer of immunity. This requirement was in striking contrast to results obtained with adoptive transfer of immunity against intestinal nematode parasites, where much smaller numbers of cells were found to be effective. Wakelin and Wilson[115] found that the threshold for transfer of immunity against *Trichinella spiralis* in mice was between 4×10^6 and 1×10^7 cells, in rats high levels of protection were transferred with between 1×10^7 and 7×10^7 cells.[116] Transfer of immunity against *Nippostrongylus brasiliensis* in rats was achieved with as few as 5×10^7 cells.[117-118] The need to use large numbers of cells against coccidial infections could be taken to imply that the lymphocytes effective in transfer were B cells. There should be a direct correlation between numbers of antibody-forming cells transferred and the amount of antibody synthesized in the recipient, and it would be reasonable to expect that, if immunity were antibody-dependent, a large number of B cells would be necessary to achieve adequate antibody levels in the recipient. However, as discussed previously, transfer of immune sera has, in general, failed to transfer significant resistance to infection, and it therefore seems more probable that adoptive transfer depends primarily upon the activity of T lymphocytes. That this is indeed the case was demonstrated in the NIH mouse — *E. vermiformis* system using conventional methods of cell separation involving adherence to nylon wool.[119] Non-adherent cell fractions were predominantly T cells (65 to 73%), adherent fractions (86% B cells) were further purified using anti-T-cell antisera and complement to remove contaminating T cells. Only T lymphocytes transferred immunity, T-depleted B cells being without effect despite an effective transfer of anti-parasite antibody responses, Figure 3b. Parallel studies[120] showed that the minimum number of cells needed to transfer immunity could be markedly reduced if cells were taken from the draining mesenteric lymph node at a time when lymphocyte proliferation was maximum (days 9 to 14). Day 11 cells, for example, transferred almost complete immunity at 5×10^7 cells, approximately 50% immunity at 1×10^7 and more than 40% at 5×10^6 cells. It follows that the earlier data showing a requirement for large numbers of immune cells in adoptive transfer probably reflected the use of unseparated cell suspensions taken at suboptimal times after immunization of donors.

Evidence that T cells were optimally effective when undergoing proliferation in response to infection suggests that their function in protection was likely to be expressed within the intestinal mucosa itself. T lymphoblasts are known to be retained preferentially in the mucosa after intravenous injection, and such localization is enhanced when the intestine is infected, as has been demonstrated in *E. nieschulzi*-infected rats and in *E. maxima*-infected chickens.[121] Given that antibody plays little or no part in immunity, then the role of T cells must be in some function other than that of providing help to B cells. An important pointer to the nature of this function has been gained by identification of the T-cell subset responsible for transfer of immunity, and by determining whether there is a requirement for other cell populations in the expression of immunity.

1. Identification of T-Cell Subsets

In the mouse the two major subsets of T lymphocytes, $CD4^+$ (T helper) cells and $CD8^+$ (T cytotoxic and suppressor) cells, are characterized by surface antigens, against which monoclonal antibodies are readily available. These antibodies can be used to enrich T-cell suspensions for one or other subset *in vitro*, prior to transfer, and can be used *in vivo* selectively to deplete an animal of a particular subset. The use of these approaches in the mouse — *E. vermiformis* model has clearly shown that effective transfer of immunity requires the presence of $CD4^+$ T cells and that $CD8^+$ T cells have little or no effect.[119] Similarly, when mice are depleted *in vivo* of $CD4^+$ cells, they lose their ability to control primary infections, whereas $CD8^+$ T-cell-depleted mice do not. These results parallel those obtained in similar studies on immunity to other intestinal parasites, such as *Giardia muris*[122] and *T. spiralis*[123] despite their different nature and habitats.

Although $CD4^+$ T cells are known to be capable of developing a cytotoxic function, this role is normally played by $CD8^+$ cells. Thus it appears that cytotoxicity mediated by the latter cells is unlikely to play a major role in protective immunity, despite the likelihood that parasite antigens are present at the surface of infected enterocytes, which would also express the class 1 MHC molecules necessary for the expression of cytotoxic T-cell function against target cells, and despite the suggestive observations that infected enterocytes are shed in sheets from the mucosal surface at the height of the infection[124] and that the normal architecture of the crypts is destroyed.[76]

B. COOPERATION OF $CD4^+$ T HELPER CELLS WITH OTHER EFFECTORS

In a number of experimental systems involving parasitic nematodes, T-helper cells mediate immunity by initiating a complex series of inflammatory events in the intestine.[125] A major component of inflammation is mucosal infiltration by a number of myeloid cell types (eosinophils, mast cells, basophils) whose production from bone marrow precursors is under T-cell control. In these systems, although immunity is transferable with T cells, these cells can only effect immunity if the recipient is capable of a normal myeloid cell response. Transfer of immunity against nematodes such as *T. spiralis* is therefore inhibited by prior irradiation of the recipient.[126] Myeloid cell responses, both intestinal and peripheral, accompany the development of immunity to eimeria[57,127] and offer a potential mechanism by which T-helper cells might function against these parasites. In the transfer system used by Rose et al.,[120] recipient mice were lightly irradiated (200 rads) before injection of cells, but this level of irradiation should not seriously interfere with myeloid responses. However, an equally effective transfer of immunity was achieved when cell recipients were irradiated at 600 rads[128] at which level myeloid responsiveness is reduced almost completely for a considerable period (Figure 3c). It appears, therefore, that the essential T-cell function in anti-eimeria immunity is independent of myeloid responsiveness, and is therefore either direct in its anti-parasite activity or involves a radio-resistant collaborative cell population.

The marked manifestations of T-cell-mediated inflammation which accompany infection (mastocytosis, goblet cell hyperplasia, villous atrophy, crypt hyperplasia, plasma cell infiltration,[110] and delayed hypersensitivity[129-132]) as well as the T-dependent antibody responses (IgG and IgA)[11,133] can therefore be seen as contingent accompaniments of immunity, rather than as primary causal events. This is not to say, however, that they make no contribution at all to overall immunity. Any alteration in the environment provided by the host for the parasite is likely to be detrimental to the survival of that organism; all of the above phenomena may therefore add to the total expression of anti-parasite immunity. What is clear, however, from the results of transfer of immune cells into heavily irradiated hosts, is that the contribution made by these phenomena is not essential to the expression of immunity, either during a primary infection or, though less certainly, to reinfection.

C. T-CELL FUNCTION IN IMMUNITY — THE ROLE OF CYTOKINES

It is now well established from many studies with intracellular pathogens that a major expression of host immunity is the release of soluble factors (cytokines), either from T lymphocytes (lymphokines) or from macrophages activated by T cells (monokines). These cytokines can kill pathogens directly, or induce killing mechanisms in the host cell itself.[134] Among the protozoa known to be subject to such immune mechanisms are *Plasmodium, Babesia, Toxoplasma, Leishmania,* and *Trypanosoma cruzi.* The cytokines implicated include the lymphokines interferon gamma (IFN-γ), lymphotoxin (tumor necrosis factor β), granulocyte macrophage colony-stimulating factor (GM-CSF), the monokines interleukin-1 (IL-1) and tumor necrosis factor (TNF-α). Of the parasites listed above, two (*Toxoplasma* and *Leishmania*) live within macrophages, in which the development of intracellular killing mechanisms under T-cell control is well documented.[135] The erythrocytic stages of *Plasmodium* and *Babesia* appear susceptible to the direct effects of cytokines, the appearance of 'crisis' forms reflecting intracellular death, resulting from the entry and activity of these molecules.[136] More relevant to the question of immunity against eimeria, however, is the recent demonstration that the exoerythrocytic stages of *Plasmodium,* the first generation schizonts in hepatocytes, appear to be susceptible to the indirect action of cytokines.[137-139] Although the mechanism underlying this effect has not yet been explained, it is presumably brought into play following interactions between these factors and membrane receptors on the host cell.

The possibility that cytokines may play a role in anti-eimeria responses is strengthened by observations that these parasites may be killed within host cells[62,63] and it has been explored by a number of workers using both *in vitro* and *in vivo* systems. Early studies[140] showed that induction of 'interferon' (probably IFN-α and -β rather than IFN-γ) by administration of 'Statolon' to chickens reduced the reproductive potential of *E. maxima.* Similar results were obtained *in vitro* with *E. tenella.*[141] More recently, lymphokines from mitogen- or antigen-stimulated T lymphocytes have been shown to affect adversely the invasion or development of *E. bovis*[24] and *E. tenella* in cultured cells.[23,25] It has now been shown that this effect is associated with IFN-γ.[142] The relevance of these *in vitro* observations to the expression of immunity *in vivo* must remain speculative, both because of the limited development possible *in vitro* and because some of the effects observed depend upon the phagocytic activity of macrophages, whose role in resistance *in vivo* is obscure. Nevertheless, recent experiments in mice have confirmed that IFN-γ does play an important part in immunity against *E. vermiformis.*[143] BALB/c mice depleted of this lymphokine by treatment with monoclonal anti-IFN-γ antibody lost their ability to regulate the course of a primary infection. The results of experiments in which the timing of anti-IFN-γ treatment was varied relative to infection suggested that the release of IFN-γ influences the development of the schizogonic cycles. At the present time, it is not known how IFN-γ might act, whether directly upon infected enterocytes or indirectly via local macrophages, stimulating them to release other cytokines. (If this were the case, macrophages would form the radio-resistant population capable of collaborating with transferred T cells — see above.) Enterocytes are known to respond to IFN-γ *in vivo,* as they express class II MHC molecules on their surface membranes when the intestine is infected or inflamed,[6] but nothing is yet known of their capacity to express intracellular anti-pathogen responses. The effects of anti-IFN-γ treatment were seen only in primary infections; no effect was seen when immune mice were treated prior to challenge.

1. Transfer Factor

An unexplained facet of immunology is the apparent ability of cell-free, dialyzable extracts of immune lymphocytes to confer immune responsiveness upon naive recipients. The term 'transfer factor' was coined to describe this activity[144] which appears to differ from effects obtained by transfer of antibodies, intact lymphocytes or known lymphokines. A role for such a factor in transfer of immunity against eimeria has been proposed by several authors, working with avian, bovine and rodent species. Successful transfer of resistance to infection has been obtained in both homologous[145-147] and heterologous systems.[148] The immunity generated is reported to be nonspecific in its effect upon the parasite. The numbers of cells used to prepare the cell-free extract are sometimes very much smaller than the numbers required to transfer immunity adoptively (10 to 100× fewer) and at present it is not possible to propose a functional explanation of this phenomenon.

VI. GENETIC INFLUENCES UPON IMMUNITY

Experience in the field has shown quite clearly that different breeds of chicken vary in the degree of resistance they exhibit to infection with eimeria. Extensive studies under controlled conditions have confirmed the genetic basis of this phenomenon and the potential for selecting resistance as a heritable trait (early papers reviewed in References 149 and 150). More recent papers have used host variation in chickens as a tool for analysis of the immune responses that correlate with resistance, following a trend that is firmly established in experimental immunoparasitology.[151] To the data available from these studies in the target host can now be added the results of work using a number of rodent eimeria.

A. AVIAN HOSTS

A series of papers by Lillehoj and colleagues has exploited the differences in susceptibility between inbred SC and FP chickens to show that they exhibited differences in both antibody and cell-mediated responses to infection (*E. acervulina, E. maxima, E. tenella*).[92,152-153] FP strain birds, which overall were more susceptible to primary infections, showed somewhat lower T-cell proliferative responses to antigen than did SC when tested against *E. acervulina* and *E. tenella* but higher responses against *E. maxima*.[152] Some correlations have been established between parameters of resistance to infection and genes linked to the major histocompatibility complex (MHC)[154] but it is clear, as would be expected from other immunogenetic studies, that MHC-linked effects are substantially modulated by the activities of other genes.[155] In a detailed comparison of MHC congenic strains and other inbred strains sharing MHC genes but with different genetic backgrounds, Lillehof et al.[156] found considerable differences in resistance, particularly to reinfection, which could be attributed to specific MHC haplotypes (e.g., B^{19} or B^{12} on the 15I genetic background). However, equivalent, and in some cases greater, differences were seen between strains sharing common MHC common alleles (e.g., B^{13}) but having different backgrounds. Interestingly, little correlation was observed between levels of serum IgG or bile IgA and resistance to infection.

Despite the data obtained from such studies, it is clear that there are severe limitations imposed by the restricted availability of inbred and congenic fowl, and this has impeded progress in identifying functional correlates between genetically determined differences in resistance and the expression of immune effector mechanisms. Rather more progress has been made using murine systems, where such restrictions do not apply.

B. MURINE HOSTS

The existence of marked mouse strain variation in response to infection has been demonstrated in relation to *E. ferrisi*,[104] *E. vermiformis*,[103] and *E. falciformis*.[105,157] Some common patterns emerge from this work, e.g., BALB/c mice appear to be most resistant in each case, but other strains show contrasting phenotypes, e.g., C3H are resistant to *E. ferrisi* and *E. vermiformis* but susceptible to *E. falciformis*. Only limited evidence has been obtained for any major effect of MHC-linked genes on resistance or susceptibility[157] (and Joysey et al., unpublished) Figure 3d. Differences in phenotype were apparent only in primary infections, even susceptible strains expressing resistance to challenge.

A number of interesting pointers to the functional correlates of strain-dependent differences emerged from the use of mutant mice in two of the above papers. Rose et al.[103] found that primary infections were enhanced in Biozzi low antibody-responder mice, in C57BL/6 beige mice with granulocyte and NK cell deficiencies, and in athymic mice on two different backgrounds. Only the latter strains, however, showed defective development of resistance to reinfection. Stockdale et al.[105] noted, as did Rose et al.,[103] that there were strong background gene effects evident in athymic strains. BALB/c nude mice were considerably more susceptible than CBA/H nudes, whereas intact BALB/c mice were much more resistant than intact CBA/H. These observations confirm the dominance of T-cell-mediated forms of resistance over genetically determined levels of innate susceptibility, as do the results of Rose et al.[158] who used a variety of immunosuppressive treatments to demonstrate the role of immune responses in determining strain phenotype. Collectively, they emphasize the importance of understanding the control and expression of immune responsiveness in eimeria infections, specifically responsiveness dependent upon T-cell function.

In a number of murine host-parasite relationships T-cell function is known to be regulated by MHC linked genes, and there are clear correlations between MHC haplotypes and resistance or susceptibility to infection.[151] In others, MHC-linked influences, although measurable, are masked by the stronger effects of genes located elsewhere in the host genome. From the studies so far carried out, this latter situation would appear to apply to the genetic control of immunity to eimeria. Thus both MHC and background gene influences regulate resistance to infection in the fowl (see above), and although MHC-linked differences have been recorded in congenic mice infected with *E. falciformis*[157] and *E. vermiformis*[159] (and Joysey et al., unpublished) these are less pronounced than the differences associated with different genetic backgrounds.

Detailed comparisons of the course of infection and of immune responses in BALB (resistant) and C57BL (susceptible) background mice given *E. vermiformis* have led to a clearer view of the basis for the genetically determined differences in resistance. Creation of radiation chimeras between MHC compatible strains of these backgrounds (BALB/b and C57BL/10 - both H-2^b), essentially replacing the bone marrow and all bone marrow-derived cells of one strain by that of the other, showed that response phenotype was donor-cell determined, i.e., that relative resistance or susceptibility was expressed through some population of cells originating from bone marrow-precursors.[159] Reciprocal exchange of immune lymphocytes between these strains showed that each developed T cells capable of transferring immunity, and the degree of resistance expressed was comparable in each strain, whether given homologous or heterologous cells.[128] Thus, genetic control of greater resistance in BALB mice must reflect a quantitative or temporal difference in response. It has now been shown that the difference is temporal. BALB mice respond more rapidly to *E. vermiformis* than do C57 mice, measured in terms of both lymphocyte proliferation and antibody response. They also generate cells capable of transferring immunity some days earlier than C57 (Rose et al., submitted).

The value of this work in the model system has been to show that genetically determined differences in resistance reflect a difference in rate of response, rather than an inability, or reduced ability to mount an effective response. From this it is possible to conclude that improving the rate of response, in both resistant and susceptible hosts, should enhance the degree of protection expressed against eimeria, even during initial infection.

VII. SUMMARY AND CONCLUSIONS

The prevalence and importance of intestinal parasites has focused a great deal of attention on the mechanisms that underlie the generation and expression of protective immunity. There seems now to be a consensus opinion that intestinal immunity, whatever the target parasite, is never the result of a single effector mechanism, operative under all circumstances and in all hosts.[4,125,160-161] The very diversity of mechanisms available to the host to deal with such parasites (Figure 1) make such a possibility extremely unlikely and existing experimental data strongly reinforce this view. Eimerian parasites are no different from other intestinal organisms in this respect and thus some earlier controversies concerning the 'cellular' or 'humoral' nature of immunity to coccidiosis can now be seen as misguided. The question 'which is the mechanism of immunity', has now to be re-phrased as 'which mechanisms can effect resistance, under which circumstances and in which hosts are these mechanisms expressed, and how are these mechanisms generated?' The objective of this chapter has therefore been to summarize and to review critically recent papers on immunity to coccidiosis, in order to marshall data that can be used to address these questions. In this concluding section an attempt is made to put together a working hypothesis of immunity, which might indicate lines of approach relevant to both the analysis of resistance in hosts of economic importance and the stimulation of immunity under controlled conditions. Such an hypothesis has to accommodate the diversity of mechanisms available, the differential susceptibilities of developmental stages of the parasite, and the quite different situations obtaining in primary and in subsequent infections.

The value of the model provided by *E. vermiformis* in the mouse has been to show that immunity can operate early enough in a primary infection to reduce oocyst output and the duration of patency. It is quite clear that the immunity is effective against the later schizogonic stages and that is is expressed through the activity of $CD4^+$ T cells. In light of the data obtained in other systems by manipulation of infection and challenge (see Section III) it appears that these cells are primed by exposure to antigens from developing schizonts, expressed directly on the enterocyte surface and therefore available to T cells in the context of induced class II MHC molecules, or released when merozoites are liberated, taken up by local antigen-processing cells and presented to T cells in the context of constitutively expressed class II molecules. This antigenic stimulation results in the expansion of clones of primed cells, release of IFN-γ and activation, directly or indirectly, of intracellular killing mechanisms. The potential for similar T-cell-mediated immunity must exist in all eimeria infections, but functional expression will depend upon (1) the biology of the eimeria species concerned, and (2) the genetically determined capacity of the host to respond at a particular rate relative to the development of the parasite.

Although cytokine-mediated protective immunity is perhaps the only form of resistance likely to have any impact upon an initial infection, it is but one of the responses that $CD4^+$ T cells are capable of inducing, many more of which may play a role in the expression of resistance against reinfection, particularly under field conditions, where infection is a continuous process, unlike the separated pulse infections used in the laboratory.

The first target of this resistance will be the sporozoite, which may be susceptible to the effects of specific IgA and IgM antibodies present in the lumen as the result of secretion across the mucosal epithelium or across hepatocytes into bile. IgG antibodies may also contribute at this stage, their entry into the lumen being potentiated by increased vascular and epithelial permeability. Antibody-mediated effects may depend upon, or be potentiated by, the trapping action of intestinal mucus, which may be qualitatively different in the immune host as a result of T-cell-induced changes in goblet cell function. All of these responses are likely to be short-term, and would wane in the absence of reinfection, but continuous reinfection would tend to maintain effective levels of antibody and altered mucus secretion.

Trapped sporozoites may be permanently damaged by prolonged exposure to the luminal environment, although clearly some escape or recover sufficiently to continue development. Subsequent development within enterocytes will be susceptible to T-cell-mediated intracellular killing as in primary infections but, in addition, IgM and IgG antibodies available within the mucosa at the time of merozoite release may interact directly with these stages to disrupt their normal entry into host cells, to bring about lysis with complement components, or to promote phagocytosis and cytotoxicity by myeloid cells within the lamina propria. A major feature of immune responsiveness to reinfection with avian eimeria (*E. tenella, E. maxima*) is disruption of sporozoite transport, from the point of entry to the enterocytes of the crypt, and a similar effect may be expected against other species which migrate within the mucosa. As the transporting cells are known to be intraepithelial lymphocytes,[65,68] a subset of T cells, it is possible that cytokine-mediated influences on the behavior of these cells, affecting subsequent parasite development should also be considered as a facet of T-cell-mediated resistance.

The potentiated responsiveness of the primed intestine would result in a number of profound inflammatory changes, structural and physiological, local and systemic, when the host is re-exposed to parasite antigens. These changes reflect the direct activities of T-cell-derived lymphokines, as well as the indirect consequences of inflammatory cell populations and the formation of antigen-antibody complexes. It is likely that many of these changes will contribute to resistance by making the local environment inimical to normal development and survival of the parasite.

Resistance to reinfection is therefore more likely to be the result of multiple effector mechanisms than is resistance during an initial infection, although the strength of any one component may reduce the relative contribution of those which act at a later stage in the development of the parasite. For example, as shown with *E. tenella* in chickens, peak biliary IgA responses correlate with highly effective immunity directed against the sporozoite itself,[51] thus minimizing the numbers of parasites against which later stages of immunity might operate. Where immunity against the sporozoite is less effective, for whatever reason, other mechanisms of resistance will play a greater role.

This working hypothesis identifies a number of important questions which need to be answered if data on immune responses can form the basis for the rational development of successful immunoprophylactic strategies. These questions will include:

1. Which antigens or epitopes are immunodominant, and which elicit protective responses?
2. Which stages of the parasite present these antigens and to what extent are they common?
3. Do T-cell-mediated and antibody-mediated responses involve the same antigens?
4. What determines the rate of T-cell responsiveness?
5. What are the mechanisms of intracellular killing and how are these mediated by T cells?
6. Are all stages of the parasite equally susceptible to immunity?
7. In what ways do MHC-related and background genetic influences regulate protective responses?
8. What is the nature of infection-associated immunosuppression?
9. How can the immune response be manipulated to maximize its effectiveness against infection?

All of these questions are relevant to the design and development of vaccines. An important consideration, which this review has attempted to emphasize, is that mechanisms and targets may exist which are not normally expressed or affected during normal infections, but which vaccine-induced immunity should attempt to potentiate and utilize.

REFERENCES

1. **Davis, P. J.,** Immunity to coccidia, in *Br. Poult. Sci. Symp.*, No. 16, *Avian Immunology*, Rose, M. E., Payne, L. N., and Freeman, B. M., Eds., British Poultry Science Ltd., Edinburgh, 1981, 361.
2. **Rose, M. E.,** Immune responses to *Eimeria* infections, in *Proc. Georgia Coccidiosis Conf., Research in Avian Coccidiosis*, McDougald, L. R., Joyner, L. P. and Long, P. L., Eds., University of Georgia, Athens, 1986, 449.
3. **Rose, M. E.,** *Eimeria, Isospora* and *Cryptosporidium*, in *Immunology, Immunopathology and Immunoprophylaxis of Parasitic Infections*, Vol. 3, Soulsby, E. J. L., Ed., CRC Press, Boca Raton, FL, 1987, 275.
4. **Befus, D. J. and Bienenstock, J.,** Factors involved in symbiosis and host resistance at the mucosa-parasite interface, *Prog. Allergy*, 57, 76, 1982.
5. **Lanzavecchia, A.,** Antigen-specific interactions between T and B cells, *Nature*, 314, 537, 1985.
6. **Bland, P.,** MHC class II expression by the gut epithelium, *Immunol. Today*, 96, 174, 1988.
7. **Kaufmann, S. H. E.,** $CD8^+$ T lymphocytes in intracellular microbial infections, *Immunol. Today*, 9, 168, 1988.
8. **Durum, S. K., Schmidt, J. A., and Oppenheim, J. J.,** Interleukin 1: an immunological perspective, *Annu. Rev. Immunol.*, 3, 263, 1985.
9. **Miller, H. R. P.,** Gastrointestinal mucus, a medium for survival and for elimination of parasitic nematodes and protozoa, in *Symp. Br. Soc. Parasitol.*, Vol. 24, *Pathophysiological Responses to Parasites*, Vol. 24, Rose, M. E. and McLaren, D. J., Eds., Cambridge University Press, Cambridge, 1987, S77.
10. **Levine, N. D. and Ivens, V.,** *The Coccidian Parasites (Protozoa, Sporozoa) of Rodents, Illinois Biological Monographs*, No. 33, University of Illinois Press, Urbana, 1965.
11. **Rose, M. E., Peppard, J. V., and Hobbs, S. M.,** Coccidiosis: characterization of antibody responses to infection with *Eimeria nieschulzi, Parasite Immunol.*, 6, 1, 1984.
12. **Mockett, A. P.A. and Rose, M. E.,** Immune responses to eimeria: quantification of antibody isotypes of *Eimeria tenella* in chicken serum and bile by means of the ELISA, *Parasite Immunol.*, 8, 481, 1984.
13. **Clare, R. A., Strout, R. G., Taylor, R. L., and Aeed, P. A.,** Bile and serum immunoglobulin levels during primary and secondary infections with *Eimeria tenella* in chickens, *Vet. Parasitol.*, 25, 33, 1987.
14. **Trees, A. J., Crozier, S. J., McKellar, S. B., and Wachira, T. M.,** Class-specific circulating antibodies in infections with *Eimeria tenella, Vet. Parasitol.*, 18, 349, 1985.
15. **Danforth, H. D.,** Use of monoclonal antibodies directed against *Eimeria tenella* sporozoites to determine stage specificity and *in vitro* effect on parasite penetration and development, *Am. J. Vet. Res.*, 44, 1722, 1983.
16. **Augustine, P. C. and Danforth, H. D.,** Effects of hybridoma antibodies on invasion of cultured cells by sporozoites of *Eimeria, Avian Dis.*, 29, 1212, 1985.
17. **Crane, M. St. J., Norman, D. J., Gnozzio, M. J., Tate, A. C., Gammon, M., and Murray, P. K.,** *Eimeria tenella:* quantitative *in vitro* and *in vivo* studies on the effects of mouse polyclonal and monoclonal antibodies on sporozoites, *Parasite Immunol.*, 8, 467, 1986.
18. **Whitmire, W. M., Kyle, J. E., Speer, C. A., and Burgess, D. E.,** Inhibition of penetration of cultured cells by *Eimeria bovis* sporozoites by monoclonal IgG antibodies against the parasite surface protein P20, *Infect. Immun.*, 56, 2538, 1988.
19. **Rose, M. E. and Long, P. L.,** Immunity to coccidiosis: protective effects of transferred serum and cells investigated in chick embryos infected with *Eimeria tenella, Parasitology*, 63, 299, 1971.
20. **Long, P. L. and Rose, M. E.,** Immunity to coccidiosis: effect of serum antibodies on cell invasion by sporozoites of *Eimeria in vitro, Parasitology*, 65, 437, 1972.
21. **Bedrnik, P.,** Attempts at transfer of immunity against *Eimeria tenella in vitro* and *in vivo, Folia Parasitol. (Praha)*, 24, 277, 1977.
22. **Davis, P. J. and Porter, P.,** A mechanism for secretory IgA-mediated inhibition of the cell penetration and intracellular development of *Eimeria tenella, Immunology*, 36, 471, 1979.
23. **Onaga, H., Tajima, M., and Ishii, T.,** Activation of macrophages by culture fluid of antigen-stimulated spleen cells collected from chickens immunized with *Eimeria tenella, Vet. Parasitol.*, 13, 1, 1983.
24. **Speer, C. A., Reduker, D. W., Burgess, D. E., Whitmire, W. M., and Splitter, G. A.,** Lymphokine-induced inhibition of growth of *Eimeria bovis* and *Eimeria papillata* (Apicomplexa) in cultured bovine monocytes, *Infect. Immun.*, 50, 566, 1985.
25. **Kogut, M. and Lange, C.,** Effect of lymphokine treatment on the invasion of cultured cells by *E. tenella, Lymphokine Res.*, 7, 31, 1988.
26. **Long, P. L., Rose, M. E., and Pierce, A. E.,** Effects of fowl sera on some stages in the life cycle of *Eimeria tenella, Exp. Parasitol.*, 14, 210, 1963.
27. **Rose, M. E.,** *Eimeria*, in *Parasite Antigens in Protection, Diagnosis and Escape*, Parkhouse, R. M. E., Ed., *Curr. Topics Microbiol. Immunol.*, 120, 7, 1985.

28. **Jeffers, T. K. and Long, P. L.,** *Eimeria tenella:* immunogenicity of arrested sporozoites in chickens, *Exp. Parasitol.*, 60, 175, 1985.
29. **Bhanushali, J. K. and Long, P. L.,** Role of sporozoite induced immune responses in protection against *Eimeria tenella: in vivo* and *in vitro* studies, in *Proc. Georgia Coccidiosis Conf., Research in Avian Coccidiosis,* McDougald, L. R., Joyner, L. P. and Long, P. L., Eds., University of Georgia, Athens, 1986, 526.
30. **Rose, M. E., Lawn, A. M. and Millard, B. J.,** The effect of immunity on the early events in the life-cycle of *Eimeria tenella* in the caecal mucosa of the chicken, *Parasitology,* 88, 199, 1984.
31. **Augustine, P. C.and Danforth, H. D.,** A study of the dynamics of the invasion of immunized birds by *Eimeria* sporozoites, *Avian Dis.*, 30, 347, 1986.
32. **McDonald, V., Wisher, M. H., Rose, M. E., and Jeffers, T. K.,** *Eimeria tenella:* immunological diversity between asexual generations, *Parasite Immunol.*, 10, 649, 1988.
33. **McDonald, V., Rose, M. E., and Jeffers, T. K.,** *Eimeria tenella:* immunogenicity of the first generation of schizogony, *Parasitology,* 93, 1, 1986.
34. **Reduker, D. W. and Speer, C. A.,** Antigens of *in vitro* produced first generation merozoites of *Eimeria bovis* (Apicomplexa), *J. Parasitol.*, 72, 782, 1986.
35. **Robertson, N. P., Reese, R. T., Henson, J. M., and Speer, C. A.,** Heat shock-like polypeptides of the sporozoites and merozoites of *Eimeria bovis, J. Parasitol.*, 74, 1004, 1988.
36. **Laxer, M. A., Healey, M. C., and Youssef, N. N.,** Production of monoclonal antibodies specific for *Eimeria tenella* microgametocytes, *J. Parasitol.*, 73, 611, 1987.
37. **Wallach, M. G., Mencher, D., Yarus, S., Pillemer, G., Halabi, A., and Pugatsch, T.,** *Eimeria maxima:* identification of gametocyte protein antigens, *Exp. Parasitol.*, 68, 49, 1989.
38. **Pugatsch, T., Mencher, D., and Wallach, M.,** *Eimeria maxima:* isolation of gametocytes and their immunogenicity in mice, rabbits and chickens, *Exp. Parasitol.*, 68, 127, 1989.
39. **Mencher, D., Pugatsch, T., and Wallach, M.,** Antigenic proteins of *Eimeria maxima* gametocytes: cell-free translation and detection with recovered chicken serum, *Exp. Parasitol.*, 68, 40, 1989.
40. **Pasternak, J. and Fernando, M. A.,** Host cell response to coccidian infection: an introspective survey, *Parasitology,* 88, 555, 1984.
41. **Davis, P. J.,** Studies on the immune response to *Eimeria tenella* in the chicken, M.Sc. thesis, Chelsea College, University of London, 1976.
42. **Henry, D. P.,** Allergy and immunity in coccidial infections, *Proc. Soc. Exp. Biol. Med.*, 28, 831, 1931.
43. **Beyer, T. V.,** Immunity in experimental coccidiosis of rabbit caused by heavy infective doses of *Eimeria intestinalis,* in *Progress in Protozoology,* Ludvic, J., Lom, J., and Vavra, J., Eds., Academic Press, New York, 1963, 448.
44. **Niilo, L.,** Acquired resistance to reinfection of rabbits with *Eimeria magna, Can. Vet. J.*, 8, 201, 1967.
45. **Rommel, M.,** Verlauf der *Eimeria scabra* — und *E. polita* - Infektion in vollempfänglichen Ferkeln und Läuferschweinen, *Berl. Munch. Tieraerztl. Wochenschr.*, 83, 181, 1970.
46. **Witlock, D. R. and Danforth, H. D.,** Surface changes induced by immune serom on *Eimeria tenella* sporozoites and merozoites, *J. Protozool.*, 29, 441, 1982.
47. **Speer, C. A., Wong, R. B., and Schenkel, R. H.,** Ultrastructural localization of monoclonal IgG antibodies for antigenic sites of *Eimeria tenella* oocysts, sporocysts and sporozoites, *J. Protozool.*, 30, 548, 1983.
48. **Speer, C. A., Thamnana, P., and Schenkel, R. H.,** Immune complexes induce the formation of an unusual micropore in second-generation merozoites of *Eimeria tenella, J. Parasitol.*, 71, 258, 1985.
49. **Speer, C. A., Wong, R. B., Blixt, J. A., and Schenkel, R. H.,** Capping of immune complexes by sporozoites of *Eimeria tenella, J. Parasitol.*, 71, 33, 1985.
50. **Whitmire, W. M. and Speer, C. A.,** Ultrastructural localization of IgA and IgG receptors on oocysts, sporocysts, sporozoites and merozoites of *Eimeria falciformis, Can. J. Zool.*, 64, 778, 1986.
51. **Rose, M. E. and Hesketh, P.,** *Eimeria tenella:* effects of immunity on sporozoites within the lumen of the small intestine, *Exp. Parasitol.*, 63, 337, 1987.
52. **Preston-Mafham, R. A. and Sykes, A. H.,** Changes in permeability of the mucosa during intestinal coccidiosis infections in the fowl, *Experientia,* 23, 972, 1967.
53. **Long, P. L.,** The pathogenic effects of *Eimeria praecox* and *E. acervulina* in the chicken, *Parasitology,* 58, 691, 1968.
54. **Rose, M. E. and Long, P. L.,** Immunity to coccidiosis: gut permeability changes in response to sporozoite invasion, *Experientia,* 25, 183, 1969.
55. **Kouwenhoven, B. and van der Horst, C. J. G.,** Histological observations with respect to the immune mechanisms in *Eimeria acervulina* infection in the domestic fowl, *Z. Parasitenkd.*, 42, 11, 1973.
56. **Rose, M. E., Long, P. L., and Bradley, J. W. A.,** Immune responses to infections with coccidia in chickens: gut hypersensitivity, *Parasitology,* 71, 357, 1975.
57. **Rose, M. E., Ogilvie, B. M., and Bradley, J. W. A.,** The intestinal mast cell response in rats and chickens in response to coccidiosis with some properties of chicken mast cells, *Int. Arch. Allergy Appl. Immunol.*, 63, 21, 1980.

58. **Ruff, M. D. and Wilkins, G. C.,** Intestinal absorption following challenge of immune chicks with *Eimeria acervulina, Avian Pathol.*, 13, 25, 1984.
59. **McKenzie, M. E., Long, P. L., and Burke, W. H.,** Concentration of [125]globulins in intestinal tissues of immunized and non-immunized chickens infected with *Eimeria acervulina, J. Parasitol.*, 72, 791, 1986.
60. **Huntley, J. F., Newlands, G., Miller, H. R. P., McLauchlan, M., Rose, M. E., and Hesketh, P.,** Systemic release of mucosal mast cell protease during infection with the intestinal protozoal parasite *Eimeria nieschulzi*. Studies in normal and nude rats, *Parasite Immunol.*, 7, 489, 1985.
61. **Horton-Smith, C., Long, P. L., and Pierce, A. E.,** Behavior of invasive stages of *Eimeria tenella* in the immune fowl *(Gallus domesticus), Exp. Parasitol.*, 14, 66, 1963.
62. **Mesfin, G. M. and Bellamy, J. E. C.,** Effects of acquired resistance on infection with *Eimeria falciformis* var *pragensis* in mice, *Infect. Immun.*, 23, 108, 1979.
63. **Leathem, W. D. and Burns, W. C.,** Effects of the immune chicken on the endogenous stages of *Eimeria tenella, J. Parasitol.*, 53, 180, 1967.
64. **Rose, M. E. and Hesketh, P.,** Immunity to coccidiosis: stages of the life-cycle of *Eimeria maxima* which induce, and are affected by, the response of the host, *Parasitology*, 73, 25, 1976.
65. **Riley, D. and Fernando, M. A.,** *Eimeria maxima* (Apicomplexa): a comparison of sporozoite transport in naive and immune chickens, *J. Parasitol.*, 74, 103, 1988.
66. **Fernando, M. A. and Pasternak, J.,** *Eimeria necatrix:* induced proteolytic sensitivity of infected chick crypt cells, *Exp. Parasitol.*, 56, 20, 1983.
67. **Danforth, H. D. and Augustine, P. C.,** *Eimeria tenella,* use of a monoclonal antibody in detemining the intracellular fate of the refractile body organelles and the effect on *in vitro* development, *Exp. Parasitol.*, 68, 1, 1989.
68. **Lawn, A. M. and Rose, M. E.,** Mucosal transport of *Eimeria tenella* in the cecum of the chicken, *J. Parasitol.*, 68, 1117, 1982.
69. **Al-Attar, M. A. and Fernando, M. A.,** Transport of *Eimeria nexatrix* sporozoites in the chicken: effects of irritants injected intraperitoneally, *J. Parasitol.*, 73, 494, 1987.
70. **Fernando, M. A., Rose, M. E., and Millard, B. J.,** *Eimeria* spp. of the domestic fowl: the migration of sporozoites intra- and extra-enterically, *J. Parasitol.*, 73, 561, 1987.
71. **van Doorninck, W. M. and Becker, E. R.,** Transport of sporozoites of *Eimiera necatrix* in macrophages, *J. Parasitol,* 43, 40, 1957.
72. **Challey, J.R. and Burns, W. C.,** The invasion of the cecal mucosa by *Eimeria tenella* sporozoites and their transport by macrophages, *J. Protozool.*, 6, 238, 1959.
73. **Pattillo, W. H.,** Invasion of the cecal mucosa of the chicken by sporozoites of *Eimeria tenella, J. Parasitol.*, 45, 253, 1959.
74. **Doran, D. J.,** The migration of *Eimeria acervulina* sporozoites to the duodenal glands of Lieberkühn, *J. Protozool.*, 13, 27, 1966.
75. **Davis, P. J. and Porter, P.,** *Eimeria tenella:* control of parasitic behavior through mucosal immunity and in-feed immunization, in *The Secretory Immune System*, McGhee, J. R. and Mestecky, J., Eds., *Ann. New York Acad. Sci.*, 409, 810, 1983.
76. **Rose, M. E. and Millard, B. J.,** *Eimeria vermiformis:* host strains and the developmental cycle, *Exp. Parasitol.*, 60, 163, 1985.
77. **Cerf-Bensussan, N., Quartoni, A., Kurnick, J. T., and Bhan, A. K.,** Intraepithelial lymphocytes modulate Ia expression by intestinal epithelial cells, *J. Immunol.*, 132, 2244, 1984.
78. **Rose, M. E. and Long, P. L.,** Vaccination against coccidiosis in chickens, in *Symp. Br. Soc. Parasitol.*, Vol. 18, *Vaccination Against Parasites*, Taylor, A. E. and Muller, R., Eds., Blackwell Scientific, Oxford, 1980, 57.
79. **Long, P. L., Johnson, J., and Wyatt, R. D.,** Pathological and clinical effects of *Eimeria tenella* in partially immune chickens, *J. Comp. Pathol.*, 91, 581, 1981.
80. **McKenzie, M. E. and Long, P. L.,** Immunization of chickens against coccidiosis with extracts of *Eimeria*-infected tissues, *Poult. Sci.*, 65, 892, 1986.
81. **Murray, P. K., Bhogal, B. S., Crane, M. St. J., and MacDonald, T. T.,** *Eimeria tenella — in vivo* immunization studies with sporozoite antigen, in *Proc. Georgia Coccidiosis Conf., Research in Avian Coccidiosis*, McDougald, L. R., Joyner, L. P., and Long, P. L., Eds., University of Georgia, Athens, 564, 1986.
82. **Clare, R. A. and Danforth, H. D.,** Major histocompatibility complex control of immunity elicited by genetically engineered *Eimeria tenella* (Apicomplexa) antigen in chickens, *Infect. Immun.*, 57, 701, 1989.
83. **Blumenthal, F.,** Zur Serodiagnose der Syphilis. Diskussion (Wassermann Reaktion und Kokzidiose der Kaninchen), *Berl. Klin. Wochenschr.*, 45, 618, 1908.
84. **McDermott, J. J. and Stauber, L. A.,** Preparation and agglutination of merozoite suspensions of the chicken coccidian *Eimeria tenella, J. Parasitol.*, 40, 23, 1954.
85. **Rose, M. E.,** Serological reactions in *Eimeria stiedae* infection of the rabbit, *Immunology*, 2, 112, 1959.

86. **Horton-Smith, C., Long, P. L., Pierce, A. E., and Rose, M. E.,** Immunity to coccidia in domestic animals, in *Immunity to Protozoa,* Garnham, P. C. C., Pierce, A. E., and Roitt, I., Eds., Blackwell Scientific, Oxford, 1963, 1973.
87. **Rose, M. E.,** Host immune responses, in *The Biology of the Coccidia,* Long, P. L., Ed., University Park Press, Baltimore, 1982, 330.
88. **Onaga, H., Saeki, H., Hoshi, S., and Ueda, S.,** An enzyme-linked immunosorbent assay for serodiagnosis of coccidiosis in chickens: use of a single serum dilution, *Avian Dis.,* 30, 658, 1986.
89. **Nolan, A., Goldring, O. L., Catchpole, J., Gregory, M. W., and Joyner, L. P.,** Demonstration of antibodies to *Eimeria* species in lambs by an enzyme linked immunosorbent assay, *Res. Vet. Sci.,* 42, 119, 1987.
90. **Catchpole, J., Nolan, A., Gregory, M. W. and Arthur, M. J.,** Observations on the effects of monensin on the production of antibodies to coccidia in lambs, *Vet. Rec.,* 118, 75, 1986.
91. **Diesing, L. and Ahmed, J. S.,** Immediate type hypersensitivity in murine coccidiosis, *Zentralbl. Bakteriol. Hyg.,* A258, 310, 1984.
92. **Lillehoj, H. S. and Ruff, M. D.,** Comparison of disease susceptibility and subclass-specific antibody response in SC and FP chickens experimentally inoculated with *Eimeria tenella, E. acervulina* or *E. maxima, Avian Dis.,* 31, 112, 1987.
93. **Davis, P. J., Parry, S. H., and Porter, P.,** The role of secretory IgA in anti-coccidial immunity in the chicken, *Immunology,* 34, 879, 1978.
94. **Rose, M. E.,** Immunity to eimeria infections, *Vet. Immunol. Immunopathol.,* 17, 333, 1987.
95. **Davis, P. J., Barratt, M. E. J., Morgan, M., and Parry, S. H.,** Immune response of chickens to oral immunization by 'trickle' infections with *Eimeria, in Proc. Georgia Coccidiosis Conf., Research in Avian Coccidiosis,* McDougald, L. R., Joyner, L. P., and Long, P. L., Eds., University of Georgia, Athens, 1985, 618.
96. **Mielke, D.,** Immunobiologische Aspekte bei Kokzidien-Infektionen, *Monatsh. Veterinaermed.,* 37, 629, 1982.
97. **Lillehoj, H. S.,** Effects of immunosuppression on avian coccidiosis: cyclosporin A, but not hormonal bursectomy, abrogates host protective immunity, *Infect. Immun.,* 55, 1616, 1987.
98. **Bhogal, B. S., Tse, H. Y., Jacobson, E. B., and Schmatz, D. M.,** Chicken T lymphocyte clones with specificity for *Eimeria tenella, J. Immunol.,* 137, 3318, 1986.
99. **Crane, M. St. J., Murray, P. K., Gnozzio, M. J., and MacDonald, T. T.,** Passive protection of chickens against *Eimeria tenella* infection by monoclonal antibody, *Infect. Immun.,* 56, 972, 1988.
100. **Rose, M. E., Wakelin, D., and Hesketh, P.,** Gamma interferon controls *Eimeria vermiformis* primary infection in BALB/c mice, *Infect. Immun.,* 57, 1599, 1989.
101. **Pierce, A. E. and Long, P. L.,** Studies on acquired immunity to coccidiosis in bursaless and thymectomized fowls, *Immunology,* 9, 427, 1965.
102. **Rose, M. E. and Long, P. L.,** Resistance to *Eimeria* infections in the chicken: the effects of thymectomy, bursectomy, whole body irradiation and cortisone treatment, *Parasitology,* 60, 291, 1970.
103. **Rose, M. E., Owen, D. G., and Hesketh, P.,** Susceptibility to coccidiosis: effect of strain of mouse on reproduction of *Eimeria vermiformis, Parasitology,* 88, 45, 1984.
104. **Klesius, P. H. and Hinds, S. E.,** Strain-dependent differences in murine susceptibility to coccidia, *Infect. Immun.,* 26, 1111, 1979.
105. **Stockdale, P. G. H., Stockdale, M. J., Rickard, M. D., and Mitchell, G. F.,** Mouse strain variation and effects of oocyst dose in infection of mice with *Eimeria falciformis,* a coccidian parasite of the large intestine, *Int. J. Parasitol.,* 15, 447, 1985.
106. **Mesfin, G. M. and Bellamy, J. E. C.,** Thymic dependence of immunity to *Eimeria falciformis* var. *pragensis* in mice, *Infect. Immun.,* 23, 460, 1979.
107. **Rose, M. E. and Hesketh, P.,** Eimerian life cycles: the patency of *Eimeria vermiformis,* but not *E. pragensis* is subject to host (*Mus musculus*) influence, *J. Parasitol.,* 72, 949, 1986.
108. **Rose, M. E., Ogilvie, B. M., Hesketh, P., and Festing, M. F. W.,** Failure of nude (athymic) rats to become resistant to reinfection with the intestinal coccidian parasite *Eimeria nieschulzi* or the nematode *Nippostrongylus brasiliensis, Parasite Immunol.,* 1, 125, 1979.
109. **Mitchell, G. F.,** T cell dependent effects in parasitic infection and disease, in *Immunology 80, Progress in Immunology IV,* Fougereau, M. and Dousset, J., Eds., Academic Press, London, 1980, 794.
110. **Rose, M. E. and Hesketh, P.,** Coccidiosis: T-lymphocyte-dependent effects of infection with *Eimeria nieschulzi* in rats, *Vet. Immunol. Immunopathol.,* 3, 499, 1982.
111. **Rose, M. E. and Hesketh, P.,** Immunity to coccidiosis: adoptive transfer with peripheral blood lymphocytes and spleen cells, *Parasite Immunol.,* 4, 171, 1982.
112. **Rommel, M. and Heydorn, A.-O.,** Versuche zur Übertragung der Immunität gegen *Eimeria* — Infektionen durch Lymphozyten, *Z. Parasitenkd.,* 36, 242, 1971.
113. **Liburd, E. M., Armstrong, W. D., and Mahrt, J. L.,** Immunity to the protozoan parasite *Eimeria nieschulzi* in inbred CD-F rats, *Cell. Immunol.,* 7, 444, 1973.

114. **Ogilvie, B. M. and Rose, M. E.,** The response of the host to some parasites of the small intestine: coccidia and nematodes, *Les Colloques de l'Institut National de la Santé et de la Recherche Médicale, Immunity in Parasitic Diseases,* 72, 1977, 237.
115. **Wakelin, D. and Wilson, M. M.,** Transfer of immunity to *Trichinella spiralis* in the mouse with mesenteric lymph node cells: time of appearance of effective cells in donors and expression of immunity in recipients, *Parasitology,* 74, 215, 1977.
116. **Crum, E.D., Despommier, D. D., and McGregor, D. D.,** Immunity to *Trichinella spiralis.* I. Transfer of resistance by two classes of lymphocytes, *Immunology,* 33, 787, 1977.
117. **Ogilvie, B. M., Love, R. J., Jarra, W., and Brown, K. N.,** *Nippostrongylus brasiliensis* infection in rats: the cellular requirement for worm expulsion, *Immunology,* 32, 521, 1977.
118. **Nawa, Y. and Miller, H. R. P.,** Protection against *Nippostrongylus brasiliensis* by adoptive immunization with immune thoracic duct lymphocytes, *Cell. Immunol.,* 37, 51, 1978.
119. **Rose, M. E., Joysey, H. S., Hesketh, P., Grencis, R. K., and Wakelin, D.,** Mediation of immunity to *Eimeria vermiformis* in mice by L3T4$^+$ T cells, *Infect. Immun.,* 56, 1760, 1988.
120. **Rose, M. E., Wakelin, D., Joysey, H. S., and Hesketh, P.,** Immunity to coccidiosis: adoptive transfer in NIH mice challenged with *Eimeria vermiformis, Parasite Immunol.,* 10, 59, 1988.
121. **Rose, M. E., Hesketh, P., and Ogilvie, B. M.,** Coccidiosis: localisation of lymphoblasts in the infected small intestine, *Parasite Immunol.,* 2, 189, 1980.
122. **Heyworth, M. F., Carlson, J. R., and Ermak, T. H.,** Clearance of *Giardia muris* infection requires helper/inducer T lymphocytes, *J. Exp. Med.,* 165, 1743, 1987.
123. **Grencis, R. K., Riedlinger, J., and Wakelin, D.,** L3T4-positive T lymphoblasts are responsible for transfer of immunity to *Trichinella spiralis* in mice, *Immunology,* 56, 213, 1985.
124. **Long, P. L.,** Pathology and pathogenicity of coccidial infections, in *The Coccidia: Eimeria, Isospora, Toxoplasma and Related Genera,* Hammond, D. M. and Long, P. L., Eds., University Park Press, Baltimore, 1973, 254.
125. **Wakelin, D.,** Immunity to intestinal parasites, *Nature,* 273, 617, 1978.
126. **Wakelin, D. and Wilson, M. M.,** Immunity to *Trichinella spiralis* in irradiated mice, *Int. J. Parasitol.,* 10, 37, 1980.
127. **Rose, M. E., Hesketh, P., and Ogilvie, B. M.,** Peripheral blood leucocyte response to coccidial infection: a comparison of the response in rats and chickens and its correlation with resistance to reinfection, *Immunology,* 36, 71, 1979.
128. **Rose, M. E., Wakelin, D., Joysey, H. S., and Hesketh, P.,** Immunity to coccidiosis: T-cell control of infection with *Eimeria vermiformis* in mice does not require co-operation with inflammatory cells, *Parasite Immunol.,* 11, 1989.
129. **Rose, M. E.,** *Eimeria tenella:* skin hypersensitivity to injected antigen in the fowl, *Exp. Parasitol.,* 42, 129, 1977.
130. **Giambrone, J. J. and Klesius, P. H.,** Chicken coccidiosis: correlation between resistance and delayed hypersensitivity, *Poult. Sci.,* 59, 1715, 1980.
131. **Taylor, R. L., Strout, R. G., Clare, R. A., and Aeed, P. A.,** Delayed wattle reactions in *Eimeria tenella* infected chickens, *Dev. Comp. Immunol.,* 10, 387, 1986.
132. **Shi, Y., Mahrt, J. L., and Mogil, R. J.,** Kinetics of murine delayed-type hypersensitivity response to *Eimeria falciformis* (Apicomplexa Eimeriidae), *Infect. Immun.,* 57, 146, 1988.
133. **Rose, M. E. and Hesketh, P.,** Immunity to coccidiosis: T-lymphocyte- or B-lymphocyte-deficient animals, *Infect. Immun.,* 26, 630, 1979.
134. **Clark, I. A.,** Cell-mediated immunity in protection and pathology of malaria, *Parasitol. Today,* 3, 300, 1987.
135. **Nogueira, N.,** Intracellular mechanisms of killing, in *Immunobiology of Parasites and Parasitic Infections,* Marchalonis, J. J., Ed., *Contemporary Topics in Immunobiology,* Vol. 12, Plenum Press, New York, 1984, 53.
136. **Clark, I. A.,** Does endotoxin cause both the disease and parasite death in acute malaria and babesiosis?, *Lancet,* ii, 75, 1978.
137. **Ferreira, A., Schofield, L., Enea, V., Schellekens, H., van der Meide, P., Collins, W. E., Nussenzweig, R. S., and Nussenzweig, V.,** Inhibition of development of exoerythrocytic forms of malaria parasites by γ-interferon, *Science,* 232, 881, 1986.
138. **Mellouk, S., Maheshirari, R. K., Rhodes-Feuillette, A., Beaudoin, R. L., Berbiguier, N., Matile, H., Miltgen, F., Landon, I., Pied, S., Chigot, J. P., Friedman, R. M., and Mazier, D.,** Inhibitory activity of interferons and interleukin 1 on the development of *Plasmodium falciparum* in human hepatocyte cultures, *J. Immunol.,* 139, 4192, 1987.
139. **Schofield, L., Ferreira, A., Altszuler, R., Nussenzweig, V., and Nussenzweig, R. S.,** Interferon-γ inhibits the intrahepatocytic development of malaria parasites *in vitro, J. Immunol.,* 139, 2020, 1987.
140. **Long, P. L. and Milne, B. S.,** The effect of an interferon inducer on *Eimeria maxima, Parasitology,* 62, 295, 1971.

141. **Fayer, R. and Baron, S.**, Activity of interferon and an inducer against development of *Eimeria tenella* in cell culture, *J. Protozool.*, 18, 12, 1971.
142. **Kogut, M. H. and Lange, C.**, Recombinant interferon-γ inhibits cell invasion by *Eimeria tenella*, *J. Interferon Res.*, 9, 67, 1989.
143. **Rose, M. E., Wakelin, D., and Hesketh, P.**, Gamma interferon controls *Eimeria vermiformis* primary infection in BALB/c mice, *Infect. Immun.*, 57, 1599, 1989.
144. **Lawrence, H. S.**, Transfer factor, *Adv. Immunol.*, 11, 195, 1969.
145. **Liburd, E. M., Pabst, H. F., and Armstrong, W. D.**, Transfer factor in rat coccidiosis, *Cell. Immunol.*, 5, 487, 1972.
146. **Klesius, P. H. and Giambrone, J. J.**, Adoptive transfer of delayed hypersensitivity and protective immunity to *Eimeria tenella* with chicken derived transfer factor, *Poult. Sci.*, 63, 1333, 1984.
147. **Klesius, P. H.. Kramer, T., Burger, D., and Malley, A.**, Passive transfer of coccidian oocyst antigen and diphtheria toxoid hypersensitivity in calves across species barriers, *Transplant. Proc.*, 7, 449, 1975.
148. **Klesius, P. H., Qualls, D. F., Elston, A. L., and Fudenberg, H.**, Effects of bovine transfer factor (TFd) in mouse coccidiosis *(Eimeria ferrisi), Clin. Immunol. Immunopathol.*, 10, 214, 1978.
149. **Jeffers, T. K.**, Genetics of coccidia and the host response, in *Br. Poultry Science Symp.*, No. 13, *Avian Coccidiosis*, Long, P.L., Boorman, K. N., and Freeman, B. M., Eds., British Poultry Science Ltd., Edinburgh, 1978, 51.
150. **Wakelin, D.**, Genetic control of susceptibility and resistance to parasite infection, *Adv. Parasitol.*, 16, 219, 1978.
151. **Wakelin, D. and Blackwell, J. M.**, Eds., *Genetics of Resistance to Bacterial and Parasitic Infection*, Taylor and Francis, London, 1988, 287.
152. **Lillehoj, H. S.**, Immune response during coccidiosis in SC and FP chickens. I. *In vitro* assessment of T cell proliferation response to stage specific parasite antigens, *Vet. Immunol. Immunopathol.*, 13, 321, 1986.
153. **Lillehoj, H. S.**, Influence of inoculation dose, inoculation schedule, chicken age and host genetics on disease susceptibility and development of resistance to *Eimeria tenella* infection, *Avian Dis.*, 32, 437, 1988.
154. **Clare, R. A., Strout, R. G., Taylor, R. L., Collins, W. M., and Briles, W. E.**, Major histocompatibility (B) complex effects on acquired immunity to cecal coccodiosis, *Immunogenetics*, 22, 593, 1985.
155. **Johnson, L. W. and Edgar, S. A.**, Ea-B and Ea-C cellular antigen genes in Leghorn lines resistant and susceptible to acute cecal coccidiosis, *Poult. Sci.*, 65, 241, 1986.
156. **Lillehoj, H. S., Ruff, M. D., Bacon, L. D., Lamont, S. J., and Jeffers, T. K.**, Genetic control of immunity to *Eimeria tenella*. Interaction of MHC genes and non-MHC linked genes influences levels of disease susceptibility in chickens, *Vet. Immunol. Immunopathol.*, 20, 135, 1989.
157. **Mahrt, J. L. and Shi, Y.**, Murine major histocompatibility complex and immune response to *Eimeria falciformis*, *Infect. Immun.*, 56, 270, 1988.
158. **Rose, M. E., Wakelin, D., and Hesketh, P.**, Susceptibility to coccidiosis: contrasting course of primary infections with *Eimeria vermiformis* in BALB/c and C57BL/6 mice is based on immune responses, *Parasite Immunol.*, 7, 557, 1985.
159. **Joysey, H. S., Wakelin, D., and Rose, M. E.**, Resistance to infection with *Eimeria vermiformis* in mouse radiation chimeras is determined by donor bone-marrow cells, *Infect. Immun.*, 56, 1399, 1988.
160. **Mitchell, G. F.**, Effector cells, molecules and mechanisms in host-protective immunity to parasites, *Immunology*, 38, 209, 1979.
161. **Wakelin, D. and Grencis, R. K.**, Immunological responses to intestinal parasite infection, in *Immunology of the Gastrointestinal Tract*, Vol. 2, Miller, K. and Nicklin, S., Eds., CRC Press, Boca Raton, FL, 1987, 1.

Chapter 15

CONTROL OF COCCIDIOSIS: CHEMOTHERAPY

Larry R. McDougald

TABLE OF CONTENTS

I. THE CONCEPT AND HISTORY OF COCCIDIOSIS CONTROL WITH DRUGS

The many species of coccidia affecting domestic animals and man cause a variety of diseases which require extensive use of anticoccidial drugs to relieve economic losses or human suffering. We have had effective drugs to prevent or treat coccidiosis of most types since the late 1940s, but new drugs are constantly needed to replace those made useless by drug resistance. The approach to chemotherapy in domestic animals varies with the type of husbandry practiced. Young chickens are reared in large flocks and are at great risk from the several species of coccidia which have short, direct life cycles and great reproductive potential. Today anticoccidial drugs are mixed into poultry feeds at low levels for continuous prevention of coccidiosis, allowing coccidiosis control to be managed from a single point in large production units, the feed mill. In animals that are subject to sporadic exposure, medication is given only to treat clinically ill animals. Treatment of toxoplasmosis in humans is difficult and often will not prevent the severe side effects of the disease. Until recently there was no known treatment for cryptosporidiosis in man, and patients with a depressed immune response were likely to die from dehydration caused by the extreme diarrhea. New drugs are finally offering hope for cure of crytosporodiosis.

The history of anticoccidial drugs was reviewed in detail.[1-3] This review will emphasize recent developments and an overview of the field.

II. METHODS OF LABORATORY AND FIELD EVALUATION OF CANDIDATE ANTICOCCIDIAL COMPOUNDS

The laboratory models for testing anticoccidial compounds are among the best in the industry. Compounds can be tested *in vitro*, in laboratory clinical tests, or in larger pen trials to simulate commercial animal production units. The advantage with these systems is that the target disease organism is used, in the animal for which the drug is intended. Chemical compounds or fermentation products can be tested *in vitro* against *E. tenella* in cell culture or in chick embryos.[4] Primary chick kidney cells are infected with sporozoites of *E. tenella* and the test materials are added to the culture medium. Any effects of the test

TABLE 1
Spectrum of Activity of Some Anticoccidial Drugs

Drug	Activity against each species of *Eimeria*				
	E. acervulina	*E. maxima*	*E. necatrix*	*E. tenella*	*E. brunetti*
Amprolium	−	+ +	+ +	+ + +	−
Robenidine	+ + +	+ + +	+ + +	+ + +	+ + +
Monensin	+ + +	+ + +	+ + +	+ +	+ + +
Ethopabate	+ + +	−	−	−	−
Sulfaquinoxaline	+ + +		+ +	+	
Zoalene	+	+ +	+ +	+ +	+
Lasalocid	+	+ + +	+ + +	+ + +	+ + +

Note: General patern of efficacy; may not apply in specific instances. −, no activity; +, low activity, + +, moderate activity; + + +, strong activity.

Data compiled from various sources.

compound on development of the parasites can be seen by microscopic examination 72 to 96 h later or can be detected electronically by the use of fluorescent-linked monoclonal antibodies. While most workers have used primary chick cells, a system employing a bovine kidney cell line has also been successful.[5]

Clinical models consist of groups of chicks 7- to 14-days old which are given a ration containing the test compound, then infected with one or more species of *Eimeria*. After 6 to 7 days, the weight gain, mortality and lesion score are recorded.[6] Some investigators use droppings scores[7] and fecal oocyst counts.

The floor pen test has become a standard tool for evaluation of candidate drugs under simulated natural conditions. For this test a facility is divided into pens of equal size, usually 40 to 50 sq. ft, to accommodate 50 to 60 birds at commercial stocking density. The birds are grown to market weight while receiving test compounds continuously in the feed. Floor litter, brooding, feeding, and watering are designed to simulate practical conditions. Coccidia can be introduced at 2 to 3 weeks of age, by contamination of the litter, feed, or water, or can be introduced by infected seeder birds.[8] The parameters used in this type of test are weight gain, feed conversion ratio, mortality from coccidiosis, and lesion scores of birds sampled from each pen.[9]

III. BIOLOGICAL CHARACTERISTICS OF ANTICOCCIDIAL DRUGS

A. SPECTRUM OF ACTIVITY

Coccidiosis is a complex of diseases in many different hosts. Thus, the activity of a drug must be defined in terms of the species of coccidia and the effectiveness in a particular host. Commonly a drug will be highly effective against some species of *Eimeria* in chickens, but lack significant activity against others. The relative activity of some drugs against laboratory strains of five species of coccidia in chickens is illustrated in Table 1.[2] These general rankings may not apply with field isolates of coccidia, where drug resistance might have developed.

B. STAGE OF THE LIFE CYCLE MOST SENSITIVE TO DRUGS

The activity of drugs must be understood in terms of specific chemical action (and thereby specific biological action) against specific developing stages of the parasite. The ionophores, quinolones and clopidol, act against the sporozoite and trophozoite, while nicarbazin, amprolium, halofuginone, arprinocid, and sulfonamides act principally against later stages.[10,11] Diclazuril acts against most stages of *E. tenella* but only the sexual stages of *E. maxima*.[12] Amprolium, arprinocid, and sulfonamides also act against sexual stages of laboratory strains.[13] Under field conditions it is doubtful whether any practical significance can be attached to the time of action.

C. MODE OF ACTION

The exact nature of the action of most drugs is poorly understood. In some cases the mode of action can be deduced from the chemical structure or from labratory studies with other systems or models. One of the best known modes of action is that of the quinolone anticoccidials. The drug binds reversibly to parts of the electron transport system in the mitochondria of sporozoites and thereby blocks any energy-requiring reactions.[14-15] Amprolium is chemically similar to thiamine, and presumably acts by blocking thiamine utilization by the parasite.[16] The polyether ionophores render cell membranes permeable to sodium and potassium ions, much energy can be expended by the cell to maintain equilibrium.[17] Sporozoites are unable to survive in cells after treatment with ionophores, probably because they lack an efficient mechanism to maintain osmotic equilibrium.[18] Arprinocid appears to interfere with the purine salvage pathway, but just how this is related to anticoccidial activity is not clear.[19]

IV. PRODUCTS USED FOR COCCIDIOSIS CONTROL

A. THE POLYETHER IONOPHORES

In 1951[20] a group of fermentation products was discovered which had interesting effects on transport of ions across cell membranes but no apparent useful activity. Later these products were found to have anticoccidial activity[21] and to improve feed utilization by ruminants. Most of these compounds are produced by fermentation of *Streptomyces* or *Actinomadura*, and the commercial production depends on large scale fermentation of these organisms rather than chemical manufacture. The first of these products was monensin, introduced in 1971 and still in widespread use in 1989. Many new compounds in this class have been isolated and characterized, and some of them may eventually be used as anticoccidial drugs.

Typically, the ionophores have broad spectrum activity, are coccidiocidal in action, and have a narrow margin of safety when given to chickens or other animals. The anticoccidial efficacy of ionophores appears mediocre in laboratory tests, but under field conditions the results are consistently superior to those of other classes of drugs. The ionophores act by facilitation of the movement of inorganic cations across cell membranes, thus upsetting the balance of physiologically important ions such as sodium and potassium. Biological activity and toxicity are closely tied to these physiological effects. Activity is most marked against the sporozoite,[22] but merozoites are also affected if they come into contact with the drug.[23]

Of all anticoccidial drugs, the coccidia had the most difficulty in becoming resistant to the ionophores, and the class of products remains important after over 18 years of intensive use. Much of the attention given ionophores has focused on toxic side effects in poultry and extreme toxicity to animals such as horses. Most of the ionophores will depress feed intake and growth if given at levels only slightly higher than the intended use level.[24]

1. Monensin

Introduced in the U.S. in 1971, this was the first of the ionophores used in chickens, and remains widely used in 1989. The use level is 100 to 121 ppm in the U.S., 100 ppm in most other countries, and 80 ppm in Japan. Most of the use of monensin is at 100 to 110 ppm. Monensin is best known for control of intestinal species of coccidia; combined feeding of roxarsone is thought to improve control of *E. tenella*. There is no required withdrawal period for monensin in most countries, but the producer usually withdraws the drug for 5 to 7 days to allow "compensatory growth".[25] The LD_{50} of monensin in horses is 2 mg/kg of body weight, so caution is recommended in feed mills where more than one type of feed is manufactured. Monensin is best known for its generally favorable influence on performance of the chickens and litter quality in addition to good anticoccidial activity.

2. Salinomycin

This product was discovered in Japan in the early 1970s and has since been marketed in all of the poultry producing countries in the world. The most common use level is 60 ppm, although the approved level in Japan is 50 ppm and in the U.S. is 44 to 66 ppm. Salinomycin has better anticoccidial activity against *E. tenella* than monensin, and generally good activity against other species. The toxicity to horses is less than that of monensin, but 30 ppm of salinomycin in feed may cause mortality in growing turkeys. In 1984 to 1989 salinomycin was used extensively, replacing monensin as the most widely used anticoccidial.

3. Lasalocid

The chemistry of lasalocid differs significantly from that of monensin or salinomycin, resulting in differences in the effects on chickens and coccidia. Lasalocid shows strong affinity for divalent cations, in contrast to the monovalent ion preference of monensin and salinomycin. Efficacy in laboratory tests is strongest against *E. tenella* and weakest against *E. acervulina*, and this is consistent with field observations. Lasalocid was approved for use in the U.S. in 1974 at 75 ppm, but control of *E. acervulina* was not adequate and the level was expanded to 75 to 125 ppm in 1977. Lasalocid is often suspected of interaction with water intake and dietary electrolytes, but the link is by no means clear. Field observations suggest that lasalocid offers improved coccidiosis control over other ionophores where monensin-tolerant *E. tenella* are present.

4. Narasin

Chemically similar to salinomycin, narasin was introduced in Europe in 1983, and approved in the U.S. in 1987. The approved use level is 60 to 80 ppm for broilers. Narasin has not been promoted extensively for use as a single entity, but will be sold as a combination with nicarbazin to take advantage of the synergistic effects between these compounds (see below).

5. Maduramicin

This ionophore has affinity for both monovalent and divalent ions, and is the only commercially produced ionophore to have a sugar moiety as a side chain. The use level of maduramicin is 5 to 6 ppm, and it is registered in Europe, South America, and the Far East. It was approved in the U.S. in early 1989. Tests against field isolates in the U.S. and in South America show that the efficacy is comparable or superior to that of older ionophores, in control of lesions of coccidiosis.[26]

6. Semduramicin

The product was still under development in 1989. The projected use level is 20 to 25 ppm. Unpublished research data suggests that this product has broad spectrum activity and will be competitive with other ionophores.

7. Nicarbazin

This product is not a single compound, but an "equimolar complex of 2,4-dinitrocarbanilide and pyrimidine". The use level of nicarbazin is 124 ppm in most countries, although it is commonly used in South America at 100 ppm. Nicarbazin was first used in 1955 making it the oldest anticoccidial in common use today, and one of the first to have truly broad spectrum activity, coccidiocidal action, and high specific activity.[27] Recent surveys show that nicarbazin is the most effective against field isolates of any of the currently available products, even though it has been used widely for many years.[26,28] The longevity of this product is a result of two major factors: (1) the mode of action of nicarbazin does not allow coccidia to easily develop resistance[29] and (2) the conservative use of the product. Because of toxic side effects in broilers, and the potential for toxicity in breeder hens, nicarbazin is used only in the starter ration, and only in the winter or cooler months. Nicarbazin alters the physiological response to heat stress in chickens and will increase losses from heat stress.[24,25,30] When breeder hens are inadvertently given nicarbazin, there is an initial loss of pigment from the shell of brown-shelled eggs, than a serious drop in egg production. Many producers using this product monitor every bath of breeder feed, to avoid contamination of hens. A quick-test is supplied by the manufacturer to help in this procedure. Nicarbazin is used almost entirely in the starter feed because of a tendency to slightly depress weight gain when given in both the starter and grower feeds.

8. Nicarbazin Combined with Narasin

When nicarbazin is combined with narasin, a synergism occurs which allows improved coccidiosis control with lower concentrations of each drug.[31]

9. Halofuginone

This product is characterized by high specific activity, broad-spectrum control and mixed coccidiostatic and coccidiocidal action.[32] It has been used throughout the world, mostly in shuttle programs as an adjunct to one of the ionophores.

10. Diclazuril, Clazuril, Toltazuril, and Tiazuril

In the early 1970s a class of compounds with distant relation to uracil was found to have potent broad spectrum efficacy against coccidia in chickens.[33] These leads were eventually abandoned because of teratogenic side effects of some of the metabolites, but the class was rediscovered in the 1980s with toltazuril, clazuril, and diclazuril.

a. Toltazuril

The optimum use level is 25 ppm when given in the water to infected chickens. This product shows excellent utility for treatment of outbreaks where some of the flock is already showing clinical signs of infection. The manufacturer recommends the use of toltazuril in place of conventional continuous medication, through the planned use of intermittent medication. The utility of this approach in the poultry industry can only be proven through field use. The disadvantages of this product are the possibility of rapid drug resistance and a long withdrawal period because of slowly depleting residues in edible tissues.

b. Clazuril

This product has a limited spectrum of activity against chicken coccidia, but is highly active against coccidiosis in pigeons[34] and will be sold only in Europe for use in non-food-producing applications.

c. Diclazuril

The optimum use level for this product is 1.0 ppm, although considerable activity is reported at 0.5 ppm.[35] Broad-spectrum activity was reported against all important species infecting chickens. The margin of safety of this product in chickens is exceptional: feeding of 50× dosage to growing chickens does not significantly depress feed intake or weight gain. The developmental stage most sensitive to diclazuril varies with the species of coccidia. For instance, *E. tenella* is highly sensitive to diclazuril in all stages of development. In contrast, *E. maxima* is killed only in the sexual stages.[12] For this reason, the conventional test for efficacy, based on weight gain and lesion scores does not accurately reflect the efficacy of diclazuril against *E. maxima*. Other measurements, such as the effect on oocyst passage, give a better picture of the overall efficacy of diclazuril against this species.[36] Drug resistance was slow to develop in strains of *E. tenella, E. maxima*, or *E. acervulina*. It was necessary to start with a low level of the drug (0.06 ppm) and gradually raise the level over a series of 18 generations in chicks, to produce strains resistant to 1.0 ppm of diclazuril. Under simulated field conditions (in floor pens) diclazuril was highly efficacious and was well tolerated, allowing good performance.[37]

B. DRUGS USED FOR CHICKENS

By far the most signficant use of anticoccidial drugs is in broiler chickens, estimated at well over $300 million in 1989. The market for drugs for poultry has increased steadily as the poultry industry has grown over the past several decades. The practice of preventive application of anticoccidial drugs is universally accepted, even in developing coutnries. The intensive use of anticoccidial drugs creates an environment which pushes the drug resistance potential of the coccidia to its limits. After introduction of monensin in 1971, the polyether ionophores became the products of choice. Despite the eventual erosion of the effectiveness of this remarkable group of products by emergence of coccidia with "reduced sensitivity", these products remain the mainstay of the coccidiosis programs around the world.

1. Programs for Coccidiosis Control in Poultry

Anticoccidial drugs are mixed into the feed at low concentrations, to be given continuously to the growing chickens. In many instances, government regulations require that drugs be withdrawn for a few days before birds are slaughtered for human consumption. In common practice, the rations given for the final 5 to 7 days are not medicated with anticoccidial drugs. In recent years the practice of using two drugs in a "shuttle program" has become widely accepted. In this plan, a strong drug is used in the starter feed, followed by one of the ionophores in the grower feed. These programs are intended to take advantage of the excellent efficacy of drugs which cannot be used continuously for one reason or another.

2. Drugs Used Commonly in Poultry in 1989

More than 40 products were introduced for control or prevention of coccidiosis after 1946, but only a few of these are used today. Most of the older drugs were replaced by more effective products or were made ineffective by drug resistant coccidia.

C. DRUGS USED FOR TURKEYS

The market for anticoccidial drugs for use in turkeys is mostly limited to the U.S., France, and the U.K. Drugs are given only for 8 to 10 weeks, even though the birds are reared to 14 to 22 weeks to reach market weight. The important species of coccidia in turkeys are *E. adenoeides, E. meleagrimitis*, and *E. gallopavonis*. Several products are listed in the Feed Additives Compendium as registered for use in turkeys, but most are not commonly used.

1. Amprolium

The oldest product in common use, amprolium, was approved for use at 0.0250% in the U.S. Although drug resistance has limited the effectiveness of this product, it is still used widely because it is extremely safe and there is general feeling that the poults are allowed to develop immunity to coccidia which will give them needed protection after 8 weeks of age.

2. Monensin

This ionophore is registered for use at 54 to 90 g/ton in the U.S. Monensin is highly effective against all important species of turkey coccidia.[38-40]

3. Lasalocid

Even though Mitrovic[40] considered the optimum level to be 125 ppm, the best use level of this ionophore will probably be near 80 to 90 ppm.

4. Halofuginone

Levels of 1.5 to 3.0 ppm of halofuginone are highly efficacious against all important species of coccidia in turkeys.[41] Halofuginone is highly efficacious and well tolerated in turkeys grown to market weight in floor pens.[42]

D. DRUGS USED FOR CATTLE

During the 1960s and 1970s the sulfonamides and nitrofurans were used against bovine coccidiosis for treatment of clinically ill animals, although the efficacy of these products was somewhat limited. In recent years, more of the modern anticoccidial products have been registered for use in cattle, and some are used in a prophylactic manner rather than for treatment of outbreaks. The practice of finishing steers in large feedlots exposes the animals to coccidiosis much in the same way as chickens are exposed in the large rearing facilities, and more emphasis must be placed on prevention. There are many species of coccidia in cattle, but the most important are *E. bovis* and *E. zuernii*.

1. Amprolium

Administration of 5 mg/kg/day in the feed for 21 days is approved regimen for prevention. Clinically ill animals can be treated with amprolium at 10 mg/kg/day for 5 days. The safety margin of this product is excellent.

2. Decoquinate

In the U.S. decoquinate is approved for use at 0.5 mg/kg/day for prevention of coccidiosis in cattle. Miner and Jensen[43] found complete control of *E. bovis* in calves when decoquinate was given at the rate of 0.5 to 0.8 mg/kg for 21 to 28 days, with no relapse after withdrawal of medication. Administration at 33 ppm in feed was effective in comparison with other products.[44] There has been no apparent problem with drug resistance with this product in bovine coccidia, in contrast to that seen in chicken coccidia.

3. Lasalocid

The polyether ionophores are given continuously to cattle in feedlots for improvement of rate of gain and/or improved feed efficiency. Coincidentally, these products are highly efficacious against coccidia. Fitzgerald and Mansfield[45] found complete control with lasalocid at 112 ppm in the feed, but Foreyt et al.[44] found good control by feeding 0.5 to 1.0 mg/kg of body weight. Further, feeding of 33 ppm in the feed prevented clinical signs and reduced oocyst counts. Lasalocid is not registered for use in the feed for prevention of coccidiosis in cattle, but can be used at 10 to 30 g/ton for improvement of gain and/or feed efficiency.

4. Monensin

When given at 30 g/ton (33 ppm) in the feed (about 1.0 mg/kg/day) monensin was completely protective against infections with *E. bovis*.[46] Fortuntely, the level of monensin approved for improvement of feed efficiency is 5 to 30 g/ton, allowing legal use at effective levels for both uses. In another test, this product was effective against *E. bovis* and *E. zuernii*.[44]

5. Salinomycin

Although not sold for use in cattle in the U.S., salinomycin is highly effective in prevention of *E. bovis* when given at 0.33 to 1.0 mg/kg/day.[47]

E. DRUGS USED FOR SHEEP

The important species of coccidia infecting sheep are *E. ovinoidalis* and *E. ahsata*. Treatment is given to clinically ill animals, and in some instances prophylactically to animals exposed to great risk of infection, such as young lambs and animals newly confined to feedlots.

1. Amprolium

Amprolium is highly effective in prevention of coccidiosis in sheep when given continuously in the feed at 100 to 200 ppm.[48] Baker et al.[49] treated lambs with amprolium in the feed at 50 mg/kg/day for 21 days with good suppression of coccidiosis. There is no reported resistance to amprolium in bovine coccidia.

2. Monensin

The ionophores are generally toxic when given in single daily doses, and are not recommended for use in this manner.[50] When given in a complete feed or supplement at 10 to 30 g/ton, however, oocyst passage and other clinical signs were eliminated and weight gains and feed efficiency were protected.[51-53]

3. Lasalocid

Like other ionophores, lasalocid is effective when given prophylactically to sheep in feedlots, at the level which would be used for improvement of feed efficiency.

F. DRUGS USED FOR GOATS

Decoquinate, at 0.5 mg/kg is the only product approved for use in goats in the U.S. This product was highly effective and well tolerated when given at 0.3 to 4 mg/kg with better results at higher levels.[54] Research reports indicate that amprolium is effective in treatment of infections at 33 or 66 mg/kg.[55] Lasalocid at 30 g/ton in complete feed was only moderately effective in prevention of coccidiosis in goats.[56]

G. DRUGS USED FOR RABBITS

The most commonly used product for prevention and treatment of coccidiosis in rabbits is sulfaquinoxaline. Other sulfonamides are also effective. Good control of *E. stiedai* was found when rabbits were given feed containing 250 ppm of sulfaquinoxaline.[57] Robenidine given at 33 ppm in feed gave some protection against natural infections.[58] The ionophores have good efficacy in rabbits, but their toxicity limits usefulness. Diclazuril is highly effective against coccidiosis in rabbits.[59]

H. DRUGS USED FOR SWINE

Severe dysentery and mortality in neonatal piglets in large farrowing facilities is often caused by *Isospora suis*. Sanitation is probably more effective than any drug treatment in this particular instance, although Sangster et al.[60] successfully used amprolium to treat farrowing sows and 3-day-old piglets.

V. DRUG RESISTANCE

Throughout the history of chemotherapy the "faithful shadow" of drug resistance has made its grim presence known.[61] Anticoccidial drugs have been no exception, as the first paper on resistance to sulfaquinoxaline was published in 1955[62] and numerous others have followed.[63,64] Drugs vary widely in their modes of action, with the result that resistant strains of coccidia emerge within weeks with some of the drugs, and in other cases may require many years. The development of resistant or tolerant strains of coccidia may in some cases be confused with the limited spectrum of activity of some drugs. For instance, amprolium and zoalene have little native efficacy against *E. acervulina* or *E. maxima*, while ethopabate, has no efficacy against *E. tenella*. When a drug such as halofuginone or robenidine fails to control a field isolate of *E. tenella*, however, there is no doubt that the strain of coccidia has become resistant as a result of prior exposure. When a strain of coccidia becomes resistant to a drug, there is usually a co-lateral resistance to other drugs of the same chemical class but not to unrelated drugs. Drug resistance is a genetic trait,[65] and tends to remain in a population for many years.

A. TECHNIQUES FOR PRODUCING RESISTANT STRAINS OF COCCIDIA IN THE LABORATORY

The coccidia of chickens and turkeys have been studied extensively and laboratory workers have used two approaches to producing drug resistant strains. The earliest workers used a system which has proved reliable with many types of drugs and disease organisms: first, a low dose of the drug is used which allows reproduction of the parasites to some extent, then after several generations the dosage of the drug is raised. Eventually the dose of drug can be raised to meet or exceed the recommended use level, and parasites reproducing in this environment are completely resistant. Despite the apparent reliability of this test in producing resistant strains, the predictive value was somewhat limited. The results of tests of this type were summarized by Chapman.[63] Another test was based on the use of high doses of oocysts/bird and administration of the recommended level of the drug.[29] If doses of 5 to 10 million oocysts/bird are given, a resistant line might be propagated if the mutation rate for a given drug resistance trait is one in 5 to 10 million. Resistance to the quinoline compounds can be selected in this way, but not to nicarbazin or the ionophores. Usually only a single isolate of one or two species is used, and the results may not be indicative of the drug resistance potential of coccidia encountered under field conditions.

B. FIELD SURVEYS OF DRUG RESISTANT COCCIDIA

Extensive surveys on coccidia isolated from commercial poultry flocks were conducted in the 1970s[66-68] and continue today. Coccidia are easily propagated from samples of droppings or litter taken from poultry farms when the birds are about 4 to 5 weeks old. The tests may be conducted with mixed isolates as they are received from the field[28,29] or individual species can be isolated for testing.[63] Coccidia have more difficulty in becoming resistant to some drugs than to others, and this is reflected in the results of field surveys. Resistance to halofuginone, clopidol, the quinolones, and robenidine was found within a year of first use in some countries, but resistance to monensin and other ionophores was slow to develop. The change in sensitivity to ionophores was apparent when coccidia from broiler farms were compared to coccidia from breeder farms where ionophores had not been used.[69] In recent years, the incidence of coccidia resistant to ionophores has increased significantly.[28,70-72] Even though there is a general cross resistance between ionophores, there is often a difference in response of a strain to each drug.[26]

C. STRATEGY TO AVOID DRUG RESISTANCE

The prolonged use of a drug leads inevitably to emergence of drug resistant coccidia. In recent years many producers have attempted to thwart the resistance-developing process by alternate use of drugs. Two such programs are called "shuttle" and "rotation". The rotation program is simply the alternate use of two or more drugs at various intervals.

Producers may use one drug for 3 to 6 months, then another for a similar period. The shuttle program is rotation practiced within the growout of a flock, where one product is used for 3 to 4 weeks then another used until the flock is sold. There is little experimental evidence to support these programs, because of the difficulty in constructing suitable models, but there is general agreement that poultry producers are able to postpone drug resistance and extend the usefulness of products in this way.[73]

VI. FUTURE APPLICATIONS FOR ANTICOCCIDIAL DRUGS

Prevention of disease is essential for maintaining a high level of productivity in food-producing industries. The poultry industry is an example of a highly efficient production system. Small deficiencies in productivity, measured only in a few grams of body weight or a few points in feed conversion can often make the difference between profit and loss. There is good evidence that inability to control coccidiosis was the direct cause of failure of otherwise well-run production companies in the 1950s. The poultry industry and other animal industries were given valuable tools for control of coccidiosis with the introduction in 1971 of menensin and the later introduction of other ionophores. By 1975 over 80% of the poultry industry in the U.S. had adopted the ionophores, and this acceptance continues today. Unfortunately, the "faithful shadow" of drug resistance has limited the effectiveness of these exceptional products. We have other new drugs, but it is unlikely that any of them will last for 20 years in intensive use. Perhaps is is unreasonable to expect that we will discover another class of compounds which will give us the years of service that we have enjoyed from the ionophores. What, then, does the future hold? For the immediate future the industry must take advantage of shuttle programs and rotation to avoid resistance where possible. In some instances old drugs can be used in a limited way if the results of sensitivity tests demonstrate that the coccidia respond to the product. New drugs must be used more conservatively, to prolong their usefulness. Considerable effort in several laboratories is aimed at the development of vaccines, either attenuated live products or genetically engineered antigens. Studies on drug resistance in other fields suggest that ways may be found to block drug resistance. Whatever the approach, we need innovative research on new ideas to help find ways to control coccidiosis and ensure the continued high level of productivity in our food industries.

REFERENCES

1. **McDougald, L. R.,** Chemotherapy of coccidiosis, in *Biology of the Coccidia,* Long, P. L., Ed., University Park Press, Baltimore, 1984, 373.
2. **McDougald, L. R.,** Coccidian and related infections, in *Chemotherapy of Parasites,* Campbell, W. C. and Rew, R. S., Eds., 1986, 159.
3. **Reid, W. M., Long, P. L., and McDougald, L. R.,** Coccidiosis, in *Diseases of Poultry,* 8th ed., Hofstad, M. S., Barnes, H. J., Calnek, B. W., Reid, W. M., and Yoder, H. W., Eds., 1984, 692.
4. **McDougald, L. R.,** The growth of avian Eimeria in vitro, in *Avian Coccidiosis,* Long, P. L., Boorman, K. N., and Freeman, B. M., Eds., Br. Poultry Science Ltd., Edinburgh, 1978, 185.
5. **Schmatz, D. M. and St. J. Crane, M.,** Parasite-specific incorporation of ^{3}H-uracil by intracellular Eimeria tenella and its use as a quantitative measure of development, in *Research in Avian Coccidiosis,* McDougald, L. R., Joyner, L. P., and Long, P. L., Eds., 1986, 134.
6. **Johnson, J. and Reid, W. M.,** Anticoccidial drugs: lesion scoring techniques in battery and floor pen experiments with chickens, *Avian Dis.,* 28, 30, 1970.
7. **Morehouse, N. F. and Baron, R. R.,** Coccidiosis: evaluation of coccidiostats by mortality, weight gain and fecal scores, *Exp. Parasitol.,* 28, 25, 1970.
8. **Brewer, R. N. and Kowalski, L. M.,** Coccidiosis: evaluation of anticoccidial drugs in floor-pen trials, *Exp. Parasitol.,* 28, 64, 1970.

9. **McDougald, L. R., Wang, G. T., Kantor, S., Schenkel, R., and Quarles, C.,** Efficacy of maduramicin against ionophore-tolerant field isolates of coccidia in broilers, *Avian Dis.*, 31, 302, 1987.
10. **Reid, W. M.,** Anticoccidials: differences in day of peak activity against *E. tenella*, in *Proc. Symp. Coccidia and Related Organisms,* University of Guelph, Ontario, Canada, 1973, 119.
11. **McDougald, L. R. and McQuistion, T. E.,** *Eimeria tenella:* anticoccidial action of drugs in birds with surgically-closed ceca, *Z. Parasitenkd.*, 59, 107, 1979.
12. **Maes, L., Coussment, W., Vanparijs, O., and Marsboom, R.,** In vivo action of diclazuril on the developmental stages of *Eimeria maxima:* a histological study, *J. Parasitol.*, in press.
13. **Mathis, G. F. and McDougald, L. R.,** Experimental development of amprolium or dinitolmide in *Eimeria acervulina* and its effect on inhibition of sporulation of oocysts, *J. Parasitol.*, 67, 956, 1981.
14. **Wang, C. C.,** Studies of the mitochondria from *E. tenella* and inhibition of electron transport by quinolone coccidiostats, *Biochim. Biophys. Acta,* 396, 210, 1975.
15. **Wang, C. C.,** Inhibition of the respiration of *E. tenella* by quinolone coccidiostats, *Biochem. Pharmacol.*, 25, 343, 1976.
16. **Rogers, E. F.,** Thiamine antagonists. Chemotherapy, *Ann. N. Y. Acad. Sci.*, 98, 412, 1962.
17. **Pressman, B. C.,** Biological applications of the ionophores, *Annu. Rev. Biochem.*, 45, 501, 1976.
18. **Smith, C. K., III and Strout, R. G.,** *Eimeria tenella:* accumulation and retention of anticoccidial ionophores by extracellular sporozoites, *Exp. Parasitol.*, 48, 325, 1979.
19. **Simashkevich, P. M. and Wang, C. C.,** Purification and characterization of hypoxanthine-xanthine-guanine phosphoribosyl transferase (HXGPRTase) from a protozoan parasite *E. tenella, Fed. Proc.*, 40, 1582, 1981.
20. **Berger, J., Rachlin, A. I., Scott, W. E., Sternbach, L. H., and Goldberg, M. W.,** The isolation of three new crystalline antibiotics from *Streptomyces, J. Am. Chem. Soc.*, 73, 5295, 1951.
21. **Shumard, R. F. and Callender, M. E.,** Monensin, a new biologically active compound. VI. Anticoccidial activity, *Antimocrob. Agents Chemother.*, 1968, 369, 1967.
22. **McDougald, L. R. and Galloway, R. B.,** Anticoccidial drugs: effects on infectivity and survival intracellularly of *Eimeria tenella* sporozoites, *Exp. Parasitol.*, 40, 314, 1976.
23. **Long, P. L. and Jeffers, T. K.,** Studies on the stage of action of ionophorous antibiotics against *Eimeria, J. Parasitol.*, 68, 363, 1982.
24. **Keshavarz, K. and McDougald, L. R.,** Anticoccidial drugs: growth and performance depressing effects in young chickens, *Poult. Sci.*, 61, 699, 1982.
25. **McDougald, L. R. and McQuistion, T. E.,** Compensatory growth in broilers after withdrawal of anticoccidial drugs, *Poult. Sci.* 59, 1001, 1980.
26. **McDougald, L. R., Solis, J., Lamas DaSilva, J. M., and Braga, M.,** A survey of sensitivity to anticoccidial drugs in 60 isolates of coccidia from broiler chickens in Brazil and Argentina, *Avian Dis.*, 31, 287, 1987.
27. **Cuckler, A. C., Malanga, C. M., and Basso, A. J.,** Antiparasitic activity of substituted carbanilide complexes, *Science,* 122, 244, 1955.
28. **McDougald, L. R., Fuller, L., and Solis, J.,** Drug sensitivity of 99 isolates of coccidia from broiler farms, *Avian Dis.*, 30, 690, 1986.
29. **Weppelman,R. M., Olson, G., Smith, D. A., Tamas, T., and VanIderstine, A.,** Comparison of anticoccidial efficacy, resistance, and tolerance of narasin, monensin and lasalocid in chicken battery trials, *Poult. Sci.*, 56, 1550, 1977.
30. **Buys, S. B. and Rasmussen, R. W.,** Heat stress mortality in nicarbazin fed chickens, *J. S. Afr. Vet. Assoc.*, 49, 127, 1978.
31. **Tonkinson, L. V., Jeffers, T. K., and Callender, M. E.,** Anticoccidial efficacy of various ratios of narasin and nicarbazin, *Poult. Sci.*, 66, 187, 1987.
32. **McQuistion, T. E. and McDougald, L. R.,** The anticoccidial action of arprinocid and halofuginone against *Eimeria* spp. in the chicken, *Vet. Parasitol.*, 9, 27, 1981.
33. **Chapel, L. R., Howes, H. L., and Lynch, J. E.,** The site of action of a broad-spectrum aryltriazine(?) anticoccidial, CP-25,415, *J. Parasitol.*, 60, 415, 1974.
34. **Vercruysse, J. and Geevaerts, J.,** Efficacy of toltrazuril and clazuril against experimental infections with *Eimeria labbeana* and *E. columbarum* in racing pigeons, *Avian Dis.*, in press.
35. **Vanparijs, O., Marsboom, R., and Desplenter, L.,** Diclazuril, a new broad spectrum anticoccidial in chickens. I. Dose titration studies and pilot floor-pen trials, *Poult. Sci.*, in press.
36. **Mathis, G. F. and McDougald, L. R.,** Anticoccidial efficacy of diclazuril in chickens, *Poult. Sci.*, 67, 115, 1988.
37. **McDougald, L. R., Mathis, G. F., Quarles, C. Q., and Seibert, B.,** Anticoccidial efficacy of diclazuril in chickens under stimulated use conditions, *Proc. 38th West. Poult. Dis. Conf.*, 39, 1989.
38. **Anderson, W. I., Giambrone, J. J., Fletcher, O. J., Jr., Eidson, C. S., and Reid, W. M.,** Demonstration of *Eimeria tenella* in Bursa of Fabricius of chickens, *Avian Dis.*, 20, 752, 1976.
39. **McDougald, L. R.,** Anticoccidial action of monensin in turkey poults, *Poult. Sci.*, 55, 2442, 1976.

40. **Mitrovic, M., Schildknecht, E. G., and Marusich, W. L.,** Lasalocid in prevention of turkey coccidiosis, *Poult. Sci.*, 58, 1154, 1979.
41. **McDougald, L. R.,** Halofuginone: laboratory titration of anticoccidial activity in young turkeys, *Poult. Sci.*, 65, 280, 1986.
42. **McDougald, L. R., Mathis, G. F., Schwartz, J., Quarles, C. L., Kennedy, T., and Grant, R. J.,** Anticoccidial efficacy of halofuginone in turkeys reared to market weight, *Poult. Sci.*, 65, 1664, 1986.
43. **Miner, M. L. and Jensen, J. B.,** Decoquinate in the control of experimentally induced coccidiosis of calves, *Am. J. Vet. Res.*, 37, 1043, 1976.
44. **Foreyt, W. J., Rice, D. H., and Wescott, R. B.,** Evaluation of lasalocid as a coccidiostat in calves: titration of efficacy and comparison with monensin and decoquinate, *Am. J. Vet. Res.*, 47, 2031, 1986.
45. **Fitzgerald, P. R. and Mansfield, M. E.,** Efficacy of lasalocid against coccidia in cattle, *J. Parasitol.*, 65, 824, 1979.
46. **McDougald, L. R.,** Monensin in the prevention of coccidiosis in calves, *Am. J. Vet. Res.*, 39, 1748, 1978.
47. **Benz, G. W. and Ernst, J. V.,** Efficacy of salinomycin in treatment of experimental *Eimeria bovis* infections in calves, *Am. J. Vet. Res.*, 40, 1180, 1979.
48. **Badiola, C. and Schindler, P.,** The control of coccidiosis in lambs by amprolium, *Proc. Symp. Coccidia and Further Prospects of their Control*, November 28—30, Prague, 1979.
49. **Baker, N. F., Walters, G. T., and Fisk, R. A.,** Amprolium for control of coccidiosis in feedlot lambs, *Am. J. Vet. Res.*, 33, 83, 1972.
50. **Leek, R. G., Fayer, R., and McLoughlin, D. K.,** Effect of monensin on experimental infections of *Eimeria ninakohlyakimovae* in lambs, *Am. J. Vet. Res.*, 37, 339, 1976.
51. **Bergstrom, R. C. and Maki, L. R.,** Coccidiostatic action of monensin fed to lambs: body weight gains and feed conversion efficacy, *Am. J. Vet. Res.*, 37, 79, 1976.
52. **Fitzgerald, P. R. and Mansfield, M. E.,** Ovine coccidiosis: effect of the antibiotic monensin against *Eimeria ninakohlyakimovae* and other naturally occurring coccidia of sheep, *Am. J. Vet. Res.*, 39, 7, 1978.
53. **McDougald, L. R. and Dunn, W. C.,** Efficacy of monensin against coccidiosis in calves, *Am. J. Vet. Res.*, 39, 1459, 1978.
54. **Foreyt, W. J., Hancock, D., and Wescott, R. B.,** Prevention and control of coccidiosis in goats with decoquinate, *Am. J. Vet. Res.*, 47, 333, 1986.
55. **Marlow, C. H. B.,** Amprolium as a coccidiostat for Angora goats, *J. S. Afr. Vet. Med. Assoc.*, 39, 93, 1968.
56. **Calhoun, M. C., Baldwin, B. C., Craig, T. M., and Huey, R. L.,** Evaluation of the efficacy of Bovatec for controlling experimentally induced and naturally occurring coccidiosis in Angora goats, *Tex. Agric. Exp. Stn. Prog. Rep.*, 18—21, 1988.
57. **Gwyther, M. J. and Dick, J. W.,** Efficacy of monensin and sulfaquinoxaline against the coccidium, *Eimeria stiedae* in rabbits, *Poult. Sci.*, 55, 1594, 1976.
58. **Peeters, J. E., Halen, P., and Meulemans, G.,** Efficacy of robenidine in the prevention of rabbit coccidiosis, *Br. Vet. J.*, 135, 349, 1979.
59. **Vanparijs, O., Hermans, L., Van der Flaes, L., and Marsboom, R.,** Efficacy of diclazuril in the prevention and cure of intestinal and hepatic coccidiosis in rabbits, *Vet. Parasitol.*, in press.
60. **Sangster, L. R., Stuart, B. B., Williams, D. J., and Bedell, D. M.,** Coccidiosis associated with scows in baby pigs, *Vet. Med. Small Anim. Clin.*, 1317, 1978.
61. **Schnitzer, R. J. and Grunberg, E.,** *Drug Resistance of Microorganisms*, Academic Press, New York, 1957.
62. **Cuckler, A. C. and Malanga, C. M.,** Studies on drug resistance in coccidia, *J. Parasitol.*, 41, 302, 1955.
63. **Chapman, H. D.,** Anticoccidial drug resistance, in *Biology of the Coccidia*, Long, P. L., Ed., University Park Press, Baltimore, 1982, 429.
64. **Chapman, H. D.,** Drug resistance in avian coccidia. A review, *Vet. Parasitol.*, 15, 11, 1984.
65. **Jeffers, T. K.,** Genetics of coccidia and the host response, in *Avian Coccidiosis: Proc. 13th Poult. Sci. Symp.*, Long, P. L., Boorman, K. N., and Freeman, B. M., Eds., Br. Poultry Science Ltd., Edinburgh, 51, 1978.
66. **Jeffers, T. K.,** *Eimeria tenella:* incidence, distribution and anticoccidial drug resistance of isolates in major broiler-producing areas, *Avian Dis.*, 18, 74, 1974.
67. **Jeffers, T. K.,** *Eimeria acervulina* and *E. maxima:* incidence and anticoccidial drug resistance of isolates in major broiler producing areas, *Avian Dis.*, 18, 331, 1974.
68. **Jeffers, T. K.,** Anticoccidial drug resistance: differences between *Eimeria acervulina* and *Eimeria tenella* strains within broiler houses, *Poult. Sci.*, 53, 1009, 1974.
69. **Mathis, G. F., McDougald, L. R., and McMurray, B.,** Drug sensitivity of coccidia from broiler breeder pullets and broilers in the same integrated company, *Avian Dis.*, 28, 453, 1984.
70. **Chapman, H. D. and Shirley, M. W.,** Sensitivity of field isolates of *Eimeria* species to monensin and lasalocid in the chicken, *Res. Vet. Sci.*, 46, 114, 1989.
71. **Rotibi, A., McDougald, L. R., and Solis, J.,** Response of 21 Canadian field isolates of chicken coccidia to commercial anticoccidial drugs, *Avian Dis.*, 33, 365, 1989.

72. **Stallbaumer, M. and Daisy, K. J.,** The effect of monensin, narasin, salinomycin and nicarbazin on field strains of chicken coccidia, *Avian Pathol.*, 17, 793, 1988.
73. **McDougald, L. R.,** Coccidiosis, in *Diseases of Poultry*, 9th ed., Calnek, B. W., Barnes, H. J., Beard, C. W., Reid, W. M., and Yoder, H. W., Jr., Eds., Iowa State University Press, Ames, in press.

Chapter 16

CONTROL OF COCCIDIOSIS IN CHICKENS: IMMUNIZATION WITH LIVE VACCINES

Martin W. Shirley and Peter L. Long

TABLE OF CONTENTS

I. INTRODUCTION

Although anticoccidial drugs, virtually exclusively, are used for the control of coccidiosis in chickens, the development of safe live vaccines as a further option is now a topic of research in several laboratories worldwide. This interest is reflected in an increase in relevant patent applications (Table 1). For reasons of industrial secrecy, it is difficult to assess just how much progress is being made towards the goal, both of live vaccines and those produced by recombinant DNA technologies. Nevertheless, the practical application of live vaccines should occur much sooner than molecular vaccines, which still remain a long-term prospect.

Dickinson et al.[1] were the first to vaccinate chickens deliberately against coccidiosis. In 1952 a live vaccine consisting of virulent strains ('Coccivac') was introduced by the animal health company Dorn and Mitchell Inc. (then a subsidiary of Sterling Drug Inc. as were the present manufacturers, Sterwin Laboratories Inc.)

A considerable disadvantage of 'Coccivac' is the inclusion of virulent strains of *Eimeria* and, unless the dose is carefully controlled, there is a possibility that the virulent strains will cause disease as they recycle through animals which have been insufficiently immunized. An obvious improvement to this type of vaccine would be the replacement of these strains with attenuated parasites that are unable, or less likely, to cause disease. Further improvements could arise by methods facilitating the more precise administration of oocysts, for example by 'in feed' delivery or by individual dosing of all birds in a flock. Both of these aspects of vaccination have recently been investigated and substantial progress towards the development of safer live vaccines has been made.

II. SPECIES OF *EIMERIA* TO BE INCLUDED IN AN EFFECTIVE VACCINE

In view of the uncertain taxonomic status of two species of *Eimeria*, there still remains some doubt as to the number of species that will be needed in a live vaccine to ensure that it is comprehensive. There is unanimous agreement that *E. acervulina, E. brunetti, E. maxima, E. mitis, E. necatrix, E. praecox,* and *E. tenella* are valid species, but the taxonomic status of *E. mivati* and *E. hagani* has been questioned, e.g., Shirley et al.[2] and Shirley.[3] From the results of electrophoretic analyses of enzymes, cross-protection studies, measurements of oocysts and prepatent times, growth in embryonated eggs and studies on gross pathological changes in infected intestine, Shirley et al.[2] considered that laboratory strains of *E. mivati* were mixtures of *E. acervulina* and *E. mitis*. Edgar and Fitz-Coy,[4] however, dispute this view.

Since the original description of *E. mitis* by Tyzzer[5] and a subsequent study by Joyner,[6] reported investigations on this species have been rare. However, the taxonomic uncertainty of '*E. mivati*' has focused more attention on *E. mitis* and recently it has been the subject of several detailed studies. Shirley et al.,[2] McDonald and Shirley[7] and Novilla et al.[8] described the characteristics of strains of *E. mitis* and Chapman,[9] who used the technique of enzyme electrophoresis, showed that the species was commonly present in commercial poultry houses in the U.K. (in 30% of isolates recovered from the field). Tyzzer[5] described *E. mitis* as producing neglible injury to the intestine and Joyner[6] referred to the species as 'benign'. Shirley et al.[2] and Novilla et al.[8] also reported the absence of any distinctive and consistent gross intestinal lesions although the species was found to be markedly pathogenic with doses of one million oocysts causing significant loss of body weight gain. The lesions in the intestine are microscopic and would not normally be diagnosed at postmortem.[10]

Protective immune responses to coccidial infection are (parasite) species-specific (Chapter 14) and a fully comprehensive live attenuated vaccine needs to contain all species. *Eimeria praecox* is extremely common in commercial poultry houses[9] but it is the least pathogenic

TABLE 1
Issue of Patents that Relate to Vaccines for Coccidiosis

Year	Number of patents issued
1963—1980	2
1981	0
1982	2
1983	1
1984	0
1985	4
1986	3
1987	4
1988	4
1989	4

Note: Search made for patents using keywords coccid., oocyst, parasitic, protozoa, vaccine, eimeria, immunization.

Source: World Patent Index, using Dialog. Produced by Derwent Publications Ltd, Rochdale House, 128 Theobalds Road, London, WC1X 8RP.

species and, arguably, may not be needed in a live vaccine. Long[11] and Gore and Long,[12] however, did show a depression in mean body weight gain of about 14% in chickens given a dose of 1×10^6 oocysts. Libby et al.[13] reported a consequence of not vaccinating chickens against all the major species of *Eimeria* when chickens previously given *E. tenella, E. necatrix, E. maxima, E. acervulina,* and *'E. hagani'* subsequently suffered clinical coccidiosis caused by *E. brunetti*. Moreover, they reported that outbreaks of coccidiosis due to *E. maxima* and *E. necatrix* occurred in flocks of broilers given a vaccine which contained only *E. acervulina, E. tenella,* and *'E. hagani'*. Libby et al.[13] recommended that all species (except *E. praecox*) ought to be included in a vaccine although Bickford,[14] implied that *E. praecox* should also be included. Shirley et al.[15] and Long and Johnson[16] similarly considered that an attenuated line of *E. praecox* would also be of value in a vaccine.

The phenomenon of antigenic diversity within some species also needs to be taken into account when the composition of a vaccine is to be determined. This will be discussed in the concluding remarks of this chapter because it remains one of the most important areas for further research because comparatively little is known about the extent of immunological differences within populations of most species.

III. USE OF VACCINES CONTAINING VIRULENT PARASITES

New vaccines which contain virulent strains will rely upon novel methods of administration to ensure a more efficient and uniform primary infection. Such methods would be expected to minimize, if not eliminate completely, any risk of disease as the progeny of the vaccine strains recycle through the flock.

A. SINGLE-SPECIES VACCINES

A vaccine for the control of strains of *E. maxima* which are partially or fully resistant to ionophores has been developed from the studies of Bafundo and Jeffers.[17] The vaccine, which contains oocysts of an ionophore-sensitive strain, is manufactured by Sterwin Laboratories Inc. and distributed by Elanco Products Company. It is given in the hatchery (via a Beak-o-Vac Machine) to 1-day-old birds which, 2 to 4 h later (when all sporozoites should have entered epithelial cells of the intestine) receive feed medicated with ionophorous anticoccidial drugs to provide protection against subsequent natural exposure with other species. Experimentally, a dose of 2×10^3 oocysts induced good protection, based on lesion scores, body weight gain, and concentrations of serum β carotenoids, against challenge with 5×10^4 oocysts given from 16 days later. Ionophores exert their anticoccidial action primarily against extracellular stages (including sporozoites) which are present within the lumen of the gut.[18] Delaying the administration of the drug until after vaccination enables sporozoites from drug-sensitive parasites in the vaccine to enter epithelial cells and initiate the immune response. Chapman[19] showed that when monensin was given in the diet for 48 h before the inoculation of oocysts it did not completely suppress oocyst production in birds given as few as ten oocysts.

B. MIXED-SPECIES VACCINES

The daily administration to chickens of small numbers of oocysts (over a period of 2 weeks) was first shown by Joyner and Norton[20,21] to induce a degree of protection greater than that arising from the same number of oocysts given on a single occasion. While these studies were done on a small scale in the laboratory and the chickens were inoculated individually by the oral route, the concept of 'trickle' vaccination has recently been the subject of potential commercial application. Davis and Reynolds,[22] for example, described methods by which oocysts (of virulent strains) could be blended within a suitable 'carrier', e.g., gum of a cold water-soluble polysaccharide, and then incorporated as a free-flowing pre-mix into poultry feedstuffs. The numbers of oocysts mixed into the feed were such that newly hatched chickens consumed about 1 to 20 oocysts per day rising with increased food consumption with age. A more sophisticated modification of this approach was provided by Davis and Harris[23] who described the encapsulation of small numbers of oocysts within sodium alginate beads. These were about 5 mm in diameter when moist and 1 to 3 mm diameter when dehydrated. These gel beadlets could be stored in the moist state, i.e., in an aqueous medium until required and then incorporated into a feedstuff. The use of materials other than sodium alginate may improve the water-retaining properties of the beadlets.

Confirmation that chickens reared under laboratory conditions or in floor pens could be effectively immunized with different species of *Eimeria* by the in-feed administration of oocysts in alginate beads has been provided by Norton and Joyner[24] and Johnson et al.,[25] respectively. These studies were done with virulent strains or mixtures of virulent strains and attenuated lines. No outbreaks of clinical coccidiosis were observed after vaccination. However, because precocious lines offer a much wider margin of safety they are most likely to be used in vaccines.

In-feed vaccination through the use of live encapsulated oocysts is an attractive proposition for the control of coccidiosis. The procedure ensures that numbers of oocysts consumed are directly linked with food intake and that the small but increasing numbers of parasites enhance immunity without causing clinical disease.However, for large scale vaccination of commercially reared chickens it will be necessary to overcome the problem of oocysts within beads being killed by high temperatures used in pelleting food (80 to 90°C) and fat-sprayed (60 to 80°C) feedstuffs. Encapsulated oocysts present in food stored in large hoppers on the outside of poultry houses could also be exposed to potentially lethal or harmful temperatures. The viability of coccidial oocysts would be expected to decrease as the ambient temperature increased, due to the more rapid utilization of the reserves of carbohydrate (amylopectin) which are the major sources of energy to the parasite.[26,27]

A method to improve the administration via the water of virulent oocysts was described by Lee.[28] Trials were done in which 4-day-old broiler chicks were given access to water troughs containing the equivalent of between 100 to 1000 oocysts per bird of *E. acervulina, E. maxima, E. necatrix,* and *E. tenella* suspended in a 1.5% aqueous solution of carageenan.

This formulation provided more even suspension of oocysts and a more uniform uptake by the birds. When the chickens were 7-days-old they were similarly exposed to a further dose of 130 oocysts each of *E. necatrix* and *E. tenella*. Judged by the results of mortality, average weight gain or feed conversion, there were no differences between the vaccinated birds or control birds given monensin throughout.

The administration of oocysts *via* water fonts is arguably one of the easiest methods of exposing young chickens to coccidial infection. Obvious advantages of giving a vaccine via the water are that chickens of any age can be vaccinated and it is possible to store the vaccine on the farm under the most favorable conditions (i.e., +4°C) until required. However, the use of an aqueous suspending medium such as carageenan is essential to prevent the rapid settling out of oocysts. Ross and Hurnik[29] showed that, when drinking, the beaks of young chickens penetrated the water in fonts to a depth of only about 10 mm.

IV. METHODS USED FOR ATTENUATION OF *EIMERIA* SPP.

Two approaches have been particularly successful for the true attenuation of species of *Eimeria* from the chicken, viz, serial passage in embryonated eggs, and repeated selection for early development (i.e., precocious development) of oocysts in the chicken.

A. SERIAL PASSAGE IN EMBRYONATED EGGS

Historically, serial passage of eimeria in the chorioallantoic membranes of embryonated eggs offered the first hope that a truly attenuated live vaccine could be developed.

Most success has been obtained with *E. tenella* and egg-adapted lines of this species have been used in small-scale studies to vaccinate chickens successfully in the U.K.,[30] U.S.,[31] and Czechoslovakia.[32] An attenuated line of *E. tenella* has also been produced by this method in Japan.[33] Egg-adapted lines of *E. mitis*[30,34] and *E. necatrix*[30,35,36] have also been obtained and used to immunize chickens. However, disadvantages of these parasites include (1) the relative instability of the trait of attenuation, i.e., rapid reversion to virulence on passage in chickens,[36,37] (2) the difficulty of producing the numbers of oocysts that would be required for use in the field (oocysts would have to be produced either by passage in eggs or by passage in chickens but, in the case of the latter, with the attendant risk of reversion to virulence), and (3) the loss of immunogenicity which usually accompanies their passage in eggs. In practice, it was often difficult to obtain populations with a satisfactory combination of immunogenicity and attenuation of virulence. However, the major limitation of egg-passage as a means of developing attenuated lines has been the total lack of success in obtaining complete development of *E. acervulina, E. maxima* and *E. praecox* in the embryonic host. In view of the success with selection for early production of oocysts, work to develop egg-adapted lines for inclusion in vaccines has largely been abandoned.

B. SELECTION FOR EARLY DEVELOPMENT OF OOCYSTS

The first evidence that repeated selection for early development of oocysts would markedly reduce the virulence of a population of *Eimeria* was provided by Jeffers[38,39] who studied the Wisconsin (WIS) strain of *E. tenella*. The basis of attenuation of virulence was that a proportion of the second generation schizonts in the selected line either did not completely differentiate or were very small at maturity. After ten generations of selection a subpopulation was established from infection with a single oocyst and the trait for attenuation was stable for at least 25 generations of relaxed passage. Crucially, despite the changes to its endogenous life cycle, the selected population induced an immune response which protected chickens against challenge with virulent strains. Jeffers[38] concluded that "these attenuated strains have potential use in a live coccidiosis vaccine which would induce protective immunity, while remaining innocuous to the bird". Selection for early development of oocysts has subsequently been extended to all species, viz, *E. acervulina,*[16,40] *E. brunetti,*[41,42] *E. maxima,*[16,43,44] *E. mitis,*[16,45] *E. necatrix,*[16,46] *E. praecox*[15,16] and other strains of *E. tenella.*[16,32,47,48] Jeffers[39] referred to the selected populations as precocious strains but, in accordance with the recommendation of Joyner et al.,[49] the term 'precocious lines' was introduced by McDonald et al.[40]

V. CHARACTERISTICS OF PRECOCIOUS LINES

With few exceptions the characteristics of all selected precocious lines are similar, i.e., compared with their parents there are changes in the prepatent time, endogenous development, reproductive potential, and virulence.

A. PREPATENT TIME

The prepatent time of each parent strain decreased with increasing selection for early development of oocysts (e.g., Figure 1). It was possible to obtain a precocious line only in a 'stepwise' manner, i.e., the prepatent period became gradually reduced. The selection of a precocious line from a cloned population of a laboratory strain of *E. praecox* also occurred in a similar manner (see Chapter 10). The developmental times of precocious lines selected in the U.K.[50] or U.S.[16] and their parents are shown in Table 2.

B. ENDOGENOUS DEVELOPMENT

Wild-type strains of each species possess an endogenous life cycle characterized by three or four generations of asexual multiplication (schizogony) before gametes are formed and the subsequent oocysts are excreted in the droppings. Typically, the final generation of schizogony is either absent or substantially depleted in the precocious lines (Table 3). In addition, faster maturation of first generation schizonts occurred in precocious lines of *E. brunetti,*[50] *E. mitis,*[51] and *E. tenella.*[48]

Selection of precocious lines of *E. necatrix* or *E. tenella* commonly did not result in the deletion of a generation of schizont. Those of the second generation usually matured faster, were smaller and contained fewer merozoites than those of their parents.[32,38,39,48,52] However, more intense selection of the Wis strain of *E. tenella* resulted in a line (WIS-F-96) which produced gametocytes directly from first generation schizonts.[53,54] Fundamental differences between the endogenous development of the precocious lines may, as Kawaguchi et al.[48] pointed out for the NIAH/P20 and WIS-F lines of *E. tenella*, arise from the use of slightly different methods of selection and variations in the intensity of selection. Moreover, innate differences between the parent strains would also be expected to contribute to the characteristics of the derived precocious lines.

C. REPRODUCTIVE POTENTIAL

The reproductive capabilities of most precocious lines, including all of those obtained in the U.K. and U.S., are substantially less than those of their virulent parents (Table 4) and they decreased with increasing selection for early development of oocysts (Table 5). In contrast, precocious lines of *E. tenella* obtained from the Jena[32] and NIAH[48] laboratory strains are not characterized by any reduction in reproductive potential. Interestingly, the Jena[55] and NIAH[56] strains are both characterized by the same electrophoretic form of the enzyme glucose phosphate isomerase, coded GPI-9. In contrast, the Houghton and WIS strains of *E. tenella* are both characterized by an alternative form of the enzyme coded GPI-1.[55] The electrophoretic profile of the Penn State strain of *E. tenella*[16] has not been examined. The significance of these differences in isoenzymes is not known, but egg-adapted lines derived from strains of *E. tenella* with GPI-9 are significantly more virulent than those with GPI-1.[3] It is possible that differences in reproduction in precocious lines might be linked with variation of GPI.

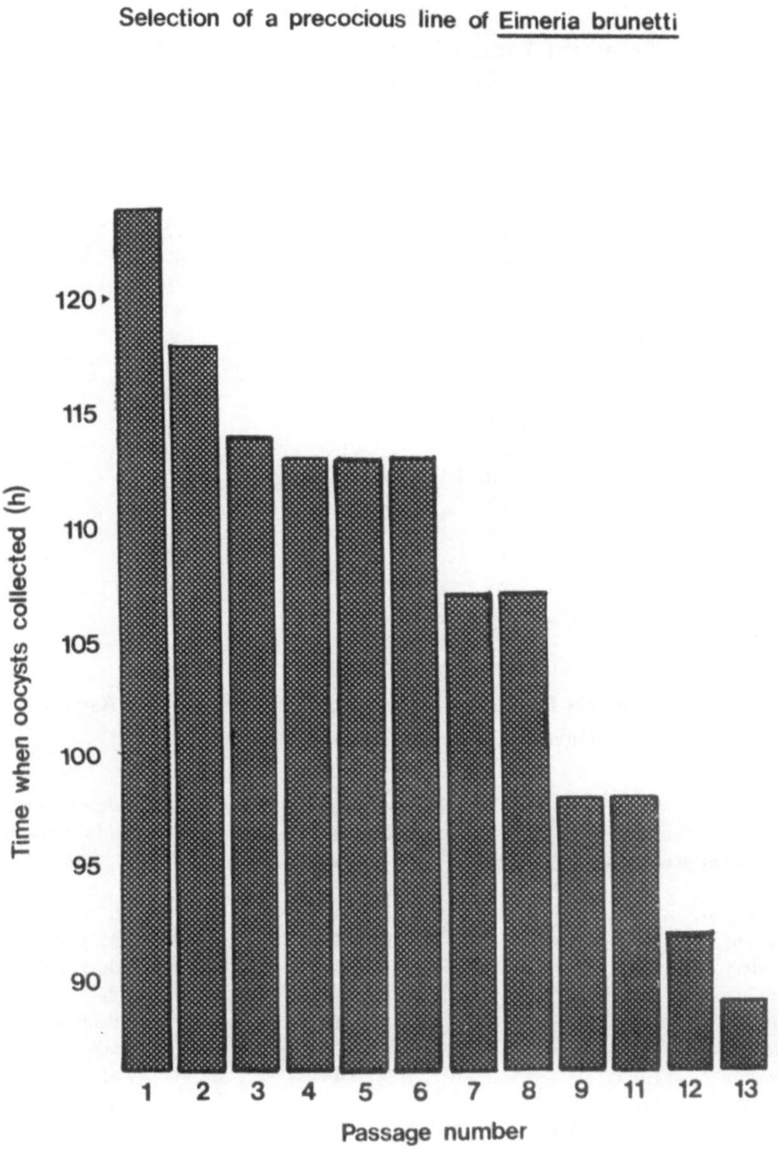

FIGURE 1. Reduction in prepatent time during selection of precocious line of *E. brunetti* (coded HP).

It should be noted that, from the practical standpoint of obtaining sufficient numbers of oocysts for a vaccine, large yields of oocysts can still be recovered from chickens if the dose given is increased accordingly.

D. ATTENUATION OF VIRULENCE

All precocious lines obtained in the U.K., the U.S., and Japan are significantly less virulent than their parents (Figures 2 and 3), a trait which increased with increasing selection for early development of oocysts (Figures 4 and 5). The virulence of a precocious line of *E. tenella* selected by Bedrnik et al.[32] was not significantly different from that of its virulent parent strain, as shown by comparisons of mortality, lesion scores, or relative weight gains of infected chickens.

TABLE 2
Prepatent Times of *Eimeria* After Repeated Selection for Early Development of Oocysts

Species	Strain	Prepatent time, hours	
		Parent strain	Precocious line
E. acervulina	H & C	97	72
	H	89	62
E. brunetti	VA	120	75
	H	120	89
E. maxima	M × 10	121	106
	MF	120	107
	C	125	110
E. mitis	FS 50	93	76
	H	91	64
E. necatrix	FS 105	138	120
	H	138	120
E. praecox	GA	83	66
	H	84	60
E. tenella	PS	121	112
	H	132	109

TABLE 3
Effect of Selection for Early Development of Oocysts on Asexual Development and Prepatent Time

Species[a] (parent/precocious line)		Number of generations of schizogony in parent strain	Generation of schizont depleted in precocious line
E. acervulina	H/HP	4	4th
E. brunetti	H/HP	3	3rd + 2nd
E. maxima	C/CP	4	4th + 3rd
E. maxima	MF/MFP	4	4th
E. mitis	H/HP	4	4th + 3rd
E. necatrix	H/HP	3	2nd[b]
E. praecox	H/HP	4	4th
E. tenella	Wis/Wis F-125	3	2nd[b]
E. tenella	Wis/Wis F-96	3	3rd[c] + 2nd[c]
E. tenella	H/HP	3	2nd[b]
E. tenella	NIAH/NIAH;P20	3	2nd[b]

[a] H = Houghton, C = Chichester, MF = Mixed Field, Wis = Wisconsin, NIAH = National Institute of Animal Health (Ibaraki, Japan).
[b] Second generation schizonts of these precocious lines developed faster, were smaller and contained fewer merozoites than the equivalent stage of the parent strains.
[c] Eliminated.

E. IMMUNOGENICITY

Most precocious lines stimulate protective immune responses (to challenge with virulent homologous strains) virtually as good as those induced by their parents, see Tables 6 and 7. With few exceptions the studies on immunogenicity have been done in wire-floored cages and criteria used to assess the immune status of the birds after challenge have included output of oocysts,[15,43] body weight gains and lesions scores.[16] Shirley and Millard[57] confirmed the immunogenicity of a complete vaccine of attenuated lines (an egg-adapted line of *E. tenella* and precocious lines of six other species) that was given to chickens kept on litter

TABLE 4
Effect of Selection for Early Development of Oocysts on the Reproduction of Different Species and Strains

Species[a]	Dose of oocysts given ($\times 10^{-2}$)	Output of oocysts per bird ($\times 10^{-6}$) after infection with: Precocious line	Parent strain	% reduction in output of oocysts of precocious line c. f. parent strain
E. acervulina	1.0	7.7	114.4	93
E. brunetti	0.5	0.5	19.3	97
E. maxima[b]	0.5	0.3	11.5	97
E. maxima[c]	0.5	0.3	13.2	98
E. mitis	1.0	2.0	82.1	98
E. necatrix	0.5	4.4	24.3	82
E. praecox	0.1	1.7	27.4	97
E. tenella	0.2	6.7	50.3	87

[a] Parent strains were Houghton strains, except:-
[b] Chichester strain
[c] MF, Mixed Field strain

TABLE 5
Effect of Selection for Early Development on Reproduction of *E. brunetti*

Generations of selection	Output of oocysts per bird ($\times 10^{-6}$)[a]
0	35.0
10	8.8
15	1.8
22	1.3
26	0.9

[a] 100 oocysts given.

Data modified from Shirley, M. W. et al., *Avian Pathol.*, 15, 705, 1986.

and challenged 4 weeks later with virulent homologous strains. McDonald et al.[47] showed that young birds kept on litter and given 200 oocysts each of a precocious line of *E. tenella* were completely protected against challenge 21 days later with virulent homologous or heterologous strains.

The immunogenicity of a virulent strain of *E. tenella* (from 'Coccivac') and two precocious lines coded WIS-F-125 (characterized by small second generation schizonts) and WIS-F-96 (characterized by the development of only one generation of schizogony) was investigated in floor pen experiments by Johnson et al.[58] The WIS-F-125 line and the virulent strain induced similar protection in chickens against challenge with the virulent strain whereas the WIS-F-96 line was poorly immunogenic. McDonald et al.,[47] however, in a detailed and quantitative assessment of the immunogenicity of the WIS-F-96 line and its parent WIS strain, found that the first generation of schizogony was, in fact, highly immunogenic. Mass-for-mass the greatest immunizing power was associated with the first generation of schizont but the greater abundance of the (also immunogenic) second generation schizonts almost certainly accounts for their importance in the induction of protective immune responses.

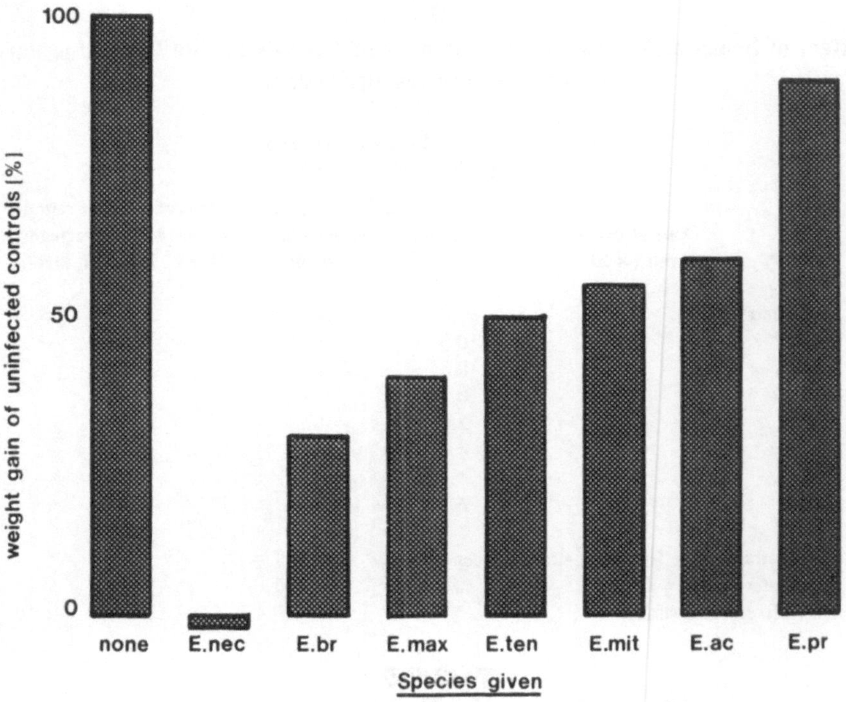

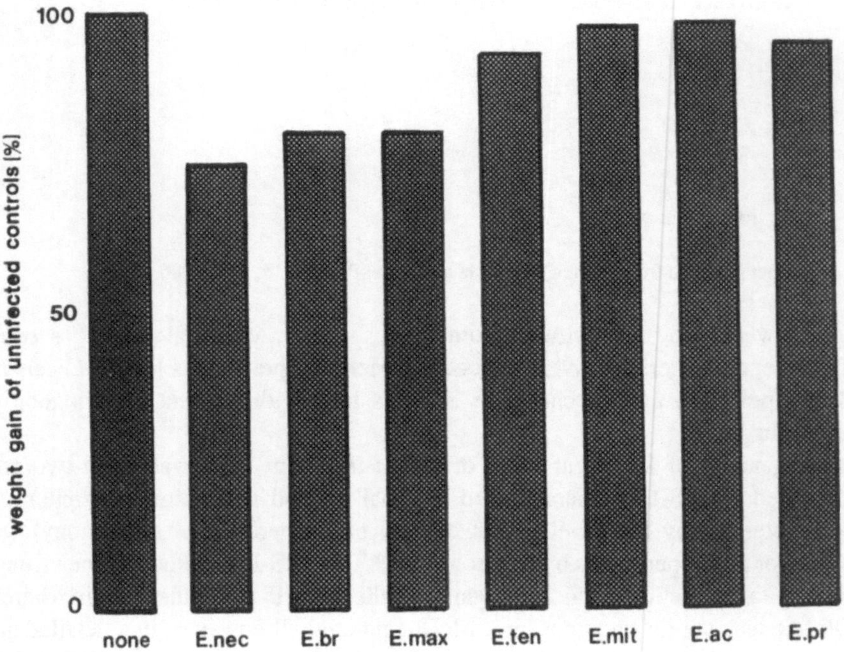

FIGURES 2 and 3. Effects on body weight gain of infection with unselected parent strains or precocious lines derived from them. Vertical axis is the weight gain of infected chickens between days 0 to 5 or 0 to 6 (depending upon species) expressed as a percentage of the weight gain of uninfected controls. Figure 2 = Parent strain. Figure 3 = Precocious line.

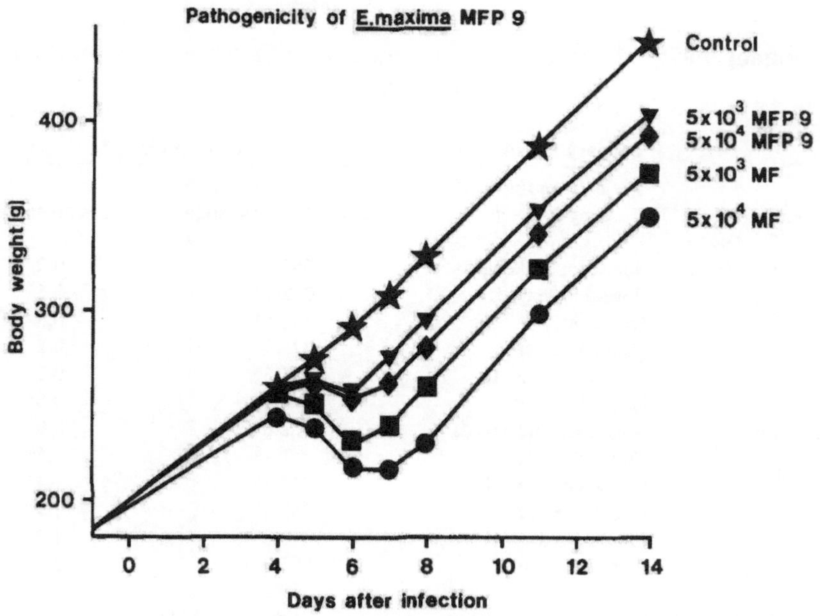

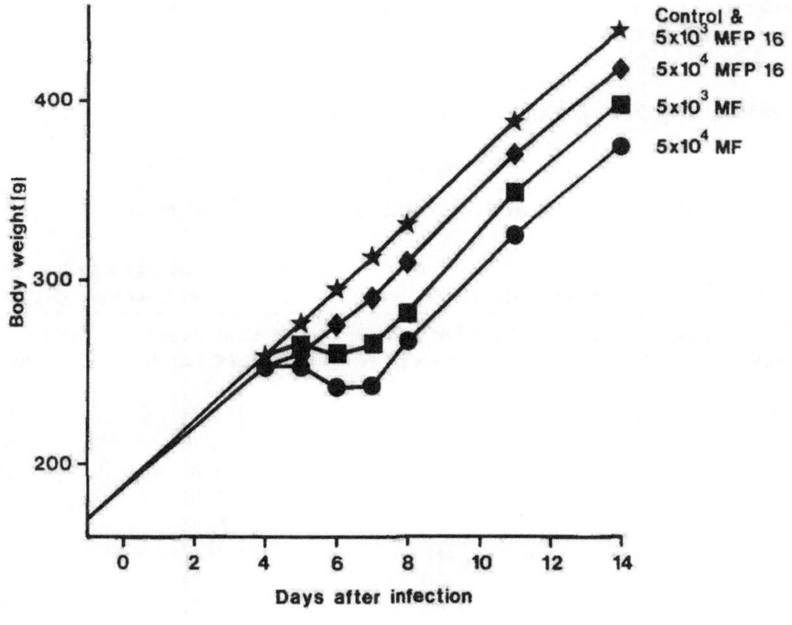

FIGURES 4 and 5. Body weight changes of 3-week-old LS chickens given *E. maxima*. (★) Control group not given oocysts; (▼) 5×10^3 precocious line; (◆) 5×10^4 precocious lines; (■) 5×10^3 parent strain; (●) 5×10^4 parent strain. Figure 4 = Precocious line after 9 generations of selection. Figure 5 = Precocious line after 16 generations of selection. Note that mean body weights of control group and group given 5×10^3 MFP16 were identical throughout experiment.

F. STABILITY OF ATTENUATION

Unequivocal evidence for the potential genetic stability of attenuation of precocious lines was first provided by Jeffers.[59] Cross-fertilization was demonstrated between populations of *E. tenella* characterized by the genetic markers of precocious or normal development, and resistance or sensitivity to the anticoccidial drug, decoquinate. Sutton et al.,[60]

TABLE 6
Immunogenicity of Precocious Line and Parent Strain of *E. acervulina* and *E. necatrix*

Species given	Primary inoculation: Parasite and dose of oocysts given	Results of challenge[a]: Mean wt. gain (g) 6 days after challenge	Lesion score
E. acervulina	None (not challenged)	356.3[A]	0.0
	None (challenged)	262.9[B]	2.5
	Precocious 1 × 10^3	336.5[A]	0.8
	Parent 1 × 10^3	328.9[A]	0.8
	Precocious 1 × 10^4	325.5[A]	0.2
	Parent 1 × 10^4	349.1[A]	0.0
E. necatrix	None (not challenged)	362.0[A]	0.0
	None (challenged)	41.4[B]	3.5[b]
	Precocious 5 × 10^3	319.4[A]	1.4
	Parent 5 × 10^3	287.6[A]	1.9
	Precocious 1 × 10^4	340.0[A]	1.4
	Parent 1 × 10^4	295.0[A]	2.1

Note: [A,B]Values with different superscripts are significantly different (p)0.05)

[a] Challenge of 1 × 10^6 oocysts (*E. acervulina*) or 2 × 10^5 oocysts (*E. necatrix*), 14 days after primary inoculation.
[b] 20% mortality.

Modified from Long, P. L. and Johnson, J. K., *Avian Pathol.*, 17, 305, 1988.

TABLE 7
Reproduction and Immunogenicity of *E. praecox*

Parasite given[a]	Primary inoculation: Dose of oocysts ($\times 10^{-3}$)	Primary inoculation: Output of oocysts per bird ($\times 10^{-6}$)	Results of challenge with 1 × 10^3 oocysts of *E. praecox* (H): Output of oocysts per bird ($\times 10^{-6}$)	Percent protection c. f. control group
H	0.1	29.7	0.21	99.9
HP10	0.1	4.3	1.40	99.5
H	1.0	152.9	0.01	99.9
HP10	1.0	33.9	0.01	99.9
None			272.44	
H	0.1	27.4	0.17	99.9
HP19	0.1	1.7	0.94	99.6
H	1.0	114.6	0.01	99.9
HP19	1.0	8.9	0.01	99.9
None			230.50	

[a] H = Houghton strain; HP10, HP19 = Precocious line after 10 and 19 generations of selection, respectively.

who used the genetic markers of precocious development, arpinocid resistance, and isoenzymes of glucose phosphate isomerase, similarly demonstrated the stability of the trait of precocious development in a line of *E. acervulina*.

G. DRUG SENSITIVITY OF VACCINAL LINES

The introduction of precocious lines of *Eimeria* into commercial poultry houses could be expected to have beneficial effects if the lines survive in the environment and are drug

TABLE 8
Drug Sensitivity of Seven Precocious Lines of *Eimeria* of U.S. Origin

	Percent protection afforded by medication		
Treatment	Body wt. gain after challenge	Food conversion	Lesion scores
Amprolium, 125 ppm	18	43	62.7
Lasalocid, 125 ppm	98	100	100
Nicarbazin, 125 ppm	89	98	86.6
Stenorol, 3 ppm	90	98	86.6
Salinomycin, 66 ppm	82	97	81.7
Monensin, 100 ppm	72	93	74.6

Note: Chickens given 25,000 oocysts each of *E. brunetti, E. maxima, E. necatrix,* and *E. tenella,* 50,000 oocysts of *E. mitis* and 100,000 oocysts each of *E. acervulina* and *E. praecox.*

From Long, P. L. and Johnson, J. K., in preparation.

sensitive. Precocious lines selected in the U.K. were from drug sensitive parents and those obtained in the U.S. are well controlled by anticoccidial drugs except amprolium (Table 8). The efficacies of the other drugs tested, based on body weight gain and feed conversion after challenge were, in decreasing order, lasalocid, stenorol, nicarbazin, salinomycin, and monensin.

VI. APPLICATION OF LIVE ATTENUATED VACCINES

A. FLOOR PEN TRIALS

Exposure of chickens to field-type coccidial infection may be initiated by the use of floor pens to house broiler chickens managed and fed in accordance with current commercial practice. In these trials chickens are exposed to naturally occurring coccidial challenges. Long and Millard,[61] for example, showed that 2-week-old birds given a mixture of embryo-adapted (attenuated) and virulent coccidia were subsequently immune to challenge.

In studies with precocious lines by Johnson et al.[25] and Parry et al.,[62] 1-day-old chickens were exposed to infection with up to seven species of *Eimeria* via an in-feed delivery method aimed at giving chickens small daily doses of oocysts. When the chickens were challenged at 3 weeks of age with laboratory (parent) strains of the same species of *Eimeria* they exhibited a high degree of immunity which was similar to that observed in unvaccinated chickens given the ionophorous drug monensin (Table 9). In another trial, chickens were similarly vaccinated but were given an 'in-pen' challenge when they were 3-weeks-old by contact with 'seeder' birds infected with field isolates of *Eimeria*. Vaccinated birds were substantially immune when 4-weeks-old and body weights and food conversion at 6 weeks were as good or better than birds medicated with monensin (Table 10).

B. FIELD STUDIES

The most comprehensive series of field trials with a live attenuated vaccine has, as far as is known, been organized in the U.K.[63] The vaccine comprises oocysts of eight precocious lines (two of *E. maxima* and one each of the other species) and an aqueous suspending agent. It has been given to over 5×10^5 young chickens (both broilers and replacement egg-laying birds) via the drinking water. The results of a series of three trials with consecutive crops of broilers are shown in Table 11. Control of disease was excellent, particularly in replacement flocks where coccidiosis was frequently diagnosed in contemporaneous flocks

TABLE 9
Performance of Chickens and Development of Immunity after 'Trickle' Immunization of Oocysts for 21 Days

Treatment	Effect of treatment against challenge infection given at 3 weeks				Performance 0 to 7 weeks	
	B.WG[a]	L.S.	B.WG[b]	L.S.	B.W.	F.C.
Trickle dose for 21 days with parent strains	216.4	1.0	196.7	2.5	2.17[B]	1.99[A]
Trickle dose for 21 days with precocious lines	213.2	1.4	115.7	3.7	2.20[AB]	1.97[A]
Monensin, 110 ppm	53.9	8.4[c]	24.9	5.6[d]	2.30[A]	1.97[A]
None	100.1	7.4[c]	37.6	6.1[c]	2.26[AB]	1.99[A]

Note: "In-pen" birds challenged with 20,000 oocysts of the parent strains of each species when 21 days old. B.W. = average live weight (kg). F.C. = feed conversion ratio. BWG = body weight gain (g) - between 21 and 7 days. LS = lesion score measured. [A,B]Values within columns with different superscripts are significantly different (p ⟨0.05).

[a] = *E. acervulina, E. maxima, E. tenella,* and *E. brunetti* given together.
[b] = *E. necatrix, E. mitis,* and *E. praecox* given together.
[c] 30% mortality.
[d] 10% mortality.

From Parry, S. H. et al., 1989, in preparation.

of the same age on the same site that were given a diet containing anticoccidial drugs. The effects of vaccination of medication on the incidence of coccidial oocysts in the litter of flocks of growing replacement egg-laying chickens are shown in Figure 6.

VII. CONCLUDING REMARKS

It is likely that any effective vaccine for coccidiosis used commercially will protect chickens against the pathogenic effects of a coccidial infection but will not necessarily induce an absolute immunity.

Oocysts not attributable to administration of the vaccine were recovered from the litter of all flocks that were vaccinated in the U.K.[63] To determine whether these oocysts were of species not present in the vaccine, or were antigenically diverse strains, a combination of procedures which included enzyme electrophoresis (see Figure 7), studies on prepatent time and cross-immunity tests was used. The presence of only *E. acervulina, E. brunetti, E. maxima, E. mitis, E. praecox,* and *E. tenella* was revealed.

The most likely explanation for the presence of wild-type parasites in flocks given the vaccine is probably antigenic diversity within several species. Comparatively few studies, however, have been undertaken on this topic since the first report by Joyner[64] who described antigenic differences between strains of *E. acervulina.* More marked variation has been reported in *E. maxima*[65] and McDonald et al.[43] found that, in chickens kept in wire-floored cages, cross protection between a vaccine line of *E. maxima* and some unrelated laboratory and field strains was poor. Similar results were obtained with a second, antigenically distinct, vaccine line,[44] see Table 12. However, 25 oocysts of these two lines, given as a mixture to chickens kept in litter pens, induced good protection against challenge with virulent heterologous strains. Shirley and Hoyle[66] found that populations of *E. maxima,* recovered from a single farm over a period of 8 months almost completely cross-protected with each other; results which suggest that antigenic drift within this species occurs only slowly, if at

TABLE 10
The Effect of Trickle Dosing with Oocysts for 21 Days on the Performance of Birds and Development of Immunity to 'In-Pen' Challenge

	Results, seven days after introduction of seeder birds[a]					Performance (0 to 6 weeks)	
		Lesion scores					
Treatment	Wt. gain (g)	U	M	C	Total	B.W.	F.C.
Monensin, 110 ppm for 21 days	227.2	2.18^{A}	2.73^{A}	2.18^{A}	7.09^{A}	1.865^{A}	1.804^{A}
None	207.7	1.81^{AB}	2.91^{A}	1.82^{A}	6.54^{AB}	1.863^{A}	1.814^{A}
Trickle dose of precocious lines for 21 days	240.6	1.25^{B}	1.50^{B}	1.92^{A}	4.67^{B}	1.863^{A}	1.782^{A}

Note: U. M. C. = areas of intestines, viz, upper, mid, and ceca. B. W. = average live weight (kg). Fc = feed conversion ratio. ^{A,B}Values within columns with different superscripts are significantly different (p ⟨0.05).

[a] Chickens were challenged at 21 days of age by the introduction of contact "seeder" birds given field isolates of *E. acervulina, E. maxima, E. mitis,* and *E. tenella* 7 days previously.

Modified from Parry, S. H. et al., 1989, in preparation.

TABLE 11
Results of Vaccination of Three Consecutive Crops of Broiler Chickens at Farm H In the U.K.

	Treatment					
	Vaccinated			Medicated with monensin at 100 ppm in the food		
	Crop number			Crop number		
	1	2	3	1	2	3
Age at vaccination (days)	8	6	7	—	—	—
No. birds placed[a]	21,500	21,753	21,600	21,583	21,250	21,696
% mortality (all causes)	4.58	5.58	4.78	4.74	6.00	6.09
Mean live wt (kg)	2.190	2.146	2.032	2.060	2.055	2.096
Mean age at slaughter (days)	48.5	49.0	48.3	46.4	49.1	49.4
Daily live wt gain (g)	45.2	43.8	42.1	44.4	41.8	42.4
Feed intake per bird (kg)	4.468	4.592	4.491	4.058	4.459	4.506
Food conversion ratio (FCR)	2.04	2.14	2.21	1.97	2.17	2.15
Incidence of gross intestinal lesions in birds found dead	0/36	0/30	0/34	3/36	2/25	6/47

[a] Equal numbers of males and females reared separately within each crop.

Data from Shirley, M. W. et al., in preparation.

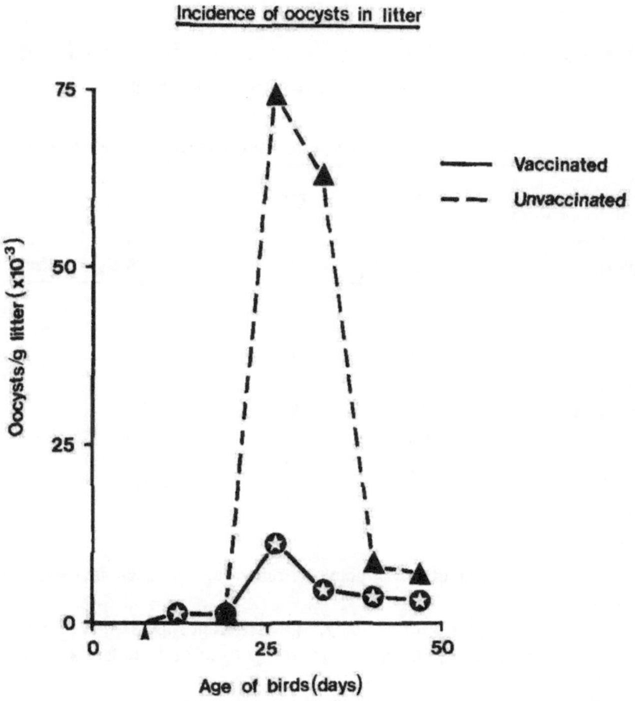

FIGURE 6. Incidence of oocysts in flocks of replacement egg-laying chickens given anticoccidial drugs (▲—▲) or a live attenuated vaccine via the drinking water (✪-✪).

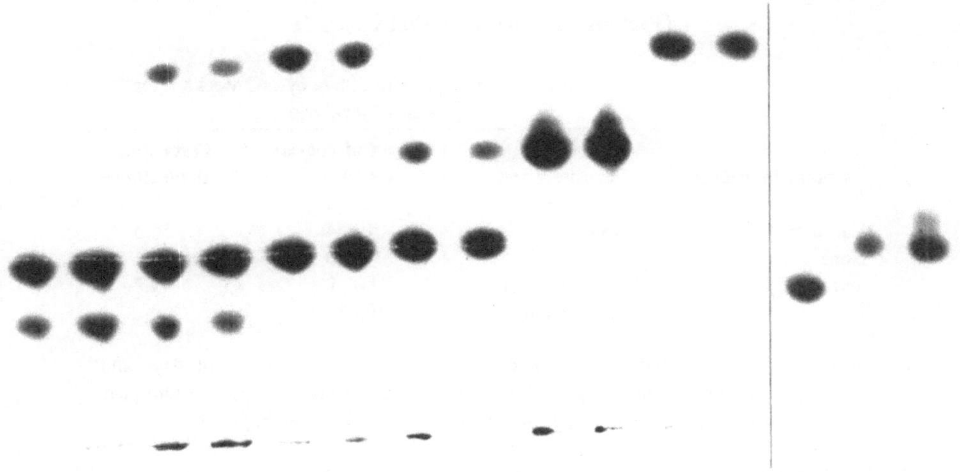

FIGURE 7. Identification of *Eimeria* spp. recovered from litter of broiler and replacement flocks given vaccine or medicated with anticoccidial drugs: electrophoretic variation of the enzyme glucose phosphate isomerase (GPI).

Lanes 1, 2 *E. acervulina* (H) - top + *E. acervulina* (W) - bottom: reference strains
Lanes 3, 4 Oocysts from vaccinated replacement broiler breeder flock = *E. acervulina* (2 strains) + *E. praecox*
Lanes 5, 6 Oocysts from vaccinated replacement broiler breeder flock = *E. acervulina* (one strain) + *E. praecox*
Lanes 7, 8 Oocysts from vaccinated replacement broiler breeder flock = *E. acervulina* (one strain) + *E. maxima*
Lanes 9, 10 *E. maxima* (H): reference strain
Lanes 11, 12 *E. praecox* (H): reference strain
Lane 13 *E. tenella* (H): reference strain
Lane 14 Oocysts from ceca from medicated replacement broiler breeder = *E. tenella*
Lane 15 *E. tenella* (W): reference strain

TABLE 12
Cross-Immunity Between Precocious Lines and Field or Laboratory Strains of *E. maxima*

Field or laboratory strain given for challenge	Precocious lines given for vaccination: MFP	CP	MFP + CP	None
	Percent protection[a]		Weight gain[b]	
Houghton	14	22	143[AB]	84[C]
Thrapston	63	21	131[AB]	50[C]
London Road	56	32	122[B]	73[C]
Watton	99	25	143[AB]	65[C]
Weybridge	99	ND	ND	ND
Chichester	13	90	152[A]	74[C]
None				147[A]

[a] Based on comparison of output of oocysts from groups of vaccinated and unvaccinated chickens.

[b] Weight gain 0 to 7 days after challenge (g). Different superscripts ([A], [AB], [B], [C]) in the two columns indicate significant differences ($p \langle 0.05$) by Student's t-test.

Data modified from References 43 and 44.

TABLE 13
Protection Against *E. acervulina*: Effect of Vaccination with the Complete Attenuated Vaccine, Against Challenge with Virulent H (Parent Strain) or a Field Isolate

Primary inoculation	Results of challenge with 500 oocysts 2 weeks after primary inoculation: Parasite given	Output of oocysts ($\times 10^{-6}$)	Percentage protection
Vaccine[a]	Isolate	19.54	55.5
None	Isolate	43.90	
Vaccine	*E. acervulina* (H)	0.28	99.7
None	*E. acervulina* (H)	109.33	

Note: Isolate: Recovered from flock of 20-day-old broiler chickens, 14 days after vaccination. Identified by electrophoretic characterization of glucose phosphate isomerase as a mixture of two strains of *E. acervulina*—see Figure 7.

Complete live attenuated vaccine, containing oocysts of *E. acervulina* HP.

all. Recent studies in this laboratory on antigenic diversity within *E. acervulina* have confirmed the findings of Joyner,[64] see Table 13, and indicate that further work on this and other species is warranted.

While vaccines against coccidiosis are likely to become available, it must be realized that none will stimulate the wider effects that are associated with some anticoccidial drugs. For example, as McDougald et al.[67] pointed out, some of the ionophores "have effects on feed consumption, performance and disease control that are unrelated to anticoccidial activity". In the future, as a consequence of increasing resistance to ionophores, the producers of poultry will almost certainly need to consider many different strategies for the control of coccidiosis. The incorporation of drug sensitive (precocious) lines in new coccidiosis vaccines could clearly have wide benefits if it results in changes in the epizootiology of field populations with the eventual dominance of drug-sensitive coccidia.

REFERENCES

1. **Dickinson, E. M., Babcock, W. E., and Osebold, J. W.,** Coccidial immunity studies in chickens, *Poult. Sci.*, 30, 76, 1951.
2. **Shirley, M. W., Jeffers, T. K., and Long, P. L.,** Studies to determine the taxonomic status of *Eimeria mitis*, Tyzzer, 1929 and *E. mivati*, Edgar and Seibold, 1964, *Parasitology*, 87, 185, 1983.
3. **Shirley, M. W.,** New methods for the identification of species and strains of Eimeria, in *Research in Avian Coccidiosis, Proc. Georgia Coccidiosis Conf.*, McDougald, L. R., Joyner, L. P., and Long, P. L., Eds., University of Georgia, Athens, 1986, 13.
4. **Edgar, S. A. and Fitz-Coy,** Differential characteristics of the lesser species of chicken *Eimeria, Research in Avian Coccidiosis,* in *Proc. Georgia Coccidiosis Conf.*, McDougald, L. R., Joyner, L. P., and Long, P. L., Eds., University of Georgia, Athens, 1986, 618.
5. **Tyzzer, E. E.,** Coccidiosis in gallinaceous birds, *Am. J. Hyg.*, 10, 269, 1929.
6. **Joyner, L. P.,** Experimental *Eimeria mitis* infections in chickens, *Parasitology*, 48, 101, 1958.
7. **McDonald, V. and Shirley, M. W.,** *Eimeria mitis*: a comparison of the endogenous developmental stages of a line selected for early maturation and the parent strain, *Parasitology*, 88, 37, 1984.
8. **Novilla, M. N., Jeffers, T. K., Griffing, W. J., and White, S. L.,** A redescription of the life cycle of *Eimeria mitis* Tyzzer, 1929, *J. Protozool.*, 34, 87, 1987.
9. **Chapman, H. D.,** The use of enzyme electrophoresis for the identification of the species of *Eimeria* present in field isolates of coccidia, *Parasitology*, 85, 437, 1982.
10. **Norton, C. C. and Joyner, L. P.,** Studies with *Eimeria acervulina* and *E. mivati*: pathogenicity and cross-immunity, *Parasitology*, 81, 315, 1980.
11. **Long, P. L.,** Studies on *Eimeria praecox* Johnson, 1930, in the chicken, *Parasitology*, 57, 351, 1967.
12. **Gore, T. C. and Long, P. L.,** The biology and pathogenicity of a recent field isolate of *Eimeria praecox* Johnson, 1930, *J. Protozool.*, 29, 82, 1982.
13. **Libby, D. A., Bickford, R. L., and Glista, W. A.,** Vaccinating chicken for coccidiosis, *Feedstuffs*, June 27, 1959.
14. **Bickford, R. L.,** Appraisal of coccidiosis vaccination for replacements, *Feedstuffs*, July 22, 1961.
15. **Shirley, M. W., McDonald, V., Chapman, H. D., and Millard, B. J.,** *Eimeria praecox*: selection and characteristics of precocious lines, *Avian Pathol.*, 13, 669, 1984.
16. **Long, P. L. and Johnson, J. K.,** *Eimeria* of American chickens: characterisation of six attenuated strains produced by selection for precocious development, *Avian Pathol.*, 17, 305, 1988.
17. **Bafundo, K. W. and Jeffers, T. K.,** Method of improving the quality of animals, European Patent 87307537.8, 1988.
18. **Long, P. L. and Jeffers, T. K.,** Studies on the stage of action of ionophorous antibiotics against *Eimeria, J. Parasitol.*, 68, 363, 1982.
19. **Chapman, H. D.,** *Eimeria tenella* in chickens: studies on resistance to the anticoccidial drugs monensin and lasalocid, *Vet. Parasitol.*, 2, 187, 1976.
20. **Joyner, L. P. and Norton, C. C.,** The immunity arising from continuous low-level infection with *Eimeria tenella, Parasitology*, 67, 333, 1973.
21. **Joyner, L. P. and Norton, C. C.,** The immunity arising from continuous low-level infection with *Eimeria maxima* and *E. acervulina, Parasitology*, 72, 115, 1976.
22. **David, P. J. and Reynolds, J. F.,** Immunisation of poultry to coccidiosis — by mixing spore carrying oocysts with the feedstuff, UK Patent 2,008,404, 1979.
23. **Davis, P. J. and Harris, H.,** UK Patent 2,144,331, 1985.
24. **Norton, C. C. and Joyner, L. P.,** Avian coccidiosis: the administration of encapsulated oocysts, *Parasitology*, 92, 499, 1986.
25. **Johnson, J. K., McKenzie, M. E., Perry, E., and Long, P. L.,** The immune response of young chickens given 'graded' or 'trickle' infection with the coccidia, in *Research in Avian Coccidiosis, Proc. Georgia Coccidiosis Conf.*, McDougald, L. R., Joyner, L. P., and Long,P. L., Eds., University of Georgia, Athens, 1986, 634.
26. **Vetterling, J. M. and Doran, D. J.,** Storage polyscharide in coccidial sporozoites after excystation and penetration of cells, *J. Protozool.*, 16, 772, 1969.

27. **Nakai, Y. and Ogimoto, K.**, Relationship between amylopectin and infectivity of *Eimeria tenella* sporozoite, *Jpn. J. Vet. Sci.*, 49, 447, 1987.
28. **Lee, E. H.**, Oral vaccine for immunisation against coccidiosis of oocysts in uniform suspension, preferably in dilute gum solution, Canadian Patent 1,204,057, 1986.
29. **Ross, P. A. and Hurnik, J. F.**, Drinking behaviour of broiler chicks, *Appl. Anim. Ethol.*, 11, 25, 1983/4.
30. **Long, P. L., Millard, B. J., Batty, A. F., and Davison, C.**, Immunisation against coccidiosis in chickens: tests under simulated field conditions, *Avian Pathol.*, 11, 131, 1982.
31. **Gore, T. C., Long, P. L., Kugut, M., and Johnson, J.**, Attenuation of *Eimeria necatrix* and *E. tenella* of U.S. origin by serial embryo passage, *Avian Dis.*, 27, 569, 1983.
32. **Bedrnik, P., Kucera, J., Jurkovic, P., and Firmanova, A.**, Results of experiments on pathogenicity and immunisation with selected lines of *E. tenella*, in *Research in Avian Coccidiosis, Proc. Georgia Coccidiosis Conf.*, McDougald, L. R., Joyner, L. P., and Long, P. L., Eds., University of Georgia, Athens, 1986, 510.
33. **Nakai, Y. and Ogimoto, K.**, Attenuation of *Eimeria tenella* by serial passage in chicken embryos, *Jpn. J. Vet. Sci.*, 49, 245, 1987.
34. **Long, P. L.**, *Eimeria mivati*: reproduction, pathogenicity and immunogenicity of a strain maintained in chicken embryos by serial passage, *J. Comp. Pathol.*, 82, 439, 1972.
35. **Shirley, M. W.**, *Eimeria necatrix*: the development and characteristics of an egg-adapted line, *Parasitology*, 81, 525, 1980.
36. **Shirley, M. W., Bellatti, M. A., and Millard, B. J.**, An egg-adapted (attenuated) line of *Eimeria necatrix*: further studies on its reproduction, pathogenicity and immunogenicity, *Parasitology*, 84, 215, 1982.
37. **Long, P. L.**, *Eimeria tenella:* reproduction, pathogenicity and immunogenicity of a strain maintained in chick embryos by serial passage, *J. Comp. Pathol.*, 82, 429, 1972.
38. **Jeffers, T. K.**, Immunization against *Eimeria tenella* using an attenuated strain, in *Proc. 15th World's Poult. Congr.*, New Orleans, 1974, 105.
39. **Jeffers, T. K.**, Attenuation of *Eimeria tenella* through selection for precociousness, *J. Parasitol.*, 61, 1083, 1975.
40. **McDonald, V., Ballingall, S., and Shirley, M. W.**, A preliminary study of the nature of infection and immunity in chickens given an attenuated line of *Eimeria acervulina*, *Parasitology*, 84, 21, 1982.
41. **Shirley, M. W., McDonald,V., and Bellatti, M. A.**, *Eimeria brunetti*: selection and characteristics of a precocious (and attenuated) line, *Avian Pathol.*, 15, 705, 1986.
42. **Johnson, J. K., Long, P. L., and McKenzie, M. E.**, The pathogenicity, immunogenicity and endogenous development of a precocious lines of *Eimeria brunetti*, *Avian Pathol.*, 15, 679, 1986.
43. **McDonald, V., Shirley, M. W., and Bellatti, M. A.**, *Eimeria maxima:* characteristics of attenuated lines obtained by selection for precocious development in the chicken, *Exp. Parasitol.*, 61, 192, 1986.
44. **Shirley, M. W. and Bellatti, M. A.**, Live attenuated coccidiosis vaccine: selection of a second precocious line of *E. maxima*, *Res. Vet. Sci.*, 44, 25, 1988.
45. **McDonald, V. and Ballingall, S.**, Attenuation of *Eimeria mivati* (=mitis) by selection for precocious development, *Parasitology*, 86, 371, 1983.
46. **Shirley, M. W. and Bellatti, M. A.**, *Eimeria necatrix*: selection and characteristics of a precocious (and attenuated) line, *Avian Pathol.*, 13, 657, 1984.
47. **McDonald, V., Shirley, M. W., and Millard, B. J.**, A comparative study of two lines of *Eimeria tenella* attenuated either by selection for precocious development in the chicken or by growth in chicken embryos, *Avian Pathol.*, 15, 323, 1986.
48. **Kawaguchi, H., Konishi, T., and Nakamura, T.**, Characteristics of a precocious line of *Eimeria tenella:* pathogenicity and endogenous development, *Jpn. J. Vet. Sci.*, 50, 445, 1988.
49. **Joyner, L. P., Canning, E. U., Long, P. L., Rollinson, D., and Williams, R. B.**, A suggested terminology for populations of the genus *Eimeria* (Protozoa: Apicomplexa), *Parasitology*, 77, 27, 1978.
50. **McDonald, V. and Shirley, M. W.**, The asexual development of precocious lines of *Eimeria* spp in the chicken, in *Research in Avian Coccidiosis, Proc. Georgia Coccidiosis Conf.*, McDougald, L. R., Joyner, L. P., and Long, P. L., Eds., University of Georgia, Athens, 1986, 502.
51. **McDonald, V. and Shirley, M. W.**, *Eimeria mitis:* a comparison of the endogenous developmental stages of a line selected for early maturation and the parent strain, *Parasitology*, 88, 37, 1984.
52. **McDonald, V. and Shirley, M. W.**, The endogenous development of virulent strains and attenuated precocious lines of *Eimeria tenella* and *E. necatrix*, *J. Parasitol.*, 73, 993, 1987.
53. **McDougald, L. R. and Jeffers, T. K.**, *Eimeria tenella* (sporozoa: coccidia): gametogony following a single asexual generation, *Science*, 192, 258, 1976.
54. **McDonald, V., Rose, M. E., and Jeffers, T. K.**, *Eimeria tenella:* immunogenicity of the first generation of schizogony, *Parasitology*, 93, 1, 1986.
55. **Shirley, M. W., Chapman, H. D., Kucera, J., Jeffers, T. K., and Bedrnik, P.**, Studies on the enzyme variation and pathogenicity of recent field isolates of *Eimeria tenella*, *Res. Vet. Sci.*, 46, 79, 1989.

56. **Nakamura, T., Konishi, T., and Kawaguchi, H.**, Isozymes of chicken coccidia, *Jpn. J. Vet. Sci.*, 48, 587, 1986.
57. **Shirley, M. W. and Millard, B. J.**, Studies on the immunogenicity of seven attenuated lines of *Eimeria* given as a mixture to chickens, *Avian Pathol.*, 15, 629, 1986.
58. **Johnson, J., Reid, W. M., and Jeffers, T. K.**, Practical immunisation of chickens against coccidiosis using an attenuated strain of *Eimeria tenella, Poult. Sci.*, 58, 37, 1979.
59. **Jeffers, T. K.**, Genetic recombination of precociousness and anticoccidial drug resistance in *Eimeria tenella, Z. Parasitenkd.*, 50, 251, 1976.
60. **Sutton, C. A., Shirley, M. W., and McDonald, V.**, Genetic recombination of markers for precocious development, arprinocid resistance and isoenzymes of glucose phosphate isomerase in *Eimeria tenella, J. Parasitol.*, 72, 965, 1986.
61. **Long, P. L. and Millard, B. J.**, Eimeria: immunisation of young chickens kept in litter pens, *Avian Pathol.*, 6, 77, 1977.
62. **Parry, S. H., Johnson, J., McKenzie, M. E., and Long, P. L.**, Immunization of broiler chickens by in feed delivery of attenuated lines of *Eimeria*, in preparation.
63. **Shirley, M. W., Roberts, B., McDonald, V., Knight, G., Geldard, H., Bushell, J., and Harding, R.**, Trials in the UK of a live attenuated vaccine for the control of coccidiosis in chickens, in preparation.
64. **Joyner, L. P.**, Immunological variation between two strains of *Eimeria acervulina, Parasitology*, 59, 725, 1969.
65. **Long, P. L. and Millard, B. J.**, Immunological differences in *Eimeria maxima:* effect of a mixed immunising inoculum on heterologous challenge, *Parasitology*, 79, 451, 1979.
66. **Shirley, M. W. and Hoyle, S. R.**, The antigenicity of *Eimeria maxima* populations obtained from commercial farms, *J. Parasitol.*, 67, 587, 1981.
67. **McDougald, L. R., Fuller, L., and Solis, J.**, Drug-sensitivity of 99 isolates of coccidia from broiler farms, *Avian Dis.*, 30, 690, 1986.

Chapter 17

CONTROL OF COCCIDIOSIS: PROSPECTS FOR SUBUNIT VACCINES

Harry D. Danforth and Patricia C. Augustine

TABLE OF CONTENTS

I. INTRODUCTION

Coccidiosis, a complex intestinal disease that is of major economic importance in domestic animals, is caused by the *Eimeria* species. The coccidia have enormous impact on the domestic poultry and cattle industries of the world because of the mortality, morbidity, and weight loss that the infection produces. A conservative estimate of the current losses to coccidiosis, based on 6- to 16-year-old data, is at least one billion dollars annually.[1,2] Control of the coccidia is primarily through the use of anticoccidial compounds; their importance is demonstrated by the direct relationship between the growth of the poultry industry and the development of the anticoccidials.[3] However, the increased costs of clearing a compound for use in food animals, which has resulted in reduced development of anticoccidials, and the resistance of the coccidia to a number of compounds already on the market, suggest that other means of control, such as subunit vaccines, may become necessary.

For a vaccine to be used successfully in the animal industry, certain basic criteria, such as cost effectiveness, a relatively long shelf-life, and ease of integration into existing commercial management practices, must be met. In addition, the immunization must give protection within a short period following vaccination and cause little adverse reaction in the animal. The greatest progress in the development of vaccines against coccidia has been with the poultry industry, so it is from these studies that the following section on subunit vaccine development is centered.

The life cycles of the avian *Eimeria* include several asexual reproductive generations, followed by the formation of sexual stages from which the oocysts eventually develop. It has been well documented that infection by *Eimeria* species induces significant humoral and cell-mediated immune responses.[4-6] Until recently, little work had been done on identifying and characterizing the antigens that are involved with eliciting the immune responses. However, with the use of molecular biology techniques and the continued interest in producing a subunit vaccine against the coccidia, the search for *Eimeria* antigens has intensified. Although only a few of the antigens that have been characterized have elicited an immune response, the investigations have yielded a greater knowledge of the makeup and complexity of this parasite.

II. SPOROZOITE ANTIGENS

The protective immune response in animals against coccidial infection is directed against the sporozoite stage, but it is generally accepted that the asexual stages produce the strongest stimulus for development of immunity.[7,8] However, because of their accessibility, the antigens of the sporozoites have been more rigorously studied than those of the other stages of the life cycle. Wisher[9] and Jenkins and Dame,[10] identified an impressive number of ^{125}I-labeled immunodominant surface antigens as well as ^{35}S methionine-labeled and unlabeled cytoplasmic or inner membrane constituents of *E. acervulina* and *E. tenella*. Each species had from 32 to 45 prominent antigens ranging in size from 9 to 350 kilodalton (kDa); approximately 15% of the antigens were found to be immunodominant. Wisher[9] further characterized up to 19 of the ^{125}I ^{35}S methionine-labeled *E. tenella* surface and internal antigens by demonstrating that they could be precipitated with immune serum. This same investigator found that sera from chickens orally immunized with two other species of coccidia (*E. acervulina* and *E. maxima*) crossreacted with a number of ^{125}I-labeled polypeptides of *E. tenella*. Murray[11] identified 15 antigens within the same kDa range as reported by the above authors from sera of birds immunized with intramuscular injections of sporulated oocyst extracts of *E. tenella*. Nine of these antigens have been shown to be immunodominant. Extracts of sporozoites or sporulated oocysts, that contained these antigens, markedly reduced lesion scores in chickens challenged with infective *E. tenella* oocysts, but the specific antigen(s) responsible for the protective response were not identified.

A number of investigators have used immune serum to screen constructs of both genomic and cDNA libraries of sporozoites of eight pathogenic avian *Eimeria* species.[12-14] At least 25 different clone isolates of mainly immunodominant surface proteins fused to beta-galactosidase were identified; these were estimated to range from 9 to greater than 200 kDa. Two of these antigens, designated cSZ-1 and 5401, have been screened for potential as immunogens.[13,14] The cSZ-1 antigen is a fragment (approximately 10 to 15 kDa) of a p240/p160 immunodominant sporozoite surface protein of *E. acervulina*. This antigen induces a measureable *in vitro* stimulation of T lymphocytes obtained from a congenic line of *E. acervulina* immune chickens, but, to date, has not elicited any type of protective immune response to a challenge infection in birds. The 5401 antigen is a 31 kDa fragment of a p200 sporzoite surface antigen of *E. tenella*. It is characterized by a region of highly repetitive amino acid sequences where nine amino acids are repeated five times. When injected subcutaneously with FCA into 4-week-old chickens, this protein stimulated a detectable serum antibody response by 2 weeks postimmunization (PI). At 4 weeks PI, the immunized birds demonstrated partial protection against challenge infection as evidenced by a decrease in intestinal lesions and an increase in weight gains when compared to unimmunized challenged controls.

Paul[15] and Files[16] used a monoclonal antibody (Mab) that inhibited penetration of *E. tenella* sporozoites *in vitro*, to identify a sporozoite surface antigen. In nonreduced SDS-polyacrylamide gels (SDS-PAGE), the protein migrated as a single band to an apparent molecular weight of 21 to 23 kDa, while in reduced gels two bands were seen at 17 and 8 kDa. The native antigen was isolated from the sporozoites by use of the Mab and inoculated into a small number of chickens. The serum antibodies elicited in the birds by this antigen were found to neutralize *in vitro* cellular invasion by *E. tenella* sporozoites and to reduce intestinal lesions in the immunized birds. The purified antigen was then sequenced and cDNA libraries screened by use of oligonucleotide probes derived from the amino acid sequence. The entire gene for this antigen was cloned, but as of yet there is no report on the ability of this genetically engineered antigen to elicit a protective immune response. Augustine and Danforth[17,18] have also reported on several monoclonal antibodies (MAbs) raised against various species of turkey and chicken coccidia which inhibit sporozoite host cell penetration. Most of these MAbs reacted with the surface of the sporozoites, were specific for only the *Eimeria* species against which they were elicited, and all recognized a 24 kDa antigen. None of these antigens have been isolated and used in bird immunization studies.

Although most of the candidate vaccine antigens that have been identified so far are sporozoite surface proteins, Danforth and Augustine[19] have recently speculated about the use of internal antigens to elicit a protective immune response. Some of the antigens of great interest are located in the refractile body (RB), and have a molecular weight of 21 to 28 kDa. These antigens are highly conserved among the *Eimeria*, and apparently play important roles in the asexual development of the parasite.[20] For *in vitro* development of *E. tenella* to occur, RB material must diffuse throughout the enlarging schizont. An MAb, that reacted with the RB, blocked this diffusion and thereby inhibited schizont development. Preliminary immunization studies with a genetically engineered RB antigen that was identified with this MAb showed reductions in lesion scores against both *E. acervulina* and *E. tenella* challenge infections. Reports by other investigators who have cloned what appeared to be similar RB antigens from *E. acervulina* and *E. tenella* cDNA libraries have shown that the antigens either induced significant *in vitro* activation of T lymphocytes that were recovered from orally immunized birds, or reduced intestinal lesion scores.[21,22] This evidence suggests that RB antigens should be considered for inclusion in a vaccine against the coccidia.

Recent studies suggest the feasibility of another area for development of anticoccidial vaccines; blockage of receptor or recognition sites for cellular invasion. Monoclonal antibodies directed against antigens on the sporozoite surface were shown to significantly inhibit invasion of cultured cells by the sporozoites.[23] A modified Western blot analysis identified interactions between specific sporozoite antigens, at 66, 40, and 23 kDa, and host cell molecules, suggesting that these sporozoite antigens may function as receptors for cellular invasion.[24] Inhibition of cellular invasion that occurs as a result of natural infection, is considered to be a manifestation of the immune response rather than a major force in protection against coccidial infection.[23,25] However, sporozoite receptor molecules may constitute prime candidates as immunogens for a vaccine, the efficacy of which is based at least partly on inhibition of parasite invasion.

III. MEROZOITE ANTIGENS

Many of the same approaches described for the identification of sporozoite antigens have been utilized to identify potentially immunogenic merozoite antigens. Jenkins and Dame[10] identified as many as 45 prominent antigens in or on the surface of second or third generation *E. acervulina* merozoites, of which approximately 15% were immunodominant. These antigens were distributed over a molecular weight range from 20 to 250 kDa. The major difference between sporozoite and merozoite antigens was that although 50% of the immunodominant sporozoite proteins were present on the surface of the parasite, over 90% of the most immunogenic merozoites proteins were surface constituents. Perhaps of greatest interest was that no major antigenic shifts were seen in a comparison of the second and third generation merozoites.

One study has identified a cDNA clone, designated as cMZ-8, that expressed an antigen of *E. acervulina* third generation merozoites.[13] The clone expressed a 150 kDa beta-galactosidase fusion protein and encoded for a 35 kDa portion of the p250 immunodominant merozoite surface antigen that contained tandem-repeated sequences.[26] The antigen was recognized by serum from chickens immunized by infection with *E. acervulina* and induced the activation of immune T lymphocytes *in vitro*. However, immunization trials with this antigen showed no significant protection against coccidial challenge.

Second generation merozoite antigens of *E. tenella* have been characterized with sporozoite and merozoite cross-reactive MAbs.[27] The sporozoite cross-reactive MAbs which were originally raised against *E. tenella* recognized 4 common antigens of 47, 60, 69, and 90 kDa of both developmental stages. The two cross-reactive merozoite MAbs raised against *E. acervulina* recognized only one antigen at 60 kDa with each stage. Screening of cDNA libraries with the merozoite MAbs has resulted in the detection and production of part of the 60 kDa antigen, but subsequent immunization trials with this antigen in birds did not show any protection against *E. acervulina* challenge.

IV. GAMETOCYTE ANTIGENS

Two immunodominant *E. maxima* gametocyte antigens, 56 and 82 kDa, have been consistently detected by chicken sera taken 11 to 14 days postinfection.[28,29] The antigens were found to be highly immunogenic, and were apparently well conserved within different strains of *E. maxima* and possibly within other *Eimeria* species. One-week-old chickens, injected intravenously with a MAb against the 56 kDa antigen, showed a 50% reduction in oocyst output after challenge with *E. maxima*. A similar reduction in oocyst output was obtained with an i.v. injection of polyclonal hyperimmune chicken serum raised against the 56 and 82 kDa antigens. cDNA clones encoding the 56 kDa antigen have been identified by immunoscreening a gametocyte cDNA expression library with the MAb. The gametocyte antigens represent a potential "transmission-blocking" component of a subunit coccidiosis vaccine.

V. PRACTICAL APPLICATIONS

The results from all of these approaches indicate that a commercial subunit vaccine for use against avian coccidiosis is not only feasible, but could be practically applied within the constraints of the poultry industry. Birds could be immunized at an early age at the hatchery by injection of a subunit vaccine, and placed on an anticoccidial until protection has been fully elicited. Evidence of protection against the coccidia has been seen within 2 to 4 weeks PI in birds immunized at 1 day of age with genetically engineered antigens, and the protection apparently lasted throughout the growout period. Several of the cloned antigens have been held at $-20°C$ for periods of 1 year without loss of effectiveness. With appropriate dosages of vaccine antigens, little pathology has been seen in the birds following immunization. Most importantly, the fact that vaccines composed of viable oocysts are already on the market suggests that a subunit vaccine would also be accepted by the poultry industry.

There are numerous problems associated with the use of any anticoccidial vaccination program when it is expanded to include the billions of commercial birds raised on poultry farms. One of the major problems is the immunological diversity among the *Eimeria* species. It has been documented with at least two chicken species (*E. acervulina* and *E. maxima*), that isolates from different geographical areas do not always crossprotect following oral immunization.[30] Subunit vaccines may have to be formulated for use in a particular area, using antigens generated from coccidia isolated within that region. Another problem involves species and stage specificity of immunity to coccidial infection. Although it is possible that a single crossreactive antigen will confer protection against all species of avian *Eimeria*, it is more likely that a multivalent vaccine that includes antigens from a variety of species and developmental stages will be needed to elicit a desired level of immunity. But, even if these problems are solved, the coccidia may possess other means for altering their antigenicity. It is not known how the coccidia may adapt to long term selective immunization pressure.

REFERENCES

1. **Fitzgerald, P. R.,** The economics of bovine coccidiosis, *Feedstuffs*, 44, 28, 1972.
2. **Biggs, P. M.,** The world of poultry diseases, *Avian Pathol.*, 11, 281, 1982.
3. **Reid, W. M., Long, P. L., and McDougald, L. R.,** Coccidiosis, in *Diseases of Poultry*, 8th ed., Hofstad, M S., Barnes, H. J., Reid, W. M., and Yoder, H. W., Jr., Eds., Iowa State University Press, Ames, 1984, 692.
4. **Rose, M. E.,** Immune responses to *Eimeria* infection, in *Proc. Georgia Coccidiosis Conf.*, McDougald, L. R., Joyner, L. P., and Long, P. L., Eds., University of Georgia, Department of Poultry Science, Athens, 1986, 449.
5. **Lillehoj, H. S., Allen, P. C., and Ruff, M. D.,** Comparative studies of humoral and cellular immune responses in two inbred strains of chickens showing different susceptibility to coccidiosis, in *Proc. Georgia Coccidiosis Conf.*, McDougald, L. R., Joyner, L. P., and Long, P. L., Eds., University of Georgia, Athens, 1986, 470.
6. **Lillehoj, H. S.,** Effects of immunosuppression on avian coccidiosis: cyclosporin A but not hormonal bursectomy abrogates host protective immunity, *Infect. Immun.*, 55, 1616, 1987.
7. **Rose, M. E.,** Immunity to *Eimeria tenella* and *Eimeria necatrix* infections in the fowl. I. Influence of the site of infection and the stage of the parasite. II. Cross-protection, *Parasitology*, 57, 567, 1967.
8. **Rose, M. E. and Hesketh, P.,** Immunity to coccidiosis: stages of the life cycle of *Eimeria maxima* which induce, and are affected by, the response of the host, *Parasitology*, 73, 25, 1976.
9. **Wisher, M. H.,** Identification of the sporozoite antigens of *Eimeria tenella*, *Mol. Biochem. Parasitol.*, 21, 7, 1986.
10. **Jenkins, M. C. and Dame, J. B.,** Identification of immunodominant surface antigens of *Eimeria acervulina* sporozoites and merozoites, *Mol. Biochem. Parasitol.*, 25, 155, 1987.

11. **Murray, P. K., Bhogal, B. S., Crane, M. St. J., and McDonald, T. T.,** *Eimeria tenella* — in vivo immunization studies with sporozoite antigen, in *Proc. Georgia Coccidiosis Conf.*, McDougald, L. R., Joyner, L. P., and Long, P. L., Eds., University of Georgia, Department of Poultry Science, Athens, 1986, 564.
12. **Clarke, L. E., Messer, L. I., and Wisher, M. H.,** DNA sequences coding for antigens of *Eimeria tenella* cloned and expressed in *Escherichia coli* in *Molecular Strategies of Parasite Invasion*, New Series, Vol. 42, UCLA Symposium on Molecular and Cellular Biology, Agabian, N., Goodman, H., and Noguiera, N., Eds., Alan R. Liss, New York, 1987, 701.
13. **Jenkins, M. C., Lillehoj, H. S., and Dame, J. B.,** *Eimeria acervulina*: DNA cloning and characterization of recombinant sporozoite and merozoite antigens, *Exp. Parasitol.*, 66, 96, 1988.
14. **Danforth, H. D. and Augustine, P. C.,** Coccidia vaccines, in *Veterinary Protozoan and Hemoparasite Vaccines*, Wright, I., Ed., CRC Press, Boca Raton, FL, 1989, in press.
15. **Paul, L. S., Brothers, V. M., Files, J. G., Tedesco, J. L., Newman, K. Z., and Gore, T. C.,** Purification and characterization of a major surface antigen of *Eimeria tenella*, *J. Cell. Biochem.*, 10A, 169, 1986.
16. **Files, J. G., Paul, L. S., and Gabe, J. D.,** Identification and characterization of the gene for a major surface antigen of *Eimeria tenella*, in *Molecular Strategies of Parasitic Invasion*, New Series, Vol 42, Agabian, N., Goodman, H., and Nogueira, N., Eds., Alan R. Liss, New York, 1987, 713.
17. **Augustine, P. C. and Danforth, H. D.,** Effects of hybridoma antibodies on invasion of cultured cells by sporozoites of *Eimeria*, *Avian Dis.*, 29, 1212, 1985.
18. **Augustine, P. C. and Danforth, H. D.,** Use of monoclonal antibodies to study surface antigens of *Eimeria* sporozoites, *Proc. Helminthol. Soc. Wash.*, 54, 207, 1987.
19. **Danforth, H. D. and Augustine, P. C.,** *Eimeria tenella*: use of a monoclonal antibody in determining the intracellular fate of the refractile body organelles and the effect on *in vitro* development, *Exp. Parasitol.*, 68, 1, 1989.
20. **Augustine, P. C. and Danforth, H. D.,** Monoclonal antibodies reveal antigenic differences in refractile bodies of avian *Eimeria* sporozoites, *J. Parasitol.*, 74, 653, 1988.
21. **Jenkins, M. C., Danforth, H. D., Lillehoj, H. S., and Fetter, R. H.,** cDNA encoding an immunogenic region of a 22 kilodalton surface protein of *Eimeria acervulina* sporozoites, *Mol. Biochem. Parasitol.*, 32, 153, 1989.
22. **Liberator, P. A., Profous-Juchelka, H., Weimer, J. L., Crane, M., and Turner, M. J.,** Cloning and characterization of a protective antigen from *Eimeria tenella*, *J. Cell. Biochem.*, 13E, 142, 1989.
23. **Augustine, P. C. and Danforth, H. D.,** A study of the dynamics of the invasion of immunized birds by *Eimeria* sporozoites, *Avian Dis.*, 30, 347, 1985.
24. **Augustine, P. C.,** Molecular interactions of cultured turkey kidney cells with specific antigens of *Eimeria adenoeides* sporozoites, *Proc. Soc. Exp. Biol. Med.*, 190, in press, 1989.
25. **Rose, M. E. and Hesketh, P.,** *Eimeria tenella*: effects of immunity on sporozoites within the lumen of the small intestine, *Exp. Parasitol.*, 63, 337, 1987.
26. **Jenkins, M. C.,** A cDNA encoding a merozoite surface protein of the protozoan *Eimeria acervulina* contains tandem-repeated sequences, *Nucl. Acids Res.*, 16, 9863, 1989.
27. **Danforth, H. D. and McAndrew, S. J.,** Hybridoma antibody characterization of stage-specific and stage-cross-reactive antigens of *Eimeria tenalla*, *J. Parasitol.*, 73, 985, 1987.
28. **Mencher, D., Pugatsch, T., and Wallach, M. G.,** Antigenic proteins of *Eimeria maxima* gametocytes: cell-free translation and detection with recovered chicken serum, *Exp. Parasitol.*, 68, 40, 1989.
29. **Wallach, M. G., Mencher, D., Yarus, S., Pillemer, G., Halabi, A., and Pugatsch, T.,** *Eimeria maxima*: identification of gametocyte protein antigens, *Exp. Parasitol.*, 68, 49, 1989.
30. **Rose, M E.,** Host immune response, in *The Biology of the Coccidia*, Long, P. L., Ed., University Park Press, Baltimore, 1983.

INDEX

D

E

F

N

O

P

Q

R

U

V

W

Z